SIR AUREL STEIN

Archaeological Explorer

Sir Aurel Stein
(1929)

SIR AUREL STEIN

Archaeological Explorer

Jeannette Mirsky

The University of Chicago Press

Chicago and London

The University of Chicago Press, Chicago 60637
The University of Chicago Press, Ltd., London

© 1977 by Jeannette Mirsky
All rights reserved. Published 1977
Paperback edition 1998
Printed in the United States of America

06 05 04 03 02 01 00 99 98   2 3 4 5

ISBN: 0-226-53177-5

Library of Congress Cataloging in Publication Data

Mirsky, Jeannette, 1903– *1987*
    Sir Aurel Stein, archaeological explorer.

    Includes index.
    1.   Stein, Marc Aurel, Sir, 1862–1943.   2.   Archaeolo-
gists—Asia, Central—Biography.   I.   Title.
DS785.S84M57   939′.6′007202[B]   76–17703

♾ The paper used in this publication meets the minimum
requirements of the American National Standard for
Information Sciences—Permanence of Paper for Printed
Library Materials, ANSI Z39.48-1992.

To

JOHN HARMON
(1934–1973)

and to

CECILE AND HARRY STARR

dear and true

friends

Contents

PART 6

The Borderlands and into Persia, 1926–36

445

PART 7

The Promised Land at Last, 1937–43

505

Preface

"The most prodigious combination of scholar, explorer, archaeologist and geographer of his generation," is Owen Lattimore's characterization of Sir Aurel Stein; his discoveries were "the most daring and adventuresome raid upon the ancient world that any archaeologist has attempted" according to Sir Leonard Woolley; "unforgettable" is Laurence Sickman's adjective for Stein.

While Sir Aurel lived (1862–1943) there was no need for a biography. A wide public followed his extraordinary expeditions as he himself told them in an outpouring of books and articles and lectures. A wider public kept abreast of his travels through newspaper dispatches telling of far-flung journeys in little-known exotic places and of his spectacular "finds." Since his death, almost unnoticed during the war, his name has fallen into the eclipse common to those no longer in the limelight; it continues to hold its place in repositories proper to scholarly endeavor: the footnotes, bibliographies, and dedications of contemporary writings. It would indeed be difficult for anyone engaged in any of the many disciplines—social, political, technological, religious, economic, ecological, art-historical, and so forth—located anywhere within the wide continental sweep from the Persian Gulf to the Pacific watershed, dealing with events occurring during the millennia between the neolithic and medieval, not to touch on Stein's pioneering efforts.

The range and importance, style and success of Stein's archaeological reconnaissances would be enough to recommend him to a biographer. What lifts his achievements into a special category is the historical vision that, to the end, motivated his questing.

For long, Europeans tended to view their cultural heritage in narrow, parochial terms as though their roots were limited to ancient Greece and the Bible lands. Thus seen, Europe is wrenched away from the mainland to become an island, distinct and apart from the rest of the Old World. Yet again and again the European imagination has been stirred by its kinship to ancient Egypt, Mesopotamia, Achaemenid Persia, and Han, T'ang, and Sung China. As the manifold roots of Europe's cultural heritage were explored, time and space were majestically extended.

Orientalist, archaeologist and explorer, geographer and topographer, Sir Aurel Stein was of that small scholarly fraternity of pioneers who created the modern paradigm which views the Eurasian landmass as one cultural field whose forces were its four high civilizations: the Mediterranean West, the Indian and Iranian, and the Chinese. Central Asia, with which his name is most notably associated, was a region where these four met and interacted.

Why was I attracted, and what attracted me to attempt this biography, devoting much of twelve years to the work? To the questions put to me, I cannot say *why* a subject takes possession of a writer but I can suggest *what* fascinated me about Stein's work. Let me try to explain it in my own shorthand.

As known to historians, the modern paradigm—it was sobering to learn—was initiated not by a European explorer but by a Chinese, when, about 125 B.C., Chang Ch'ien went overland and established contact with the West. Thereafter for fifteen hundred years an artery of transcontinental routes laced East and West together. Then Columbus ventured across the Atlantic. Europe's overseas epic began. Sea lanes replaced the overland routes.

There is a subtle but significant difference between maritime commerce and land traffic. Whatever the dangers of oceanic venturing, a ship in a foreign port remains a familiar home-base, whereas a caravan, however welcome, is an uninvited guest in a succession of alien entrepots.

Because Spain and Portugal controlled the southern approaches to Asia, English, Dutch, and Russian sailors went northward hoping to find the shortest distance to Cathay. In a manner of speaking, the Arctic Ocean replaced the Himalayan passes and the Taklamakan desert of Central Asia as the inhuman region separating the main centers of the Old World. The long parade of ships that attempted the Northwest and Northeast Passages was the subject of my first book, *To The Arctic!* It was the earlier story, still unfamiliar, that brought me to Aurel Stein.

This book was to have been short. Originally the biography of Stein was to have been supplemented by a volume of letters, including his unpublished "personal narratives." When this became impracticable, I had to retrace my steps to combine the material from both in one book.

In drawing on the letters and unpublished personal narratives, I have taken the liberty of letting the story run along without denoting ellipses; the pages filled with dots looked unseemly. Out of Stein's vast printed outpouring, I have utilized whatever books and articles would carry the story ahead fully and honestly and so obviated the repetitiousness of an archivist's bibliography. After some thought I decided to omit the diacritical marks used on Oriental words—while consequential for the specialist, they are but of passing interest to the general reader. To understand is, I hope, to forgive.

Institutions and individuals have sustained me physically and spiritually. It is a pleasure to look back and, with a sense of accomplishment, express my gratitude to the Rockefeller Foundation, to the National Endowment for the Humanities, to the Stein-Arnold Fund, and to the Lucius N. Littauer Foundation. Their generous responses to my proposal made possible the long period of research and writing. Princeton University twice honored me by naming me a Visiting Fellow in the Department of East Asian Studies, where I became the beneficiary of the friendship, scholarship, and active encouragement of Professors Frederick W. Mote and Marion J. Levy, Jr., and, at different times, of Visiting Professors A. F. P. Hulsewé, Joseph Fletcher, and Denis Twitchett.

My research began in New Delhi in 1965. I am deeply indebted to Sri. A. Ghosh, Director-General of the Archaeological Survey of India, for expediting my search of its archives; to Dr. B. Ch. Chhabra, Chief Epigraphist, learned in the ancient history of India, who remains a valued friend; and to Dr. Krishna Devi for his vivid vignettes of his work in the field with Stein. At New Delhi that same year I was fortunate in many conversations to have Dr. Sourindrath Roy, Joint-Director of the National Archives of India, expatiate on details in his book, *The Story of Indian Archaeology*. There, too, my search was guided by Mr. M. L. Ahluwalia, the Archivist, and the Assistant-Director, Miss Dhan Keswani. My sincere thanks go to all who enabled me to complete my investigation in the months spent in New Delhi. At the Magyar Tudományos Akadémia in Budapest, the Director, Professor L. Ligeti, a pupil of Paul Pelliot, was most cordial. My examination of their Stein material was facilitated by Dr. L. Bese, an outstanding Mongolist, and Dr. Hilda Ecsedy, an expert in Turkic and Tibetan. There I was presented with more than a dozen reprints of Stein's articles, a prized and much used gift. I thank them for their help.

In London, at the British Academy where boxes filled with thousands of Stein's letters to his family had been sent from Vienna, the Secretary, Sir Mortimer Wheeler, spoke brilliantly on some of the persons and problems referred to in the letters. He and Sir Laurence Kirwan, Director and Secretary of the Royal Geographical Society, the officials charged with disbursements from the Stein-Arnold Fund, honored my work with a grant. It was also an expression of friendship. My warm thanks go to Dr. Richard D. Barnett, former Keeper of the Department of Western Antiquities, for supplying me with copies of Stein's letters to his father, Dr. Lionel D. Barnett, Deputy-Keeper of the Department of Oriental Antiquities, a scholar who held Stein's respect, confidence and affection. Quite by accident, at Oxford, at the Bodleian Library, I found the great cache of Stein's papers deposited there. The courtesy and efficiency shown me by Dr. R. W. Hunt, Keeper of Western Manuscripts, and his assistant, Mr. D. S. Porter, enabled me to accomplish my search thoroughly and speedily; to both I am grateful.

Individuals, some friends of long standing, some newly-made along the way, at different times and in different ways, have become part of the book. Gladly and with gratitude I acknowledge their help. Dr. Gustav Steiner of Vienna gave me family pictures and his blessing on my portrayal of Stein, the man; Professor Martin B. Dickson, an encyclopaedia on the history, peoples, and languages of Iran and Central Asia, answered my questions with patience and an animated interest; Brigitte Schaeffer shared the tedium of translating Stein's German letters and articles and by her response to their contents transformed the sessions into delightful occasions; Julia Davis, a most talented veteran writer and an innocent on Central Asian matters—a model reader for the manuscript as it was being written—gave salutary suggestions and comments; Professor Mark Leone was an ideal guide to the history and the recent developments in archaeology; James and Lucy Lo, literati and master-photographers, whose unique, comprehensive record of Tun-Huang frescoes and sculptures was made at the temple-caves, enriched my understanding of the site's importance in art history; Sylvia Massell and Elizabeth Beatson counseled and encouraged me during the years of work-in-progress; Alfred L. Bush, of Princeton University Library, Collection of Western Americana, provided a safe deposit for my precious microfilms, and Linda Oppenheim, at the Library's Information Desk, led me to a Hungarian biography which established the importance of Stein's uncle, Dr. Ignaz Hirschler; Nancy Dupree was a delightful and informed companion during my stay in Afghanistan; Professor William Hung, formerly of

Yen-Ching University, Peking, detailed the events leading up to the debacle of Stein's Fourth Expedition; Professor Yu-kung Kao, a scholar to delight the mind and soul, translated the Chinese characters on Stein's antiquities' flag; Reyner Unwin, of George Allen & Unwin, Ltd., who published my *The Great Chinese Travellers* in England, extracted the necessary permission from the Stein executors to have the Bodleian materials microfilmed; Lewis Bateman, a friend, with his sharp, professional eye was a valuable aid in the chore of proofreading; and last, but far from least, I owe much to my staunch friend, Marion Bayard Benson, whose faith in and concern for me never failed.

1

Beginnings

1

In Budapest, on 26 November 1862, their third child, a boy, was born to Nathan and Anna Hirschler Stein; he was baptized Marc Aurel. The infant was twenty-one years younger than his sister and nineteen years younger than his brother Ernst Eduard. Consider the baptism and the name, for both invite consideration and reveal the plans and hopes of the parents that did so much to form their youngest child.

Why?—the question presents itself—why did the parents have their sons baptized in the Protestant faith while they and their daughter remained Jews? Since the christianization was highly selective within the family, it cannot have been a straightforward religious matter; it suggests that the political and social climate of that period and region be examined. It must be remembered—to many people the Nazi persecution of the Jews was an aberration historically unique and monstrous—that anti-Semitism, endemic in Christian Europe, was sporadically punctuated by virulent outbursts. Of that story it is only necessary here to indicate those aspects relevant to why only the sons were baptized. Stein's parents, who grew up and were married in the first half of the nineteenth century, knew as did all Jews in Central Europe, that their religion barred them from participating in Western culture. Jews were not only legally confined to ghettos (the word itself had its origin in the district in Venice set apart for the Jews, an area often walled around in which its inhabitants were locked at sundown and on Sundays); they were not permitted in schools, universities, and professions, and at every turn they were harassed by legal disabilities. Stein's father, a small businessman, and his mother, a sister of Professor Ignaz Hirschler, were intellectually ambitious; they were responsive to the profound changes in Jewish life and outlook. The potent message of eighteenth-century Enlightenment had found its Jewish spokesman in Moses Mendelssohn (1729–86), who in brilliant, soul-searching arguments invited Jews to share in the civilization common to educated Europeans, a civilization founded in a belief in progress and in the ideals of reason and morality, tolerance and peace—a belief and ideals philosophically and rationally congenial with his religious faith. Soon after Mendelssohn's death in 1786, his appeal to reason became a possibility when French troops marching under the revolutionary banner of "Liberty, Equality, and Fraternity" unlocked the ghetto gates. Suddenly, the walls sealing off the medieval world in which the Jews had been held, tumbled down. Their freedom, alas, was short-lived. Reaction set in: the restrictive laws were reinstated. (Not until 1867, five years after Aurel was born, did the Austro-Hungarian Jews gain political freedom; legal recognition came in 1896.) But the changes fostered by the Enlightenment had a life and movement of their own—there were stirrings inside the ghetto. The Jewish tradition that enjoined systematic schooling on its members, the attitude of pious Jews that equated the unschooled with the godless and the ethical with the intellectual, gave strength and urgency to Moses Mendelssohn's message: the righteous life was no longer tightly bound to ritualistic obligations but could also be attained by significant social achievements.

Baptism, as the elder Steins saw it, was the key that unlocked the ghetto and proffered to those who had been baptized access to the scholarly riches of the world outside. In its noblest meaning, they were giving their sons the key to freedom. In the 1820s, while they were still young, of an impressionable age, Anna and Nathan were furnished with examples that would give firmness to their decision. Baptism as the way to circumvent the frustrations forced on Jews by their religion had been taken by many: especially noteworthy were Heine and Abraham Mendelssohn. The former had gone through baptism in the morning so that he could receive his LL.D. the same afternoon (1825), and the latter, a son of the great Moses and the father of the musical prodigy Felix, was baptized in 1822 in time to avoid having his thirteen-year-old son ritually inducted into the Jewish community. Those who were baptized shrugged off charges of apostasy—in his *Confessions*, written in 1853–54, Heine had defined Protestantism, which had translated the Old Testament into native languages and spread it around the world, as "a mere pork-eating Judaism."[1]

And the name given at the christening of the Stein's younger son—what of it? Marc Aurel, the then-in-vogue French form of Marcus Aurelius, belongs to the period when neoclassicism dominated European sensibilities.

Contemporary with the Enlightenment and consonant with its spirit of rationalism, the neoclassical idealized the serenity and grandeur of Imperial Rome. Beginning with the discovery and excavation of Pompeii and Herculaneum in the eighteenth century, neoclassicism flowered as the illustrations in monumental quartos presented the personages and palaces of antiquity through romantic eyes. In the lengthy procession of emperors, the voice of one came to men across the intervening centuries: Marcus Aurelius's *Meditations*. The Roman emperor's private thoughts, the soliloquies of a Stoic, randomly set down in Greek—still the language of philosophy—expressed the ethics of the school in which he has been reared; this pagan ruler's book, lofty in tone, dark in mood, extolling man's reason for decrying the folly of appetite and impulse, was widely read and highly influential. The virtues he valued—duty and responsibility, simplicity, frugality, and diligence—were the very ones valued by Jew and Protestant alike.

Yet, after sifting the easily identifiable qualities contained in the names chosen, a tantalizing, unresolved residue remains. Marc had instant Christian association with the apostle and evangelist and was therefore eminently suitable for baptism. But what of Aurel? Did its French form effectively disguise its pagan association, an association palpably questionable for christening? Marcus Aurelius, a firm believer in law and order, human as well as divine, is also remembered as a leading persecutor of Christians who, by refusing to perform their civic duty to sacrifice to the deified emperor, himself, were in his eyes members of a disloyal, traitorous sect. For the Steins the question is further complicated because Mark was Nathan's mother's father's name; yet though the older son Ernst Eduard was always called by his first name, Marc Aurel—Marc, soon reduced to an initial, appears on official documents and less and less on his own writings—was to his family and himself Aurel.[2]

Of Aurel's formative years, a little can be salvaged from the few lean facts known. Thus, for example, there is meaning in the statement that as a child he spoke both Hungarian and German, for the use of Magyar as well as German within the Stein household signals the family's allegiance to nationalistic aspirations. In Hungary, language had become a political instrument: Magyar had received a massive infusion of new words early in the nineteenth century which, without altering its structure, made it a vehicle for a new literature and ideology. Smoldering Hungarian dissatisfaction with Austrian domination, sparked by the 1848 revolution in France and fanned by the inflammatory speeches and writings of Kossuth, flared into armed rebellion. The next year, when the uprising had been put down, the role of Magyar was acknowledged: the Austrians decreed that only German was allowed in the schools and in government, a regulation that stayed in effect until 1860 when new agitation was quieted by making Magyar the official language in Hungary.

This much had been gained when Aurel started school. If German was the passport to achievement in whatever career he might choose— as well as the language of the letters which the family regularly wrote to any absent member—for Aurel, Magyar was more than proof of patriotism. It belonged to his happy, carefree childhood before he had learned to read and write, when he, the "unexpected" baby, had held the center of the stage, cherished by his doting father and mother—she called him "my precious Aurel"—by his caring, avuncular brother, and by his learned, important Uncle Ignaz; when he had been fed sweets and smothered in embraces; when his precocity had earned him delighted hugs and kisses; when home and the world were one and the same. All this was lost when, at ten, he was taken away from home and sent to the renowned Kreuzschule in distant Dresden. Separated from oneness with the family, letters, endless letters filled with the words "love," "hugs," "a thousand kisses" had to substitute for his family's enveloping intimacy. Magyar miraculously, instantly evoked that lost Eden. Later, the occasional Hungarian traveler he met in India is mentioned as though he were a long-lost cousin; to hear Magyar, to speak it gave him—to use his own word—"joy." In 1897, twenty-five years after he had first left home and was living and working in India, after he had mastered Greek, Latin, French, and English at the Kreuzschule and had gone on to do graduate work in Sanskrit and Persian at the universities of Vienna, Leipzig, and Tübingen, he attended a meeting of the Hungarian Academy of Sciences while on a visit to Budapest. "Bishop Szasz presented me to Tisza K" (the Prime Minister), he told Ernst, reporting not only what had happened but his own reactions, "who congratulated me that my Hungarian was still as good as the Bishop had claimed. This honor gave me great joy."[3]

Aurel was not the only child to be thus exiled from his family. Each child handles such a separation in his own way. But for him it may have been then that the possibility of being that anomaly in his urban world, a nomad without a firm, fixed habitation began to settle around his heart. It was an old pain when, in 1889, in faraway Lahore he received word of his father's death. To Ernst he poured out a grief compounded by his feeling of isolation. "In all this sorrow I cannot but think of our dear home which is now totally deserted. . . . I have the dearest memories of each book and picture. The less I was able to live in our home since my boyhood years, the more closely I felt attached to that sacred place, and now, where the thought of home actually makes life bearable, there exists a void."[4] It could be said that at ten he became a traveler learning to keep a lifeline to family and friends by a steady flow of letters. To return to his home was his deepest, dearest wish. All his travels were a search into the past: his peers rank him as one of the world's greatest archaeological explorers.

As we look at Stein's mother's photograph and read some of the letters that he treasured since his school days in Dresden, and from Stein's own references to her long after her death, the lack of precise facts about her becomes less significant. Anna Hirschler Stein emerges out of the shadows. Did she have her picture taken for her son to carry with him when they were to be separated? Her modish gown is in the style of the 1870s; she was, most probably, late in her fifties and obviously had dressed for the occasion. She stands, small-boned, erect, her hands resting on the back of the tufted chair in the photographer's studio—a conventional pose—holding a large folded fan and giving an impression of movement such as that of a bird momentarily alighted. Did the fan serve as an outlet for activity: opened, fluttered, closed in a quick, short rhythm? Or was it a kind of marshal's baton, an insignia of command, used to emphasize directions and dicta—her large watchful eyes, the tense smile on her lips and a determination in her stance speak of authority and discipline. Not quite a bluestocking—like her younger brother to whom she was very close, she was artistically gifted and widely read—she was a mother molding a scholar, a mother whose maternal love wanted the world for this child, the gift of her middle age.

Most probably she died in 1888: in a letter of 8 June 1889, Aurel mentions her in a reference to Ernst's wife, whom he had married in 1886: "your dearest Hetty given to you by a kind fate in order to lighten the last years of our dear parents." His mother, then, lived to see her Aurel get recognition as a promising scholar: in quick succession he had received his Ph.D. from Tübingen (1883), a grant from the Hungarian government for postdoctoral studies in England (1884), and had published his very first paper, "Old Persian Religious Literature"—the first in a lengthy bibliography—in a Budapest periodical (1885). That his mother was still alive at the end of 1887 when he left for India can be inferred from the requests he made to Ernst in that 8 June letter: "I am certain that all the works from the artistic hand of our dear Mama [did she paint, embroider, decorate china?] are well taken care of by you. Would you send me a few of the smaller pieces when you send other small relics? Of the books I ask for the collection of German classics which my dearest Mama pointed out I would inherit because they have the monogram 'A.S.' " A secret, special bond this monogram: "A.S." was interchangeably Anna or Aurel; a mystical bond that united them in their love of what the classics stood for—all that was good and noble in the past. The scholarly pursuit of learning the mother left to her brother Ignaz; her role was rather to prepare Aurel to be at ease in the great world to which his talent and training would take him: she taught him the importance of dress, deportment, and manners.

The bent of her purpose, sheathed in affectionate complaints, admonitions, and presents, comes out in the letters she wrote her twelve-year-old Aurel. "I learned from Ernst that Mr. and Mrs. Lamm visited you and you did not even mention it although I am sure they were very kind to you and praised you to Ernst. Did you get the suit? Does it fit? Please take care of it so it is still nice when we are in Töplitz where I shall buy you everything you need, even shoes. . . . Now my dear child, I ask you to be very industrious so that I can be satisfied when I spend the holiday with you. Many, many kisses till then from your loving mother."[5] In October of the same year: "My dear son. I want to tell you that I really long for a letter from you and cannot understand why I have had no reply either to my letter or to the big parcel of presents I posted to you. I would like to hope that you are well and shall perhaps have some news today. . . . How did you spend the three days of the holiday and what did you use them for?" And again: "I am constantly waiting for letters from you, for although I received your New Year's greetings and do admire its faultless writing, one feels that this was prescribed for you. Once a year this is all right, otherwise I'd prefer reading your own thoughts. Deutsch brings this letter to Dresden with our love and a box of goodies, and in addition I put in collars and some of your shirt material to be used to mend them. . . . I also gave three thalers in groschen coins. I saved this in your savings purse for you and I know you will make the best use of your money and not squander it on just foolishness but I do not want you to deny yourself what you need when you want to give joy to somebody. It was Papa who bought the little purse for the coins and I cherish the hope that all this will give you joy." In a brief note written when Aurel was already fifteen, his mother invites him to join her in an act implementing her recurrent themes of frugality and joy-through-giving: "Although I wrote you yesterday, I must pay for another letter to prevent your selling the collection of minerals. Kornel asked me to give him one like it as a present for his 13th birthday and you know very well that I am his godmother and in any case would want to give him a nice present."

How successful she was in inculcating such virtues and manners in Aurel and how conscious he was of her teachings, of this, he himself offers testimony. More than half a century later, when he was sixty-eight, on a letter from an American friend who playfully reported Boston's reaction to Stein's visit there (November 1929–April 1930)—"You must have behaved yourself your very prettiest and done all the nice little things your mother taught you when a small boy"[6]—Aurel added a gloss in pencil: "How intuitive! Yes, it was my dear mother who taught me to think of the feelings of others."

It is an exaggeration but not an untruth to say that Stein had three fathers. First, of course, there was his elderly "Papa," a grandfatherly figure, who seemed to be summoning up his own courage when he addressed his small, distant son as "my fine brave boy" and showed his own apprehensiveness that at Aurel's "tender years" he "dare undertake the long tiring excursion."[7] Then there was his uncle, Professor Ignaz Hirschler, who furnished him with an incomparable example in fortitude and intellectual commitment. And, lastly, there was his brother Ernst, so much his senior yet still his sibling. From Stein's references to them, the part each played in forming his qualities of heart and mind is clear. What makes their joint efforts quite special is the fractioning of the parental role, that ambiguous, ambivalent mixing of the loving father and the strict disciplinarian, the begetter of the usual love-hate relationship. Thus, while Aurel had no single figure who in the name of love demanded conformity and strict obedience, he did have, inbuilt and functioning, a trust in the essential goodwill of the older man, be it embodied in his professors, established scholars, or administrative heads, a respect that cut across ethnic lines to include Indian and Chinese as well as European scholars.

At the end of 1887 Stein took passage for India, passing through the Suez Canal, then a relatively new wonder, to first set foot on Asian soil. He wrote "my dearest, best Papa" almost daily beginning in February, 1888. His language borders on the sentimental. During the first decade after his transplantation to India, his words sound with his desperate effort to annihilate the vast distance separating him from home and reinstate the immediacy of contact. Stein's body of published writings is notable for its clarity and precision; its color comes from the extraordinary places he traversed and the "finds" he rescued from oblivion. But in these Sturm und Drang communications, the complexity of the German syntax is further embroidered with the late nineteenth century's addiction to a flowery style. Thus, in the same letter in which he requested the books with the "A.S." monogram, he also asks for the copy of the famous Brockhaus *Lexikon* because it "gave so much pleasure to our dear Papa during the past winter. Among the most precious relics I also count the small bolster from our dearest Papa's bed as well as that from the green divan on which so often I saw his dearest head rest during his afternoon nap. Also the black walking-stick which dear Papa leaned on during his last years."[8] Papa, who long ago had bought him the "little purse for the coins" had continued to care for Aurel's savings. Now he thanked Ernst for his offer to take care of his small capital: "I have according to your wishes informed the Commercial Bank to put my savings account as well as my checking account at your disposal as formerly it was at Papa's. . . . I will keep the savings from my May salary here so as to have a small deposit with the Agra Bank for unexpected expenses, something I did not have before. I could not deny myself the pleasure of sending some money for our dearest Papa who waited for it so longingly and sent everything I had at all times. Alas, that it was his last trip to the bank that took him outdoors on the 9th [May]."

The word "longingly," which Aurel wrote so baldly, needs an explanation. The books and pictures, evidence of the family's devotion to culture, had been bought before their third child upset the budget. The father sent Aurel to the Kreuzschule, but by the time of his postgraduate studies, age had forced the father to retire; it was Ernst who financed his brother's training. Frugality, so often implied or mentioned, was not just a philosophical ideal; it had its basis in reality. In their last years, the parents, as was customary in those days, lived with their married son. That Papa was not an old-age pensioner "longing" for a monthly allowance from India is clear from the consolation Aurel expressed to Ernst in that letter of 8 June 1889: "Our dear Papa was granted the favor of seeing you happy in your marriage and me materially secure and of knowing that I could expect positive results from my position in Lahore." Possibly, to hazard a guess, well-meaning friends must have repeatedly asked Papa how a man could make a living out of dead languages, and, perhaps, he himself had wondered how Aurel would support himself. The money that came regularly from India justified everything: his monthly walk to make the deposit in the bank was a kind of triumphal march.

Two and a half years after Stein lost his "dear Papa," he was notified of his Uncle Ignaz's death. He wrote to Ernst in December 1891: "I feel as if I had lost my father for a second time." His letters during that period were filled with plans for a way to escape—the word is not too strong—and return to Europe. The sad news tempered his discontent; home was now invested with desolation: "I do not even want to think how empty our *Vaterstadt* has become. What can ever replace the strength always offered to us by his fatherly advice?" Ernst had sent him a sheaf of laudatory obituaries; their presentation of his uncle's life made him curious to know how the man himself had viewed his extraordinary career. "I hope the fragments of the autobiography Uncle Ignaz was dictating during his last year will be soon available."

Professor Ignaz Hirschler, 1823–91, was a pioneer in a new branch of medicine: a doctor who specialized in the treatment of diseases and affections of the eyes. For much of his working life, he was the only eye doctor in all Hungary. Until late in the nineteenth century, the living human eye was as unknown and inaccessible as the North Pole; its physiology invited discovery and exploration. In an outpouring of articles, Hirschler described conditions which, it is hard to believe, waited for instruments and techniques of the twentieth century to handle. His reputation was established long before Aurel was born. He was a graduate of the celebrated medical school in Vienna where he remained on for some years as an associate of his professor, also assisting him in teaching; in 1847, to learn what other specialists were doing, he spent three years in Paris, working and teaching in the medical school there. Returning to Budapest—a fervid nationalist he missed the 1848 uprising—he applied to the university for a teaching post. He was refused; he was a Jew. Undeterred, he trained students in a hospital for poor children; aroused, he founded an association that agitated for Jewish emancipation and in 1867 succeeded in winning political rights; he raised funds to send Jewish youths abroad for graduate study; for years he served as the head of the Jewish community. After he had retired, 1881, the government that had refused him a professorship honored him with a medal. This did not silence him any more than the rebuff had; he organized another group whose goal was to secure civil rights for the Jews—finally, five years after his death that too was attained (1896).

Is there a hint that despite the loving ties between Anna Stein and her brother the baptism of her sons did alienate their families? It might seem so from Aurel's asking Ernst to "let it be known in the appropriate quarters that I would appreciate one of Uncle Ignaz's books, the Latin edition of Horace."[9] However, no shadow of disagreement came between uncle and nephew as the boy responded to the older man's love of antiquity and his informed interest in archaeology; from the continuing exchange of letters it is clear that Aurel received the support of his uncle's Jewish friends, Goldziher, the eminent Arabist, and Vámbéry, the daring traveler and Orientalist: the alternate solutions to the religious problem were mutually respected. The model his uncle provided—he walked with confidence among the Hungarian fashionable and powerful and moved easily in the literary and scientific circles—was the leavening influence in Aurel Stein's drive to achieve. As Ignaz Hirschler had been a pioneer and explorer in his profession, just so his nephew would become in his discipline.

And Ernst? Steady, reliable, alert to duty, circumspect, orderly, a trifle overanxious—the adjectives pile up; a practical man, a man devoted and faithful. He gives no sense of inner excitement, high ambition, or special talent; nor, happily, any trace of envy for those marked for renown. He supported Aurel's postgraduate studies; later, when at times Aurel felt out of step, he cheered him on. He was a spectator—a parade needs spectators along its line of march—and he willingly accepted that role. One prerequisite of the spectator is that he stand still: Ernst became the necessary fixed point, the reviewing stand. While Aurel was a student, his way ahead was clearly marked and Budapest, where his parents and Uncle Ignaz lived, was home—essential to his well-being. But when in 1888 he arrived in Lahore to begin the unmarked path of a career—he felt his time-consuming position offered neither financial security nor scholarly opportunity; and when, in fairly quick succession his mother, father, and uncle died, then wherever Ernst lived was home.

Little can be salvaged of the facts of Ernst's life. At forty-three he married; Hetty, his wife, a beauty and an heiress, was a sensible thirty-two; they had two children, Thesa and young Ernst. Sometime after Papa's death he became the well paid manager of a coal mine in Jaworzno, a town in the booming, industrialized part of Galicia near Upper Silesia. Aurel addressed the envelopes "Herr Wohlgeboren, Herr Generalsecretar"; he began, "My dearest Ernst" and often closed "with love to dear Hetty and a thousand kisses and yet another thousand." The brothers corresponded regularly and often: pages filled with clear cursive writing detail the minutiae of daily events—illnesses, servant troubles, news of relatives and friends, vacation plans, the children, and remarks about investing Aurel's savings or explanations why they were omitted. It is possible to reconstruct the rhetoric and content of much of what Ernst wrote; both were mirrored in Aurel's letters, for both conform to the model set long ago by Papa who insisted that the writer comment on the news received before giving his own. Only to Ernst did Aurel feel free to complain, to confide his hopes and plans and stratagems—he was eager to get a position in Europe or, if that failed, to better his situation in India—to relate how he lived, whom he met, where he traveled, and to give the progress of his first major project, the editing and translation of Kalhana's *Rajatarangini: A Chronicle of the Kings of Kashmir.*

Ernst had the immense satisfaction of seeing the beginning of his brother's success. Over the years, through the letters he had followed Aurel's changed attitude toward India: at first Ernst underlined the words and phrases charged with homesickness. These grew fewer and fewer as India, the land of exile where Aurel felt imprisoned by his job, became the land that offered unique opportunities, the land that brought to his heart the friendships it craved—with the Andrewses, the Allens, the Arnolds—which were to last all the days of their lives, and a particular place that became home, his beloved Mohand Marg, the high alpine meadow in Kashmir where, whenever his duties and the season permitted, he would pitch his tent and work. From his camp there almost the whole of Kashmir was spread before him, "from the ice-capped peaks of the northern range to the long snowy line of the Pir Panjal—a little world of its own enclosed by mighty mountain ramparts,"—" 'the land in the womb of the Himalaya.' "[10]

Best of all, Ernst saw the changes in his brother with his own eyes when Aurel had home leave in 1893 and again in 1897, and he was there to rejoice when he returned from his first spectacular archaeological expedition into Chinese Turkestan, June 1901, the hero of scholars and the public alike. At last, Ernst could relax: his young brother's long, worrisome apprenticeship was finished, and before him was an assured future, financially as well as intellectually. He had found his road and marked it well.

In October of that same year, Ernst had a stroke. (It was Hetty who sent him the news, sometimes heartening, sometimes disturbingly ominous.) There is a touching letter on 26 March of the following year, 1902: "I thank you for acknowledging the wish I expressed last month," Aurel wrote Ernst, and his formality carries the solemnity of his message. "You deny me the fulfillment of what my conscience considers a duty and which would give me the most heartfelt satisfaction. However, you do understand that everything I have and can acquire or earn belongs to you and your family the instant you might need it. I am thankful for your promise and only hope that no unfortunate events will make it necessary for my solicitude to be remembered. . . . I beg fate to allow you to get well and allow me to see you again." Half his wishes, fate granted him: Aurel had a farewell visit a few weeks before Ernst died, October 1902.

In such a manner Aurel Stein was formed. What kind of man, of human being was he? Short of stature—his passport gave his height as 5'4"—with his mother's slight frame; wiry, a man of exceptional strength and stamina, he took delight in strenuous physical activity. It is told that when he was in his sixties on a tour in the rugged hill country along the Northwest Frontier of India, a young soldier native to that region, who was detailed to accompany him, reported to his officer, "Stein Sahib is some kind of supernatural being, not human; he walked me off my legs in the mountains; I could not keep up with him. Please do not send me to him again, Sir."[11] The one weakness in that otherwise tough physique was a recurrent dyspepsia, perhaps the price Stein paid for his "idea that time given to such grossness as feeding is time ill spent."[12]

Time, for Marc Aurel as for Marcus Aurelius, was a most precious commodity—every moment of "the hair's breadth of time assigned a mortal" should be spent wisely, nobly, fully. How the emperor found relaxation is not clear, but happily Stein included long daily walks or rides (he was an accomplished horseman). Once he tried riding a bicycle but gave it up because the "shameless thing" did not watch where it was going, and he decided he could "not spare the time to have a broken leg."[13] On occasion he enjoyed picnics, tea, or dinner with friends, and, when it seemed advisable (depending on the guest list: he was a master at manipulating the "mysterious processes of officialdom," as one of his best and oldest friends recalled with admiration and affection),[14] the parties and formal gatherings of which the British cantonments in India had a wide assortment. Most agreeable of all was an evening of good conversation to which, according to all in his wide circle of friends, he contributed pleasantly.

As with time, so it was with money: he was frugal in his tastes and modest in his habits; the price of any article or service, whether spent or earned, was considered and noted; he preferred second-class accommodations on boats and trains to the extravagance of first-class passage. Daydreaming about the future was, he wrote Ernst on 8 September 1892, "a pleasant pastime and does not cost anything." But he was not stingy. He had learned at his mother's knee that generosity, not self-indulgence, gave pleasure; generosity thoughtfully directed to relatives and friends, to associates and servants, be it in presents sent from distant places calculated to arrive in time to celebrate an anniversary, public recognition for the efforts of a colleague, acknowledgement of indebtedness for help or information given, or energetic efforts to secure pensions or positions for those who had served him well—all reveal this side of his ledger sheet.

He was methodical. He routinely went into *purdah* (the Urdu word for the seclusion enforced on women), to spend ten to twelve uninterrupted hours a day writing the reports for each of his many expeditions. He had no home. Living arrangements were temporary and centered on his books and papers: the books carefully chosen for study, reference, and refreshment (he thanked Ernst enthusiastically for the volumes of Sainte-Beuve sent him) and masses of papers carried with him in capacious brief cases, each clearly labeled "Personal correspondence," "Proofs" (books or articles in press), "Journal," "Work-records," "Map data," "Accounts," "Photographs," and so forth. As long as he had his books and papers with him, wherever he was he felt at home; a bed, a table— or any adequate writing surface—a chair and lamp—and his home was furnished. Eventually a fox terrier became a fixture in his life; always called Dash, the last being Dash VII, he trotted along on his expeditions and when "in purdah" on the evening walk.

However simple his living requirements were, one element was vital: a view. His letters are filled with descriptions of the views his tent commanded. Nature is always mentioned and always sketched in: the panorama of mighty mountain ranges and their glaciers, of forested hills and flower-carpeted meadows; of the vast Taklamakan desert, that waterless Mediterranean of Inner Asia, whose waves are gigantic sand dunes cut and shaped by wind, along whose shores are isolated oases, settlements connected with one another by the ancient international caravan routes which had seen armies and pilgrims and traders move between East and West. In much of his travels he was conscious that he was moving through a pristine world where men passed but did not linger. His letters describing this terrain, which he covered on foot, by camel and horse, move at the same slow pace as they attempt to convey something of what he saw and felt.

Stein was methodical in the larger sense—systematic or organized might be better words—in the thorough way he planned, prepared, and concluded his expeditions—but that is part of the unfolding story.

Stein never married, nor is there any hint of any relationship with anyone of either sex. He was not a misogynist: his sister-in-law Hetty and

Helen Mary Allen, the wife of his lifelong friend, knew him to be devoted as well as appreciative of their qualities and that he enjoyed the companionship of women. Since his parents' union was the basic fact of his childhood, why, it must be asked, did he never marry? His parents' marriage inevitably formed his image of marriage, an image which he firmly believed could not be compatible with what he decided was to be his life's work. "His work as the first consideration under all circumstances governed his life," Andrews explained in a context quite different but pertinent and plausible. "He planned well ahead and never allowed anything to interfere with the furtherance of his plans."[15]

His plan, the grand design for his life, was collocated while he was still at the Kreuzschule: Alexander the Great, the youthful world-conqueror, became his hero and provided the incentive to become immersed in Greek and Roman histories and later to master the ancient languages and history of Persia and India. For the remainder of his long life Stein was faithful to solving the problems posed by his boyhood idol's achievements: in various expeditions he traced the routes of the Macedonian's Eastern campaigns, trying to verify the exact places where critical battles had been fought; his very last reconnaissance through the wastes of Gedrosia when he was over eighty was to settle the route by which Alexander had led his troops in their wretched, near-disastrous retreat from the Indus Delta to Persia. By his expeditions and by his words, Stein acknowledged the spell Alexander had cast over him. (It would seem likely, whether he was conscious of it or not, that his schoolboy's imagination had also been influenced by Schliemann's excavations. That millionaire, amateur archaeologist who, with Homer in one hand and a spade in the other, announced in the 1870s that he had unearthed Priam's Troy and the golden capital of Agamemnon, metamorphosed the classics, making them history and history a subject to be deciphered at the end of the archaeologist's spade.)

Later, when he was a graduate student, two other men added further dimensions to his grand design. One was Hsüan-tsang, the seventh-century Chinese pilgrim who took the long overland route to India in order to visit the sites holy to Buddhism; his great travelogue, *Record of the Western Regions,* at once accurate, richly informative, and recognized for the light it had cast on early Indian geography, made Stein call him his Buddhist Pausanias. (Sir James Frazer had written, in his introduction to his translation of Pausanias, that without him "the ruins of Greece would be a labyrinth without a clue, a riddle without an answer.") During Stein's first expedition into Chinese Turkestan, Hsüan-tsang, the Buddhist Pausanias, was promoted to patron saint.

The other man in whose tracks Stein followed was the late thirteenth-century Venetian trader, Marco Polo. His book, translated by Sir Henry Yule, described the kingdoms and marvels of the East. From these three, Alexander the Great, Hsüan-tsang, and Marco Polo, Stein formed a triangle whose sides connected antiquity with the Middle Ages, East and West, international trade with a universalistic religion, Buddhism. Stein's role was to perceive their interconnections, and through the work which lasted throughout his long, long life, he made many archaeological reconnaissances to verify, reconciling record and site, and thus giving them life and meaning for future workers.

Stein chose the world within the triangle he had formed, a world whose promise and excitement and discovery filled his days and years and brought him a sense of fulfillment. Marriage must not interfere; he put even the thought of it aside; he remained single; he had no regrets.

2

Of the influences that directed Stein to India, two can be singled out. The first was the importance given in the latter part of the nineteenth century by Central European scholarship to philological studies. A chair in comparative philology graced every well-known university. Comparative philology, as linguistics was then called, was born in 1786 at a meeting of the Asiatic Society of Calcutta, where Sir William Jones (1746–94), who had founded it the previous year, said: "The *Sanskrit* language, whatever be its antiquity, is of a wonderful structure: more perfect than the *Greek,* more copious than the *Latin,* and more exquisitely refined than either, yet bearing to both a close affinity."[1] This address and his subsequent work in his tragically brief life ushered in a grand era of comparative philology when Sanskrit and Persian became the keys to unlock the prehistoric world of Indo-European, its parent language whose considerable progeny flourish in many modern languages.

The other bridge to India was furnished by the life and labors of the Hungarian Csoma de Körös (1784–1842), a legend in his own time and eventually a patriotic symbol stimulating his countrymen's interest in Oriental research. His romantic pilgrimage, his death in the distant Himalayan foothills captured the imagination of more than one Hungarian who in one way or other and for one reason or other made their way to the East. A penniless student who rose to be an underpaid teacher in his college in Transylvania, Csoma's harsh frugality was a lifelong assertion that money and scholarship must not bed down together. Even when given a room in the Asiatic Society's building in Calcutta, Asian as well as European scholars marveled that he subsisted entirely on tea enriched with dollops of butter, Tibetan style, and small amounts of plain boiled rice; that he slept on a mat on the floor flanked by boxes of books, rarely emerged from his cell-like room, and had but one suit which he wore day and night.

He started his journey as a dream-possessed wanderer who expected to find in the wilds of Central Asia Greater Hungary, the supposed cradle of the Hungarian people whom popular belief connected with the Huns. He became an ascetic scholar whose special talents and unusual mentality were directed to Tibet, an involvement that postponed his quest for twenty years while he explored and explained the "recondite Buddhist lore of Western Tibet."[2] Stein was at pains to place Csoma's pioneering work in its proper perspective. "In the narrow circle of Western scholars devoted to the study of Buddhism, his memory was greatly respected as that of the first European who had acquired a systematic, scholarly knowledge of the Tibetan language and opened access to the extensive literature composed in it. His published grammar and dictionary of Tibetan were acknowledged as standard authorities on the language. But subsequent labours showed that most of this Tibetan church literature was but translated from the Sanskrit and, as further research succeeded in recovering more and more of these sacred texts of Buddhism, Tibetan studies in Europe slackened for a time and with them also the interest in this remarkable man who had been their founder."

In 1819, when Csoma was thirty-six, he began his travels. On foot he made his way, with long detours to avoid regions where cholera and the plague were raging, through Turkey, Egypt, and Persia to Bukhara and thence via Kabul and Lahore to Kashmir. In the years spent on the road, hardships and privations were dismissed—Csoma merely listed localities and dates. And so he came to the mountain kingdom of Leh, the chief settlement of Western Tibet, a gateway to Central Asia. There he was solemnly warned that "the road to Yarkand was very difficult, expensive and dangerous for a Christian." In the hope of finding another way, Csoma returned to Kashmir (1823), where quite by chance he met another traveler, William Moorcroft, who told Csoma of his own experience with the heartless rapacity of the hill tribes beyond Leh. Moorcroft, who was to perish the next year on his way to Yarkand—did he die of natural causes or was he poisoned?—possessed a rare combination of fearlessness and sensitivity. Moorcroft's appreciation of Kashmirian weaving began the European vogue for cashmere shawls; it was his intuitive perception of Csoma's unusual qualities "that [enabled] the wandering scholar, who previously had been content to pass through some of the most fascinating and then practically unexplored parts of Central Asia without adding to our knowledge of their people, geography, or archaeology," to settle down to Tibetan studies. These occupied him for the next twenty years.

"Your opinion of Csoma's character accords with mine," Stein wrote in a letter to Ernst 27 September 1896. "He became the discoverer of Tibetan literature through his chance meeting with Moorcroft. Who knows where his urge to travel might have taken him! His ascetic life and his iron determination remind me of Anquetil-Duperron [1731–1805], the discoverer of Parsi literature. His etymologizing which has often been called childish, is probably the result of insufficient methodological training. Certainly there was no lack of opportunity in Calcutta to become familiar with modern comparative linguistic studies, but Csoma was probably too old to warm up to new methods."

Stein's equating of Csoma's inner rigidities with those of Anquetil-Duperron is spelled out by Waley in his short, appreciative study, "Anquetil-Duperron and Sir William Jones." Waley describes Duperron's old age as touched with a distasteful "sordidness": choosing to live on "four sous a day, explaining 'Nulla corporis lintei lotio mutatio,' a Tacitean way of saying his underclothes are never washed or changed."[3] Also, as Csoma translated Buddhist Tibetan renderings, so Duperron used a contemporary Persian translation to render the Sanskrit Upanishads. While his work invited Sir William's scorn, it influenced European thought: in 1813 his rendition fell "into the hands of Schopenhauer and soon permeated his philosophy."

Was Stein's a personal judgment of Csoma, or was it a youthful reaction? Fifteen years later, when he himself had made his way into Central Asia, he had professional as well as patriotic reasons for claiming kinship with him. Both had left home in pursuit of a dream and had willingly accepted the isolation and immense toil it had imposed. But unlike his heroic, misinformed countryman, Stein had reached Central Asia with results far richer than he had dreamed of. In 1912, while encamped on his beloved Mohand Marg, he looked down on the valley of Kashmir into which the earlier Hungarian traveler had made his way and also looked back on the hundred years that separated their presences there—he was intent on reconciling Csoma's legend with reality. He sought to redress the judgment passed on Csoma's quest, arguing that it was not quixotic but premature. He used an occasion both appropriate and felicitous: the Hungarian Academy of Sciences' honoring the memory of one of its members, Dr. Theodore Duka, of whom more will be said on a later page. Csoma de Körös, Duka, and Stein—three Hungarians associated with work in India; Duka, who had served the Indian Medical Service for a quarter of a century, had, on his retirement, piously collected unpublished material for his "biographical sketch" of Csoma. To set this sketch properly Stein wrote: "It was natural that those serious philologists who, during the last third or so of the nineteenth century, did so much to clear up the true affinities of the Hungarian language by systematic research on the lines of comparative grammar, were inclined to think of Csoma's aims and achievements with something like pitying indulgence."[4]

Condescension was a gesture alien to Stein. Gladly, scrupulously, he always acknowledged his indebtedness to those travelers and workers who had marked the path he followed. But for Csoma de Körös he felt something special and, musing on the parallel between Csoma and himself, he saw the hand of fate in the guise of the warning given to Csoma at Leh. The warning had "frustrated a design which seemed destined to anticipate, as it were, our archaeological activities a century later."[5]

There was neither fumbling nor hesitation in the course of Stein's education: from boyhood on, with no interruptions, his work was clearly marked by his talents and interests. At the Kreuzschule in Dresden, which he attended from 1872 to 1877, his most noticeable talents were an exceptional facility in languages and the ability to study harder, longer, and better than most of the other students. Even his childhood "hobby," a fascination with old topographies, had a future. From the Kreuzschule he returned home to attend the Lutheran grammar school, the "gymnasium" which prepared him for the university and where he began his Oriental studies. During his years there he became familiar with the Indian, Persian, and Central Asian collections in the library of the Hungarian Academy of Sciences, especially with the report of Lajos Lóczy, a geologist-geographer who, on an East Asian expedition, mentioned the cave temples at Tun-huang. He spent happy years, reading voraciously and living at home. His ambition took form then—to prepare himself so that he might obtain a teaching post in Oriental studies at some university, an ambition that remained strong during his first decade in India. After the gymnasium he went, as was customary for students in Central Europe, to different universities to listen to the lectures of the Orientalists and to choose from among them the professors under whom he would work.

Two men guided his studies, giving them depth and focus. Rudolph von Roth (1821–95), professor of Indo-European languages and the history of religions at the University of Tübingen, and George Bühler (1837–98), professor of Indian philology and antiquities at the University of Vienna, an authority on Indian paleography. Both introduced Stein to the finest level of contemporary scholarship; their heuristic methods prepared their students to see problems in interrelated disciplines. Roth was co-author of the monumental Sanskrit-German dictionary (known as the St. Petersburg Lexicon, its seventh and final volume was published in 1875), which Basham considers "probably the greatest achievement of Indological scholarship in 19th Century Europe."[6] Roth had handled the Vedic portions (of which the Rig Veda, the oldest Indian religious text, composed about 1500 B.C., was in a Sanskrit already then fairly archaic) and had an unrivaled knowledge of Sanskrit manuscript materials.

It would be wrong to assume that, because the scholars of that world were fairly esoteric, they were stodgy; there was in all their research a Sherlock Holmesian aspect. It is told, for example, how Roth came to possess a unique Vedic manuscript. From a traveler's garbled account of a Brahmanic cult in Kashmir, he made an informed guess that its existence indicated an unknown version of a text of the Arthava-Veda (a collection of magical incantations combined with a number of hymns of a prephilosophical, speculative nature). Through scholarly, quasi-official channels, he persuaded the British authorities in India to try and locate such a manuscript in Kashmir, that then inaccessible earthly paradise. After many years, it came to pass that in 1875 the maharaja of Jammu and Kashmir sent the manuscript to Sir William Muir, the lieutenant-governor. His Excellency was nonplussed when the object of negotiation was handed to him—a messy bundle of grimy, tattered birch-bark leaves, 287 pieces held loosely together by a cord passed through a hole in the center of each leaf. Help was at hand. An urgent wire brought George Bühler, then professor of Oriental languages in Bombay, to the viceregal mansion. One look at the manuscript made from the trimmed and smoothed inner bark of birch trees and Bühler knew that what it needed was a washing. Reassuring Sir William that the ink used would not be affected, he proceeded to launder the bundle in the lieutenant-governor's bathroom. The manuscript was restored by its bath and Sir William was impressed by Bühler: the scholar entrusted the bundle to a native bookbinder, and a week later the pages, clean and properly ordered, were sent to Roth. (On his death it became one of Tübingen University's greatest treasures.) This was the man and this the university which in 1883 granted Stein his Ph.D.

Bühler, unlike most other continental Sanskritists, lived and worked in India before being called to Vienna where Stein studied under him. It might have been that Bühler's stay in India gave Stein the idea that he too might achieve a professorship by taking the same circuitous path via an Indian college; what is certain is that he was following Bühler's example when he headed straight for Kashmir to find a particular manuscript whose existence Bühler had established.

The story of this search begins far away and long ago with the earliest visit of a European to Kashmir in the mid-seventeenth century and was concluded by Stein's masterly edition of Kalhana's narrative poem, *Rajatarangini: A Chronicle of the Kings of Kashmir*. This twelfth-century chronicle relates only the history of Kashmir, not of the rest of India; yet, as Stein explained, it aroused keen interest as "being practically the sole extant product of Sanskrit literature possessing the character of a true chronicle."[7] Europe first learned of this "histories of the ancient Kings of Kachemire" from a Persian compilation; subsequently, with the increasing desire to learn more of India's history, the need to find Kalhana's chronicle grew. However few the searchers were, they included distinguished names: Sir William Jones, who died before he could carry out his declared intention; Horace Hayman Wilson, Boden Professor at Oxford who held the first chair in Sanskrit (1832) and whose "Essay on the Hindu History of Cashmir" (1825) introduced European scholars to the first cantos of Kalhana's epic; William Moorcroft, who before this *Essay* was published had arranged for a transcription to be prepared for the Asiatic Society's library; the polymath General Sir Alexander Cunningham (1814–93), father of Indian archaeology, who, as a young officer stationed in Kashmir after the first Sikh War, using coins he had collected, fixed some dates in Kalhana's account (*The Ancient Coinage of Kashmir, With Chronological and Historical Notes*, 1846) and by iconographic reading of ruined Hindu temples, identified localities and thus began recreating Kashmir's ancient topography (*An Essay on the Arian Order of Architecture . . . , 1848*); and finally, Bühler, the expert on ancient Indian scripts who made a special trip to Kashmir in the summer of 1875.

Why did Bühler go to Kashmir? Why did Stein call it his "memorable tour?" In the half century between Moorcroft and Bühler, workers in epigraphy, linguistics, and numismatics had begun deciphering the mystery of India's past to separate the historical from the wonderful and legendary. Bühler's firm knowledge of Indian scripts convinced him that many of the errors and much of the confusion in the *Rajatarangini* were due to faulty transcriptions. He agreed with Wilson, who long before had characterized them as being so defective "that a close translation of them, if desirable, would have been impracticable." The first prerequisite, Bühler saw, was to obtain a script free from clerical errors. The copies available were all in the Devanagari script: therein, he was certain, lay the fault. This script, in which Sanskrit and modern Hindi are written, did not become common in Kashmir until the second quarter of the nine-teenth century when it replaced the S'arada script used by Kashmirian scholars. (The differences can be likened to those between German writ-ten in Gothic and German written in Latin script.) Bühler concluded that the numerous corruptions in all the Devanagari transcripts made the recovery of a S'arada text imperative, and not just any S'arada text but the particular one from which Moorcroft's transcription had been made. "The versified colophon attached to this transcript, informs us that the original Manuscript from which the Annals of the kings of Kashmir were copied had been obtained from the learned Kashmirian Pandit S'ivarama [who] is praised in the colophon as the representative of that family which alone in Kashmir had always preserved a copy of the Royal Chronicles."[8] There was one, and only one, true manuscript: the *codex archetypus*.

Bühler set out to find it. In Kashmir he traced it to the grandson of the pandit who had permitted Moorcroft to have it copied. There his good fortune deserted him. Bühler was permitted only a glimpse before the owner took the manuscript away. The brief look at the peculiar, ex-tremely cursive writing made it clear that it would present difficulties for anyone not familiar with S'arada script. Bühler's brilliant deductions, his fondest hopes, his pilgrimage to study the archetype all came to naught; he left India having failed to get that manuscript.

Stein heard the story from Bühler himself. Though he already knew its plot from the *Report* Bühler wrote (1877), he shared the sense of defeat Bühler still felt. It might well have been then that Stein thought that if he could get to India, he might be the fortunate one to secure that S'arada archetype which his teacher had located and whose transcription waited to be done. It was a task worthy of the preparation he had re-ceived. How was he to get to India? He did not know. The day for a wanderer like Csoma de Körös had passed with the passing of the East India Company. It became clear to Stein that the road to India began in England; and to get there became his first objective.

After receiving his Ph.D. in 1883, with laudatory commendations from both Roth and Bühler, Stein received a stipend from the Hungarian government to cover two years of postdoctoral research in Oriental languages and archaeology at the Universities of London, Oxford, and Cambridge (1884–86). He lost no time in getting to London, and there one of the first men Stein called on was Dr. Theodore Duka (1825–1908), a corresponding member of the Hungarian Academy of Sciences and known to his Uncle Ignaz. Duka was one of the well-born, well-educated Hungarians whom the abortive revolution of 1848 had cast up on the political-asylum shores of England. After eluding the Austrian police and eschewing the Hungarian plotting-pots of Paris, Duka entered medical school. There his close friendship with a fellow student, the son of a general who had served in India, secured Duka an appointment as surgeon to the Honorable East India Company's Establishment in the Bengal Presidency, 1854. With the company's demise following the Great Mutiny, he joined the Indian Medical Service and held the high rank of surgeon-colonel when he retired in 1877. *In Memoriam, Theodore Duka,* which Stein wrote out of "motives of *pia memoria"* (he wrote it in 1912, taking time out from his pressing labors of preparing for his Third Central Asian Expedition), was his way of discharging an obligation to a man "whom I had revered not merely [as] the scholar, soldier, and man of truly noble character, but also the kindest paternal friend."[9]

Between the two, the young scholar abroad for the first time and the retired, older man settled after years of Indian service, there was an instant mutual appreciation and respect which made for a fine friendship whose nature was implicit in their backgrounds and pursuits. Although a British subject, Duka was still a Hungarian patriot. His time and energy were servant to "his desire to secure credit for Hungary's share in the advancement of science and his eager wish to stimulate interest for Oriental research in the country of his birth." And Stein, the epitome of this desire and wish, now speaking Magyar, now English, was made welcome to Duka's hospitable home, large and comfortable and located in the part of West London jokingly called "Asia Minor" because so many retired Indian and colonial officials of distinction had settled there.

Luck was with Stein in the year of his grant. Had he come to England in 1883, he would have missed Duka who was then in India collecting material for his biography, *The Life and Works of Alexander Csoma de Körös,* to mark the centenary of his hero's birth; had he come a year later, Duka would have been preoccupied with arrangements for the International Medical Congress to be held in Budapest. But 1884 was propitious. Duka, engrossed in his biography, found Stein a congenial companion, conversant with Csoma's life and helpful in evaluating his contributions. A frequent guest in the Duka home, Stein there met two men each of whom had won a knighthood and an appointment to the powerful India Council: Sir Henry Rawlinson (1810–95) and Sir Henry Yule (1820–89). Duka's early faith in his young friend steadied into a lasting friendship. Stein wrote that Duka's last letters "which followed me far into Central-Asian deserts showed no sign that his life's warm sun was rapidly sinking. But my last letter written in March 1908 [from a camp in the midst of the Taklamakan Desert] was destined not to reach him." Thus Stein ends his filial *In Memoriam*.

Stein made a clear distinction between "paternal friend" and patron. For Duka he wrote an obituary addressed quite properly to Hungarians; but to Yule and Rawlinson he dedicated reports of his Central Asian expeditions addressed to the international scientific community. The superb two-volume *Ancient Khotan*, published in 1907, the account of the scientific results of his first probings of Chinese Turkestan's past, Stein inscribed: "To the Memory of Colonel Sir Henry Yule, the Great Elucidator of Early Travel and a Pioneer in the Historical Geography of Central Asia, this account of the antiquarian results of my journey on which his works were my best companions, is dedicated with deep respect and admiration for the scholar, the writer, and the man."[10] Similarly, the scientific and artistic results of the extended third expedition were presented in 1928 in the four-volume *Innermost Asia: Detailed Report of Explorations in Central Asia, Kan-su and Eastern Iran;* its dedication ran: "To the Memory of General Sir Henry Rawlinson, Bart, whose labours in the field and study illuminated the ancient history of Asia, this record is inscribed in grateful remembrance and sincere admiration for the explorer, the scholar, and the man."[10] Though the lofty, sonorous, orotund phrasing of these dedications might be typical to that era and proper for such majestic statements of achievement, the wordings are precise and pertinent, carefully chosen and not inappropriate. At the peak of his fame, Stein was paying homage to two men, much, much older, famous and important, who had responded to him when he was a young, untried foreign student.

Giants old and young they were—a difference in time, not kind. The older men who had realized their potentials recognized the yet-to-be-realized potential of the younger man; they could appreciate better than others that what separated them was not a difference in age but the considerable advances made since their youth in Oriental studies. Yule and Rawlinson had gone out to India as soldiers when India was still administered by the East India Company; by roundabout ways they had come to their chosen work; they were amateurs who, writing their own curricula, had become scholars. Vastly different from Stein whose academic training had fitted him to make straight for his goal before ever he set foot in India. Asia, they knew, was where he belonged; once there he too would find his life's work.

It is possible, though difficult, to be brief about Yule and Rawlinson, important in Oriental studies and crucial in Stein's life. In 1824, Rawlinson, a beardless fourteen-year-old military cadet, had gone to India. He made the voyage out in the company of the newly designated governor, an Oriental scholar, who in leisurely shipboard talks stimulated the boy to think of the East as a vast primer wherein lessons were waiting to be read. Whether his youth facilitated his learning languages, or whether his linguistic talent revealed itself early, is a moot point; within a year Rawlinson had become familiar with several Persian and Indian vernaculars and was acting as interpreter. The studies he began then continued and dominated his life. In 1833, as one of the officers sent to Persia to help train the shah's army, he won his first acclaim for a spectacular feat: a practiced horseman he made a 750-mile ride in 150 consecutive hours to warn the British minister at Tehran of the arrival of a famous Russian agent in nearby Herat. (This was the beginning of the shadowboxing between the British and the Russians as, extending their Asian frontiers, they approached each other in Afghanistan.)

In Persia, inevitably, Rawlinson was attracted to one of the wonders of the world, the long-known, undeciphered inscription at Behistun. On a blood-red cliff rising wall-like 500 feet above a cluster of springs, a favorite camping site for caravans plying from ancient times between Babylon and Ecbatana (Hamadan), Darius the Great (522–486 B.C.) had had his might depicted in a bas-relief. Under this scene he proclaimed his greatness in the three principal languages of the Persian Empire—Babylonian, Old Persian, and Elamite. All were written in cuneiform, the name given the wedge-shaped, or arrow-headed, script. When his workmen had completed the 1200-square-foot complicated sculpture, they smoothed the surrounding surface leaving the rock-face sheer to make certain that no man should be easily tempted to desecrate Darius's words and scene—only a wayward trickle of water, obeying its own laws, blunted the letters in the Babylonian section. To read this ancient imperial message was a task congenial to Rawlinson's linguistic and athletic prowess. Lowered by ropes from a ledge above, risking his life as he dangled hundreds of feet above the ground, he began the laborious copying task in 1835. Nine years later, his patience worn thin by the interruptions of official duties, Rawlinson resigned from the military; two years later he had completed the first stage: stroke by careful stroke, all three inscriptions were in notebooks.

He began with the Old Persian section. Though he could identify it from similar inscriptions known from other ancient Persian sites, it was still unintelligible. To his considerable power of ratiocination and command of the vernacular, he added a bold intuitive skill: twenty-five hundred years after they had been carved, Rawlinson could spell out the words of the Achaemenian ruler. Mastery of the Old Persian encouraged him to try the more difficult Babylonian section. With the cooperation of two other brilliant linguists, the Irishman Edward Hincks and the Frenchman Jules Oppert—cuneiform's "holy triad"—Babylonian gradually yielded to their determined, adroit efforts. These were the very years when Henry Layard, inspired and encouraged by his friend, Paul-Emile Botta, who in 1842 at Khorsabad unearthed the palace of the Assyrian king Sargon II, started excavating at Nineveh (1845–51). The Louvre and the British Museum house the monumental sculptures sent to Europe, evidence of the splendor and the power of the Assyrians described in the Bible. Layard chanced on the palace of Ashurbanipal II, Sargon II's great-grandson, one of the earliest collectors and antiquarians—his enormous library preserved ancient cosmological myths, literary pieces, and sophisticated astronomical and mathematical calculations. Like the pyramids, this was real, unimpeachable evidence of the past—but doubts, suspicions, skepticism, frank disbelief began to cloud the decipherers' work.

For them the year 1857 was fateful. In *The Sumerians,* Samual Noah Kramer tells the story of the great test.

It was a mathematician and inventor and not a professional Assyriologist who brought matters to a head. W. F. Fox Talbot, who did research on integral calculus and helped lay the foundation for present-day photography, was also an amateur Orientalist. . . . Having obtained a still unpublished copy of an inscription of the Assyrian king, Tiglath-pileser 1 (1116–1076 B.C.), he made a translation of it, and dispatched it sealed to the Royal Asiatic Society on March 17, 1857, suggesting that the Society invite Hincks and Rawlinson to prepare independent translations of the same text and send them in sealed, so that the three independent translations might be compared. The Society did so and also invited Jules Oppert, who was then in London. All three accepted the invitation, and two months later the seals of the four envelopes containing the translations were broken by a specially appointed committee of five members of the Royal Asiatic Society. . . . All in all the verdict was favorable for Assyriology as then practiced; the similarities between the four translations were reasonably close and the validity of the decipherment vindicated.[11]

And so archaeology and decipherment could go forward together in fruitful partnership. Archaeology, then as now, is more than the thrill of finding buried treasure. An understanding of what the most common-place unearthed material may mean, what information is locked up in discovered inscriptions—this, the reason for the digging, lags far behind in popular appreciation. The relationship can be likened to that between exploration and cartography: Columbus crossed uncharted seas and found a world utterly unknown to Europe; he died without appreciating what he had found. That step waited on astronomical observations and calculations which gave the new land's longitude—the east-west measurements; only when this had been done came the certain knowledge that between the new land and the Cathay Columbus had sought lay another vast ocean. Like the cartographers, Rawlinson, Hincks, and Oppert gave scholars the tools to reach back into a lost, buried world, to announce the Sumerian world a cradle of civilization where, among its other inventions, there was writing, writing by which man could transmit personal and social experiences and the cultural creativity of the ancient Mesopotamian world "to generations yet unborn."[12] The conception of archaeology as a hand-maiden to the scholarly community guided all of Stein's work in the field.

Yule differed from Rawlinson and addressed himself to different problems. The Scottish home in which he grew up was tilted toward the East: his father, a good Persian and Arabic scholar, possessed miscellaneous Oriental knowledge. The word "miscellaneous" prefigures the rich texture of Yule's own work. As a boy he read William Marsden's *Marco Polo,* the latest and best version of a journey written just before 1300 that had informed and perennially fascinated Europe with its "wonders and marvels of the East." Certainly, the seed was planted then, and just as certainly at the same time and from the same book came a clear notice of what was entailed in attempting a critical edition of such a work. Marsden (1754–1836), who became an Orientalist and numismatist, had gone out to Sumatra for the East India Company and there had been impressed by the truthfulness of Polo's Sumatran details and amused that his medieval mind had identified the rhinoceros as the mythical unicorn. Feeling in need of scholarly guidance, he put his problem to the librarian at St. Mark in Venice whose reply outlined the prerequisites needed to do justice to the great medieval traveler's account. Among other competences, a study of Polo required a "full and precise acquaintance with the geography of the Middle Ages; with the travels of those days; with Oriental history; with the languages prevailing in early and modern times amongst the peoples; with the manners, the natural history, and the rare productions of those countries; and, at the same time, with the Venetian dialect of Italian."[13] This was, the good librarian felt, a necessary beginning to counteract the disbelief that had long clouded Polo's account. By creating a caricature, Marco of the Millions, part clown, part ruffian, Venetians expressed the hilarious contempt felt for a liar of such magnitude. Had not Marco Polo described black stones that burned, Indian nuts the size of a man's head, and paper as valuable as gold and silver coins? Centuries passed before Europeans saw coal, coconuts, and paper scrip.

It took Yule time, courage, and self-discovery to realize that he was a writer and a scholar. Before he retired from the army in 1858—he served as an engineer from 1853 until 1857—he had published his first book, *Mission to the Court of Ava,* an account of his visit as member of a commission to that kingdom in northern Burma. Perhaps it was his wife's illness that dictated his retirement, or perhaps that was an acceptable excuse; perhaps, in England, he was moved by the dynamic ideas then shaping historical scholarship. Lord Acton, at that time a student at Göttingen, the center for the new critical method, jotted down its relationship to the pulsations of the romantic movement: "Romantik. Its services to history. The alternate sympathy and detachment by which we understand and are able to judge times, ideas and men"; and: "Romantik enlarged the horizon of culture. Everything was brought into it, Antiquity, the East, Literature, Language."[14] Yule found his ideal topic, the blend of the romantic and the historical—Marco Polo's journey overland to Cathay, magical in name and image, which occupied a large part of the European imagination.

The time was doubly ripe for such a historical investigation: the Italian revolution of 1860 opened the archives—even those of the Vatican—and Yule combed those of Florence, Pisa, Palermo, and Venice. His search went far beyond Italy: an extensive correspondence with scholars, librarians, archivists, and helpful consular officials in nearly all parts of Europe. He was richly rewarded in the information he sought. In 1866 he published *Cathay and the Way Thither, Being a Collection of Medieval Notices of China.* It was a great book; time has only made it greater.

Yule had a tidy mind. This book was a historical reconnaissance, an exercise in surveying, collecting, understanding various motivations, organizing and placing the heterogeneous materials the librarian of St. Mark, long before, had deemed essential. He was, it must seem, clearing the way to focus directly on his main target: Marco Polo. In 1870 he published *The Book of Ser Marco Polo the Venetian, Concerning the Kingdoms and the Marvels of the East.* Yule's books were Stein's guides and companions in his explorations of Central Asia.

In 1885 Stein interrupted his grant to return to Budapest for his year's compulsory military training. Usually such a year is a wasted one for young men intent on and engrossed in professional studies. However it happened, he did his stint at the Ludovika Academy, the Hungarian army's mapmaking school. Its director, Captain Karoly Kuess, an outstanding cartographer, introduced the best methods then known to military surveyors. Stein's boyhood hobby in ancient topographies took on a new meaning—he understood the theoretical concepts and the technical competence that went into the making of a map. A map, like a book, expresses a particular idea; unlike a book, it is a graphic statement. At that time the best way to construct a map expressing the form, extent, and location of a portion of the earth's surface demanded a knowledge of surveying using triangulation and trigonometrical calculations. The techniques Stein learned at the Ludovika he put into practice on his Central Asian expeditions. On all three, his small party included a two-man team of native surveyors (trained by the Indian Survey Department) whose cartographic work he directed and coordinated. The results were presented in the *Memoir of Chinese Turkestan and Kansu,* published in 1923; the magnitude of such work in a region largely unknown and rarely traversed—carried on in addition to the archaeological program— can be gauged by what was accomplished in just one area—50,000 square miles in the northern Nanshan Range. The *Memoir* is an adult fulfillment of a schoolboy hobby.

When the year's service was completed, Stein hastened back to England. Much of 1886 was spent studying the coin collections at the British Museum in London and the Ashmolean Museum in Oxford. Stein had become aware of the way ancient coins were evidence of historical events. Numismatics, expertly used, had helped establish chronology, political power, and trade where written documents did not exist. His first scholarly writing, "Zoroastrian Deities on Indo-Scythian Coins," was published in the *Oriental and Babylonian Record,* 1887.

As Stein's grant-term was drawing to a close, Rawlinson suggested to the India Council that the young Hungarian was eminently qualified to fill the dual vacancy of registrar of Punjab University and principal of Oriental College at Lahore, newly founded. Endorsed by Yule, the

suggestion was quickly implemented. Thus, with no delay, Stein left his
student days behind him and, saying farewell to his family, was on his
way to India. Rawlinson and Yule were patrons in the grand tradition:
aloof by virtue of their age and position; receptive to talent; generous,
gracious, encouraging, and helpful. They created Stein's image of the
British gentleman-scholar—the bearing, the manners, the self-sought,
self-taught preeminence. If it meant hanging from a cliff or burrowing
through archival warrens, they had added significantly to the written
record that, whatever its script and however discovered, transcended
time and distance to add integrally to the sum of man's experience. They
were his models whose embryonic likeness he would perceive in the
younger men he met at Lahore—Andrews, Allen, and Arnold. Perhaps
his great patrons were in Stein's mind when, on a small plateau, under
a roof of heavenly deodars, he took the oath to become a British subject.
Perhaps, too, at a much later date and in a different way they provided
him examples: both men, nearing the end of their long productive lives
when he met them, were still immersed in work, busy, active. They only
retired when death came, not before. This too Stein would remember.

There was another kind of learning in which during his gymnasium and
university years Stein became proficient. He grasped—whether on a conscious or an intuitive level it is impossible to say—the fundamentals of
how the real world works. Perhaps his aptitude was formed by his dual
citizenship in the Lutheran and Jewish worlds (how else explain Uncle
Ignaz's extraordinary and anomalous position?), perhaps by daily encounters with the omnipresent bureaucracy of the Austro-Hungarian Empire, perhaps in response to his mother's admonitions—it would be un-
wise to try to pin it down. But somewhere, somehow, from someone,
Stein learned an "organizational language." Its syntax was simple: however large the organization, however impersonal its mode and however
governed by policy or tradition, administration was in the hands of individuals. The proper approach—that is, the behavioral expression of the
organizational language—was to engage the attention of a particular individual whose position gave him influence or authority. It could be
through an introduction, chance meetings, casual visits, a note of congratulations or inquiry, a reprint, an invitation to a lecture. There is
never a suggestion of fawning or sycophancy; he did not connive or rise
at another's expense. Rather Stein impressed men who held strategic
posts with his intelligence and quiet good manners; he invited them to
consider projects of cultural significance and, enlisting their administra-

tive cooperation, assured them that given their good offices he would carry the project through to a successful conclusion. The idea he proposed—*his* was the initiative—was well thought out and carefully presented, precise in plan and method, lofty in aim, and, so that no official feel slighted, his idea was made with due regard for the proper channels. Because he asked to be allowed to contribute to a noteworthy enterprise, officials paid attention to his request; they knew that its success would bring them credit, and they also knew that he would always publicly acknowledge their sympathetic assistance. Such an approach took time. Stein planned well ahead knowing that a seed needs to germinate before sprouting; it worked surprisingly well and flowered and flourished more often than not.

Stein's approach can be discerned from remarks about the grant that took him to England. That it did not just "happen" he indicates in a letter written four years later (1888) while on his very first Indian vacation. As soon as college was over, he fled the "grey oven of Lahore" for Srinagar, the "Venice of India," as the metropolis of Kashmir was called. In a comfortable, rented tent, pitched in a grove of old chinars, the majestic Oriental plane whose dense foliage creates a cool shade, he was within walking distance of the town's reading room. Looking through "the newly arrived English 'Times,' quite unexpectedly and deeply moved [I] read of the death of the respected Hungarian Minister of Education. My mind is filled by the news. I see him as the patron without whose active support I could never have finished my studies as easily nor achieved my present position. . . . As long as I live I will remain deeply grateful to the memory of this cultured man who was responsive to what was fine and of worth and who always met me with a friendliness that went to the heart."[15]

His letters describing this trip appeared in a Munich newspaper (1889) as a long travel piece, "Holiday Trip to Srinagar." At his suggestion— he knew the value of making his name known—Ernst had arranged for their publication. They were dedicated "To the memory of his dearly beloved father who departed this earth in the distant homeland on 10 May 1889, in deep sorrow, by the author."

The father image is one imposed by Stein's use of the word *patron*— not a chancy word for a philologist who knew that it derived from *pater* and that his attitude to his patron was the reciprocal: filial. As if to underscore this, there was the subtle but significant change in Stein's approach once he was in India. Patron, word and symbol, disappears and is replaced by the bureaucratic language of title-*cum*-name. Thus: "At six P.M.

I was received, after first announcing myself, by the European-educated Governor, Dr. Suraj Bal . . . [who] fully understood my intention to acquire Sanskrit manuscripts and expressed the wish to be of service to Dr. Rattigan, my esteemed chief and Vice-Chancellor of Punjab University who had recommended me to him." However, once introduced and recommended, Stein performed brilliantly.

Following this opening ploy, the governor acknowledged his presence by sending an equerry bearing a load of choice fruits and vegetables, an "attention doubly valued by me for it leads me to hope for the best in the promise given to influence the Pandits." Within two days the governor had arranged a meeting with the pandits in his house. "Rajah Ram Singh, the Maharajah's brother, honored it with his presence and I was introduced to the prince who knows Sanskrit. After the most educated of the Pandits had greeted me in that language, we had an animated discussion on a topic suggested by me which gave them confidence in my intentions. I had hardly returned to my tent when it was surrounded by Pandits ready to show me their manuscripts. . . . Professor Bühler's visit of thirteen years ago is well remembered by all. Many manuscripts which were then hidden have since gone into the possession of others; thus I have reason to hope for valuable acquisitions. The Pandits do not get along well with each other which means that each must be received separately in the tent and, since the meeting, an even greater number appeared with manuscripts to be examined by me." So great was the crush that Stein was forced to rent a larger tent "with three separate rooms in which everything is examined in privacy and the best kept with much bargaining."

Four days later he quotes Goethe's *Sorcerer's Apprentice:* " 'Master I am in deep trouble; I cannot rid myself of the spirits I summoned.' " Pandits came from morning until night. They advanced with circumspection and looked suspiciously around, and only when certain that they were alone would they drag a manuscript from its hiding place under the coat. In this hush-hush and tiptoeing, Stein was helped by Pandit Damodar whom Bühler had known and recommended: "a thoroughly learned man who seems to shake Sanskrit verses and prose out of his sleeve. He takes pains, spending long hours, to explain these artistic products of his poetic talents until I am forced to erect a dam against the flood of didactic erudition. He has much humour and an inexhaustible supply of proverbs and anecdotes. Another younger Pandit surprised me with his exact knowledge of the old geography of Kashmir and the true historical sense accompanying it which he displays in the many questions relating to it. I wish I had such teachers in our Oriental College."

Most manuscripts were modern copies of well-known works—the pandits were impressed with Stein's familiarity with the field—but he did secure three, each at least four hundred years old, written on birchbark; these treasures brought in from outlying villages had escaped Bühler's eagle eye. A fortnight of Stein's precious vacation slipped by in this scrutiny of manuscripts. For most of them he made short notes after a hurried examination so as to concentrate on those of the greatest importance. With each came the hope that it would be the desired copy of the chronicle. He punctuated the long hours spent in his office-tent with short excursions, wandering through the lovely, deserted, and overgrown Mughal gardens and visiting nearby Hindu temple ruins, sites now converted to Muslim shrines. Everything charmed him; even the problems. "The Sanskrit characters in Kashmir differ considerably from those customary in India and are difficult to read; however I have accustomed myself to it as well as to the Sanskrit pronunciation of the Pandits." Stein was doing what he wanted to do, in the place he had wanted to be, and he was being immensely successful. Only one flaw marred the quiet excitement of this first visit to Kashmir: the archetype he had hoped to find was not forthcoming; nor could he stay there and wait for it to appear.

Next in his planned tour of the princely kingdom of Jammu and Kashmir was a visit to the palace of the maharaja "to devote the last ten days of my vacation to the library there which until now has not had a scientific examination." In 1871, when Bühler had published his catalogue of Sanskrit manuscripts in private libraries, this collection had not been available to a European. In attempting to penetrate it Stein was, as he might have said, "in the tracks" of his great professor, and he took pains in approaching the project. "Mr. Tupper, Secretary of State for the Punjab, informs me in a gracious note that in line with my request he has written to the State Council of Jammu urging them to open the library to me so that I may catalogue it. In his opinion, his suggestion will have the desired effect."

After the delights of the "Venice of India" and the sense of accomplishment his weeks of work in Kashmir gave him, it was hard for Stein to leave Srinagar, "to tear myself away from my learned friends and everything else that had fascinated me. The arrival of autumn the day before yesterday [1 October] contributed to my elegiac mood when saying farewell. A fierce storm drove rain through the valley and when the heavy, low-lying clouds lifted, the heights surrounding the city were covered with a brilliant new snow and overnight the lush green of the slopes had turned a reddish-brown." For the alpine nature of Kashmir, Stein felt doubly grateful: its mountain-girt loveliness had insured its historical isolation, a seclusion that fostered its remarkable tenacity of tradition, preserving local lore, religious cults, and social customs. It was to his pandits' fine scholarship that he owed this view into the past; but it was his own scholarship and his manner of assuming that as men devoted to the same intellectual pursuits they belonged to a common brotherhood that won him friends among their number.

He began the roundabout way to Jammu. "At five in the afternoon I boarded the boat accompanied by three younger Pandits who had become especially close to me during my stay here. The buildings of the town as well as the surrounding hills were glowing in the rays of the setting sun as I glided down river under the seven bridges and took my farewells from the city of happiness. I felt choked with emotion. At one after another, at different landings, my young friends left me; the last was Pandit Govind Kaul whom I had come to value for his erudition based on a truly historical sense. It consoled me that I could truthfully call out 'Auf Wiedersehen' since he decided to accept my invitation and join me in Lahore so that we can collaborate. By the time he had mounted the broad steps to Ram Jiv's Temple where he left the boat, it was already dark. Soon the last houses of the town had fallen behind. Lying in the quiet boat it was long before sleep came—my mind was filled with sad thoughts."

Stein had fallen irrevocably in love with "the green paradise of Kashmir"; in Pandit Govind Kaul he had found the Kashmirian scholar whose erudition was to be essential to the translation and editing of the *Rajatarangini* on which they would work together over the next decade. Their commitment to exacting scholarship brought with it a mutual respect; their joined efforts a lasting friendship.

Anticipating Stein's success in obtaining the S'arada manuscript he had vainly sought, that part of the story can be told here. The manuscript's owner who had coldly refused Bühler had since died; three of his heirs had cut the manuscript, divided it, and were resolved to maintain the policy of their predecessor. But Stein was not to be denied: "More than a year passed in repeated endeavors and negotiations, which proved fruitless but were instructive to me in a small way of the methods of eastern diplomacy."[16] In this stalemate, he approached a pandit, member of the Kashmir State Council, whose son was a pupil at Oriental College. Late the following year, 1889, the several parts "of the codex archetypus of the Rajatarangini became once more for a time, reunited in my hands." In a footnote (ah! the sweet gems in scholarly footnotes) Stein confided that this codex, "obtained with so much trouble, was nearly lost on my voyage to England in 1890. The box which contained it, was dropped overboard in the Ostend harbour through the carelessness of a Flemish porter, and recovered only with difficulty. Fortunately, my collation of the text was complete and safely packed elsewhere. Happily, too, the soaking with sea-water left no perceptible trace. . . . The owners, when they received back in 1892 their respective parts, had no inkling of the *abhiṣeka* [lustration] their household talismans had undergone."[17]

To go back to the second part of Stein's "Holiday Trip." A week after leaving Srinagar and his pandit friends, he arrived "at the capital of Kashmir which, situated at the foot of the last slope towards the Punjab plain has the same unfortunate heat as Lahore. Except for this I cannot complain. The letters of recommendation of the Resident [Colonel W. F. Prideaux] and the State Secretary had a fine effect: I was received most cordially by the Maharajah and his brother, Prince Amar Singh, have been put up in the palatial *dak* bungalow [accommodation for travelers] as guest of the court, and been allowed an elephant as part of the royal transportation."[18]

Armed with an official order opening the manuscript collection to him, but proceeding with grace and decorum, Stein arranged to have the list of manuscripts, along with several sample bundles of these, brought from the temple where they had been kept to the new library. "I saw to my delighted surprise that the collection gathered from all over India by the late prince had been underestimated by me: it contains no less than about 8000 items including many old and important manuscripts. I spent from six in the morning until about two P.M. looking through the manuscripts brought out to me and taking notes. Then I had an audience with the Maharajah who received me in the open Dunbar [public audience]. Surrounded by his councillors and the entire court, he asked me to describe his father's treasures and the state of European Sanskrit studies. I had the seat of honour on his right, Prince Amar Singh on his left, and before us, on the carpet, the entire court sat. Twelve of the most learned Pandits had been summoned at the Maharajah's express wish and I had to discourse in Sanskrit with them and also recite Vedic verses from the books printed in Europe which I presented to him. In half an hour the audience was over. On my entrance and departure the guards fired a salute which, an undeserved honour, made my elephant restless."

The one defect in an otherwise complete triumph Stein took pains to correct before leaving Jammu. Allowing a few days to pass while he looked through the manuscript bundles and had extracts made by three pandits assigned to work under his direction—time enough to let the pandits react to him and report to their priestly superiors—Stein then paid a courtesy call on the high priest of the temple where the manuscripts had been kept. "I finally succeeded in chasing the last clouds from this holy man's brow who, at the beginning, had been upset at the 'desecration' of the treasures entrusted to him although the Prince's command could not have been seriously opposed." With a good rapport established and the cataloguing well started, Stein left Jammu in time to resume his duties at Oriental College.

This maiden trip inside India reveals qualities for which there does not seem to have been any preparation. His demonstration of scholarly excellence could have been expected—was he not the highly recommended pupil of Roth and Bühler? So, too, his exemplary proficiency in adjusting to Kashmirian speech and calligraphy and his innate courtesy and respect shown to the pandits. What was new and unexpected was his debut in camping—there is no other word—on this his first tour. At twenty-six, a Central European with no tradition of the wilderness-frontier, city-born and city-bred, his years spent in classrooms and libraries, not a sportsman, military man, or alpinist with periods spent "roughing it" (and long before camping became a commonplace served by an industry offering a vast paraphernalia of equipment and portable gadgetry), Stein crossed a rugged mountain area whose geography as well as its past he wanted to explore. He adopted what seemed right for his needs from the experience of English officers in border travel but mainly experimented to find his own style of cross-country living. Well versed in the reports left by earlier travelers, he was alert to every associated detail: standing on a pass with the valley below hidden under heavy rain clouds, Stein knew that from here Hsüan-tsang had been "able to see the snowy peaks in the northeast." He planned to travel light—"my suitcase serves as a table, my camp-bed as a seat"—to live, whenever possible, off the country, buying chickens, eggs, and milk as he could; to limit himself to one attendant who cooked his food and set up his tent; to hire guides, bearers, and pack animals locally as the need arose; to converse with whomsoever he met so as to learn whatever was noteworthy about the area and the march ahead; and to take delight in climbing rocky mountain passes and tramping through mighty forests and alpine meadows.

There is a fairy-tale quality to the account of his 1888 trip, despite a precise accounting of mundane details as he moves through an enchanted land. "13 August: Right after our departure at six A.M. we climbed a steep mountain with large glittering snow-fields below the highest peak. . . . As we moved along the ridge, the wind which in the morning blows up the valley and in the afternoon down the valley, drove patches of heavy fog toward the peaks. They soon overtook us and shrouded everything. Toward the east, the Alatopa valley along with everything else was hung with clouds; while to the left of the ridge we could look into the Chand valley and hear the roaring of the Hillen brook some 2000 feet below. Luckily the fog around the ridge was not too dense, otherwise it would have been difficult to keep to the narrow trail. I estimated the altitude by points given on my map: 11,500 feet. The flowers still grow abundantly; the scent from the masses of mint was almost overpowering. There are only gnarled trees and dwarf firs and a low, creeping bush. Pir Baksh [the guide], Piru for short, is an interesting companion who as a hunter in the retinue of officers has seen large parts of this alpine world. He has been in Astor [a garrison town in Northern Kashmir] and Ladakh where the valleys are over 10,000 feet high. He knows Gujrat and has admired the railroad there. But by and large he has remained a simple hunter. He knows the valleys and mountains around Barrangalla like the inside of his pocket and points out many places where years ago, sahibs, sometimes with their ladies often spent months in tents surrounded by ice and snow. (Memsahib, a corruption of Madame Sahib, is used for European ladies; Begum is the title for native ones.) It is strange how many hardships English ladies can take when participating in the sport of their husbands. Hunting is only possible here in spring and winter.

"Piru and others can only imagine Europeans as English officers. Indeed, they are almost the only ones who visit Kashmir; the country is so difficult of access that civil servants, limited in their furloughs, prefer Simla or Murree. . . . All military matters are well known to these people—the first question addressed to me by my bearers, by the buffalo hunters we meet usually concerns the length of my furlough and whether I am a captain- or lieutenant-Sahib. . . . All distances deceive in the fog and thus the march that brought me to this quiet place at 1:30 P.M. seemed doubly long. The area is covered by a short, scanty grass; there is a penetrating cold. My tent, which I took the precaution to surround with a stone wall is very comfortable. I hear the gay chatter of the bearers sitting around the camp fire. Nothing reminds me of the fact that I am living in the clouds. Tired and full of expectation for tomorrow's march, I turn in early."

3

During the decade when Stein held the post of registrar of Punjab University and principal of Oriental College, his letters are filled with Sturm und Drang. Not the passive type who waited for things to happen to him, Stein felt he was floundering when unavailingly he tried all "appropriate action" to realize his goal: to be a university professor like Roth and Bühler. The decade, then, was the length of time it took him to free himself from this goal and accept the reality of unknown and untried possibilities; to create, as Yule and Rawlinson had created, his own empire; and to appreciate that just by being in India he had been given, as it were, a blank check drawn on the subcontinent's storehouse of wonders and mysteries. And Ernst, reading his brother's words and reading between the lines, was a concerned and, when asked, a helpful spectator before whom this drama was freely acted out.

It is doubtful if Ernst ever really had a clear picture of Lahore where his brother lived and worked. Hetty was less bothered by the larger issues—she concentrated on his living arrangements and the women he met in India. Ernst, it would appear from Aurel's reassurances, tended to think of Asian cities in terms of the inhuman climate, the dread diseases, and an animallike poverty, cities where even the commonplace was barbaric and the ordinary filled with danger. In what other terms could he think of what was non-European? Had not medieval cartographers peopled Asia with such tribes as the Cynocephalae, the Dog-headed men, or the Unipeds depicted happily with their one monstrous foot held aloft as a parasol? And how could Ernst have formed a favorable idea of Lahore when, although Aurel described it and sent him pictures, at every opportunity, like an arrow ready to fly, he sped away from it?

Yet in England Lahore was not thought of as a grim, distant outpost. It was known as an important, lively city, a large, beautiful city with a great past. Set along one bank of the Ravi, a tributary of the mighty Indus, its slight elevation was man-made, the result of layers left by a succession of early inhabitants (as Budapest sits atop settlements from neolithic to Roman times). The Moghul emperors, Janangir and Shah Jahan, had built splendid palaces; their imperial presences lingered on in the magnificent gardens they had laid out close to the city, the lovely Shahdara and the extensive Shalimar. When in 1849 the British annexed the Punjab, Lahore became the province's administrative center and, because of its strategic and commercial importance, a key rail junction. By the time Stein arrived there the city had educational and cultural activities—a university, public library, museum, and the Mayo School of Art; imposing administrative buildings, official residences—each large house set in its well-kept garden in the best English tradition; clubs and hotels. A sizable British-officered army cantonment close-by added the manpower to insure a lively social calendar.

Aurel gave Ernst a map. "The house I rented with Mr. Andrews," he wrote 23 October 1894, "is situated in the oldest, rather quiet part. You can find it on the map—on the western edge of the Anarkalli, near the Punjab Govt Office, on the right, directly under the inscription 'Tomb of Anarkalli.' " But for Ernst, a map of Lahore was like an X ray, informative but lifeless. To flesh it out, Aurel sent him photographs, carefully identifying friends and acquaintances, buildings and monuments; snapshots taken at picnics, garden parties, and even some of the well-attended tennis matches showing the court surrounded by spectators. What did Ernst running a coalmine in Galicia and innocent of the newly invented game of lawn tennis make of strange ladies and gentlemen intent on strange rites?

Because of this discontinuity in their lives, Aurel concentrated on persons and matters and procedures which since they were related to his own well-being were comprehensible to Ernst and Hetty. An artless mixture of news giving an understanding of Stein during his troubled decade of search. Only Ernst heard his brother's lieder, now expressing an elegiac homesickness, now a bravura of expectations. Other themes come very soon: a monotonous beat of complaints at time eaten up by official duties; codas concerned with saving money for the future; lyrical passages singing Kashmir's beauty and bucolic tunes of camping trips seeking sites of archaeological and historical interest. Most of these themes join to form the motif of the *Rajatarangini* which in Stein's hands had three movements. The first was his critical edition in Sanskrit, edited with Pandit Govind Kaul in 1892; the second, the *Memoir of the Ancient Geography of Kashmir*, in 1899, "since for a full comprehension of Kalhana's Chronicle a minute study of the ancient geography was indispensable";[1] and the third, the masterly two-volume *Rajatarangini, Translated, with an Introduction, Commentary, and Appendices*, 1900, dedicated "To the Memory of George Bühler."

His letters, written without ulterior motive, free from reticence and before Stein became famous, are a treasure trove. That they survive is a small but praiseworthy miracle. They were carefully kept by Ernst and, after his death, by Hetty; and, after her death (she died in October 1934 "almost thirty-two years to the hour after Ernst"), by their daughter Dr. Therese Steiner. Each in turn safeguarded them through changes in residence and the dislocations and upheavals of two world wars and the Nazi occupation of Austria, until finally, in the 1960s, they were given to the British Academy to which in 1921 Stein had been elected to membership.[2]

At the end of the letter of 31 May 1889 bewailing the "catastrophe which robbed us of our good Papa," comes the laconic announcement, "My Pandit [the father of a pupil at Oriental College] reports from Srinagar that after much trouble he managed to obtain permission for a comparison of the Rajatarangini which I was unable to see last summer." Then, in the same paragraph, Stein adds that he went "as was the wish of dearest Papa to the Lt. Gov's swimming school where there are seldom any people."

People. Names. Quick identifications. At the university, at the hotel where he lived, at various functions—small, medium, large, professional, personal, social—he met many people whom in his letters he introduced to Ernst. Of all those in Lahore, "by far the most interesting person was Lockwood Kipling" (1837–1911). He had come out to India over twenty years before to serve as architectural sculptor to the Bombay Museum of Art; subsequently becoming curator of the Lahore Museum. His son Rudyard, at the beginning of *Kim*, called it "The Wonder House," because it housed masterpieces of "Graeco-Buddhist sculptures done, savants know how long since, by forgotten workmen." The novel describes it as it was then: "hundreds of pieces, friezes of figures in relief, fragments of statues and slabs crowded with figures that had encrusted the brick walls of the Buddhist *stupas* and *viharas* of the North Country and now, dug up and labelled, made the pride of the Museum."[3] Stupas, whose form derived from the earth heaped up over a burial, were elaborated into sacred commemorative mounds; viharas were the cluster of Buddhist monasteries and temples.

The curator's office of this Wonder House "was but a little wooden cubicle partitioned off from the sculpture-lined gallery." Its curator, "the white-bearded Englishman," his son portrayed as a man erudite and sensitive, widely known for his knowledge, imbued with reverence for the faith that had inspired this art and appreciative of the messages carved into the images by devout artists long forgotten; a man, moreover, capable of responding similarly to a contemporary exemplar, an attitude expressed movingly in his meeting with the aged Tibetan lama whom he honored as a "scholar of parts." Clearly stated in that meeting and subsequent parting is the conviction that a friendship was possible between the well-bred Englishman and the Tibetan lama—disparate in all external matters—who shared a common, inner passion. If Stein had need of an example, Lockwood Kipling must have given him the freedom to value such Indian scholars as Pandit Govind Kaul.

Under Kipling's tutelage, Stein studied the collection and became familiar with Greco-Buddhist art. Gandhara sculpture, an art of the highest quality then hardly known outside the Punjab, would be increasingly appreciated for its special importance. Its "greatest contribution," in Rowland's words, "to the art of Asia was the invention of the Buddha image."[4] Stein's thoroughgoing familiarity with the museum's treasures would unexpectedly provide him with the opportunity to gain the ear of power and the sanction to begin his Central Asian expeditions in which, among other investigations, he would trace the transmission of the Buddha image as it went overland from Gandhara to China.

On quite another level, Lockwood Kipling played a role in Stein's life: it was he who introduced Stein to Fred Andrews soon after the latter became vice-principal of the Mayo School of Art (1890). The two younger men quickly became friends. Their friendship lasted fifty-three years, ending only at Stein's death; it was an important friendship and, as the years passed, increasingly significant in Stein's work. Both Kipling and Andrews were influenced by the Arts and Crafts Movement, which with William Morris as its spokesman considered art in holistic terms—with no valid distinctions separating the fine arts from the applied arts and thus elevating social purpose, virtuosity, and taste to the same level as great talent. The movement's high social and moral basis gave its adherents a missionary fervor, a double-edged emotion that encouraged them to value the traditional Indian handicrafts (something

Moorcroft had responded to innately) even as their inner vision was loyal to the medieval European idiom proclaimed by John Ruskin. Of the two, Kipling had been part of the movement's beginning and with a beginner's openness could see beauty in the anonymous Gandhara sculpture, whereas Andrews came out to Lahore formed by the first show of the Arts and Crafts Exhibition Society (1888) eager to bring the Indian genius into the Arts and Crafts formulas. Andrews lacked the prophetic quality of William Morris and his architectural training, but he did share his response to the wide range of handicrafts, his proficiency in many minor arts, and his preoccupation with detail. It was these traits that he brought to his friendship with Stein. Baldly stated, Fred Andrews became Stein's third hand—an intimate part of his life and valued as a man values his right hand. Stein, on his side, was a kind of outrigger for Andrews, giving him stability and keeping his talents on their true course.

Their first outing together cemented the friendship. Soon after they met, they made a short summer tour in the Salt Range, hills lying between the Indus and the Jhelum. In his reminiscences of Stein, Andrews remembered that their reason for going was to see "fragments of sculptured stone that had been found. Many of these stones had been carried away by the Public Works Department and used in the construction of a local railway bridge following a time-honored practice common in many parts of India."[5] Who suggested the tour? Stein to see the sculptures or Andrews to see the desecration wrought? Andrews was vigilant to try to stop such vandalism being a member of the "Anti-scrape" as the Society for the Protection of Ancient Buildings which Morris had founded in 1877 was popularly called. Andrews recalls what might have been the sealing bond on that brief expedition. "As an enthusiastic amateur photographer [a fairly new and limited skill at that time], not a very successful one, I photographed the various sites and finds, and at the same time initiated Stein into the mysteries of the art and craft of photography. He has often referred with undue gratitude to these his first lessons in the art in which later he became so proficient." It is clear that the innocuous pictures Aurel sent to Ernst were meant, like the letters that followed, to inform him while he practiced and perfected this new, valuable skill.

8 April 1891: "Your fears about the fantastic heat have fortunately been groundless so far. We are having very pleasant weather: the evenings especially are delightful and the air is filled with the perfume of oranges and lillies now in bloom. . . . Last Friday we had a costume ball.

I took care of the obligation to wear fancy dress by transforming the cuffs and lapels of my tails into a Windsor court uniform by means of colored silk which cost only a few rupees. . . . I am owed about 230 rupees from January for the salary of the Pandit, travel expenses, etc. My stipend of 3,000 rupees [for the Jammu cataloguing] is supposed to be made in the course of the summer. . . . I am in correspondence with Dr. Hoernle, the Secretary of the Asiatic Society of Bengal, and hope the latter will accept my entire *Raja[tarangini]* edition on favorable conditions."

23 August 1891: Back in Kashmir, he was about to visit "the sacred lakes at the foot of the Haramuth glacier. The first one is sacred to Siva and his faithful Nanda [a bull, the god's mount]. We reached it after an hour and a half climb over high moraines. The massive ice-wall in which the glacier ends has a grandiose appearance. On the shores of this quiet lake I met a small band of Dards ["a third group of Aryans parallel with Iranians and Indians"].[6] It was only a short distance to the higher, more sacred lake which the Kashmirians consider one of the sources of the Ganga [Ganges] and a goal of their annual pilgrimages. I breakfasted on the shores of this desolate lake. Those who are not wealthy enough to bring the ashes of their loved ones to the real Ganga, bring them here and scatter them into the lake whose shores are strewn with bones. Returned at 5 P.M. wet through from crossing icy mountain streams but invigorated by the long march in the fresh mountain air."

25 August 1891, still Kashmir, Vangath: "Last night at Pir Bakhsh's suggestion the Kulis and Gujars [tribal people] who in the summer months pasture their flocks in the high valleys, gave me a real seranade. Some of the Kashmirian songs were very melodic and reminded me of Hungarian songs."

28 August 1891, Srinagar: "I found my tent which had been pitched under the same three Chinars where I had been three years ago. Pandit Govind Kaul brought manuscripts to see and possibly to buy. The next day I took a ride with G. K. along the eastern shore of the lake to visit several villages and springs mentioned in the *Raja*. On my return I met Mr. Andrews. He is living on three boats with his wife and child. The next morning, with Andrews I made a short excursion to the temple of Pandrathan which I had visited in 1888 and then went to the palace— Raja Amer Singh had invited me. My report about the Vangath temples [the ruins being despoiled for pieces of sculpture] interested him and he declared he was willing to spend something to have them restored. . . .

All in all the Raja was very gracious; he was enthusiastic about my suggestion to build the new library in Jammu (to house the Sanskrit manuscript collection) in the form of a temple."

11 November 1891, back in Lahore at his university post: "I am very busy and work from 7 in the morning until five in the afternoon without interruption. Saturday night Dr. Hoernle came from Amritsar [the great Sikh city, the site of the Golden Temple] where he had evaluated a coin collection for the government. He was my guest for 48 hours and I had the pleasure, almost unknown for several years, of being able to talk *con amore* with someone in my own field. Dr. H. has been in India for 25 years and has many relatives who are serving as missionaries in the country. He has a good position in Calcutta but I do not envy him the pleasure of teaching Bengali babus [native clerks who write English]."

9 December 1891, Lahore: "My edition [the Sanskrit *Rajatarangini*] is approaching its last third. Today I made arrangements with Harrassowitz [an outstanding publisher in Germany who was to handle the European distribution] concerning royalties—I receive 75% of the sale price of all copies sold in Europe and America."

22 December 1891, Swabi, Christmas vacation: "This morning I left Hazru and crossed the alluvial plain of the Indus; at 11 A.M. I reached the 'Father of Rivers' and with ponies and camels crossed it in a fairly primitive boat. In midstream I was received by the Tahsildar of Swabi, at the Assistant Commissioner's suggestion. He is to escort me across the border. I landed near the old village of Und, the Udabhauda in the *Raja*. There I was on foreign soil; the language is Pashtu, an Iranian dialect, and utterly incomprehensible to my people. I was put up very comfortably in the bungalow of the civilian authorities. Along the border, the administrative offices are housed in small stations equipped for defense purposes if needed. They remind me vividly of the stations excavated along the Roman wall in Northumberland. [Two days later he had recrossed the British border.] The Tahsildar had contacted the Khudo Khel tribe to announce my proposed visit to the ruins of Ranigat and while we were still in Swabi the khan's messenger had told me that he had no objections to my visit and would assume responsibility for our safety. There has been no trouble with the Khudo Khel for five years.

"General Cunningham, the first European to visit these ruins [1848], thought that he recognized in them the mountain-fortress of Aornos, whose conquest by Alexander is one of the best known episodes in the accounts of the Macedonian's Indian campaign. I think this identification is not tenable for several reasons though the Ranigat, as it is called by the inhabitants, is a place of particular interest. . . . It is quite obvious that the archaeologist was preceded by covetous sahibs: I found over a dozen magnificent torsos mostly portraying the Buddha as teacher, prince and ascetic while their heads probably grace a mantle somewhere in the Punjab or in England. Only a few of the plundered items found their way to the museum at Lahore, and yet these pieces constitute almost the finest of the great collection of Graeco-Buddhist sculptures. . . . Gradually the tribal elders appeared with their 'pages,' rather wild looking figures most of them carrying rifles stolen from the British, though some carried ancient flint locks or the broad Pathan sword which looks like a large chopping knife. I lost no time and started working with my surveyor, a Sikh, Gurditt Singh, on a plan of the site. We had some initial difficulties because the chiefs insisted on examining all instruments but refused to help with the measurements. I finally succeeded in distracting them by the rather exotic spectacle of making tiffin. I returned that evening to Mangram because the Jirgah did not want to be responsible for our safety in the village. I spent Christmas eve in good mood."

25 December, Mardan: "In Shahbazgarhi I saw the famous rock inscription of King Asoka's Edict of Tolerance." Basham calls Asoka (who began his reign about 265 B.C.) "the greatest and noblest ruler India has known, and indeed one of the great kings of the world. . . . The keynote of Asoka's reform was humanity in internal administration and abandonment of aggressive war."[7] Conquering by righteousness, not force, won for Asoka numerous victories including those of the Hellenist kings of Syria, Egypt, Macedonia, Cyrene, and Epirus—so his edicts engraved on rocks and pillars widely scattered throughout his vast empire claimed. Stein, reading the northernmost of these pronouncements—"the oldest surviving Indian written documents of any historical significance"—thus had a firsthand confrontation with Asoka in whose reign "Buddhism ceased to be a simple Indian sect and began its career as a world religion." For him, reading the edict *in situ* made his Christmas day memorable.

13 January 1892. Back in Lahore, the antiphonal mood came to the surface: "How vividly the wish to spend the Christmas season under your roof is in my heart I need not tell you. Perhaps it will be fulfilled before I establish permanent residence in Europe again. After eight years' service European officials are eligible for a furlough of up to two years' duration. Although by that time I may not succeed in securing a government position which will take my services to the university into consideration, I must hope for this. Mr. Fanshawe, our chief secretary to Govt is favorably disposed and has taken upon himself to have the Lt. Gov-

ernor, who retires at the end of February, express his satisfaction with
my services in an official document and bring it to the attention of his
successor. . . . As soon as I shall have finished the text, I'll start working
on the Jammu catalogue. During the holidays I will get started on the
historical and topographical notes which will form volume 2. . . . I work
ten hours a day and feel fresh and well because the weather is cold and
from the relaxation I get riding, playing tennis and walking. . . . I often
see the elder Kiplings and they always receive me graciously. Their
famous son stopped in Lahore on his way from Australia to Europe."

26 January 1892. Matching what Ernst and Hetty wrote of their servant problems brought out a rare picture of Stein's living arrangements:
"In my household, if it can be called that, a delightful change has taken
place. Since returning from Kashmir, I could see that I would have to
find a new servant. My old one whom I kept out of habit and because
of the hardships he endured on my tours, has become increasingly stupid
since he devotes himself entirely to opium. However, a replacement
was not easy to find. My perseverance seems to have paid off: my new
Kanzamar, a Muslim, makes a very favorable impression and appears to
have been a happy choice. Formerly I had resigned myself to do without
my beloved 'orderliness' in my rooms. Jakkir Uddin, my new major
domo has wrought a real revolution: now everything is so proper, clean
and in its place that it looks like a barracks ready for inspection. He
keeps my books in their right places while cleaning. For this he gets 16
rupees a month and it is well deserved. . . . I can hardly express my feelings in being able to say that next year I'm coming home." The last
four words were underlined by Ernst, a kind of Amen.

3 February 1892: "I beg you not to worry about my alleged liver
problem. So far I haven't taken even notice of the existence of my liver
and enjoy an excellent digestion. This is very handy since on tour I cannot be choosy in matters of diet. I am still very fond of cold baths which
many people here cannot indulge in because of their livers. . . . I expect
to be finished with the *Raja* by the end of March. Harrassowitz expects
to sell 50 copies at the beginning; my share will be 16 marks per copy.
In India, the subscriptions of the local governments will absorb a goodly
number and bring in a tangible profit, especially since the Bombay Educational Press asks only a 10% commission. I have not decided on the
sale price but it should be about 16 rupees."

10 February 1892: "I am overwhelmed with work of all kinds—teaching duties and exacting administrative work—but I feel well. . . . I cannot detail to what degree I will practise greater economy and would

1. Nathan Stein, Aurel's father

2. Anna Hirschler Stein, Aurel's mother

3. Aurel Stein while a student at Tübingen

4. Aurel Stein during his year of military service (1886)

1

2

3

4

5. Harriet Stein, Aurel's sister-in-law (1926). To Aurel's English friends she was "Hetty," to him she was "Jetty," the German form

6. Ernst Stein, Aurel's brother (about 1890)

7. Aurel Stein in the dress prescribed for the investiture ceremony

5

6

8. Stein on horseback, with Dash, shortly before starting on the Third Expedition (1913)

9. Camp of the servants at Mohand Marg (1912)

10. Fred Andrews (the Baron) and Mrs. Andrews (the Baroness) with Stein and Dash III in Srinagar (1916)

11. Standing: Fred Andrews and P. S. Allen (Publius).

Seated, left to right: Mrs. Andrews, Miss Satow (cousin to the Allens), Helen Mary Allen (Madam), and Maud Adam, her sister

12. Lockwood Kipling (1893)

10

11

12

13. Aurel Stein (1919)

14. Book plate designed for
Stein by Fred Andrews

14

13

15. Stein with Thomas Arnold (the Saint; later the Hierarch)

16. Stein in flight togs for aerial photography

17. Stein in front of his tent at Macedonian citadel wall, Dura-Europos (1929)

15

16

17

18. Stein's caravan marching over high dunes on the way south from Tarim River to Keriya River

19. View south toward snowy range from ridge above Shalgan Davan, circa 17,700 feet; Stein's plane table in foreground

18

19

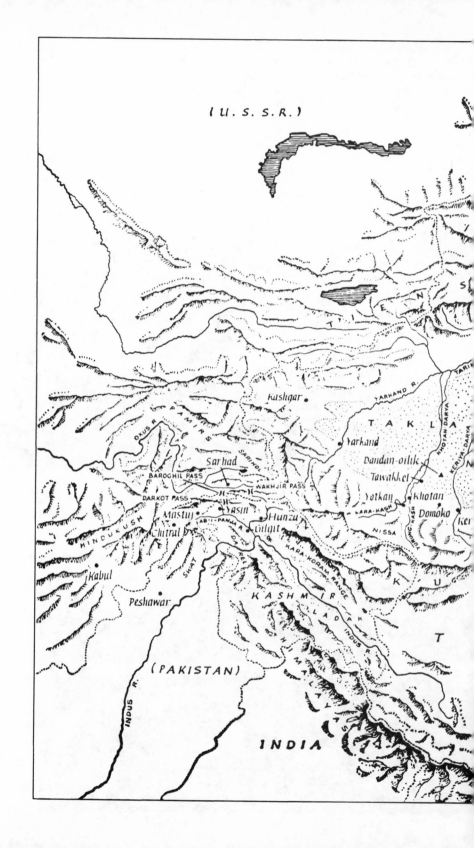

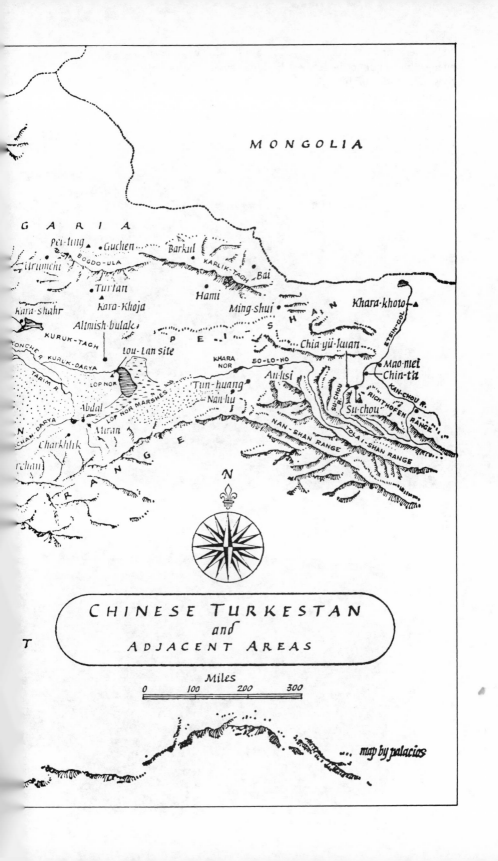

MONGOLIA

GARIA

Pei-tung ▲ • Guchen···· Barkul
Urumchi BOGDO-ULA KARLIK-TAGH Bai
Turfan Hami
kara-shahr kara-Khoja Ming-shui Khara-khoto ▲
Altmish-bulak PEI- SHAN Chia-yü-kuan
KURUK-TAGH lou-Lan site Mao-met
TONCHE R KURUK-DARYA KHARA Chin-tu
TARIM R. NOR SO-LO-HO Au-lisi KAN-CHOU R.
LOP NOR An-lisi SU-CHOU R. RICHTHOFEN
Abdal LOP NOR MARSHES Tun-huang Su-chou RANGE
Nan-hu NAN-SHAN RANGE TOLAI-SHAN RANGE
CHARCHAN-DARYA Miran
Charkhlik
rchan RANGE

N

CHINESE TURKESTAN
and
ADJACENT AREAS

Miles
0   100   200   300

map by palacios

20. Cave-temples at Tun-huang; the temple to the left rises nine stories high to accommodate a colossal seated Buddha (photograph courtesy of Mr. and Mrs. James C. M. Lo)

21. Stucco head of colossal Buddha found at Miran site

20

21

22. Stein's "daftar" household at Dal Lake Cottage

23. Group of Stein's party at Turfan. Left to right: Ibrahim Beg, R. B. Lal Singh, Stein with Dash III, Musa Akhun, Afraz-gul Khan, Jasvant Singh

22

23

24. Haramuth, one of the mighty peaks Stein always viewed with pleasure

25. Stein's summer encampment on Mohand Marg

26. Surveyor Muhammad Ayub Khan with Dash VI (1935)

27. Stein's breakfast table on Mohand Marg

24

25

26

27

28. Hassan Akhun, Stein's
head camel man

29. Pasa, of Keriya, hunter
of wild yaks and guide to
Stein

30. Turdi, the Dak-man from
Khotan

28

29

30

31. P'an Ta-jên (Pan Darin)
with his two sons at Urumchi

32. Wang Tao-shih, Taoist priest at the Caves of the Thousand Buddhas

33. Chiang Ssŭ-yeh, Stein's Chinese secretary and helpmate

32

33

rather refer you to my remittances of the next few months as tangible proof."

17 February 1892: "Sir James Lyall, who will be leaving us soon has, on the suggestion of our director of curriculum ordered 50 copies of the *Raja* to be bought by the Punjab Govt at a price of 16 Rupees for volume 1. This should amount to a tidy sum. The text book committee has asked ex officio to subscribe to further copies for school libraries and I will try to get additional orders from other local governments. With the expected Kashmir subscription, the number of copies sold on publication should come to 200; Harrassowitz has not been idle and, from a report I received today, has distributed 500 copies of my prospectus. He has said that after a few days he had a satisfactory number of orders. . . . The proceeds will thus, hopefully, be a nice contribution to this year's savings."

20 April 1892: The heat had settled on the Punjab; it was 108° in the shade, and the wind when it blew was heavy with dust. "I had occasion to speak with Mr. Sime, Director of Public Instruction, about a vacancy in the Education Department, who explained that two school inspectors are soon to retire. He will suggest my name when a vacancy occurs. But there will be another applicant who will give me competition not through his qualifications but because of his family connections. Meanwhile I will try to secure support in the secretariat where I have friends among the civilians. I have also tried to get assistance from London and asked to have Sir H. Rawlinson write to the Governor. Otherwise I wait patiently for my opportunity and hope to be able to exploit it should a friendly fate offer it to me."

12 May 1892. He reports to Ernst about a letter that Rawlinson *did* write: "I have a very high opinion of Dr. Stein's acquirements in the field of Oriental research and think that he well deserves the support and encouragement of the Government of India. If indeed he would be employed in a leading position in one of our great Educational Establishments in India where his extensive knowledge of the various languages of the Sanskrit family might be turned to account in directing the studies of the young, I feel assured that it would be advantageous to the Public interests, and I have no hesitation therefore in writing this note and requesting you to convey it to him as a testimonial on my part to his literary merits." Stein was "deeply touched" by Rawlinson's recommendation. It "will be of use to me as well as a dear memento of this truly great explorer. At his advanced age—he is almost 80 years old— I can hardly hope to see him again."[8] Despite all his efforts Aurel was

forced to conclude that "since no actual vacancy occurs at present, I can gain nothing more than this testimonial."

From April, when he began his efforts to secure an appointment as inspector of schools in the Education Department, through June, when he received Rawlinson's encouraging recommendations, he waited for an answer. His hopes, his mobilizing of influential sponsors netted him a polite "no." In the letter to Ernst (3 September 1892) he reveals his mood: "I have been reading the memoir of St. Arnaud, Marshall of France (1798–1854) and found a passage which clearly expresses something I have long been thinking—the advantages of a sharply delineated career which at all times offers a tangible goal. 'On ne pense qu'a avancer, a monter, a gagner un grade de plus, et en pensant a ce point unique, on y tend avec plus de rigeur.' " The mood was still with him a few days later when he told how much he "enjoyed reading St. Beuve whose charm remains undiminished. I often think of our dearest Uncle Ignaz who originally introduced me to him." In his hour of rejection, he combined two father figures, finding therein strength and consolation. He goes on to explain why he so desired the position: "A major attraction of the Inspector post would be the fact that I could exchange my sedentary life with the much pleasanter one of camp life. Everybody in the Plains wants to be on tour and that's what an inspector must do the greater part of the year. I would gain knowledge of the country and people. Kashmir where all together I have barely spent nine months is more familiar to me than the Punjab and its inhabitants."

24 May 1892: "I hear from Roth very rarely but it is probably my fault for not having time to correspond about things that interest him. I cannot indulge in Vedic and Avesta studies and must restrict myself to helping colleagues and patrons in Europe by procuring manuscripts and other materials when the opportunities arise. I have done this quite often, for example for Burkhardt in Vienna who is working on the Kashmir language, a study for which I could never spare the time. . . . Max Müller is going to hold his Oriental Congress in Oxford this September and I suggested to Col. Prideaux that the Maharaja should send a number of copies of my work to the Congress. I am planning to submit a report of last year's tour illustrated with photographs."

17 July 1892: "Rain, rain rain. Everything is green again and large stretches of Lahore are under water. The first rain [of the monsoon] brought an enormous number of frogs which fill the night with a deafening noise. Lizards, ants, frogs, and so forth are constant visitors in the bed—and bath-room."

6 August 1892. Happy and comfortable in Kashmir, Stein saw "Colonel Prideaux who received me very graciously, obviously satisfied with the *éloge* I gave him. He agreed with my suggestion to present six copies of the *Raja* to the Congress as a courtesy of the Kashmir Govt. I wrote to Professor Bühler to ask him, if he attended the Congress, to accompany the gift with a few words about the purpose and execution of the work. The Maharaja also received me in his new chalet and expressed satisfaction with the acknowledgements made to him. With Raja Amer Singh, who is very *simpatico*, I discussed the arrangements for this year's tour. As in the previous year, 500 rupees are given to me for excavation, etc. as well as the power of authority for the preservation of artifacts. My mention of him in the Preface seems not to have been without effect, especially as the "Lahore Civil and Military Gazette" took pains to mention his interest in my work. I believe I have also won a patron in Mr. Lawrence, Settlement-Commissioner for Kashmir: He and his wife invited me urgently to be their guest at Srinagar this Christmas. If there is any ice for skating an invitation like this would be very tempting."

The tour of 1892 began with an effort to locate the castle of Lohara that "played an important part in Kashmir history as the ancestral home and stronghold of the dynasty whose narrative fills the last two cantos of Kalhana's work," so Stein begins Note E in the *Rajatarangini*.[9] In pursuit of the castle's site, Stein toured the "well-populated and fertile mountain district formed by the valleys of the streams which drain the southern slopes of the Pir Panjal Range between the Tattakuti Peak and the Toshamaidan Pass." By mid-August he was in the high mountain country that separates Kashmir from the western Punjab.

14 August: "I made a resolution to learn Kashmiri this year and enjoy the fact that I can use the little I have learned so far."

16 August: "The path [into the valley of Mandi] seemed to be very old. In several places where it leads over steep rocks overhanging a wild mountain stream, steps are cut into the rocks. Now I understand why all conquerors from Mahmud of Ghazni [971–1030, the most successful of the Turkish chieftains who from Afghanistan raided deep into India] to the Sikh, Rangit Singh [1780–1839, the great Sikh leader, the "Lion of the Punjab"] who tried to penetrate into Kashmir over the Toshamaidan Pass were brought to a standstill before the fortress of Lohara. . . . I could not miss this chance to present to the Divan [State Council] a petition to free the valley's inhabitants from forced labor. This was accomplished favorably in no time at all. In the Punjab, under European ad-

ministration, his Honor, the Lt. Gov., would have been only able to refer the petition to the following Tuesday!"

"The stout castle of Lohara, built of rough, uncut stones solidly set in a framework of wooden beams, had been reduced by neglect, heavy monsoon rains and equally heavy winter snows to a shapeless heap of stones. Of its former might, all that remained was a lingering tradition." Rubble and legend and the fortress's impregnable location identified it as the bastion that had kept Kashmir inviolate for seven hundred years repelling all would be invaders.

20 August: "From Loharin valley I returned to the open grassy slopes of Maidan [13,000 ft.], choosing this route along the northern slope of the Pir Panjal because it is one of the oldest and most important connections between the Punjab and Kashmir and plays a large role in the *Raja*. . . . Once a Kashmiri pretender invaded the Happy Valley by this Pass and reached Srinagar in two and a half days. To make certain that this route permits so speedy a march, I am very happy to have furnished proof of Kalhana's truthfulness. If times were different I could imitate Prince Sussala [seventh century] and fill my pockets with gold in the suburbs of Srinagar tomorrow." At Srinagar Stein did fill his pockets: "Colonel Prideaux told me that the 1500 rupees subvention was available. I also received pleasant prospects for the second volume," he wrote to Ernst on 27 August 1892.

The second part of the tour was in search of the shrine of Sarada, once Kashmir's most important pilgrimage site. Stein was familiar with the shrine's legendary origin which dictated the stages of the pilgrimage. It began with an ascetic who practiced austerities that he might behold the goddess Sarada. Divine guidance led him toward the Sarada forest where, as promised, Sarada showed herself in her true form and vanished. The second step took the ascetic to a spring; when he bathed in it half his body became golden, a sign of liberation from darkness. The third step at the foot of a mountain was the setting for the dance of the goddesses. And so, advancing step by step, the ascetic was rewarded and invited by the goddess to her residence.

The first reliable information directed Stein to a difficult pass and then through equally difficult gorges to the Kisanganga Valley; there he finally located Chandra Pandit whose family were hereditary guardians of the shrine and who agreed to accompany him. His guide took him along a path that brought him to the left bank of a roaring mountain stream at least 150 feet wide.

10 September 1892: "I had to cross over to the right bank on a rope

bridge. It consists of three ropes, the lower one in the middle for your feet, the other two for your hands. The ropes are held in position by Y-shaped poles and the entire dangling structure is attached on both banks to high, wooden cross-beams—like a suspension bridge. Only in the center where it swayed in the wind was it unpleasant. The rope under one's feet seems to disappear; directly below is the roaring water. However, with Chandra Pandit in front and Pir Bakhsh behind, I got across without any problem." At a turn of the path the shrine came into view: with a magnificent amphitheater of high peaks behind, it stood on a prominent, terracelike foot of a spur of a high pine-clad peak. Immediately below, where the Kisanganga was joined by another stream, was a sandy beach where the pilgrims performed their ablutions. The legend of the ascetic and his golden bath has a basis in fact: the Kisanganga "drains a mountain region known as auriferous to the present day."[10]

Stein describes the site further in his letter to Ernst of 10 September 1892: "I made camp by the old temple which is well preserved. This sanctuary, probably the one farthest north in Kashmir, is still visited by the Hindus. In the winter the snow is said to reach as high as its twenty foot walls. I am happy to have found this place and the fact that I found it at all was by reading Kalhana with great care and following his description. I spent all day surveying the hill. In the covering thicket I found remains of old walls which makes my identification correct. Fortunately, I did not have to return by the rope bridge because the villagers had made a crude raft for the District Officer and moved it back and forth across the river by a rope."

Colonel Prideaux, the Kashmir resident who had helped Stein in his pursuit of ancient geography, asked Stein for a short memorandum on the Toshamaidan Pass. Because Kashmir was part of Britain's sensitive Northwest Frontier, he needed information about alternate routes for military purposes. Stein, happy to reciprocate for the many favors he had received, gave Prideaux a full statement. "I had occasion to use this route across the *Pir Panjal* the end of August last when returning from a visit to the *Loharin* valley. In previous years [I had] visited several of the best known passes leading from Kashmir south (the *Pir Panjal, Banihal, Hajji Pir Passes*)."[11]

A bare five years after Stein began his exploration of Kashmir, he disclosed what might be called the "Stein method"—never to use the same pass twice: the true explorer's approach. When later he would make his expeditions into Central Asia, he would choose a different pass for each crossing from India.

27 November 1892, Lahore: Stein opened the presents sent from home for his thirtieth birthday: ties "to wear on formal occasions which could not be bought here except for much money; elegant gloves which made me especially happy that you found the proper size which is unavailable in India; and beautiful handkerchiefs which must have been selected by dear Hetty." Thirty years old! and he had not attained the recognition he dreamed of. He pinned his heart on his sleeve in the accepted romantic idiom. "Spent my birthday in the shadow of the Imperial Tomb [the lovely mausoleum of the Emperor Jahangir] in the company of St. Beuve and Horace," the books given him by Ernst and Uncle Ignaz. He might well have been a tiny figure in the overwhelming isolation of a Caspar David Friedrich painting—the canvases he saw during his schooldays in Dresden foretold and formed Stein's way of acting out his loneliness.

With the Sanskrit edition of the *Rajatarangini* in print, 1893 was the year Stein concentrated on getting a furlough—he wanted to visit Ernst and press his candidacy for a professorship at some European university.

1 March 1893: "I assure you I know how to appreciate the advantage of a two-year furlough in 1895 as well as a short visit home this summer. Yet I have little hope that both will come through. I only have a claim to a four-month, non-accumulative furlough every two years. No provision was made for the furlough every European servant has the right to expect: this is because the Registrar's position originally was not a full-time one but an extra job given to officials or professors occupied with their own duties. This was not true in my case and so I must somehow manage to secure for myself those privileges due to every civil servant. . . . I have written to Roth and Bühler detailing my academic aspirations and have attempted to approach Weber [Albrecht Weber, 1825–1901, professor of ancient Indian languages and literature at the University of Berlin] by sending him a copy of the *Raja*. . . . I am enclosing a check for forty pounds for copies sold at the Oriental Congress. The sum for the seventy-five copies sent to various government agencies has not yet been paid."

11 April 1893: "I would never have dreamed that Dr. Duka would suggest me for election to the *Academia* even though he had a pious admiration for our uncle. . . . I will certainly be grateful to him for this even if, as is to be expected, I fail to get the candidacy."

17 May 1893: "I am happy to be able to return your good news with good news of my own even if they do not represent all I strived to ac-

complish for my future. Without Mr. Gordon Walker's untiring efforts in my behalf, the compromise would not have been reached. . . . There are two important points: First, the reconfirmation of the registrar every two years has been changed to a five year period—this makes the position less insecure. Secondly, my claim to a furlough has been granted albeit with certain restrictions. . . . As regards the pension, I had better give up hope and rely on savings. Or, if the chance, to get a regular teaching post somewhere, outside India, it is always possible to invent reasons of health which would forbid further stay in Lahore. . . . The manuscript of the catalogue will be finished and in print by July. My honorarium, to be fixed by the State Council and the British Resident, will respect the advice of the Bengal Asiatic Society. Col. Barr has assured me he is going to press for a liberal sum. Moneys owed by Jammu have come—amount to 1200 rupees now. . . . I work all week from morning to night. Luckily the heat is still bearable and my cold is gone. . . . I hope to leave Bombay on July 21st which brings me to Brindisi about August 4th. Round trip ticket second class, costs 750 rupees, 200 less than the same passage on Lloyds."

24 May 1893: "I am using my holidays (4) to finish my contribution to the Roth Festschrift." [It would be "The Saka Kings of Kabul" and would be published in 1899.]

6 June 1893: "Good news! the Senate has unanimously accepted the new regulation without change. It only remains for me to apply for permission for the six-month furlough to begin the end of July. . . . I send a copy of Cook's *Unicode* which is very useful in sending telegrams and saves money. I have a copy at my disposal. . . . That my election to the Academia fell through does not surprise me. The choice of a foreign member obviously means honoring foreign scholars with a European reputation. Strictly speaking I am neither a foreigner nor by any stretch of the imagination a man with the necessary qualification. . . . I wrote Dr. Duka some weeks ago and thanked him for his consideration."

20 June 1893: "I want to devote the first part of my furlough being with you and your family. Later in the fall I will make a brief visit to our home town. . . . Christmas again with you as the steamer which brings me back to Indian soil leaves January 7th."

Back in India after his European vacation, Stein enjoyed the last few days of his freedom before returning to Lahore. He visited the Gwalior region, known for its superb sculptures of Hindu divinities and mythological scenes. He was especially conscious of the influence a group of famous temples set within the confines of a rockbound fortress had exerted on later Islamic architecture. Then, swinging back to Lahore, he stopped off at Agra and New Delhi.

22 May 1894: "It is 33 degrees centigrade in my room!. I have been asked by Col. Barr, the resident in Kashmir, to spend about five days in Jammu in order to make the necessary notes on the history of the Sanskrit collection. . . . I am also thinking ahead to my Kashmir vacation and have advised Pir Bakhsh about the tours planned for the first half of August. How I would love to extend my stay there until I have completed the second volume of my *Raja*—six to eight months would do it— but that is an unrealizable dream. . . . I hope to have a copy of the [Jammu catalogue] ready, all except the index and preface in time for the Congress of Orientalists; but I don't know who will present it. Bühler keeps stubbornly silent which is very unpleasant for me."

15 August 1894: "On the Dudh Kuth Pass. Twelve thousand feet high! Cooler than Srinagar which I left on the 10th. Came by boat to Bandipur which I had not visited since 1889 and found it changed. There are the main headquarters of the road to Gilgit and Chitral for the British troops guarding the Russian border. Large supply depots and barracks show the care supplies and transport across the northern passes require. . . . I had suspected a frontier fortress here mentioned in the *Raja*, and my suspicion was confirmed when, in digging, I found traces of walls. I am taking advantage of the opportunity to learn Kashmiri and regularly take lessons both on the march and at camp during the evenings from Pandit Kashi Ram. Though not a scholar like Govind Kaul, he is more reasonable—satisfied with 15 rupees per month—a fine person and a good walker."

22 August 1894: "I am sitting under pine trees while working on a memorandum for a project I discussed with Mr. Lawrence in Gulmarg. Its objective is to free me from the Lahore misery for a considerable time so that I can complete the *Raja* commentary. Mr. L., who has been writing a report on the ethnohistory of Kashmir (a sort of administrative district monograph), is convinced that my work on the commentary is very necessary and that it would be possible to persuade the Kashmir Council—with the consent of the India Govt.—to borrow my services from the university for a specific period so that I could complete the task without interruptions. At present economies are being made in the Kashmir budget and my full Lahore salary could cause difficulties. For my part, I would consider the freedom for scholarly work worth a finan-

cial sacrifice but only up to a reasonable limit. . . . If Simla agrees, the matter is settled (they are the bosses). Following Mr. L.'s direction, I have submitted an official request from here: my condition is that during the period of "special duty" I receive half my Lahore salary, that the Govt give me free quarters in Srinagar where I will work, and that my claim to furlough and other matters will not be jeopardized.

"Maclagan, under secretary with the Indian government, is in Kashmir on vacation. He hopes the matter can be carried through and thinks he can rely on the government's request for the list of archaeological monuments. This was to have been done by Mr. L. but as a non-expert he does not feel up to the task and would like to have the assignment turned over to me since I have the material from my *Raja* work. . . . Anyhow the whole matter could not be arranged before the fall of 1895—the correspondence with Simla alone will take months. I can await the result in peace."

Maclagan, later Sir Edward, whose friendship with Stein antedated that with Fred Andrews, recalled that "in Stein's early days, the idea that the Indian Government should maintain an archaeological explorer"[12] would have been considered preposterous. It was this attitude that compelled Stein to seek posts in the educational field. Stein "carried out his official duties with all possible faithfulness and tenacity, but he was always straining at the collar and every moment he could snatch from routine duties was devoted to his favourite line of research. For a long time he was constantly asking for periods of 'special duty' or for 'extension of leave' and anyone who knows the intense aversion of the official mind to special concessions will realize the difficulties he had to encounter. From constant applications of this type Stein became a real expert at manipulating the mysterious processes of officialdom."

17 October 1894: "In the night ride across the Wullar Lake, a small storm made me worry for the safety of my manuscript. It seemed as if the Goddess of Wisdom, represented by the waters of Kashmir, was unwilling to let me abduct the manuscript. This is what happened 1200 years ago to the Chinese pilgrim Hsüan-tsang, who had to leave his Sanskrit manuscript in the angry Indus River."

30 October 1894. He rented two rooms in the Andrews house. Its location "fits my decision to remove myself from social commitments. Since Pandit Govind Kaul has arrived I am busy working on the arrangement of the manuscripts which are to go to Europe. Only two pages of the index for the catalogue are lacking; thus the first copies should be

ready to send by mid-November. . . . Hungarian papers for which I longed, arrived; with news from home I feel even more isolated than is really the case."

28 November 1894: "I am busy [at the Durbar] distributing complementary copies of the *Catalogue*[13] which just arrived from the printer. I received the first copy on my birthday. Eighty are going to Harrassowitz of which seventy will be distributed to libraries, academies and scholars. *Viva reclamatio!* . . . Last week my Kashmir plans were updated without coming any closer to realization. The Indian Govt. was willing to ask the Kashmir State Council whether it would like to have the work done at its own cost. But the university would first have to be consulted. I discussed the matter with Dr. Rattigan, the vice-chancellor, the next day and used the opportunity to inform him of my further plans. He expressed, as was to be expected, scruples associated with his own retiremen. He had to admit, however, that for me it would be more important to complete so noteworthy a task than facilitate the vice-chancellor's duties. He promised to seek the consent of Dr. Sime while he was still in office. Dr. Sime, whom I saw the next day, took the whole problem more easily although he too paid me unwelcome compliments about my quasi-irreplaceability. I will have my friends in the Punjab secretariat use their influence on Rattigan and can thus hope to prevent the matter from getting bogged down. It shows once again that in India no favorable turn of Fate levels my road to success."

And so 1894 passed and it was another year.

20 February 1895: "From Grierson [1851–1941, George Abraham, a noted linguist], Secretary of the Bengal Asiatic Society, I hear that the Society will undertake the honorarium for the *Catalogue*—4000 rupees—as suggested by Bühler to me. Since Kashmir finances are in trouble, I consider it diplomatic not to pursue my claim seriously until the matter of my furlough has been settled. I have received many thank-you letters from recipients of the gift-volume. . . . The low rate of the rupee shows that my hope of making a fortune in India is not being made easy by Fate."

1 May 1895: "I am making preparations for the summer trip to Chitral. It will be especially hard because I am taking a number of books and the transportation will be difficult. All the mules have been requisitioned for the troops."

21 May 1895: "Mrs. Andrews is ill and I stayed home to give Andrews moral support. . . . I am hoping to camp quietly on a meadow on the Pir Panjal. I would be only a day's journey from Srinagar. The Chitral matter

has had a happy conclusion: Indian diplomacy has for a long time pro-claimed all territory south of the Hindu Kush as in the British sphere of influence via India."

4 July 1895, Mohand Marg: "Roth died in Tübingen. I last saw him in his beloved garden along the Necker River. I am gratified that I could give him joy before he died by bringing him some manuscripts. . . . I work ten hours a day but am as fresh as I could possibly want to be."

7 August 1895: "I will leave my beloved meadow on 6 September. It is in full glory and bloom: the meadows are covered with foot-high flowers and look like the most magnificent carpet in the world—the main colors are blue and yellow. I will quit this alpine world with many regrets."

Rested, refreshed, Stein started early in September in search of a sacred shrine Kalhana had referred to and which had long escaped identification. "There the goddess *Sarasvati* herself is seen in the form of a swan on a lake situated on the summit of the *Bheda*-hill which is sanctified by the *Ganga*-source."[14] Portrayed as a beautiful young woman, Sarasvati, the wife of Brahma, is the patron of art, music, and letters and is the divine being who invented the Sanskrit language and the Devanagari script. She holds a lute and has a swan in attendance. Stein tells of the search. "Neither Professor Bühler nor myself had succeeded in tracing any information whatsoever among the Pandits of Srinagar. . . . I first obtained an indication of the right direction . . . when examining an old miscellaneous codex acquired by Bühler during his Kashmir tour. . . . Among the local names mentioned, only one was known to me . . . but this locality alone would not have sufficed had not a reference in Abu-L-Fazl's account of the 'mirabilia' of Kashmir mentioned 'a low hill on the summit of which is a fountain which flows throughout the year and is a pilgrimage site for the devout. The snow does not fall on this spur.' . . . There remained now only the task of tracing actually in the direction indicated the site of the Tirtha [shrine] and any local traditions attaching to it. This I was able to accomplish in September 1895 on a short tour."[15]

Stein's search brought him to a remote, idyllic spot. The path led along a well-cultivated ridge, through charming forest scenery following a mountain stream to an area where small grassy meadows were hugged by thickly wooded spurs of mountains. The bend of the stream held a hillock, and on its summit he found a man-made pool—55 feet square, its corners oriented to the cardinal points. As at other ritual baths, stone steps surrounded the tank of limpid water wherein the pilgrims had made their ablutions. Though badly decayed, they still showed reliefs, the carving of a superior kind and of ancient workmanship. Scattered about were other fragments of carved figures and scenes and traces of old walls. An old Gujar, patriarch of a group that brought their buffalo there for summer grazing, knew that the water in the tank never froze and always remained at the same level though springs bubbling up fed it copiously. Stein, always precise and methodical in his survey of ancient sites unfortunately had not equipped himself with a "thermometer and hence was unable to take the temperature."[16] Save for that, he was satisfied that this was the long-sought shrine, the sacred basin "sanctified by the Ganga-source."[17]

29 October 1895: "I am happy that you are satisfied with my saving program. I have been able to avoid many former expenses by giving up getting involved in what calls itself society in Lahore. I don't lose anything that gives me real satisfaction and gain time and comfort. . . . In 1898 I would like to effect a move to Europe. [Underlined by Ernst.] Lacking anything better I could hope to get a *honorar-extraordinariat* [a paid but untenured] professorship and could take the preliminary steps in the summer of 1897. In February of 1898 my second five year term ends and, if all goes well at the university, I can count on getting severance pay."

In summary, it might be said Stein's tension resulted from two very different rhythms: the official one that tied him to his post and its nagging, time-consuming duties and the personal one—his own work and response to nature—epitomized in his beloved Mohand Marg. An understanding of this inner turmoil becomes apparent in his letters, which subtly meter his changes in mood.

It is possible to see 1896 as a watershed in Stein's view of his future. True, there are the same laments at being tied to sterile administrative duties and his continued efforts to return somehow, somewhere to scholarly work in Europe. But these yield to new notes. He found important new friendships—his family, as it were away from home; he received official blessing to take part in unexpected, exciting borderland excursions. The opportunities India offered widened. All this translates into contentment, or perhaps where previously he strained at his position, he now began to strain at geography—his horizon moved beyond the limits of Alexander the Great's Asian venture. His growing interest in Greco-Buddhist art led him outside political boundaries to Greater India—the vast hinterland of Central Asia, homeland of tribes like the Kushan and Saka who established dynasties in the Gandhara region, and the high road to India for pilgrims who carried Buddhism back to China. While reconstituting the ancient geography of Kashmir he first came to value the travelogue written by Hsüan-tsang, the great seventh-century Chinese pilgrim. It is impossible to pinpoint when Hsüan-tsang joined Marco Polo in giving Stein a desire to explore eastward; but in a letter of 2 February 1899 he told Ernst that "an archaeological exploration" into Chinese Turkestan "has been in my thoughts since 1897."

26 February 1896: "If I could obtain a paid position in Budapest I would give up my hope of a position in Germany which, at best, is nebulous. Goldziher [one of Europe's outstanding Arabists, a close friend of Uncle Ignaz] writes that he had brilliant offers from foreign universities which he was forced to decline because of private matters. To be sure, Germany's shortage of Arabic scholars equals that of Sanskritists. . . .

You will be glad to hear that I will soon have a chance to practice French. M. Foucher, the young French scholar should arrive here about mid-March. Judging from his letters he must be a very pleasant person. . . . I have had a very interesting piece of work explaining data found in an 8th century travelogue [by the pilgrim Ou-k'ong] published recently by Sylvain Lévi and Edouard Chavannes. This work is most closely related to the *Raja* and was thus unavoidable. In the next mail I will send the manuscript to Lévi who, I hope, will take care of its translation and publish it in "Le Journal Asiatique" since the travelogue was published there."

4 March 1896: "The Indo-Scythian essay will get started before the end of this [month] and I hope to complete it during May. ["The Saka Kings of Kabul" in *Festgruss,* for R. von Roth]. The essay explaining the Chinese information about Kashmir got much longer than I had expected but should go off the next mail."

11 March 1896: "What irony it would be if the devaluation of the rupee were limited to *my* stay in India. . . . As regards Bühler, you are not giving him enough credit when you assume he was exploiting me in the Kashmir topography. His influence in scholarly circles rests above everything else on his just recognition of others' results. He does not need to appropriate their work. I will discuss my Kashmir data in a separate contribution to his encyclopaedic work. As early as 1893, Bühler had identified me as his co-worker in his prospectus for the Ancient Geography of the Indian northwest."

25 March 1896: "I have finished my report about the newly discovered travelogue of the Chinese pilgrim and since it got rather long—30 to 40 pages—I had to give up the idea of having it translated for Le Journal Asiatique. I therefore sent it to Bühler with the request to present it to the Vienna Academy for publication. ["Notes on Ou-k'ong's Account of Kashmir," vol. 135 (Vienna: Akademie der Wissenschaften, 1896).] My next project will be the essay about the Huns."

1 April 1896, To Hetty: "I am depressed by the knowledge that there does not seem to be any way, either through scholarly work or devotion to my official duties, to get ahead. At least there are hopes in Europe, although I must be prepared to have similar disappointments there. . . . More and more I see that I must settle down somewhere; otherwise I have all the disadvantage of being a man without a country. In scholarly pursuit one runs the danger of becoming a cosmopolite. Even so it will be hard for me to say farewell to India. It means giving up many attractive projects for which here I only need the leisure but which in Europe

are quite impossible. I will also miss the kind of friends I find here. I have always felt very comfortable here socially; in Germany as well as Hungary, it will be difficult in this respect. At least dearest Ernst has prepared me for this in the way he presents the situation."

13 May 1896: "The enclosed letter from Bühler shows him at his nicest and should please you. . . . My friend Foucher and his aunt, Mme. Michel, left Lahore two days ago and after six weeks of the most pleasant and refreshing company, I feel very lonely. They will arrive in Kashmir only a few days before me. F has made a thorough study of the sculptures in the museum and read Sanskrit with one of Pandit's colleagues. F got a grant for his Indian trip through the help of Sénart and just before his departure he was nominated Maître des Conférences at the École des Hautes Études, taking the place of Professor Lévi who went to the Sorbonne. Apparently the French know that one cannot become an Indologist without having seen the land and its people. F will be able to give the gentlemen among my colleagues an idea of how little free time I have for scholarly research in Lahore. Mme. Michel is a very practical woman and a good traveler. But her presence forces F to travel on the most comfortable roads and do without visits to remote ruined cities. Mme. M is apparently quite well-to-do. It is an advantage to my friend that she does not speak a word of English and I have acquired considerable fluency in French."

19 June 1896, camped on Mohand Marg: "I enjoy the freedom and work eleven hours a day. After dinner I take down Kashmiri tales from the mouth of a peasant-bard and am thus collecting valuable material which I will put to good use in Europe. Hatim, the story-teller, is known throughout the Sind Valley and to my great pleasure was brought up here by the authorities.

"To explain the enclosed letter from Foucher, let me mention that Dash is the name of a frisky little dog which I brought along as a protection against rats and other enemies. F and Mme. have a good deal of enjoyment with Dash and Dash climbs all around as though he were a mountain goat although in Lahore he never saw a rock. . . . I received the first issue of Bühler's *Outline:* to put your mind at ease my name is listed among the contributors."

13 August 1896: "It was good to know that you approve of my idea to write down Kashmiri fairytales. Its primary purpose was to produce phonetically correct texts which up to now have been non-existent for Kashmir. I also think that translating them into German as I proposed to do will be useful. The tales themselves are not wholly original since

the people in the Valley are better at borrowing and adapting than inventing. But even if the tales themselves are well-known and spread all over India, they are a welcome addition for a certain humorous idiom unfamiliar to the Indian spirit. Since Hatim is on vacation I use the time formerly spent with him handling my correspondence."

27 September 1896: Stein was back in the Happy Valley having "left Mohand on the 21st. At Manigam I was greeted by F and his aunt. F has decided to carry tents and servants with him during his Indian stay—certainly the pleasantest—and most expensive—way to travel. I envy F his interesting tour. . . . Since 1887 I have felt drawn to the border regions and now they interest me all the more. Before leaving Mohand Marg I completed a short essay on the sixty new inscriptions written in unknown characters which Major Deane had sent to me last year. I had obligated myself to do the job and mailed it to the Asiatic Society of Bengal. Let the Turkologists look for the key—these writing systems are certainly not Indian."[1]

4 November 1896: "Major Deane, who is in charge of the Swat district, has made arrangements for Foucher and myself to go there over Christmas."

Major Deane, one of the "Lords of the Marches," as Stein with a historical flourish called officers commanding key posts on the Northwest Frontier, had had his curiosity whetted by strange inscriptions carved on rocks in Swat. Stein's eagerness to explore monuments of the past in the northern borderlands happily coincided with the government's decision that control of Swat was vital to India's security. Obviously, Swat's mighty mountain barrier could be penetrated from the north and east: traders using anciently known passes went regularly between Russian and Chinese Turkestan and India. Britain and Russia, each increasingly suspicious of the other's expansion into Turkestan, kept a careful watch; both kept their eyes on the weakness of Imperial China.

Swat was a district that had a powerful and lasting appeal for Stein. He responded to views that charmed the eye—the distant presence of snowcapped mountains whose glaciers unfailingly provided the blessing of water for its lush, irrigated rice fields. To its natural beauty was added a richly textured history. The fertile valley drained by the Swat river (with the territory of Buner to the southeast) was the ancient Udyana, a country famous in Buddhist tradition. That alone would have been sufficient for him. But there was more. Earlier, the region had witnessed stirring events in the strenuous campaign which brought Alexander the

Great (ca. 327 B.C.) across the snowy Hindukush to the Indus and his invasion of the Punjab. One of Stein's ambitions was to find the exact location of the "Rock of Aornos," the mountaintop fortress which the Macedonian had successfully invested. And still more. Where the Swat and Kabul rivers meet, not far from the open plain of Peshawar, had been the ancient kingdom of Gandhara—an inscription of Darius I (about 519 B.C.) claimed it as a satrapy of his vast empire. Taxila, one of its cities, was even then known as a seat of learning.

Influences and incursions had left their imprints on the northwestern borderlands. By Stein's time the solid blank that had settled on that region following the presence there of Alexander the Great was splattered with hints obtained by inscriptions and coins and by events pieced together from Chinese, Greek, and Roman historians. Scholars are still trying to sharpen the indeterminate outlines that give the movement of tribes, the names and dates of rulers belonging to dynasties who held power briefly and vanished utterly, evidences of the rise and spread of Buddhism—a tangle of happenings that affected the Northwest Frontier in particular but the rest of Asia as well.

After Alexander's death, after the bloody sequel when his generals fought for possession of his empire, one, Seleucus Nicator, about 300 B.C. secured the Asiatic provinces—from the Aegean to the Indus, from the Oxus to the Persian Gulf. Some fifty years later the Seleucid Empire, a cultural conglomerate, began to break apart. In the east, the Greek governor of Bactria (the land lying between the Hindukush and the Oxus) declared his independence; similarly and almost simultaneously, the Iranian governor of Parthia (a large, ill-defined area running south from the Caspian steppe) established his own dynasty. In no time, the Greeks, mainly mercenaries whose business was war, fought their way across the Hindukush: by the early second century B.C. they had added the Indus Valley and the Punjab to their kingdom. Then civil wars ensued; Greek against Greek; usurpers against heirs—the former took the Kabul Valley and Taxila, the latter retained the Northwest Frontier.

The story of the Greeks in Bactria and India invites the imagination: facts are few. At best there is a lean outline based almost entirely on their stunning coins, some of the most splendid portrait coins ever minted. Chance discoveries of magnificent hoards not only established the existence of these forgotten Greek dynasties, they provided the key to unlock ancient inscriptions. Greek legends on one side had, on the other, characters in Kharoshthi, the vernacular used then and there in northern India. The coins indicate that the political fragmentation continued: soon the Greeks lost Bactria to the Parthians, retaining only a minor kingdom in the Kabul-Punjab region.

Meanwhile, in far-off Central Asia, a people whom the Chinese called the Yueh-chih were badly mauled by the Hsiung-nu (Huns?) and driven out of their tribal territory—this started a tremor felt by other nomadic peoples of the steppes. Whether it was caused by tribal rivalries or, it may be, by changes in the rainfall pattern, the Yueh-chih fled from their territory on China's northwestern border to the west. So they came into Transoxania, the home of the Scythians whom Alexander the Great had wisely refused to battle. The nomadic steppe tribes were wagon-dwellers who moved their encampments about within their territories to provide pasturage for immense herds and flocks; their mounted horsemen were their striking arm whose mobility made them irresistible against settled communities. (Both China and Europe would taste the power of the Huns and Mongols.)

About 160 B.C. the Yueh-chih, speakers of an Iranian language, having been shunted off and driven away by other tribes, bore down on the Scythians who, fleeing across the Oxus, in turn attacked and occupied Bactria. The Yueh-chih followed. In short order they took over Bactria (about 130–100 B.C.) and sent the Scythians southward. The latter, known in India as the Sakas, overwhelmed the Greeks there. Maus (about 80 B.C.), the first Saka king of Taxila, continued minting coins using legends in Greek and Kharoshthi. With the Saka rise to power the rule of the Greeks in India—save for isolated petty chiefdoms—ended.

To return to the Yueh-chih: after they had been dislodged by the Huns, after the vicissitudes of their long march, they settled in Bactria and, according to tribal custom, divided the country into districts, one for each clan. Even as they were settling down, in distant China, the great Han emperor Wu-ti (140–87 B.C.) had the idea that if the Yueh-chih returned to their original territory together they could crush their common enemy, the Hsiung-nu. To that end he sent an envoy, Chang Ch'ien, to find the Yueh-chih and invite them back. The mission so conceived and undertaken is one of the world's great epics, written by a contemporary, Ssu-ma Ch'ien, called the Father of Chinese History. Chang Ch'ien, he related, was almost immediately captured by the Huns and held prisoner for ten years before he could escape and make his way westward—a long and difficult journey, going from tribe to tribe until at last he located the Yueh-chih. They were happy to see him but declined the emperor's invitation to return: they liked Bactria.

The overland trail Chang Ch'ien had pioneered (138–125 B.C.) quite soon became the Silk Road which brought the Chinese Empire into contact with the Romans. Bactria enjoyed an unexpected prosperity when it found itself the middleman in the lucrative exchange of Chinese silk for Roman gold. Of Chang Ch'ien's remarkable mission, what mattered was not that he failed in his attempt to woo the Yueh-chih back but that by it China discovered Europe.

To take a hop, skip, and jump in the history of Yueh-chih—it happened that under the overlordship of one of their clans, the Kushan, they were fused into a rich and formidable power. The grandson of the chief who had united them was the great Kanishka. (Even his dates are disputed, but A.D. 78 seems likely to have been the time of his reign.) Kanishka, ruler of an immense empire—from northern India deep into Central Asia—became the patron of Buddhism, though it was a later form. Mahayana Buddhism, the Great Vehicle, replaced Hinayana Buddhism, the Little Vehicle. Now the Buddha who had been the teacher, a man among men, whose wisdom, self-mastery, and enlightenment set down guidelines for mankind, was transformed into the divine Buddha, a god whose superhuman qualities had marked his miraculous acts in all his previous incarnations. Previously, the Buddha had been symbolized by the Wheel, the Footstep, the Parasol; now the deified Buddha was given human form. On Kanishka's coins the Greek or Zoroastrian gods of earlier dynasties were replaced by the divine Buddha.

Under the Kushans, the thriving oases of Central Asia spaced along the transcontinental caravan routes attracted merchants from far and wide; the Chinese recorded that the kingdom of Khotan was where they met Indians, Iranians, and Greeks. In that vast hinterland, the isolated settlements (so liable to fall apart into petty kingdoms) were united by Buddhism as well as by international trade: everywhere stupas and viharas were built and beautified, each different as local artisans gave their cultural imprint to the basic forms.

And still they came: wave after wave of mounted nomads out of the steppes of Central Asia, raiding, probing for internal weakness, conquering, settling down, enjoying peace, property, and the pursuit of sedentary pleasures and inevitably becoming the target for the next wave of covetous nomads. The fifth century brought the Hepthalites, or White Huns. They occupied Bactria and "like the earlier Greeks, Sakas and Kushanas, they crossed the mountains and attacked the plains of India."[2]

The Northwest Frontier never recovered from Hunnic ravages, and when, about the mid-sixth century, they were annihilated, the region, put on a cultural back burner, simmered with tribal feuds.

Stein's mind was filled with this pre-Islamic history when he left Lahore to meet Foucher at Peshawar.

21 December 1896: "On the Swat River, Fort Chakdarra. Mme. Michel was my companion in the tonga [a two-wheeled cart] and we visited the ruins of Tahkt-i-Bahi, the niches of its chapels stripped of their sculptures. An excellent military road brought us to the Malakand Pass where last year there was fighting and this year has a well-fortified and planned camp. At Major Deane's house we were cordially received by Mr. Davis, his deputy. Our plans are to visit the ruins of cities, stupas, monasteries, to cover all the hills in this archaeological paradise. But even months would not suffice for a thorough investigation. If only I could return!"

27 December 1896, Fort Chakdarra, Swat: "I spent six fantastic days. Just enough to acquaint me with the most significant sites in which this region is so rich. That the archaeological excursions are physically strenuous heightens my enjoyment of them. I have daily rides of at least twenty-five kilometers to reach the monasteries, stupas and fortifications and return to the safe walls of the fort by evening. The cautious political administration insists on providing an escort, etc. Thus I have not lacked a considerable following which came in handy in transporting photographic equipment, etc. The troops are Dir [a nearby princely state] and Swat recruits from the valley itself.

"I feel I am on classical soil and enjoyed every minute. As extensive as are the sites, they have unfortunately suffered considerably from the barbaric digging for sculptures. In spite of Major Deane's assurances, every officer with a taste for the classical products of the old sculpture of Udyana has people dig in the monasteries and around the stupas for statues and reliefs. You can imagine what unspeakable destruction accompanies these robberies. Foucher and I often felt like Jeremiah mourning the ruins of this modern vandalism."

Stein's furlough in 1897 was spent in a determined effort to secure a European appointment. After his stay with Ernst, he visited Budapest where Uncle Ignaz's friends worked in his behalf.

29 April 1897, Budapest: "Found Vámbery [Armenius, 1832–1915, Professor of Oriental languages in the University of Budapest] at home and had a long conversation with him about my problem. He is in favour of trying to procure an extraordinary professorship for me in Budapest.

How he wants to manage that he did not explain in detail. Thus the matter seems quite illusory to me." Stein remained in Budapest the month of May, giving a lecture, attending a meeting of the Akademia, calling on influential members—doing everything suggested to court ranking bureaucrats.

1 June 1897, Budapest: "Am going to Vienna and have made my farewell call on Bishop Szasz. The creation of a chair for Indo-Iranian Philology, involves the separation of this discipline and is to be applied for by the faculty this fall and the decision should be had by March. I visited his Excellency Berzeviczy who was once Secretary of State and is considered to be the future Under-Minister and was received very well. I hope to meet Bühler tomorrow afternoon." In September, the crucial time, he was back in Budapest again trying to contact the professors whom Vámbery considered most influential.

3 October 1897, Trieste: "I left Budapest knowing I had done everything possible. Vámbery's petition was not considered because the faculty meeting had been postponed. It will be taken care of at the first opportunity. V and others think my cause at the university is as good as won." He sailed for India the next day. He had done his best. Not only in Budapest, but he had zigzagged across Europe—Vienna, London, Paris, Bonn—seeing, meeting, probing, testing.

In London he had met Sylvain Lévi (1863–1935), who had come from Paris. Of the same age, Lévi, also a Sanskritist, was about to start on a trip to India and Japan.

20 October 1897: "I reached Bombay on the 18th and the same day Lévi arrived and we spent the evening in a Sanskrit symposium. He gladly agreed to an archaeological jaunt to Kathiawar. We are on our way to Girnar, one of the most sacred pilgrimage sites in India where I hope to see much of interest. . . . Our reception by the Nawab of Jumnagarh, in which kingdom the temples are, had already been arranged for and thus I can expect a most pleasant stay. I know Lévi is grateful to me for my guidance during his first days in India and I could not wish for a more delightful companion. . . . News received at Aden tells me that I can move into my old home (Mayo Lodge) but instead of Andrews, Professor Allen will share the lodgings with me."

25 October 1897, Verawal, Kathiawar: "Verawal is on the Indian Ocean coast. Since my last letter I have spent unforgettably beautiful days in Jumnagarh where we were received as guests of the Nawab and where a piece of old India has been preserved. . . . In Kathiawar [Peninsula], where over 300 princes and nobles share in the rule, much of old

Hindu custom and usage has been preserved. I envy Bühler the years he spent here. What I must painstakingly reconstruct from sources in Kashmir, continues to live on here undisturbed. . . . My Sanskrit opened up avenues of communication and so I did not lack for courtesy and attention. Not even in Jammu was so much hospitality shown to me. I was especially glad of all this for Lévi's sake who thus became acquainted with the essence of India at its most pleasant. Days such as these compensate me for what awaits me in Lahore."

11 November 1897, Lahore: "Frankly I feel in better shape than at any place during my last European trip. I bought a good horse very inexpensively and enjoy rides with Mr. Allen in the evenings. There is much to do but it does not prevent me from enjoying a Punjab winter— sunshine, flowers and pleasant company. . . . I was invited by the new Governor of the Punjab, Sir Mackworth Young, who knew me when he was Vice-Chancellor, to be one of a small circle of friends who play tennis in the afternoons. I hope this connection will be useful before I say goodbye to the Punjab. Sir Charles Roe remains Vice-Chancellor here until his retirement next April and always shows himself to be a fatherly friend and patron to me. . . . Had good news from Major Deane: fighting in the Afridis region seems to have stopped."

9 December 1897, Lahore: "Thanks very much for the clipping on Sven Hedin's travels. [The Swedish explorer (1865–1952), then on his first extended expedition, 1893–98, was crossing Asia along untraveled routes from Orenburg, just east of the Urals, to Peking.] Kugiar, where the manuscript Hoernle edited was found, is far, far away from the Keriya River where Hedin mentioned seeing ruins. . . . Let me assure you: I would not undertake a scholarly project without ample preparation and serious attention for personal safety. Let me also put your mind at rest by reminding you that I have gained considerable experience during my ten years in India."'

27 December 1897, Malakand, Swat: "I am writing from Maj. Deane's hospitable house. It is wonderful to be able to roam through the Swat valley in any direction. My prediction that the summer fighting would calm down by Christmas has happened. Nothing seems to change except that the fortifications of the camp on the pass have been reenforced. The fortress of Chakdarra, where I camped with Foucher a year ago, has been greatly enlarged. I intend to complete my archaeological notes from last year and work them into an article for the 'Journal of the Asiatic Society.' Major Deane is the most pleasant of hosts and also an interesting histori-

cal type: the frontier politico who knows how to keep half-barbarian tribes under control."

What was Deane's "good news"? What lay behind Stein's extended mention of Sir Mackworth's official position and his demonstrated friendliness; and why did he characterize Sir Charles (revelatory words) as a "fatherly friend and patron"? What kind of tension during November and December expressed itself in the unusual testiness he showed at Ernst's solicitude? What Stein dared not tell Ernst was that Deane had mentioned the lively possibility of a military expedition being sent into Buner and had invited Stein, if he could secure the necessary permissions, to accompany the occupying force. The prospect was heaven-sent: Buner, with Swat, comprised the ancient Udyana! Stein mobilized instantly to deal with the complex situation—dealing with both the Punjab administration and the university authorities. He not only asked for the project to be sanctioned, he needed it to be acted on swiftly, within days and weeks, not months.[3]

By the very beginning of 1898, Stein could tell Ernst his exciting news.

3 January 1898, Lahore: "I had not dared mention my plans because the chances against it seemed formidable. Since last summer's uprising in the Swat valley, there was talk of having the border troops pay a visit to Buner whose inhabitants had supported the Swatis. Buner, on which no European has ever set foot, is the site of most of those mysterious inscriptions collected by Major Deane and published by Sénart and myself. It is also the last territory in which the sculptures of the Graeco-Buddhist art were saved from the vandalism of amateur collectors and their native agents. It was clear that unless controls were exercised the monuments would suffer total destruction during the occupation. It seemed equally clear that the inscriptions, heretofore only known from faulty copies should be studied on the spot for in most cases one did not know which end was up and which was down.

"Considering all these scholarly factors, I could not deny myself and followed Major Deane's advice to ask the government to make my visit possible to further archaeological research. My friend in the Punjab Government secretariat supported me and the governor granted the expense which would arise from my "deputation," the government reimbursing the university for my salary. . . . It was only ten days ago that I was certain that the occupation of Buner would come about—and this was another reason I did not mention it before. During my stay in Swat, Major D. received official notification of the intended occupation. . . . I returned to Lahore to make the necessary preparations and go back to Swat tomorrow. The·occupation should take two to three weeks and is intended mainly to make a moral impression. I beg you not to have the slightest worry—I will be properly escorted on all trips. I am convinced that the chance presenting itself to further my scientific work will open a valuable perspective for the future and will not be repeated for another half century and perhaps never."

9 January 1898, Kingargalli, Buner: "I accompanied the division to Sanghao at the foot of the mountains. The next day the Tange Pass was taken and the way to Buner open. The position was very strong but the opposition very weak: in European terms the Buners lacked rifles. As soon as the 20th Regiment had climbed a neighboring peak, the Buners had to abandon their 2000 ft. mountain. But prior to that three batteries fired away at the rock for five hours. An interesting spectacle which I observed in absolute safety from the artillery position. The effect of this bombardment was total. That evening three regiments occupied the village from where I am writing and it seems as if open resistance in Buner is over. I myself came over the Pass a few hours later and spent the following two days visiting the extensive ruins in the immediate environs. As Major D. had predicted, the population submitted at once. I feel better than ever: it was an endless joy to climb the mountains on which my ancient buildings are found."

16 January 1898, Camp Barkili, Buner. ". . . There was no resistance after the taking of the Tange Pass. This is both good and bad for me; on the one hand, I am permitted to roam with my escort over the territory of the tribes that have submitted; on the other hand the period of occupation has been cut to four or five days when the last troops will be back on British soil. This is what makes a *thorough* investigation of the numerous ruins impossible and forces me to work with greater haste. I regret that soon a thick veil will once again fall over this fascinating mountain region. Who can say when the Buddhist ruins will be visited again by a European. At any rate, I am satisfied that I was the one to examine this new region even if superficially! . . . The people of Buner had fled to the mountains. We stayed in curious quarters in the occupied villages and I learned more about the Buner household than would ever have been possible in the Punjab. . . . I feel better than ever, especially since last summer. I would appreciate it if my friends in Germany were informed of my small expedition. Could you do me a favor and send a notice to the 'Algemeinerzeitung'—I have absolutely no time to write any one. . . . I received a letter from Vámbery telling me that the faculty

has voted 28 to 4 for my appointment. I feel convinced that it will be a *fait accompli* within three months. The Minister has announced he would approve the nomination."

25 January 1898, Swat: "You will be happy to know that I have returned safely to British soil. I could have used as many weeks as days in Buner. I am in no hurry to return to Lahore and shall spend four or five days here writing my report for the Archaeological Survey and have the maps drawn for it under my supervision. ["Detailed Report of an Archaeological tour with the Buner Field Force," Bombay, 1899] . . .

Unfortunately I could not reach Mt. Mahaban where probably Alexander's Aornos should be sought: a long ride on the last day brought me closer to it than any European thus far."

Buner was far, but not far enough. To the identification of Aornos—reconciling topography with historical accounts—Stein would return again and again. This is his first mention of seeking the "Rock of Aornos," the site of one of the Macedonian's most celebrated victories. According to the legend related by the Greek historian, Arrian, "even Herakles, son of Zeus, had found [the lofty acropolis] to be impregnable."[4] When all the particulars of this fortified eyrie were told to Alexander "he was seized with an ardent desire to capture it, the story about Herakles not being the least of the incentives." As long as Mt. Mahaban was out of bounds, it remained a candidate; in 1904 when Stein reached it, his survey "furnished conclusive evidence against the location of Aornos on that range.[5] He would take up the task later when the opportunity presented itself.

In addition to the Swat and Buner expeditions, the winter of 1897–98 would remain memorable: it was then Stein began two of his closest and most valued friendships—with Thomas Walker Arnold and Percy Stafford Allen. They met in Lahore when the men joined the faculty of Government College, Arnold as professor of philosophy, Allen as professor of history. No friendship is like another in an individual's repertory; each has its unique quality, each its own grammar. With the passage of time and the way each of the men developed, the differences in Stein's friendship with Andrews, with Arnold, and with Allen become clearer. At the risk of oversimplifying, it is possible to say that Stein found a dependable lieutenant in Andrews, that he took delight in Arnold's *joie de vivre,* enthusiasm, and aesthetic response to elegance in architecture and art, and that he felt a kinship to Allen's commitment to a noble intellectual task. Stein could count friends, good friends, by the score, but these three were part of the fabric of his being.

In Victorian times such intimate friendships had the problem of maintaining the proper reserve while feeling warm affection: a man's given name was kept for family use ("My dearest Ernst"). But conventional problems have conventional solutions. How the recipient is addressed measures the relationship between the two parties. Does a letter begin "Dear Mr. Stein," or "My dear Mr. Stein," or "My dear Stein," or "My dear General"? The first used the conventional, formal mode; the second indicated a degree of acquaintanceship; the third a personal tie; and the fourth, a nickname, a close, intimate relationship. The nickname could only be used between initiates.

Andrews tells how the four friends got the nicknames by which they addressed one another (or referred to one another) for the rest of their lives: That winter, the Arnolds, newly come to Lahore, were almost daily visitors at Mayo Lodge where Allen, still single, had joined the Andrewses and Stein. Andrews was making much of his having become a member of a volunteer body known as the Punjab Light Horse.

> Volunteering being in the air, it was, I think, Allen with his delightful sense of humour, who suggested that there should be a Mayo Lodge contingent of volunteers (non-military) with Stein as C.O. [Commanding Officer]. As we were all desirous of contributing such help as we could in promoting Stein's work, this seemed appropriate enough. Eventually, we each acquired a designation or *nom-de-guerre*. Allen was Publius, Arnold was The Saint [Mrs. Arnold was the Saintess], my wife and I were the Baron and Baroness, and Stein, of course, B.G., Beg General; names by which we always thereafter addressed one another. To these was added Madam (Mrs. Allen) when she came out to Lahore to lend lustre and grace to our party, and Madam she still is to those of the band who survive.[6]

Of the three couples, it was with the Allens that friendship expressed itself in an extraordinary correspondence. Stein's first letter, sent from Mohand Marg in August 1898, when Publius had gone home to England to marry Helen Mary Allen (a cousin), was a thank-you note for "the dainty volumes of Scott. I have been reading them over my lonely dinners." By 1902 he was confiding his fears about Ernst's health; by 1904 he could discuss Hetty. "The condition of my poor sister-in-law is ever causing me deep anxiety. Her inexpressible grief [since Ernst's death] forces her to reduce all contact with things of the present to a minimum, and she suffers bitterly being separated from her boy [Ernst, then 13], a necessity she has had to recognize at last. He has just been sent to my old school in Dresden."[7]

Publius had become a kind of brother; as it had been with Ernst, the exchange of letters was regular and, once begun, continued for life. Publius's fortnightly letters followed Stein on his most distant travels, and when a batch arrived—they might have been three months on the way—they were read in seriatim so that Stein could share and savor the Allens' doings as they happened. This correspondence was Stein's lifeline. Because of it neither time nor distance could create a hiatus in the relationship. At a deeper level the Allens' manifest affection, concern, and pride in his life preserved Stein from being a solitary man, a lonely wanderer like Csoma de Körös. He was, because of it, an explorer with an intellectual commitment, a traveler in strange and distant worlds absent for a stated period and for a particular purpose.

Publius and Madam were themselves engaged in exploring a world as richly fascinating: forced to return to England because of Madam's health, they began their lifelong work—the monumental task of collecting, dating, and editing the vast correspondence of Desiderius Erasmus, "the intellectual hero of the 16th century."[8] The two quests balanced one another. The Allens uncovered Erasmus's letters in libraries, archives, and private collections all over Europe; a formidable correspondent, the letters would fill eleven volumes, the first published in 1906, the last, forty years later, carried through by Madam after Publius's death. It was the high quality of the scholarship shown in the *Letters* that won Publius the presidency of Corpus Christi College at Oxford. Stein, who made a rosary of anniversaries, wrote the Allens at Christmas 1926: "For close on 30 years your unfailing friendship has been the brightest thing in my life, a constant source of comfort and encouragement."[9]

Though Stein's friendship with Arnold lacked the steady heartbeat of letters, it never faltered. Their first meeting came when Arnold, after ten years' teaching at the Anglo-Muslim College at Aligarh, moved to Lahore. Stein wrote to Ernst on 19 February 1898: "A capable Arabist . . . I am glad that I finally have someone whose interest in ancient India is as great as mine. I cannot consider him a replacement for me at Oriental College, since by his own admission this charming but unpractical scholar is unsuited for administrative work." This initial response to Arnold's charm—a dedicated scholar by nature expansive and empathetic—held to the very end.

As Stein's first mention of Arnold was to Ernst, his last was to the scholarly community. *In Memoriam Thomas Walker Arnold, 1864–1930* appeared in the Proceedings of the British Academy (1932). Between first and final mention both men had achieved recognition; Arnold was knighted in 1912. Unlike the steady pulse of Stein's friendship with Publius, that with the Saint was like a geyser—spasmodic displays of steam evidenced a hidden, constant fire. That Arnold reciprocated Stein's friendship was a fact tested again and again and most clearly stated in a mournful scene Stein recorded. "An hour spent in the spring of 1929 by a sunny Dalmatian shore when, after a happy reunion successfully achieved after years of separation, that dear friend expressed the wish that his obituary for the Academy should come from my pen."[10] A year and a month later Arnold's heart—as his doctor had warned—stopped.

The two men matched one another in many ways. Arnold, too, had an insatiable appetite for learning, his memory a storehouse of striking detail, his conversation broadly informative and enthusiastic. Arnold's charm never palled, and in his last sad notice Stein tried to spell it out. "His genial personality, his bright outlook on life which no experience could dim, his delightful humour, his feelings of genuine sympathy for those in need of help or encouragement."[11] In writing the obituary Stein fulfilled Arnold's wish; in writing his own will Stein expressed his wish—the bequest he left for scholarly research was to be called the "Stein-Arnold Exploratory Fund."

At another part of the spectrum was Major-General Lionel C. Dunsterville (1865–1946). As a dashing junior officer he had, when in Paris on furlough, met an Hungarian nobleman whom he invited to visit him at Lahore. Sigismund de Justh, hearing of a countryman, called on Stein. "In this way," Dunsterville recalls in his autobiography, "a kindly Providence furnished me with a lifelong friend."[12] Thus he accounts for the unusual meeting of a professor and a "disreputable subaltern."

3 March 1898, Lahore: "Today the Andrews are leaving for England. I lose dear friends and advisors in them. Andrews wants to secure a position in England because in the Education Department here life was made rather unpleasant for him."

17 March 1898, Lahore: "I have found dear friends in Mr. Arnold and his wife. He is an enthusiastic Orientalist and book-collector. Mrs. Arnold, who is a good deal older, is widely traveled and is very knowledgeable in Japanese and Chinese art. She has circled the globe no less than five times. I must give Lahore credit for the fact that I have never lacked stimulating friends. . . . I sent little Ernst some stamps."

7 April 1898, Lahore: "On Sunday I invited the Allens and a few other friends to a picnic breakfast in a beautiful old garden not far from the Ravi. We had delightful hours in the shade of the mango trees on the terrace of a decaying estate of a Sikh prince. It began to get warm

quite early and by midday we had returned. . . . I hear nothing from Budapest and assume that my file wanders slowly from office to office."

5 May, 1898, Lahore: "Yes, you are right if you wish that I have as good friends in the home town [Budapest], as I have here in India. I can hardly wish for more stimulating or dearer companions—house companions—than my friends."

29 May 1898, Srinagar: "I arrived here at the very beginning of the beautiful weather. . . . Moved into my old accustomed camp grounds. The Reception Department of the Government has put excellent tents at the disposal of me and my men, making the equipping for Mohand Marg simple. . . . On 2 June I move to my alp. Here I feel as though I had returned to my second home and don't even want to think of a time when such summers might be denied me."

5 June 1898, Mohand Marg: "I was truly happy to set foot again on my beautiful alp. Pir Bakhsh, who is with me again for a short time, put my tents in order in their accustomed sites and today after unpacking my library, it seems as if I had never left my alpine paradise. It also felt good to see the honest joy with which my old acquaintances in the village greeted me. I had no difficulty in getting 30 coolies—people almost fighting for the privilege of going with me. Maybe they hope that I will have the Maharaja give me Manigam as an estate. That wouldn't be bad at all. . . . The position of epigraphist to the Indian Government, a vacancy for which I could apply with a fair hope of success, is one of the purely scholarly positions in India and I confess openly I would rather have this than a professorship in Budapest. I do not need to hide the fact that my premature departure from my Indian research would cause me pain. If I could secure a position in India which would leave me more freedom for scholarly work than my job in Lahore, I would be delighted to stay in a country of my scholarly interests. I want to hold out in Lahore for as long as possible, provided I have freedom for my summer months as in the past years. This makes my Lahore job and work not too onerous."

11 September 1898, Mohand Marg: "I admit that my last year in India has been so good to me that I find it difficult to even think of saying farewell. Since last October I only had to work in Lahore five and a half months, while the marvelous times I enjoyed in Swat, Buner and Kashmir would never be available to me in Europe. Bühler thought that I, like himself, would regret giving up India. As long as the conditions are acceptable I will not hurry for there are steps which cannot be undone and often I am plagued by doubts wondering if in Budapest I would not long to return to India. . . . My memoir on the maps of Kashmir is ready and

should be a small volume by itself. The Buner Report is at the Lahore Government Printing shop. . . . I have my alp all to myself since the cold drove the Gujars away. Delicious grapes from the Maharaja's vineyard are delivered to me—I enjoy them all the more because they are probably stolen." In this letter to Ernst there is no mention of a lengthy, carefully worded official communication he had sent the day before to the chief secretary to the government of the Punjab.

How radically Stein's inner orientation had changed: his hopes were no longer magnetized by Budapest and Europe. Work on the *Rajatarangini* was coming to a close while tours into Swat and Buner had made him eager for newer, more exotic fare. Just then, an article in the *Journal of the Asiatic Society of Bengal* (1897, no. 4) by A. F. Rudolf Hoernle, a paleographer, whom Stein knew and whose work he respected, gave definite shape to Stein's daydreaming. Hoernle had written of his several years' study of manuscript fragments known to have been found near Khotan, an important oasis on the southern rim of the Taklamakan Desert in Chinese Turkestan. They posed an interesting problem and suggested an intriguing complex of historical events: the script was Indian Brahmi but the language was an unknown, non-Indian one.

The origin of Brahmi is a matter of scholarly debate. Basham, summarizing the arguments, says: "it may have begun as a mercantile script suggested by the shape of Semitic letters, or by vague memories of the Harappa script ca. [2000 B.C.], but by the time of Asoka, though still not completely perfect, it was the most scientific script in the world."[13] From these puzzling manuscript finds, from the inscriptions Major Deane had found and Stein had studied, from the numerous ruined Buddhist establishments in Swat and Buner which he had explored with Foucher, from the routes taken by the Chinese pilgrims which connected the Indian subcontinent with the oasis cities of Central Asia—from hints and hard evidence Stein saw a pattern emerge. A mysterious past was waiting to be discovered.

Though the state of Indian archaeology did not invite work in distant, little-known regions, Stein launched his ambitious plan.

The proposal, dated Kashmir, 10 September 1898, marks the decisive change in Stein's life. As a plane, its momentum irresistible, leaves the ground and, from a lumbering machine that had taxied to its starting-point, becomes a wingéd thing designed and perfected for swift flight— so Stein's proposal lifts him into the life for which his talents and training had prepared him. The proposal has the cogency of a masterful brief.

"I have the honour to request that you will be pleased to submit the following application to His Honour the Lieutenant-Governor for favourable consideration, and, in the event of his approval, to arrange for its subsequent transmission to the Government of India.

"The object of my application is to secure the assistance of the Local and Supreme Governments towards a tour of archaeological exploration planned by me for *Khotan,* Chinese Turkestan, and the ancient sites around it." Briefly Stein gave reasons for such a project. "It is well known from historical records that the territory of the present Khotan has been an ancient centre of Buddhist culture . . . distinctly Indian in origin and character. . . . The discoveries of ancient manuscripts, coins, sculptures, etc. made during recent years show beyond all doubt that these old sites, systematically explored, promise to yield finds of great importance for Indian antiquarian research." To support this statement, he summarized "the principal objects of archaeological interest brought to light from that region as well as ancient notices regarding it." These finds, he went on to explain, resulted from "the casual search of native treasure-seekers whose statements require verification. It can safely be asserted that the antiquarian objects obtained in this manner will acquire additional value when their origin has been properly authenticated.

"The object of the tour proposed by me will be, therefore, to explore the ancient sites at and around Khotan from an archaeological point of view, to search for such data as will throw light on their history, and to make collections of ancient remains on which full reliance can be placed. I beg further to point out that since my plan was first formed, it has been announced that the Imperial Russian Academy of Sciences has arranged to send three savants for the exploration of *Turfan* (Southern Chinese Turkestan) where ancient manuscripts have been found. I am also informed that Dr. Sven Hedin's explorations, referred to in the Appendix, are likely to be resumed."

Having deftly insinuated national rivalry into the argument, Stein turned to the logistics, financing, and administrative matters. He considered the distance to be covered and the conditions of travel along the route from Lahore to Khotan via the Kara-Korum, estimating it would be necessary to allow six weeks for the journey out and the same time for the return. "For the explorations in and around Khotan at least three months would have to be allowed as the minimum period during which they could be carried out efficiently. The whole of the proposed tour would thus necessitate my absence from Lahore during six months"—he had already sounded out the university authorities and had their support for a furlough, especially since part of his absence would coincide with the annual summer recess. "Starting from Lahore the end of May, I should find the summer route over the Kara-Korum open, and thus be able to reach Khotan about the middle of July. The next weeks could be devoted to the examination of ancient remains at and near Khotan, while the cooler months would be available for the exploration of the ruined sites in the outlying desert tracts. Commencing the return journey towards the end of October, I should hope to reach Leh and Kashmir before the close of the usual travelling season." In Stein's mind the crossing of the mighty mountains that separated Lahore from Khotan was lengthy but not difficult; to him what was difficult was the financing.

He estimated the expenditures for the expedition: the "out-of-pocket expenses of the journey from Lahore to Khotan and back, including the purchase of baggage animals, camp equipage, etc., would amount to about 4,000 rupees. Assuming my absence from Lahore would extend from the 1st June to the 30th November, . . . extra expenditure would be required for little more than three months (excluding the University vacation). If these arrangements would be similar to those approved by the University authorities for the periods of my deputation to Kashmir in the summers of 1895, 1896 and 1898, the expenditure caused to the University and to be refunded would amount to about 800 rupees. In addition it would be necessary to provide a certain sum for the cost of exploring and excavating certain sites, and for the purchase of antiquities that may eventually be offered for sale. For these purposes an allotment of 2,000 rupees seems to be indicated. The total expenditure . . . would amount to 6,800 rupees." At a time when economy was the only word applied to Indian archaeology by government officials, this sum, so picayune in retrospect, loomed ominously large in Stein's mind. How he must have wished he were a Schliemann with a personal fortune to spend or that Abraham or Moses had wandered into Central Asia so that he could tap the moneys contributed for Biblical explorations.

His proposal continued: "In view of the important bearing which the proposed explorations are likely to have on Indian historical research, I beg to express the hope that the Governments of the Punjab and India will be pleased to render them possible by meeting between them the total cost. . . . I beg to point out that there is good hope of some learned societies, such as the Asiatic Society of Bengal, and the Royal Geographical Society of Great Britain, being induced to contribute towards the cost of these explorations. Their contributions may amount in the aggregate to from 1,000 to 2,000 rupees. Whatever sums may be received from these sources would be paid to the Government as a set-off.

"In return for the material assistance I have ventured to ask from the Local and Supreme Governments, all proceeds of my search will be handed over to the Government of India for the British Museum, except such as may be given, with the approval of Government, to the learned societies in return for their contributions (if any).

"In addition to the above assistance I beg to request that the Government of India will be pleased to obtain for me through the Foreign Office, the necessary passport or permission from the Chinese Government enabling me to travel in the territory of Chinese-Turkestan. It is my intention to undertake my tour in the summer of 1899 in case all preliminary arrangements can be completed within the available time. As, however, unforeseen circumstances may cause a delay or necessitate a postponement, I beg to request that the passport be made available both for the years 1899 and 1900.

"I beg further to point out that the assistance of the Local authorities at Khotan appears essential for the success of the proposed explorations, I hope, therefore, the Government of India will be pleased to arrange for the issue, by the Chinese Central, or Provincial Government, of such instructions to the Amban of Khotan as will assure to me full permission to survey and otherwise explore all ancient sites within his territory; to carry on excavations of such sites, and to retain objects of antiquarian interest brought to light there; and to acquire such objects if offered for sale."

To whet the intellectual curiosity, to involve the imagination of men who, though bureaucrats, were university-trained and proud of being responsive to historical studies, Stein added his Appendix A, "Note on Central Asian Antiquities," an admirable summary of voices from the past—voices long silent and, ghostlike, recently heard.

"The attention of all scholars interested in Indian antiquities has been directed to the old Buddhist sites of Central Asia as a new source of important materials for their researches," Stein wrote, recalling the birch-bark manuscript sent by Captain Bower for the Asiatic Society (1890) a treasure that he had acquired in Kashgar, a large oasis city in Sinkiang at the western end of the Tarim Basin. Hoernle's exhaustive examination of that famous manuscript had shown that it was written sometime in the mid-fifth century (A.D.) and was far older than any other Indian document previously known. It also raised questions relating to Indian literary tradition as well as the influence of Buddhism in Central Asia. "The great scientific importance of the find was fully appreciated by the Government of India whose enlightened patronage enabled Dr. Hoernle to prepare and publish a complete edition of the manuscript which has become a standard work for the study of Indian palaeography." Hoernle received another important collection of ancient manuscripts two years later, also Indian in writing and language, but written on paper—paper was not known in Europe until the twelfth century. At the same time, from the same place, similar manuscripts were sent to St. Petersburg. Hoernle had established that these as well as the "Bower Manuscript" came from a ruined Buddhist monastery near Kuchar, also in Chinese Turkestan; all were in the Brahmi script.

Orientalists were still engaged in studying these valuable finds when they were told of another remarkable discovery from the same region: "fragments of a birch-bark codex sent to St. Petersburg early in 1897 were found to exhibit the Kharoshthi writing of the extreme northwest of India [whereas the Brahmi script is used everywhere except the northwest] which had previously been known only from inscriptions and coins." These fragments contained portions of a sacred Buddhist text in Pali, the form of Sanskrit spoken at the time of the Buddha and thus the language used in Buddhists' canonical writings. "On palaeographical evidence [these] could not be assigned to a later date than the commencement of our era." This, the oldest Indian manuscript, was said to have been found somewhere in the region south of the great Taklamakan Desert. Another set of remarkable old documents, the "Macartney Manuscripts," had been found in the desert itself to the north of Khotan and were in an Indian script but a Turkic language. "As if to show still more strikingly the importance of the desert region of *Khotan* for Indian antiquarian research, Sven Hedin, the well-known Swedish traveller, has recently communicated interesting notices of the ancient sites visited by him near *Keria*, east of *Khotan*, sites known from Chinese sources and now buried below the advancing sand of the desert. The description of the wall-paintings traced by him in some of the sand-buried ruins, leaves no doubt as to the early date and Buddhist character of the Keria finds."

Stein emphasized that such manuscript discoveries were the merest token of what systematic research could reveal of the spread of Indian culture and religion into Eastern Turkestan. "There is ample evidence for it in the accounts of the Chinese pilgrims to the western regions. *Fa-hsien*, who reached Khotan about 400 A.D. found the country prosperous and rich and the people since early times attached to the Buddhist

faith. Magnificent convents were provided for the priests who used Indian books and the Indian language. . . . The fullest description of Buddhist Khotan, however, we owe to the great pilgrim Hsüan-tsang who
visited the country on his return from India A.D. 644. He found there
Buddhism still in a very flourishing condition, and the people in possession of a high culture. He already indicates the Indian origin of the latter,
and records traditions which point to Kashmir as the place from where it
was introduced. There were about a hundred monasteries. A number of
these are accurately described to us in Hsüan-tsang's memoirs, and their
positions could in all probability be traced without difficulty. Hsüan-tsang
already notices in the territory . . . sites which had been abandoned owing
to the advance of the desert, and which are, therefore, likely to have preserved remains considerably older than that period. . . . [Within a radius
of sixty miles of Khotan were such buried sites.] One of them, the city
of Pinia, has been described by Hsüan-tsang and visited by Marco Polo
shortly before the sands reached it."

A month after sending in his proposal, learning that the lieutenant-
governor of the Punjab was inclined to recommend the project to the
government of India, Stein brought up his heaviest reenforcement. At his
suggestion Hoernle wrote to the Department of Agriculture and Revenue,
which passed on financial matters: "I have now gone through nearly the
whole of our collection of Central Asian antiquities. . . . There are numerous points of detail on which no information can be got from the native
treasure seekers. . . . Some coins which I have deciphered seem to show
that in the first centuries B.C. and A.D. there was a Scythian or Turki
(Elighur) Kingdom which included both Khotan and Kashmir on the
two sides of the Karakorum Range. Towards the end of the first century A.D. it was broken up into two parts: the Northern portion (Khotan)
being annexed by the Chinese, the Southern (Kashmir) by the famous
Indo-Scythian King, Kanishka. I have bilingual (Chinese and Indian)
coins, issued by the Chinese administration of Khotan after 73 A.D.
Much to corroborate this and to fill in details may be discovered by European exploration."

One aspect of the finds bothered Hoernle. "Regarding a certain series
of block-prints, myself and others in Paris and London were growing
suspicious that they might be absolute fabrications. . . . My own idea was
that they might be forgeries only to the extent that ancient blocks of type
had been discovered in Khotan and that from these the crafty treasure-
seekers had prepared reprints, which they sold to collectors as genuine
ancient prints. . . . Some of the block-prints may be forgeries to *that* ex-

tent. But even this is not certain. Anyhow, this . . . can only be determined by an explorer on the spot. . . . Khotan and the Southern portion of Chinese-Turkestan seems to me distinctly the proper sphere for British exploration. It by right belongs to the British "sphere of influence," to use the modern term; and we should not allow others to secure the credit which ought to belong to ourselves. . . . and I would submit that it is for India to take it up, and for the Government of India to have the credit of it."

Hoernle's supporting letter ended with a statement whose irony escaped him. "It should be borne in mind that India is practically inaccessible to European savants, who alone, as a rule, study such things; and that a certain amount of precariousness is inseparable from deposits made in Indian Museums. For these reasons the bulk of any collection that may be made should, in my opinion, be transferred to the British Museum."[14] This attitude and such reasoning Curzon, soon to be named viceroy, would find unacceptable.

From Simla, where Stein spent the Christmas recess, he wrote a chatty letter to Hetty without mentioning the pending proposal. "Oxford plays a role in the intellectual atmosphere of my Mayo Lodge. My friends and their young wives represent the most agreeable achievement in education and culture as found at Oxford, that ancient bastion of learning. I do not lack for a pleasant and heartwarming circle of friends in Lahore. They are equally well developed physically and intellectually and are *generous* in the finest sense of the English word. Mrs. Allen [newly arrived] is a very charming and highly educated young lady."[15] His account of life at Mayo Lodge, at once high-minded and idyllic, is Stein's way of counting his blessings. His contentment was further enriched when he met the Hoernles. "He would like me as his successor at the Madresa in Calcutta. Mrs. Hoernle reminded me that we had met at the 1886 Vienna Oriental Congress and she had not forgotten me. She did everything she could to make my stay as pleasant as possible. Hoernle is engrossed in scholarly work on Central Asian antiquities." The letter ends on a note of pride: "The Senate granted me a six-weeks' furlough." This was, as he pointed out, "not a momentary change but the result of ten years' work and for that reason gives me satisfaction."

27 December 1898, Simla: "I received a letter written by the Lt. Governor of Bengal to Dr. Hoernle offering me a job at Calcutta. Its conditions are as follows: I will be taken into the Indian Educational Service as principal of the Calcutta Madresa with a salary of 800 rupees per

month. . . . Further advancement up to 1500 per month will depend on vacancies in the department. In addition I will have free quarters in the Principal's house—this represents an additional 250 rupees; also a part of the servants, garden and lighting are free. The main fact is that the job has a pension and is a more satisfactory type of work—no teaching, only administration. Dr. H has been on 'deputation' several times for scholarly work and thus has established precedents for me. Also it has almost a four-month vacation as against two and a half at Lahore. . . . The best thing for me is to accept the job. Dr. H. thinks I can get along on half my salary."

Stein gave up the amenities of Mayo Lodge to obey the dictates of thrift and prudence: he decided to accept the Calcutta offer.

5

"On New Year's eve I received the welcome news that the Home and Finance Depts. had sanctioned my Khotan tour. Since then I have heard that the Foreign Office has also agreed to arrange matters with the Chinese authorities, and that the project may hence be considered as sanctioned. I am all the same anxious to keep the matter as quiet as possible as there is always the chance of a Russian expedition being sent in the same direction, and premature news might send them moving."[1] Success in his momentous project Stein told first to Andrews.

Filled with items at once personal and businesslike, the long letter is a prototype of the hundreds that would be sent—and carefully filed—over the coming decades. It reveals much of the nature of their friendship: confidence easily given, a rich dividend of the years when they were housemates in Mayo Lodge, and Stein's reliance on Andrews to handle technical matters, run errands for his forthcoming expedition, and attend to his requests intelligently, promptly, and carefully. Still at the "My dear Andrews" stage, Stein's letter proceeds to give his news chronologically. First he refers to the Buner photographic plates which Andrews had taken with him to have processed in England—"I thank you most heartily for your kind help in the matter and only wished that the *Report* were not already finished so that I might do it 'in public.' " Then comes a detailed account of his acceptance of the principalship of the Calcutta Madresa [a Muslim College], adding, since Andrews would appreciate its full significance, "I am to act as advisor to Government in archaeological, numismatic and kindred matters. You may well imagine that I should do my utmost to develop *this* side of my prospective work." And further: "My appointment will in no way interfere with the Khotan tour, except that it makes it possible for me to postpone the start until July or August. This would be a great advantage as regards the work in the desert which in the winter is less difficult."

Next he described Christmas at Simla. Not until he comes to New Year's Eve does he mention the Khotan expedition, plunging immediately into topics already discussed and the search, already begun, for items essential for the Central Asian venture. "I have not yet received the sample collection of condensed food but it has reached Karachi safely. I am greatly obliged to you for getting the firm to send so representative a set of articles. I have tried the films and find them quite satisfactory." He outlined the kind of camera required to suit his different needs and trusted Andrews to find it.

"There is yet another request which I must trouble you with. Sven Hedin's account leads me to believe that my work in the desert would be made far easier by the use of a portable well-sinking apparatus. Along the dried-up beds of rivers, water can be struck in the sand at a depth of not more than from 9 to 30 feet. But of course digging for it after a day's march is a very troublesome and not always a promising proceeding. Now I remember from my military course of instruction that portable well-sinking apparatus of the *Norton* pattern has been used with great advantage in desert tracts during the Abyssinian war and by French expeditions in the Sahara region. I should very much like to find out what would be the approximate weight of a Norton well apparatus capable of penetrating to a depth of 30 feet, and what it would cost. I should require an apparatus as light as possible yet fit for the purpose. It would have to be taken over the passes by mules, and it would be wise to keep each load down to about lbs. 150, though, of course, 200 lbs. might be carried at a stretch. . . . I may add that the soil to be bored into consists of sand with layers of clay." Such was the precision and forethought of Stein's preparations.

"I am afraid," he continued, "I am really making too free use of your kindness by the many requests I make. I am afraid you will find that distance is no protection from me." With technical matters finished, Stein relates his other news. "I think, I wrote to you that I have been relieved for 6 weeks of my work as Registrar to complete my Rajatarangini Introduction. In order to use them I have gone into 'Purdah' and have pitched my camp in the delightful old garden where Maharaja Sher Singh was murdered. . . . I am established here in great style. All around is peace and quiet, and I often wonder if I am only 3 miles from the nearest City Gate. . . . Please do let me know if I can be of any use yet to you here. It is hard always to ask for favors and do nothing in return." Time and events would rectify the balance of indebtedness.

Not until 2 February 1899 did Stein break the news to Ernst. "To be brief, the Government of India and of the Punjab have approved my proposal concerning 'an archaeological exploration' to Khotan in Chinese Turkestan and have budgeted the required sum of 7000 rupees for travel expenses and purchase of antiquities, etc. . . . The affair is not yet public. The reason for the *discretion* is that I heard of an expedition planned by the Russians to Turfan in the area of Aksu (northeast of Kashgar) and since the route along the Khotan River is as Sven Hedin's beautiful book shows, quite comfortable, I don't want to draw the Russians into my work-area through a premature announcement. Because Khotan is considered as lying in the 'British sphere of influence,' this fact has been useful in gaining the consent from the gentlemen on the Indian Mt. Olympus. . . .

"The plan, now in the stage of realization, offers me the possibility of a field of research extremely interesting to the Indian archaeologist. Lord Curzon, the new Viceroy, himself traveled in Turkestan and will certainly be interested in the matter which will further aid me with the Government in Calcutta. Once I am in the Government service many other things can be accomplished.

"Let me, I beg, assure you that after a thorough investigation of all details I will be in absolutely no personal danger. The trip over the Karakorum has become a tour for ladies. Khotan itself, in the 'environs' of which 'my' sites are situated, is a district under careful Chinese administration. Sven Hedin's book shows (as do other publications) that the personal safety of Europeans in Chinese Turkestan is as great as in and around Kashmir. Concerning the physical hardships of individual stretches of the sandy region, I have the advantage of Sven Hedin's experiences. . . . In consideration of a possible visit with Mr. Macartney, the English political agent in Kashgar, I may possibly have to take along my tails. The world where civilization and the mail does not reach has become very small. . . . I am busy."

16 March 1899, again to Ernst: "I received your letter of the 22nd and know that you are well but unfortunately with reference to the Khotan tour also indulge your fears which in my honest conviction are, as things stand, unfounded. It is not a matter of an adventurous journey into an unknown territory, but a simple archaeological examination of sites which would have been available to a dozen travelers if they had the interest in such things. Khotan itself can be more easily reached today

than Kashmir was in the 40s. The most important finds have been made in the immediate vicinity of the cultivated region (e.g. the old birch bark manuscript in Paris came from a Buddhist vihara about 20 kms. from Khotan in the *hilly region*). The ruined cities which Sven Hedin saw are situated along the old riverbeds which carry water until the fall and from where later, when the snow melts, water can be dug out at a depth of 5 to 7 feet. . . . Lord Curzon is coming here the end of this month for a week. I know his private secretary, Mr. Lawrence, very well and in any case expect to see him."

6 April 1899: "Lord Curzon spent a week here to confer with the "Politicals" of the Afghan border, the Keepers of the Marches. . . . The most agreeable thing was that the Punjab Government asked me to serve as the Viceroy's cicerone in showing him the museum's archaeological collection. The day before yesterday I had the opportunity to introduce myself personally to Lord C. He showed in his great book on Persia how interested he is in Oriental archaeology and geography. Thus I was not surprised when he devoted a long time to the Indo-Graeco sculptures. I had half an hour to present my explanation and found in Lord C an expert questioner and listener. This was the most stimulating *collegium ambulatorium* I have experienced so far in India. I handed him my re-prints of various articles of mine and found that Lord C was acquainted with my roving expeditions on the borderlands as well as my Khotan project; at the end I had the satisfaction of having Lord C promise quite spontaneously to do what he could to open my new area of research. Mr. Lawrence, the private secretary, had obviously prepared the ground well. [In acting as a guide to Lord C] I have done the Punjab Government a personal favor for the museum has no trained director and the explana-tions of the sculpture would have been missing. I had taken care to make certain that Lord C was informed of this fact and that I was in no way responsible for the collection's rather unsatisfactory presentation. Accord-ingly, His Excellency thanked me for my service as a personal rather than an official favor. It may interest you to know that Lord Curzon who, after the premier, holds the highest office in the British Empire is only 40 years old and has traveled in Persia, Russian Turkestan, the Pamirs and China. Personally he is, with all his dignity, very charming."

19 April 1899, to Hetty: "Possibly the notification of my nomination to the position in Calcutta will arrive shortly and make it necessary for me to go there directly. Fortunately I am ready to leave within a few hours: my books and possessions are all packed without arousing any attention. I left Mayo Lodge because Mrs. Allen has typhoid fever. To

make room in the house I had to clear out so that she could be moved into my large bedroom, the coolest in the house. In about 5 hours all my possessions were packed and tinned, not an easy job, but necessary. . . . When I leave Lahore I leave with the most pleasant impressions."

27 April 1899: "Just received notice that my appointment to Calcutta has been confirmed by the Secretary of State. Please address letters for the time being to Madresa, Calcutta. In haste."

With Curzon's appointment as viceroy to India early in 1899, archaeology sailed out of the doldrums in which it had been caught for almost a generation; it took direction and strength under the steady wind of his energetic concern. His visit to Lahore in April of that same year was a personal gesture. Before leaving England he had learned of the sterile, unimaginative policy for archaeological research laid down by the government of India. Considering the vastness of India, everywhere planted with monuments and ruins of monuments whose age and meaning might help reveal its past, considering too that archaeological matters were a hodge-podge, the sad result of being administered by each state and province and thus outside viceregal direction, it is clear that Curzon had his personal priority and worked with dispatch and effectiveness to revitalize a moribund situation.

In a report of his grand tour he wrote: "I cannot conceive any obligation more strictly appertaining to a Supreme Government than the Conservation of the most beautiful and perfect collection of monuments in the world, or more likely to be scamped and ignored by a delegation of all authority to provincial administrations."[2] Then he let loose salvos of anger—righteous, patriotic anger. "The continuance of this state of affairs seems to me little short of a scandal. Were Germany the ruling power in India, I do not hesitate to say that she would be spending many lakhs [hundred thousands] a year on a task to which we have hitherto rather plumed ourselves on our generosity in devoting Rs.61,000, raised only a little more than a year ago to 88,000."[3]

There was a modern ring—unusual only in that it was uttered by a head of state—when he wrote. "When I reflect on the sums of money that are gaily dispensed on the construction of impossible forts in impossible places, which are to sustain an impossible siege against an impossible foe, I do venture to hope that so mean a standard may not again be pleaded, at any rate in my time."

Curzon's dismay at the state of archaeological research gives an added dimension to Stein's complaints. His years at Lahore fell within a particularly bleak period, a period when lack of activity in archaeology was sententiously excused as due to a lack of funds. Were not all beaureaucracies alike? Certainly that was his meaning when he thought of Budapest where "my file wanders slowly from office to office."[4] In that firm belief he planned his campaigns and mobilized his contacts so that he might pit his puny strength against the dead weight of official inertia. That he managed to do as much as he did is the measure of his enthusiasm and enterprise. He might get a "deputation" now and then, but to change the system was beyond his power, even beyond his daydreams. It took a man fired by the need for scholarly work and clothed in viceregal panoply to get the bureaucratic inertia into motion. During the lean years when Stein hoped for an appointment in Europe, he demonstrated through his work and wits his worth to India. Curzon's establishment of the Archaeological Survey of India "for the first time on a sound and secure foundation"[5] found Stein known and respected in high circles, ready, willing and set to go.

The "sound and secure foundation" had been lacking in an earlier Archaeological Survey which had never outgrown its antiquarian origin: the Asiatic Society of Bengal founded by Sir William Jones at the close of the eighteenth century. Sir William had dedicated the society to study the "Antiquities, Arts, Sciences and Literature of Asia." A private venture, its members were military and civil servants who used their free time to pursue their intellectual interests; their wide range of studies reflected, it might be said, Jones's own versatility. A genius in languages—he had taught himself Hebrew, Arabic, Persian, and a bit of Chinese and was familiar with the important European languages—he served as puisne judge of the Calcutta Supreme Court, having studied law. Certainly, in whatever direction his intellectual questing was attracted, he made signal contributions, opening up insights into a world heretofore unknown.

Columbus-like, Sir William discerned a vast continent of languages, thus beginning exciting explorations in comparative philology. He also provided the "sole firm ground in the quick-sands of Indian history"[6] when he identified Sandracottos, the Indian monarch mentioned by Greek historians, with Chandragupta Maurya, a contemporary of Alexander the Great. By "fixing the location of the classical Palibothra," he gave a "starting point from which future investigations of ancient Indian geography could proceed."[7] The impulse toward ancient history and geography thus begun was continued: Stein's masterly editing of the *Rajatarangini* and his recovery of the ancient geography of Kashmir were a significant part of the ongoing work.

Though Jones was the founding father of the Asiatic Society, he did not work alone or in a vacuum: other men stationed in many parts of India carrying on their duties in the civil or military services reported on wonders they saw—of temples sheathed in sculptured figures and marvelously decorated caves carved out of the living rock. Or they described and pondered on ancient coins they had collected; or they faithfully copied inscriptions written in scripts long forgotten, such as King Asoka's rock-carved edicts. The harvest of wonders grew, a harvest garnered by men pursuing their avocations when and as their official tasks permitted. They joined the Asiatic Society because they craved intellectual fare, and they found nourishment in the problems to which they addressed themselves. Though they were amateurs, their efforts were cumulative because their *Journal* informed others of their findings.

Synthesizing some of the fragmentary material thus accumulated waited on the erudition, the informed imagination, and, in the case of James Prinsep (1799–1840), on a command of numismatics and epigraphy. Assay master of the Calcutta Mint, he served the society as secretary from 1832 until his untimely death. What rich and exciting years those were. Writing eloquently of Prinsep's achievements in *The Story of Indian Archaeology,* Roy considers the most remarkable to be his deciphering of the two scripts used on the Asoka edicts as well as the earliest written documents. "Within the incredibly brief space of three years [1834–37], the mystery of both the Kharoshthi and Brahmi scripts [were unlocked], the effect of which was instantly to remove the thick crust of oblivion which for many centuries had concealed the character and the language of the earliest epigraphs."[8] Prinsep's task was made more difficult by the coexistence of the two scripts: Brahmi was found throughout India, except in the northwest where Kharoshthi was used, until the end of the third century A.D. (It continued on for several centuries in Central Asia.) Brahmi is read from left to right, whereas Kharoshthi, related to the Aramaic alphabet of Achaemenid Persia and like all Semitic-derived scripts, reads from right to left. Prinsep provided the key whereby Hoernle deciphered the "Bower" manuscript and other fragments found in Central Asia.

Prinsep's insights into Kharoshthi came from coins unearthed in a stupa in neighboring Afghanistan. They furnished a kind of bilingual Rosetta stone carrying the names of rulers inscribed in Kharoshthi with their Greek equivalents on the obverse. As they cleared up one problem, the Begram coin-finds posed another; they "brought to light for the first time the names of a considerable number of Graeco-Bactrian and Indo-Scythian dynasts of whom history had absolutely no knowledge."[9] Two other excavations of stupas in the Indus-Jhelum region (the vicinity of the Salt Range), yielded coins and inscriptions of the Kushans, another unknown dynasty. Who were these proud kings with strange names? Whence had they come? How extensive were their kingdoms? When did each reign? And, always present, what was their connection to Buddhism?

Now ruined, ravaged, and forlorn, the mighty stupas rising above the surrounding landscape, ignored by Muslims and Hindus alike, had been sacred to Buddhism. As a Christian church's altar is above the saint's body interred in the crypt, so the stupa took its form and sanctity from the tumulus raised over the Buddha's ashes. On his conversion to Buddhism, Asoka had had the ashes disinterred and redivided so that stupas might flower throughout his empire. Their presence, utterly unexpected, posed the questions: had Buddhism existed in India and when? To understand the bewilderment, it must be remembered that Europe's acquaintance with Buddhism, particularly its origin in India and its powerful role there during a forgotten millennium, had only begun in the 1820s.

Buddhism as a living religion had been encountered outside India, from neighboring Tibet eastward to Burma, Indo-China, China, and Japan. To the westerners, missionaries, and officials, Buddhism was another manifestation of Asian idolatry. Its historic role began to be understood when, in 1827, Henry Thomas Colebrook (1765–1837), with Jones, one of the early students of Sanskrit, declared in a lecture that Buddhism was a heretical offshoot of Hinduism. He based this on the fact that the canon of Hinayana Buddhism (today called Theravada Buddhism) was written not in Sanskrit, the archaic, learned, sacred language of the Vedas, but, as the Buddha had preferred, in Pali (about the first century A.D.). Pali was a vernacular dialect spoken by the Buddha so that his message could be widely understood, whereas Sanskrit was the property of the priestly caste. With Buddhism's spread, Pali became internationalized. Colebrook's claim that India was the birthplace of Buddhism constituted a minor revolution.

As an earthquake is recorded thousands of miles from its epicenter, so the Indian origin of Buddhism was located by Chinese writings. These were first translated at the Collège de France, for several decades the only place in Europe for Far Eastern studies. In 1814 a chair in Chinese was established there and was held by the extraordinary J. P. Abel Rémusat

(1786–1832), who taught himself to read Chinese. Among his translations were the travels of Fa-hsien, who about A.D. 400 had journeyed overland to India to visit the holy places and seek information on monastic rules. Rémusat's translation "was remarkable for a period when ideas about Buddhism were very vague and little was known of the geography of Central Asia and the history of India."[10] Stanislas Julien continued this inquiry of the transmission of Buddhism to the Far East; he brought out (1853–58) three volumes of pilgrims' accounts, the second being the biography of Hsüan-tsang by the monk Hui-Li and the third Hsüan-tsang's own *Record of the Western Regions (Hsi Yu Chi)*, a travelogue, a precise account of his sixteen-year journey (629–45) across the width of Central Asia and up and down the length of India.

Hsüan-tsang, Stein's trustworthy guide to ancient Kashmir and soon to be his patron saint, in his years spent in India mastered Pali and the various schools of Buddhist thought. On his return, he was appointed head of the Monastery of Extensive Happiness where he deposited the treasures he had brought back with him: 150 tiny particles of the Buddha's flesh, icons of the Buddha in holy acts, and 657 separate volumes of scriptures to be translated. As translator and interpreter his influence was profound. In widely spaced intervals of leisure, he entertained his monks with memories of his pilgrimage: the pleasures and perils of the road, the kingdoms and cities he had visited, the terrible terrains he had crossed, the shrines prayed at, the erudite teachers with whom he had studied, the strange people he had met, and the stranger customs he had witnessed. From this material the monk Hui-li assembled his master's biography.

The stories flooded the acolyte's imagination with images and incidents of his pious master moving in an alien world, supernaturally protected from the assaults of bandits and the blandishment of kings. At one time Hsüan-tsang was called "Tripitaka"—meaning "Three Baskets" which hold the entire canon of Hinayana Buddhism. "By the 10th century and probably earlier Tripitaka's pilgrimage had become the subject of a whole cycle of fantastic legends."[11] As Hui-Li's biography begat these legends, so these legends begat the folk-novel *Hsi Yu Chi* (bearing the same title as Hsüan-tsang's own travel book, it was translated by Waley and called *Monkey*). Anonymous for three hundred years, it delighted adults and children alike and built itself a beloved, enduring mansion in the Chinese imagination. Hsüan-tsang or Tripitaka was, from childhood on, a friend to all men and women of all ranks and all regions. At a crucial moment his name would, like the magic lamp of Sinbad the Sailor, reveal a treasure to Stein.

To complement his translations, Julien brought out a study (1861) which provided an intricately wonderful key that unlocked Sanskrit names from their Chinese forms. ("Méthode pour déchiffrer et transcrire les noms sanskrits qui se rencontrent dans les livres chinois.") Thus, far-fetched as it may seem, Julien's Chinese researches furthered those in India. "In describing the ancient geography of India," Cunningham wrote, "I would follow in the footsteps of the Chinese pilgrim Hwen-Thaang [*sic*], who, in the seventh century of our era, traversed India from west to east and back again for the purpose of visiting all the famous sites of Buddhist history and tradition."[12]

General Sir Alexander Cunningham is the colossus of the pre-Curzon period of Indian archaeology. Recommended to the East India Company by Sir Walter Scott, he sailed for India in 1833 as a nineteen-year-old cadet and, until his final retirement in 1885, proved himself to be a man of amazing energy and high intelligence. He had a long life—unlike Jones and Prinsep—and though, like them, he began as a dedicated amateur, during the last twenty-five years of his stay in India he served as the full-time archaeological surveyor, initiating a professional status for archaeologists. A mere two years after he arrived, he began digging on his own at a stupa at Sarnath, a site sacred to the Buddha's first sermon. Sent on official duty into little-known Kashmir, he continued Prinsep's numismatic analysis of Greco-Bactrian and Indo-Scythian dynasties. Like Rawlinson, Cunningham was irked and impatient at official duties that took him away from inquiries into India's past; as early as 1848 he began agitating for an organized, funded, countrywide survey of ancient sites. Not until the British government took over the governing of India did his proposal find a listener in the first viceroy. It was then that Cunningham, then Colonel, was appointed archaeological surveyor (1861). His title defined his task: to survey and describe the existing monuments of Northern India—a kind of tidy inventory of past glories. To a modest salary and field allowance a financial inducement was added: "a share in the antiquities to be discovered by him."[13] Perhaps he was too tireless, his results too prodigious—five years later the government wearily ended the survey. Cunningham returned to England.

The belief expressed in this short-lived survey was that only objects deemed attractive, unusual, or architecturally significant *to Western taste* were to be measured and recorded. Excavation, conservation, reconstruction were dismissed as unnecessary and expensive; still unformulated were the concepts of context and stratigraphy. Nevertheless, Cunningham's results, written up and circulated, were impressive and engrossing; the government could not long refuse to take up its honorable responsibility and continue them. In 1870, the Archaeological Survey, this time as a separate and distinct department in the government of India, was created and Cunningham named its director-general. On his retirement fifteen years later he summed up what he had accomplished. "I have identified many of the chief cities and famous places of ancient India, such as the Rock of Aornos, the city of Taxila, and the fortress of Sangala, all connected with the history of Alexander the Great. In India I have found the sites of celebrated cities . . . connected with the history of Buddha. Amongst other discoveries I may mention the Great Stupa of Bharhut [built ca. 150 B.C.] on which most of the principal events of the Buddha's life were sculptured and inscribed. I have found three dated inscriptions of King Asoka. . . . I have traced the Gupta style of architecture [fourth and fifth centuries A.D.—the earliest extant freestanding Hindu temples]."[14]

Expressed in a long envoie was his conviction that he had completed the major part of the task assigned to him. It was time for him to retire. He proposed that the finishing touches of the inventory-taking should be done under local authority—each state, each province setting its own goals, setting its own pace. Why he did so is hard to understand. Was he too old, too tired? Or did he believe he had no successor? Whatever the reasons, at his suggestion the department he had headed for the government of India was dissolved and the director-general's post eliminated.

Then followed the bleak years, the bleak years of Stein's tenure at Lahore; the decade when he had to circumvent apathy and indifference, to find through friends or by the flattery of laudatory phrases a lever to move the governments of the Punjab and of Jammu and Kashmir to furnish him with a modicum of help—petty sums to survey the ancient sites in Kashmir or to accompany the army into Buner. Having survived the bleak years Stein hoped that Curzon's coming would change the official attitude toward archaeology.

Stein arrived at Calcutta to assume the principalship of the Madresa on 6 May 1899; six months later, he wrote Ernst, 1 November 1899: "I must blame myself for having pushed the Calcutta position without first having acquainted myself with the climate and the living conditions. It is now clear to me how grateful I must be that a happy chance took me to Lahore first. I must also be grateful that the Khotan expedition gives me a chance to escape Calcutta and everything connected with it reasonably soon and with an advantage. By the fall of 1901 my prospects will probably have clarified themselves."

Complaints continue but without despair. Whereas he had been like a man trapped in a room looking for an exit—an appointment to a European university—now his laments were limited to the discomforts of the Calcutta climate and his living arrangements. "This is the worst season in Calcutta and this year especially bad: such conditions have not been experienced here for a long time—104 degrees in the shade. Thus I am only too glad to escape this steambath tomorrow night," he wrote on 9 May. He missed Lahore which had become a second home to him; he missed Mayo Lodge where he had lived happily in a shared domicile with his friends; he missed the familiar avenues of governmental approach; most of all he missed his summers on Mohand Marg. "I have found it advisable to choose Darjeeling as my next vacation retreat. The governor and the government have been up there for the last ten days and it is, of course, important for me to establish personal relationships with the gentlemen in the secretariat. From Darjeeling I hope, despite the approaching monsoon season, to be able to visit Sikkim, perhaps to find a temporary substitute for Mohand Marg." His reaction to Calcutta was that "as a city it is rather imposing but for my taste too European"; to the Madresa that the job "is much easier than at Lahore and there is no lack of capable assistants."

The afternoon of the last school day he left Calcutta for Darjeeling. 14 May 1899: "It is green and cool but the monsoon clouds announce approaching rain. In spite of the rain and fog on my first evening I saw the grandiose panorama of the Kinchenjanga and the other mountain giants of Sikkim and Tibet. Unfortunately I am separated from the actual alpine regions by steep valleys and a considerable distance. As a summer resort, Darjeeling is not suitable for work and in addition is very expensive and overcrowded. . . . The main thing is that I am able to introduce myself to the gentlemen of the Government and was charmingly received everywhere. My Khotan plans have just been transferred to the Bengal Government from Simla [the Punjab Government] and are to be submitted to the Governor, Sir John Woodburn, soon. After lengthy consultations with Mr. Slack, Secretary in the Revenue Department, and with his consent, I submitted a new memorandum in which I suggested a 'deputation' for a full year. If I succeed in this and if the results are good I shall find ways and means to escape the hot months in Calcutta, one way or another. Now I am finally in government service and with that the way is open. Perhaps I may yet secure the position I have wished for: archaeological sphere of activity under the central government."

Though just arrived, Stein was calculating his vacation time, matching it to the Calcutta climate. "Thanks to the monsoon, July and August are relatively cool. September is again hot but I can get away for three weeks in October. Thanks to the Ramadan holiday [the ninth month of the Muslim lunar year during which daytime fasting is strictly observed] which this year falls in January and follows Christmas, I will have about five weeks at the best time of the year for an archaeological tour. In Lahore I had the reputation for a certain facility in the exploitation of all available furloughs. I plan to secure for myself the maximum of freedom in Bengal." Stein ended this letter from Darjeeling telling of his "evening walk to the well-cared for grave of poor Csoma de Körös. The lovely cemetery on a mountain slope would please Hetty very much."

22 May 1899, Darjeeling: "I was received by Sir John Woodburn, a charming 53-year old gentleman; he welcomed me most kindly to Bengal and showed the greatest interest in my Khotan trip. . . . Yesterday I received written confirmation that Sir John had approved my deputation for *one year* starting at the end of the summer vacation in 1900. This corresponds exactly with my wishes. I can use the vacation for preparations in Kashmir and thus will *de facto* have a full 14 months at my disposal. The total cash expenses amount to 14,000 rupees and are carried jointly by the Central Government and the Governments of Bengal and Punjab. My salary of course continues independently.

"I cannot put a high enough value on this security for a real research tour. It is supposed to supply me not only with scholarly results but improve my position in India considerably: if I am successful, the way to new deputations is assured. I have every reason to be glad about what I have achieved so far; the concessions made by the Government are no small matters here and indicate a special confidence in view of the fact that I just entered Government service. . . . The postponement of the tour is welcome for several reasons—it allows me to see my work on the *Rajatarangini* wholly ready; to work systematically on my travel preparations; and last, but not least, to get the Madresa straightened out and thus prove my usefulness to the Gentlemen."

28 May 1899, Santakphu Bungalow, Sikkim: "I am glad you are satisfied with my success in Calcutta. Mr. Pedlar, the Education Director, is a scientist (chemistry) and thus understands scholarly interests—a rare thing in the Education Dept. He is very charming and is my direct superior. . . . Up here it is very cool; the bungalow is small but for me adequate, having been accustomed to a tent. The three-day march brought me along the ridge of the chain which forms the border between British Sikkim and Nepal. The magnificent rhododendrons, the many ferns,

orchids and so forth make the Singalila Range a paradise for botanists but still it cannot compare with the floral glory of Mohand Marg. The view is grandiose and this morning, for the first time, I saw Mt. Everest surrounded by other giants. There is a good path along the crest, the so-called Nepal Frontier Road, built for strategic reasons, flanking Nepal, a country still barred to Europeans."

Back in Calcutta, 13 July 1899: Stein writes of "news which moved me very deeply and will remain a sorrow for a long time. My Pandit, Govind Kaul, my faithful co-worker since 1888, succumbed to a fever in Srinagar. From the confused communication I conclude that it was typhoid. A single earlier letter had reported the Pandit's illness but without preparing me for this great loss. The poor man was only about 50 years old. On my departure from Lahore nothing led me to conclude that it would be a separation for life. Since he could be of no further help to me on the *Rajatarangini,* I had arranged that the Asiatic Society would have him work on a Kashmiri-Sanskrit dictionary. It would have been a monument to his erudition. You will understand that his premature death, a man who was my daily co-worker for ten years and for whose knowledge and character I had true respect, is affecting me very profoundly. *Requiescat in pace!* I must try to help his young son for whom I obtained a position as assistant in the accounting office of the Kashmir Government, secure a salaried position. Meanwhile I feel obliged to give material assistance to the family." And again, a week later, 20 July; "the loss of Pandit Govind Kaul lies heavily on my heart. He died much too early and I deeply feel the gap he has left in my Indian life."

Some twenty years later Stein titled the preface to *Hatim's Tales* "In Memoriam Pandit Govind Kaul." Even though he had previously acknowledged the great help his friend had given to the text of the *Rajatarangini* (1892) and to the *Catalogue of the Sanskrit Manuscripts in the Raghunatha Temple Library* (1894), and, after his death, to the translation of the *Rajatarangini* (1900), Stein seized a last occasion to pay full tribute. He sketched in Pandit Govind Kaul's background: born in 1846, son and grandson of scholars of high repute, "universally respected and of commanding social and political position in the Brahman community of Srinagar, he had a thorough literary training in Sanskrit as well as a first-hand acquaintance with the geography and history of Kashmir. His outstanding qualities had commended him to Bühler."

In August 1888, a few days after his arrival in Srinagar, Stein had met him. "I was quick to notice Pandit Govind Kaul's special interest in

antiquarian subjects. . . . I was equally impressed by his dignified personality which combined the best qualities of the Indian scholar and
gentleman. A short archaeological tour which we made in company to
sites around the Dal Lake, helped to draw us together in mutual sympathy and regard. So it was a great source of satisfaction when . . . [he]
accepted my offer of personal employment and agreed to follow me to
Lahore for the cold weather season." So began their unbroken, eleven-
year association. "Whenever Pandit Govind Kaul was by my side, whether
in the alpine peace of my beloved Kashmir mountains or in the dusty toil
of our Lahore exile, I always felt in touch with past ages full of historical
interest for the historical student of India."[15]

Thus Stein spoke of "my best Indian friend" twenty years after his
death when he himself had been knighted and stood at the pinnacle of
his fame and success. Sir Aurel wanted to record that the first sure step
on the ladder that had led to his fame and success had been made with
the "friendship and help" of Pandit Govind Kaul.

28 August 1899: "The vacations in Bengal are too scattered to be of
any use for a trip to Europe." His return from Khotan was two years off,
but he was already figuring out how to visit Ernst. "From Kashgar, Europe
can be easily reached in about three weeks thanks to the Russian RR which
goes as far as Ferghana. Since the route by the Karakoram takes about
seven weeks, I would lose no time returning to Calcutta. Also it is doubtful if the detour would be more costly and I could travel alone instead
of with a caravan. It would be wonderful to see you again in the spring
of 1901. It would also be advantageous to report on my trip in London."

Looking far ahead did not stop Stein from using a four-day holiday
for a trip to "Gaya, in Bihar, an important Hindu pilgrimage site. I hope
to find much of interest in the ruins of Bodh Gaya, situated near the
town" (20 July 1899). Gaya, north and west of Calcutta, easily reached
by train, had once been part of the kingdom of Magadha, a center of
Mauryan culture and power; there meditating under a bodhi tree the
Buddha had attained enlightenment. When, two hundred years later,
Asoka made Buddhism the state religion, one of the most important pilgrimage sites was to the Tree of Wisdom at Gaya.

24 July 1899, Gaya, Bihar: "My short visit to Bihar, the ancient
Magadha, succeeded beyond all expectation. I was graciously received by
Mr. Oldham, the [tax] collector, to whom I had been recommended.
In his hospitable bungalow I found the Indian atmosphere and way of
life which I love. I am surrounded by the quiet and space of a compound.
I made many stimulating observations in Gaya which complement my

experiences in Kashmir. Yesterday I spent unforgettable hours in the ruins of Bodh Gaya, the spot where the Buddha found his enlightenment. Extensive excavations around the site of the Bodhi tree brought forth an overwhelming mass of sculptures, some going back to the third century B.C. I studied this memorable place from early morning until evening without having seen enough."

Hungry to see more Buddhist sites in Bihar, he proposed to the government almost immediately on his return that he be sent on a ten-day tour to investigate old sites in the Hazaribagh district, a chain of hills west of Gaya. Stein's arithmetic was simple and functional: he figured that if this archaeological tour was authorized and counted as a "tour of duty" with a per diem allowance, and if it was added to the scheduled fourteen-day vacation, he could stay away from Calcutta for most of October. "The chain of hills," he explained to Ernst, 24 August 1899, "contains a number of important Buddhist ruins and since the climate in October is pleasantly dry, this small tour would benefit my physical well-being as well as my archaeological knowledge."

What he had referred to as his "certain facility in the exploitation of all available furloughs," worked as well in Bengal as it had in the Punjab. On schedule, as he had planned, he began the Bihar tour.

10 October 1899, Giryek, Bihar: "Writing from the first lap of my archaeological pilgrimage, from sacred Magadha, which I reached yesterday, heading for Parasnath, the holy mountains of the Jains. From Giridih, a coal-mining district whose sandy hills and numerous shafts remind me vividly of Jaworzno, I continued to the foot of the Parasnath by 'push-push,' a curious carriage consisting of a palanquin mounted on two wheels and pushed by four coolies. Progress was slow because the road rises sharply and the coolies who belong to the half-civilized Kolarier tribe are not too strong. In Madhuban where the Jain temples and rest-houses are situated I was, thanks to the thoughtful arrangements of the district authorities, made welcome. From there a good path led to the 4000 ft. crest—the air cool and delightfully dry. Four days were spent making a tour of the pilgrimage sites and then yesterday I reached Newadah where Mr. Oldham had organized the extensive preparations for the main part of my tour: a native landowner put two ponies and a carriage at my disposal and another offered an elephant for the jungle covered hills. This morning I drove to the eastern end of the chain of hills which bear many monuments to the life and teachings of the Buddha. Ruins of stupas crown the hills. The Chinese pilgrims described every spot in this sacred area and it gave me great satisfaction to follow Hsüan-tsang here as I had in Kashmir and Swat."

17 October 1899, Camp Kurkihar: "The long walking tours in the jungle covered hills of the Rajgir Range have given me a rich harvest of archaeological experiences. I spent four interesting days on the site of the ancient capital, mentioned before 500 B.C. Excursions into valleys now desolate and difficult of access enabled me to determine several holy sites mentioned by the Chinese pilgrims which had been missed by previous visitors. The elephant put at my disposal proved very useful in the jungle; my shoes and clothes are in a dreadful state, but I am feeling very well. Yesterday I camped in a delightful palm grove in the Jethian valley; today I have a comfortable camp near a village where almost every house contains old sculpture or building materials."

His Bihar tour succeeded better than he had expected. Using the Chinese pilgrims as guides—when they had visited Gaya it had been a lively center for Buddhist monasteries and stupas—he found and mapped routes which had gone unrecognized. He returned to Calcutta via Gaya.

1 November 1899: "Again at the Oldhams for one day which I used to visit Bodh Gaya. The road to it leads through the most friendly landscape I have ever seen in the Indian plain; the ride to the site, so rich in memories and traditions, and the coolness of the morning air will be forever in my memory." The last night he made his camp "at the foot of the picturesque granite hills where King Asoka had his famous grottos carved out of the living rock to house pious anchorites—all this almost 2200 years ago." (Before leaving for Khotan, he wrote up his findings: *Notes on an Archaeological Tour in South Bihar and Hazaribagh*.)[16]

This excursion was part of his preparation for Central Asia. In Kashmir, the traditional knowledge of the pandits had substantiated the accounts left by the Chinese pilgrims and Kalhana. In Bihar he had only the Chinese travel notes to guide him to former glories swallowed up by jungle. Practice had developed his boyhood hobby in old topographies into a kind of sixth sense—a talent given muscles by repeated exercises in reconciling geographical descriptions left by earlier travelers with whispers heard by on-the-spot examination. From his first initiation into Buddhist sculpture in the museum at Lahore, Stein had been readying himself to find and interpret Buddhist landmarks scattered in Central Asia.

Before he could start on his Khotan expedition, much remained to be done: in addition to his work at the Madresa, he faced a considerable load of writing.

21 September 1899: "My Preface is finished in manuscript. The final copying goes to the printers in the next mail. The 150-page Introduction

will be ready by the end of November. Bühler was quite correct when in 1897 he said this would be the hardest part of my work. All that remains is the Index which will go to the printer the end of December. In April, the 2 volumes with 1100 pages should be completed."

Back from Bihar. 30 November 1899, Calcutta: "I am *very* well and concentrate on finishing the Index. Two-thirds is finished and I hope to get it off my neck by 23 December when I leave for the Christmas vacation."

Christmas plus Ramadan added up to five or six weeks which he had already decided to spend away from Calcutta. A month earlier, 25 October 1899, he had notified Ernst: "I expect to spend that time in the Punjab for which I feel truly homesick. After a tour on the Indus, I would like to go to Lahore for a few weeks with friends and undisturbed work."

The new year, the new century found him among his friends. "You can imagine the feeling of well-being that was mine in the circle of warm friends and the accustomed environment after the unpleasant conditions of Calcutta," he reported happily to Ernst on 10 January 1900. "My host, the Allens, and all my old acquaintances exhausted themselves in care and attention to me and my native friends and students gave me numerous proofs of devoted attachment. Although I was much in demand socially, I succeeded in putting my visit to some practical purpose." His friends in the Punjab government, knowing of his unhappiness at Calcutta, had been on the lookout for ways to bring him back—this time to a position he would enjoy: inspector of schools. "I would find the inspectorship rather pleasant because it provides an independent kind of activity and permits periods of stay in one or another of the hill-stations during the hot months. Probably the salary would be higher—1000 to 1250 rupees—but the main attraction would be the better climate and the much more agreeable living conditions. You will see in Bühler's biographical notice, that he was inspector of schools for thirteen years and found the work to his liking."

Meetings and conferences; he was assured that "if the Bengal Government does not object, I shall get the first vacancy in the Punjab in May 1901. . . . It would give me the Rawalpindi district, one of great historical importance, between the Jhelum and the Indus." Filled with hope he began his Indus tour. 10 January 1900: "I am writing these lines at the ruined site of Taxila, which Alexander the Great saw."

17 January 1900, Peshawar: ". . . only the best. Lovely days in the Harro Valley where I examined various historical monuments of Buddhist origin never described before and where I bought a number of

interesting coins from the Greek and Indo-Scythian periods very cheaply.
. . . I went in quick marches to the Indus where I tracked down a Kharoshthi inscription, from the start of the Christian era, and last night actually took possession of the large stone which now, under guard, awaits
my return. I came to Peshawar to meet a Turki servant found for me by
a friend. He comes from Khokand (Ferghana) and also speaks Persian
which is very desirable. I shall test his usefulness on this tour." The
Lahore meetings and conferences were bearing results. "I had a letter
informing me of Sir Mackworth Young's readiness 'to take my claims
into sympathetic consideration' in filling the expected vacancy [of in-
spector] upon my return from Khotan if I get the Bengal Government's
consent by October. This will be my next task in Calcutta. I hope to be
able to return to the Punjab where I have always felt physically and intellectually happy. I cannot tell you how happy I am to be back again on
the old and dear soil where nothing reminds you of the tropics. I can
see only too clearly that the leisure in Calcutta was bought at too dear
a cost."

24 January 1900, Abbottabad: "I spent lovely days in old Gandhara,
then moving along the right bank of the Indus I visited a series of old
sites which had a special interest for me: they were the places Major
Deane found his inscriptions; I examined several Buddhist ruins not
previously described. The Kharoshthi inscription I acquired is important
and is on its way to London. The rock weighed over 200 pounds. Crossed
the Indus where it leaves the mountains and then marched through the
hilly Hazara country, reaching Abbottabad yesterday under deep snow.
I enjoy the wintry landscape, so rare for this place. Tomorrow I continue
on to Jammu to visit the Maharaja; and on the 27th I will be in Lahore
to stay about a week. My Khokandi servant is proving himself and I find
simple Turki conversation not too difficult.

1 March 1900, Calcutta: "Today I have a small victory to report: The
Bengal Government gave me permission to apply for the expected vacancy
in the Punjab. There are no reservations; it went very smoothly. I could
hardly have expected so easy an acceptance of my request. Just consider:
my eight months' active service in Calcutta will be rather costly for the
provincial government: it must bear the largest share of the Khotan expedition expenses and will probably lose me to the Punjab on my return.
The way to the Punjab is now open and I have good reason to be appointed Inspector of Schools in the Rawalpindi division—a most interesting region and cooler than Lahore. . . . The enclosed letter from Vámbery
shows that I am not missing anything tangible in Budapest. If Vámbery

who is always so sure of himself is hesitant, it emphasizes that I have nothing to expect there."

15 March 1900, Calcutta: "I am very busy with an enormous correspondence to acquire the proper equipment. In Calcutta no useful camping equipment is available." To Andrews went a steady flow for items with queries as to performance and prices. "There is also much official business concerning my expedition. To get permission for the return trip via Russian Turkestan in May and June of next year, I must apply now as it is handled through the London Ministry to St. Petersburg. I count on being successful in this since the Foreign Office wants to implement a provision made in the recently arranged concessions concerning travel in areas of mutual interest. Also I must arrange for my European leave following my deputation. I should have about 7 weeks in Europe if all goes as planned."

He received the passport for Chinese Turkestan. The British minister in Peking, who at Lord Curzon's request had secured the passport, wrote: "In the application made to the Yamen [the official bureau] it was not thought advisable to ask for the special facilities asked for by Dr. Stein in paragraph 12 of his letter of September 10, 1898, to the Government of the Punjab. Dr. Stein will probably find no difficulty in executing the surveys he mentions: as to the excavations and purchase of antiquities, it is considered that any reference to them would hinder rather than assist his objects."

Enclosed with this covering letter were the passport, a translation of it, and Chinese visiting cards "bearing his name which he may find useful." The translation reads:

The Tsungli Yamen [Foreign Office] issues a passport to the English scholar Stein.

The Yamen are in receipt of a letter from H.B.M. Minister, Sir Claude Macdonald, stating that Dr. Stein proposes to travel with some servants from India to the New Dominion [Sinkiang] and the Khotan neighborhood, and requesting the issue of a passport.

The Yamen have accordingly prepared a passport which has been stamped and issued by the Metropolitan Prefect.

The Yamen call upon local authorities upon the line of route to examine Dr. Stein's passport at once whenever he presents it for inspection, to afford him due protection according to a Treaty, and not to place any difficulties or obstacles in his way.

This passport is to be surrendered for cancellation on return from the journey. If lost, it is to be held waste paper.[17]

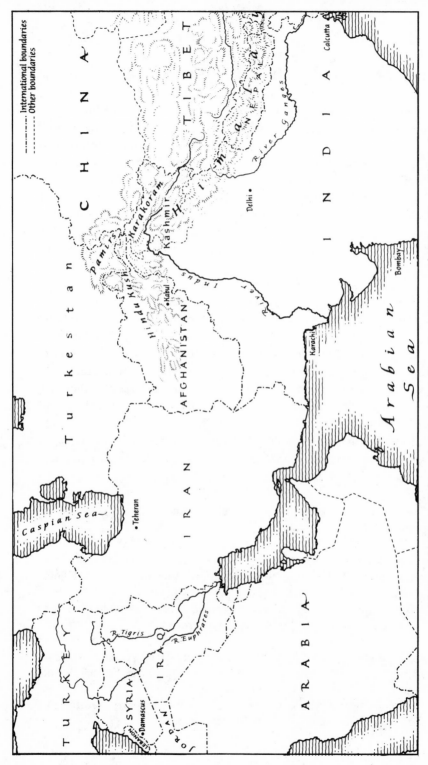

The world wherein Stein explored the past. From Jean Fairley, *The Lion River: The Indus* (© Jean Fairley 1975), front endpaper. Reprinted by permission of The John Day Company and Penguin Books Ltd.

2

The First
Central Asian Expedition
May 1900–July 1901

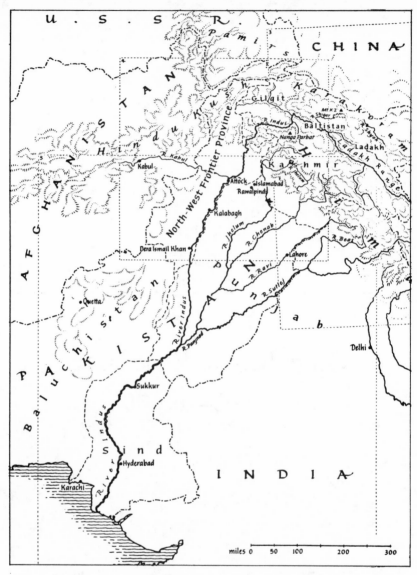

From India across the top of the world to China. From Jean Fairley, *The Lion River: The Indus* (© Jean Fairley 1975), back endpaper. Reprinted by permission of The John Day Company and Penguin Books Ltd.

6

Stein's proposal for a "tour of archaeological exploration to Khotan, Chinese Turkestan," was the blueprint for his expedition. In it he had stated where he wanted to go and why, how he intended getting there, the cost, and what he expected to accomplish. The blueprint signals Stein's act of discovery: Khotan's ancient kinship with India as evidenced by Mahayana Buddhist elements in the Tarim Basin. His discovery was a synthesis of many seemingly disparate pieces of information, and thus, in its deepest sense, his expedition was the scientific experiment designed to verify the conclusion he had already reached. On this, his maiden tour beyond the mountain confines of the Indian subcontinent, there would be no surprises, only a mounting satisfaction at the wealth of the evidence unearthed.

Ancillary, but still essential to the main search, was a piece of detective work. Stein was eager to learn the truth about the "old books" written or blockprinted in unknown characters which, in increasing numbers during the last four or five years, had been sold from Khotan to European collectors in Kashmir and Ladak as well as Kashgar. "In regard to these acquisitions the suspicion of forgery had before presented itself to competent scholars, but evidence was wanting to substantiate it, and in the meantime these strange texts continued to be edited and analysed in learned publications."[1]

Why, it is obvious to ask, why did experts like Hoernle hold them to be real and struggle to decipher them? Had they been items of precious metal or of art, they would have been looked over carefully, even suspiciously, on the chance that they might be fakes. But sandy, scorched, disheveled documents! Perhaps scholars could not imagine anything as sophisticated as forged manuscripts coming from a region which they considered a cultural vacuum. Paleographers and linguists, they could have been emotionally stirred by a hope that these strange texts from the mysterious East, like the unexpected netting of a coelacanth long thought to be extinct, might yield knowledge of primordial languages and scripts. Emotional commitments are tenacious; proofs, not arguments are needed to dislodge belief. Proofs were what Stein would have to obtain. Until now, as he had said, "no part of Chinese Turkestan had been explored from an archaeological point of view, and it struck me that however much attention these and other future discoveries might receive from competent Orientalists in Europe, their full historical and antiquarian value could never be realized without systematic researches on the spot."[2]

To call it Stein's "discovery" is not to claim priority for him. Often such discoveries are made almost simultaneously by more than one: independent in origin, they indicate the commonalty of information attained in the particular field. (One classic example, of course, is the work of Darwin and Wallace.) Thus, even as Stein was making his proposal, word reached him that both Russian and German scholars were readying expeditions to Central Asia. The differences between him and competent Orientalists in Europe were simple but significant: the latter were Europe-based and without his hard-won, wide, firsthand iconographic knowledge of the major Buddhist sites in India. Primarily, they went to add to their museum collections—as if they were studying animals in a zoo—whereas Stein, though he also collected art objects and manuscripts, broke out of the confines of art history and paleography and sought to restore lost chapters of history. On one level there was a competition for artifacts, and inevitably this was reduced in the popular imagination to a "race"— as though there were but a single goal and the three nations were off on a hundred-yard dash. Like the exploration of the moon, the getting there was impressive. Once achieved, however, it was recognized that the field was too vast, the areas to be studied too individualized, and the problems too many, too complex for any single person to command the whole.

Time would confirm that, of those who first dug in the Tarim Basin, Stein revolutionized the puzzle-solving: his act of discovery formed the new paradigm within which later scholars would work.

In three separate publications Stein detailed his first expedition: *Preliminary Report of a Journey of Archaeological and Topographical Exploration in Chinese Turkestan* was written almost immediately upon his return to England (1901); two years later *Sand-Buried Ruins of Khotan:*

Personal Narrative of a Journey, appeared; and in 1907 the splendid two-
volume *Ancient Khotan.* Though all deal with the same experience, their
emphases differ: the first, a succinct account of the expedition, is wonderfully fleshed out in the second book. Stein's Personal Narrative—his
"P.N." as it came to be called—compiled from his diary, field notes, and
letters to family and friends, told the *story* fully: plans and procedures,
obstacles, lessons learned, troubles and triumphs; what had happened and
what might occur, a recital of the day finished and always the challenge
of the desert. Finally, *Ancient Khotan* was concerned with scholarly discussions and precise inventories of the sites excavated and was richly
illustrated with the finds recovered from the desert sands. By writing
the Personal Narrative Stein freed himself from having to inject into his
scientific book "the character of the obstacles which had to be reckoned
with during my work in the Taklamakan," which he felt, was "a necessary preparation and complement to *Ancient Khotan.*"[3]

He began his planning almost as soon as he sent off his proposal. It is
axiomatic that well-planned expeditions rarely have "adventures," and
Stein planned well. His many tours within India and its borderlands had
made him expert in traveling, working, and living in out-of-the-way areas.
But these tours had either been inside India or under the aegis of the
British military, of fairly short duration and among peoples whose languages he commanded. But Chinese Turkestan? Khotan was within the
Chinese Empire and he knew no Chinese. Furthermore, his work would
take him beyond the oases that rim the Taklamakan deep into the desert
itself, a desert few people had crossed. Hsüan-tsang and Marco Polo were
trusted guides from the past, and, most recently, Sven Hedin; but none,
past or present, had attempted to stay and work in the desert itself. Long
before the concept of "living off the country" was advocated, Stein had
practiced it, knowing that wherever there were human settlements food,
clothing, shelter, and transport could be found. Within the Taklamakan,
however, there were no living habitations. Thus, in addition to carrying
the equipment essential for his work—writing and photographic materials and surveying instruments—he had to prepare for fieldwork in
the Taklamakan.

The Taklamakan needs an introduction. It lies in the immense westernmost province of China, in Sinkiang, "New Dominions," thus named
when the Chinese reoccupied it (1877) after Yacub Beg, a Muslim from
Kokand, had tried to detach Kashgaria from China and had maintained
his rule there for thirteen years. Before Yacub Beg's seizure, Sinkiang
was known by the names of its two parts: Dzungaria, the northern, and

Kashgaria, the southern. The two parts, divided by the *T'ien Shan* ("Celestial Mountains"), are fairly equal in size but utterly different in nature. Geographically, Kashgaria is the Tarim Basin with the T'ien Shan on its north, the Pamirs on its west, and the Kunlun on its south. A bird's-eye view of the Tarim Basin would show a series of concentric rings: the outermost is its encircling snow-covered mountains; next the thin circle of arid foothills and boulder-strewn alluvial fans; then its fringe of fertile oases, such as Kashgar, Yarkand, Khotan, whose existence depends on the waters of the Tarim River and its tributaries; and last, the innermost huge center, a sea of sand dunes, the Taklamakan Desert.

Almost two years after he had completed four expeditions, totaling almost seven years, covering some 25,000 miles within Sinkiang, Stein described the parts he had visited on horseback and foot. Such "quasi-archaic methods of locomotion," he called them as the new age of aerial photography was beginning, "provided the right means of acquiring familiarity with a region vast in extent."[4] Crossing the Taklamakan, Hsüan-tsang told of encountering armies of demons; making his way across the lofty Pamirs, Marco Polo had called them the "Roof of the World," but Stein spoke in sober cartographic terms of Kashgaria's "vast basins, elevated and drainageless."[5] Their "longitudinal rim is well defined in the north by the big rampart of the T'ien Shan . . . and in the south by the snowy K'un-lun ranges which divide those basins from Tibet"[6] and which, continuing eastward, join the Nan-Shan range. This last "forms the watershed towards the drainage of the Pacific Ocean. In the west [the basin] abuts on the mighty mountain mass of the Pamirs, . . . which connects the T'ien Shan with the Hindukush and on its western flanks gives rise to the headwaters of the Oxus."

Despite the demon-possessed sea of sand and the sky-piercing mountains, men in ancient times had already found their way past such barriers to connect East and West. However localized such topographical knowledge was, in such comings and goings, Sinkiang was strategic: it was the threshold leading to awesome transcontinental passes. Traveling from India, Stein the scholar saluted the watershed marking the drainage areas of the westward-flowing Oxus from those of the eastward-flowing Tarim—a barely visible height of land leading on one side to the classical world, India and Persia, and on the other side to the civilization of China; while Stein the geographer observed that the "line of greatest elvations, culminating in peaks rising to over 25,000 feet, stretched to the *east* of the watershed."[7] His route, which had already served traders and invaders, passed from the high, plateaulike valleys of the Pamirs, drained by the Oxus and its main tributaries, into the "tortuous, arid gorges [which are] the western margin of the huge trough": the Tarim Basin. The Tarim, fed by the melting ice and snow of the encircling mountains, in turn feeding the oases' irrigation canals, flows eastward for hundreds of miles only to die in the marshes of the Lopnor, a salt-encrusted seabed. So, too, die the rivers that drain the Kunlun: sooner or later all debouch in the sea of sand, the Taklamakan.

Stein speaks of "tame deserts": those found in Arabia, America, and South Africa that are deserts in their sense of solitude and emptiness, but "tame" because in them whole tribes can wander about for long periods of time sure of finding water at least at certain regular seasons."[8] How different is the true desert, "the dune-covered Taklamakan and the wastes of hard salt crust or wind-eroded clay of the Lop desert which stretch almost unbroken for a length of eight hundred miles from west to east. In them the absence of moisture bans not only human existence but also practically all animal and plant life."[9] The lack of atmospheric moisture (so ideal for archaeology) is due to the basin's position remote on all sides from the "seas and their life-giving vapours."

The scenery of this most formidable of the earth's dune-covered wastes is special. "Whether the traveler enters it from the edge of cultivated ground in the oases or from jungle belts along the river beds, he first passes through a zone of desert vegetation, mostly in the shape of tamarisks, wild poplars, or reeds, surviving amidst low drift sand . . . and 'tamarisk cones,' hillocks [built up] in the course of centuries . . . to heights reaching fifty feet or more. Farther out in the Taklamakan there emerge from the dunes only shriveled and bleached trunks of trees, dead for ages. . . . These too finally disappear among utterly bare accumulations of sand, in places heaped up into ridges rising three hundred feet or more."[10] The Taklamakan is like a mummy's shroud: to recover the lineaments of its living past, Stein would have to cut through the sand wrappings—a task as heroic as any designed for Hercules.

Like ocean swells, the dunes move, the effect of northeast winds that rage over the desert during much of the year, that are also constantly abrading the soft clayey soil unless it is already covered by dunes or anchored by desert tamarisks and poplars. At the ancient sites ruins of buildings or what were once orchards and arbors often rise above the wind-eroded bare ground on islandlike terraces: these preserve the original level while around them the ground has been scooped out lower and lower. Using the cultivated fields of the oases as a model, Stein knew how these long-dead settlements looked when canals had carried water to them. In the past as in the present, as though reading a palimpsest, "the traveller sees everywhere the same fields . . . slightly terraced for irrigation, the same winding lanes lined with white poplars and willows; the same little arbors or orchards inviting him with their shade and their plentiful produce."[11]

To prepare for work in the Taklamakan, Stein ordered special equipment. To solve the water problem, he had a pair of galvanized iron water tanks made in Calcutta. The capacity of each, limited by the carrying capacity of a camel, amounted to a mere seventeen gallons. To house himself in below-zero Fahrenheit temperatures, he had his tent given an "extra serge lining."[12] But not even stout British serge could stop the icy Turkestan winds; when the inside was warmed by the Stormont-Murphy Arctic Stove, "fed with small compressed fuel cakes (from London) steeped in paraffin" the intense cold did not "permit discarding the heavy winter garb, including fur-lined overcoat and boots," which he wore in the open. Too cold for reading or writing (the ink froze), Stein would get inside the heavy blankets where Yolchi Beg—Sir Traveler, as the Turkish attendants called his terrier Dash—"had usually long before sought refuge, though he too was in possession of a comfortable fur coat of Kashmirian make, from which he scarcely ever emerged between December and March."

But as yet all this was far in the future. Preparing for the coming winter's work in the Taklamakan, plans had been made, supplies and equipment ordered so that he would have everything needed when his expedition was ready to start. Since Kashmir was a gateway to an established route to Chinese Turkestan, Stein was pleased that he could use his old campsite within walking distance of Srinagar as his base of operations. He had the items bought for him in England sent there, and there he arrived in April 1900. It was a place whose resources he knew: it was "haunted at all hours of the day by versatile Kashmir traders and craftsmen who provide for the Sahibs' camping requirements,"[13] and whom Stein knew were capable and intelligent, and able to fashion unfamiliar items according to specifications. They made the "mule trunks and leather-covered baskets, or 'Kiltas,' in which stores, instruments, etc. were to be packed. Fur coats and warm winter clothing of all sorts to protect myself and my followers against the cold of the Pamirs and the Turkestan winter; bags to carry provisions and all the other paraphernalia which my previous experience had shown to be necessary for a protracted campaign in the mountains."[14] He was especially concerned that his eight hundred photographic glass negatives, extremely fragile, of considerable weight and bulk be packed so that they would be safe and easily accessible.

Stein was among other things, a master at the minor art of packing. Or this talent, so unlooked-for, so invaluable can be seen less as a minor talent and more as further evidence of his ability to fill each hour of each day in carrying out his multifarious activities. It is out of fashion to admire so tidy an ordering of tasks and organizing of time. Yet everything indicates that, once he was free to do what he most wanted to do, his energy, unobstructed by the rubble of self-pity and frustration, flowed strong and steady and sure. "It had taken two years, and bulky files of correspondence," he wrote at the beginning of *Sand-buried Ruins of Khotan*, "but at last I had secured what was needed—freedom to move and the means requisite for my journey."[15]

With the preparations well started, he turned to complete a task ten years in the making: writing the Introduction to his translation and commentary of the *Rajatarangini*. Retiring to a quiet spot within sight of his beloved Mohand Marg (still, in May, inaccessible under a thick blanket of snow), he settled down to work. From morning until night, without interruption, he sat at his writing table, his masses of notes close at hand. The Introduction, with notes, tables, and appendices, fills 144 printed pages; nowhere is there a hint that it was done hurriedly. "Undisturbed by intrusion of any kind . . . three short weeks afforded [the] leisure for concentrated work which, after the preceding 'rush,' seemed almost as enjoyable as if it had been a period of rest."[16]

Again it was his disciplined orderliness that allowed him in so brief a time "to sum up and review the results of labours that had extended over so long a period and so wide a field." Emotionally, it must have been hard to write "finis" to a study whose every point he had discussed with Pandit Govind Kaul: he mourned that his friend had died before he could see the completed work. Finishing the book also ended a period in his own life. The plan to find the true copy of the *Rajatarangini* was conceived while he was still a student of Professor Bühler, to whose memory he dedicated the two volumes; and, looking back, Stein must have seen that all his efforts to make the finished product worthy of his teacher had become the vehicle that brought him to the threshold of his new, self-made career: as a pioneer in Central Asian research.

May was ending when he returned to the bustle of last-minute sorting, testing, packing, and settling of accounts. He could switch effortlessly from the fuss and hullabaloo of the traders and craftsmen to enjoy quiet, leisurely farewell visits with his pandit friends. The end of the preparation period took shape when, two days before the date set for starting, Stein was joined, as arranged, by three men who would make up his small party. Ram Singh, a Gurkha, a trained surveyor sent by the surveyor-general of India to assist with the expedition's mapping program, who brought with him his cook and personal servant Jasvant Singh, a wiry little Rajput. Both men knew the region to which they were going, since they had accompanied Captain H. H. P. Deasy on his recent mapping of the sources of the Yarkand River where it rises in the Kunlun Mountains. The third man, Sadak Akhun, had been sent all the way from Kashgar to be Stein's cook and *karawan-bashi,* the local word for "general factotum"; he turned up in his "fur-lined cap and coat of unstained azure, and red leather boot-tops of imposing size" giving the expedition "a touch of Central-Asian colour."[17] At the very last moment when the little flotilla of boats carrying men and equipment was about to cast off to thread its way through the lagoons and lakes toward the foot of the mountains, the Srinagar postman triumphantly ran down to the boat and handed Stein the last of his supplies—a case of medicines. Its arrival, delayed for months because of the South African War, held pills and antiseptics, each properly labeled and its uses indicated. Such a kit was a necessity in regions where every European traveler is considered a doctor and the giving of medicines is a gesture of friendship.

The expedition began on schedule. "On the morning of the 31st of May, sixteen ponies were ready to receive the loads which were made up by our tents, stores, instruments, etc. Formidable as this number seemed to me, accustomed as I was to move lightly on my wanderings in and about Kashmir, I had the satisfaction to know that my personal baggage formed the smallest part of these *impedimenta.*"[18]

From Srinagar to Kashgar, the first objective, the fastest way was via Gilgit and Hunza, a journey facilitated by the permission granted by the government of India for Stein to use the Gilgit Transport Road, newly built to supply the isolated mountain garrisons of the Northwest Frontier. With expert help from the official in charge, Stein had drawn up a fairly accurate schedule so as to prepare the distant posts for his coming and alert them to his needs. The Gilgit Transport Road: the word "road" is perhaps misleading. At that time and in that region it signified no more (but no less) than that the way to Hunza had been made "fit for laden animals, including camels, during 'hree summer months"[19] and that coolies were no longer needed for transport of supplies. Stein would be on that road for the next month.

The long overland journey began pleasantly along an open valley filled with the fresh green of young rice shoots. The road made long zigzags to climb the range dividing Kashmir from the next valley, and at about 9,000 feet it entered a narrow field enclosed by a pine forest from which the snow had recently disappeared: its damp ground was "strewn with the first carpet of alpine flowers."[20] The shelter, a rude wooden hut, was doubly welcome (rest houses spaced at each day's march were part of the Gilgit Road construction), for with dark a storm broke bringing fresh snow. "I started before daybreak on the 1st of June. A steep ascent of some two thousand feet" led to an open ridge exposed to all the winds and still deep in snowdrifts that obliterated all traces of the road. Snow and an icy wind heralded the "bad repute which the 11,000 foot Tragbal enjoys among the Kashmirian passes."[21]

The descent was wild. The "Markobans," "hardy hillmen, half Kashmiri, half Dard," who owned the ponies, favored taking the short winter route. The path plunged steeply into a ravine in whose soft snow the ponies slid helplessly but safely to the bottom. Then, too late, they found that the snowbridges over the tumultuous stream had begun collapsing, forcing them onto narrow, treacherous snowbanks. Gingerly picking their way, they came to a stretch where the water had washed away the entire snow vault. What were they to do? Impossible to negotiate the sheer slaty bank; depressing to consider retracing their way back to make the descent by the lengthy downward zigzags. The Markobans decided to try a narrow ledge of snow still standing. The first pony "though held and supported by three men, slipped and rolled into the stream, and with it Sadak Akhun who vainly attempted to stem its fall. Fortunately neither man nor pony got hurt, and as the load was also picked out of the water, the attempt was resumed with additional care." At the worst places they piled up stones one by one and led the ponies across. Heavy rain started. Men and animals were thoroughly soaked. Impatient at having spent seven hours to cover the eleven miles to the rest house, and because the grimy cabin looked uninviting and the afternoon was still young, Stein decided to push on the fourteen miles to Gurez, where better accommodations and supplies were available.

Gurez, the first Dard village, was a miserable collection of wooden cabins scattered along the slopes of the narrow valley of the Kishanganga, the "Black Ganga." The river took its name from the black sand of its bed. Crowded between high, pine-covered mountains and under a gloomy, rainy sky, the place looked forbiddingly somber. At an elevation of 8,000 feet, with short summers and scant sunshine, its fields yielded poor crops, severely limiting its population. The scene spoke eloquently of their having crossed a "well-defined ethnographic boundary."[22] The Dards, who inhabit the valleys between Kashmir and the Hindukush, are distinct in language and physical type from the Kashmiris. "There is little in the Dard to enlist the sympathies of the casual observer. He lacks the intelligence, humour and fine physique of the Kashmiris. . . . But I can never see a Dard without thinking of the thousands of years of struggle these tribes have carried on with the harsh climate and barren soil of their mountains. They . . . have seen all the great conquests which swept over the North-West of India and have survived them, unmoved as their mountains. Gurez was once the chief place of a little Dard kingdom which often harassed the rulers of old Kashmir. But I confess, when I approached it at the close of my fatiguing double march, this antiquarian fact interested me less than the comfortable shelter which I found for my men and myself."[23]

Rested, the baggage dried out, they prepared for crossing the Burzil Pass that separated Gurez from Astor. With fresh ponies and the delicate theodolite and photographic cameras transferred to the safer backs of coolies, they headed (3 June) for the pass. Where avalanches had swept down over the road, they led the ponies across; two days of marching through high meadows recently freed from snow, past huts the Gujar herders used in summer, brought them to the telegraph office that maintained the line across the pass. At 10,000 feet, they saw the temper of the winter: "Raised high above the ground, and enclosed with heavy palisaded verandahs and sheds, the building looked more like a fort than an office."[24] Five miles beyond, at the rest house at the foot of the snow-covered pass, they were in the midst of true winter scenery.

"The only condition to be observed was an early ascent before the snow should become soft. I therefore got up at one o'clock, and an hour later my caravan was plodding up the snow-filled ravine. . . . Of the road no trace could be seen."[25] A steady two-hour climb brought them to the thirty-foot-high telegraph shelter-hut that guided the dak (mail) runners carrying the Gilgit mail during the winter. Even in June ten feet of snow surrounded it. The hard snow offered good going. By the time the first rays of the sun lighted the higher ranges to the east, they had gained the 13,500-foot summit. It had taken three hours from the rest house to cover six miles. In the shelter of a dak hut, Stein ate a hurried breakfast. With sunlight the going became heavier; the descent was tiring. Twelve hours after starting he reached Chillum Chauki—just two miles below the snow line, it was the first rest house on the Astor side. There the ponies brought from Gurez, the first laden pack animals to make the crossing that season, were relieved by fresh ones sent up from Astor.

The march leading to Astor was easy, though having crossed the watershed between the Jhelum and the Indus "meant the entry into sterner regions,"[26] slopes of decomposed rock dotted with cedars and juniper, while the stunted cultivation testified to the unfavorable conditions of soil and climate. "All the more cheerful it was to behold, by the side of the little terraced fields of more than one hamlet, an oblong sward carefully marked off with stones—the polo grounds of the villagers. Polo is the national game of all Dard tribes"; to enjoy their pastime they sacrifice valuable soil. Astor, the chief place of that hill district, was a group of villages "spread out over a mighty alluvial fan."[27] Above the naked rock hills enclosing the valley Stein beheld the "icy crests of the great mass of peaks culminating in Nanga Parbat, that giant of mountains" 26,500 feet high. The orchards and fields in a setting of rocky ridges and deep ravines cut by wild mountain torrents were like simple songs honoring the valor and sweetness of the Dards. Stein found the heat oppressive—it was almost as warm as Kashmir though the elevation was 9,000 feet—and he was glad to reach the cool shade of the rest house.

The next morning (8 June) "the march led down towards the Indus. The valley became bleaker and bleaker as the road descended." Stein pushed on another twelve miles beyond the next rest house when he heard that "Captain J. Manners Smith, V.C., C.I.E., the Political Agent of Gilgit and adjacent hill tracts, 'Warden of the Marches' "[28] was encamped in the vicinity on a hunting trip. The landscape changed as "the road, rising gradually to 5000 feet above the tossing river, took me through a charming forest of pines. . . . It was a pleasure to behold once more green moss and ferns along the little streams which rush down through the forest. But . . . at the turning of a cross spur, there spread a grander view before me. Through a gap . . . appeared the broad stream of the Indus and beyond it range after range towards the north." The fatigue of a long double march could not dull his elation. "Father Indus was greeted by me like an old friend. I had seen the mighty river . . . where it breaks through the rocky gorges of Baltistan, where it bursts forth into the Yusufzai plain, and in its swift rush through the defiles below Attock. But nowhere had it impressed me more than when I now suddenly caught sight of it amidst these towering mountains."[29]

Captain Smith invited Stein to spend the day in his camp. Sending his party ahead, he rode to a "charming spot where the ground was carpeted with wild violets, forget-me-nots . . . and where a bright little stream added to the attraction of the scene." In this wild, idyllic spot the well-fitted tents of the captain and his family were placed. "Anglo-Indian ladies know how to carry true refinement into camp life even at the most distant points." In the captain's retinue were leading men from Gilgit and Hunza: the older men talked of religious traditions that had yielded only recently to the coming of Islam; petty headmen related customs of an earlier time. Stein was delighted when "my host's pretty little children" appeared. "They were worthy representatives of the British Baby which in the borderlands of India has always appeared to me as the true pioneer of civilization."[30]

Hurrying to catch up with his party, Stein passed through desolate mountains of reddish-brown, a veritable oven baking under a hot sun, to come at last to the great Gilgit Valley. From its first village "every alluvial fan on the left bank was green with carefully terraced and irrigated fields,"[31] and side valleys showed fertile fields and tended orchards. At the Gilgit agency's headquarters, Stein lodged in a comfortable set of rooms. The men he met, whether in charge "of the troops supplying the local garrisons, or of the Commissariat, the Public Works, or the hospitals of Gilgit . . . knew and liked these hills." He was impressed that this last Anglo-Indian outpost on the Northwest Frontier offered a well-built hospital for men and zenana for women, a clean bazaar, substantial school buildings, and small, comfortable bungalows for the officers. He stayed three days while fresh ponies were collected from distant grazing fields; busy days repairing defects in his outfit and collecting additional warm clothing and foodstuffs. As always, Stein does not let the slightest courtesy go unmentioned. "Mrs. W., the only lady left in the station, kindly offered threads of her fair hair for use in the photo-theodolite. How often had I occasion to feel grateful for this much-needed reserve store when handling that delicate instrument with half-benumbed fingers on wind-swept mountain tops!"[32]

The fortnight (15–29 June) spent on the road to Hunza made Stein appreciate the steady, surefootedness of the hill ponies and the extraordinary stamina of the Kanjutis, the term applied to the men from Hunza and adjacent Nagir. On the eighteen-mile march to Nomal, "a green oasis in the otherwise barren valley," the road narrowed to a path winding along precipitous spurs many hundred feet above the river that flowed from

Hunza. From Nomal to Chalt, the limit of Gilgit territory, the path, following the river as it cuts through deep gorges with almost perpendicular walls, "is carried in long zigzags over the projecting cross-ridges, and more than once traverses their face by means of galleries built out from the rock."[33]

Rafiks, the local name for such galleries, are fastened to the sheer cliffs by branches of trees forced into fissures of the rock and covered with small stones. Elsewhere natural narrow ledges are widened by flat slabs packed over them. In some places the rafiks "turn in sharp zigzags on the side of cliffs where a false step would prove fatal, while at others again they are steep enough to resemble ladders. To carry loads along these galleries is difficult enough, and . . . for ponies, sure-footed as they are, wholly impassable."[34] Even his terrier, Yolchi Beg, so nimble on the rocks of Mohand Marg, was fearful and allowed himself the indignity of being carried. Rafiks alternated "with passages over shingly slopes and climbs over rock-strewn wastes."[35] To negotiate this terrain, the "baggage animals were left behind [at Chalt] and coolies taken for the rest of the journey up to the Taghdumbash Pamir."

The first Nagir village was Nilth. Hunza and Nagir, "two little hill states which divide between them the right and left sides of the valley jointly known as Kanjut,"[36] had long maintained their independence. Their fate was decided in 1891 when the British successfully stormed a thousand-foot cliff. In that fight, which began and ended at Nilth, Captain Manners Smith had won his V.C. Until the British came, parties of Kanjutis had plundered caravans as far away as the Pamirs and had raided the lower valleys for slaves to be sold in Turkestan. Unable to pursue their traditional economic pursuits, their life in those valleys became a continuous struggle. The old exciting way of life was replaced by earning money catering to garrisons and convoys; peaceful contacts introduced new words for new activities taken from Punjabi, Arabic, and Persian. Stein was particularly interested in this rapid spread of foreign words; he took time to collect linguistic lists. The language of Hunza, Burisheski, "has no apparent connection with either the Indian or the Iranian family of languages, and seems an erratic block left here by some bygone wave of conquest."[37] Reaching Aliabad, the next village, required an interminable twenty-six miles of constant ups and downs on absolutely barren rocky slopes. In the evening's fading light the "steep walls of rocks rising on either side fully five or six thousand feet above the river with the icy crests of Rakiposhi [over 25,000 feet] in the background formed a picture worthy of Gustave Doré. By the time I had cleared the worst parts of the road along sliding beds of detritus, it had got quite dark."[38] Yet it was another two hours before they "emerged on the open plateau which bears the village and lands of Aliabad." But the sight in the morning was thrilling: Rakiposhi, crowned with snow and ice "and without a speck of cloud or mist. To the north, mighty peaks also above 25,000 feet in height frown down upon the valley, while eastward I could see the range along which my onward path was to lead. The two days saved by the double marches between Gilgit and Hunza were used for a short halt at Aliabad."[39] Stein needed the time to "dispose of arrears of correspondence." Hunza, to which mail was forwarded from Gilgit every second day, would be his "last link of regular postal communication" in the long journey ahead.

It was a time of critical changes in Hunza. The mir's wazir, who called on Stein the first morning, assured him that arrangements for the next stage of the journey had been attended to. The wazir, hereditary chief advisor and executive of the mir, the ruler, gave Stein a tour and explained the efforts made to raise agricultural production. Elaborate watercourses "winding along the foot of the mountains often in double and treble tiers" husbanded and utilized the glacier-fed streams. What to Stein was most curious was the little kingdom's accommodation to the two powers on which she bordered: the Chinese and the British Empires. High valleys in the Yarkand River drainage system long occupied by Hunza people fell "within the natural boundaries of Chinese Turkestan";[40] this explained why the Kanjuti chiefs sent presents to Kashgar, receiving in return presents "considerably in excess of those sent." If the former was in the nature of tribute sent, the latter suggested a kind of payment by the Chinese to safeguard their far-western territory against Kanjuti raids. This exchange of gifts—however each side interpreted its political significance—continued "even after the enforcement of British sovereign rights . . . after the occupation of 1891."

Two Hunza men familiar with the Pamirs were hired to guide the party as far as Sarikol; the commandant of the mir's bodyguard was detailed to collect and command the sixty coolies hired for transport. "Swelled by these numbers, my caravan looked alarmingly large as it moved off on the morning of June 20th."[41] A short day's march brought them to Baltit, "the chief place of Hunza and the Mir's residence. Rising on a cliff from an expanse of terraced fields and orchards, the Castle of Baltit looks imposing enough with its high walls and towers. Below it, closely packed on the hillside are the rubble-built houses, some two hundred in number, of the Hunza capital." From the castle's living quarters atop the high, massive foundation walls, Stein saw below a fine polo field and, across the valley, a glacier hanging from an icy peak. The mir's apartment had a few pieces of European furniture; but its Yarkand carpets, Chinese silks, and gaily colored Kashgar prints showed that goods from Central Asia had made their way to Hunza over the Sarikol passes more easily than had Indian articles before the opening of the Gilgit road.

How did invaders without guides, without rafiks, without supplies of food, without knowing their final destination, find their way through these passes? If winter offered easy routes up frozen riverbeds, how did they survive the cold? If they waited for warmer months, then the river route was impossible and they faced an exhausting scramble over wave upon wave of arid rock mountains. Stein, with all the advantages on his side, found the next two days trying. The path led over high rugged spurs and across the "débris of an enormous old landslip"; the campsite was dismal. "The black glacier-ground sand, which the Hunza River brings down and deposits in large quantities, rose in thick dust with the wind which blew down the valley at evenings. Drink and food tasted equally gritty; it seemed a foretaste of the Khotan desert." He was doubly glad when the dak runner who had been following him, "at nightfall brought a long-expected mail." The next day's march was a hair-raising nightmare of dizzy rafiks and exhausting clambering over shingly slopes and rock-strewn wastes. At its end he reached Little Guhyal, that part of the Hunza Valley so named because its inhabitants were Wakhis, who had immigrated there from Wakhan, or Guhyal, on the Oxus.

With great ceremony, the headmen of the valley's villages escorted Stein to a charming campsite in an apricot orchard. He was delighted to hear "at last the language of Wakhan, which had attracted my attention years before I first came to India, as a remarkably conservative descendant of the ancient tongue of Eastern Iran."[42] The Wakhis numbered some thousand souls; they were taller than the men of Hunza and had the "characteristic eagle-nose of the true Iranian."[43] From happy conversation with his hosts Stein learned that the original immigration from the Oxus Valley was clearly remembered and that the connection was maintained by occasional marriages between the people of Wakhan and Sarikol. Why, Stein wondered, had the Hunza "more warlike and so pressed for land acquiesced in this invasion?" Was it, perhaps, the very peacefulness of the Wakhis? Their sign of manly power was not a weapon but an implement: on state occasions every Wakhi carried "a long staff with a small heart-shaped shovel of wood at the end, used for opening and damming up the irrigation courses that brings fertility to the laboriously cleared terraced lands."[44]

The presence of the Wakhis of Little Guhyal announced that the party was approaching the Pamirs, "a point of contact not only of great geographical divisions, but also of equally great language families." Stein knew that "close to the Kilik Pass is the point where the watersheds bounding the drainage areas of the Oxus, Indus, and Yarkand Rivers meet; and it is plain that as far as history can take us back, these areas belonged to the dominant races of Iran, India and Turkestan."[45]

Changing coolies delayed the next morning's start: "Time is always lost until everyone settles down to the load he fancies."[46] Fortunately, the day's march to Pasu was short and, after the previous day's experience, easy. Pasu's few straggling homesteads with their green fields and orchards were at the foot of the mighty Batur glacier whose trickle of water made human existence possible. On 24 June, the party crossed the glacier, a river of ice a mile and a half wide protected from the sun by an "extraordinarily thick crust of rock and shingle."[47] The crossing, Stein was told, was quite easy although some years the condition of the ice makes it "difficult even for men and altogether closes the route to animals." Obliquely referring to Russian hopes of controlling the Hunza Valley route, Stein mentioned that the glacier was an obstacle "no skill of the engineer can ever completely overcome [and] makes one realize the great natural defenses of the Hunza Valley route against invasion from the north,"[48] but he added that even so harsh a natural barrier was no guarantee against foes, for the path is further "guarded at a point of great natural strength by a rude gateway, or Darband." The Darband was necessary because the opposite bank of the river was "easily accessible to the people of Nagir, the hereditary enemies of Hunza."

After the glacier the path descended by a long rafik built on "an almost perpendicular rock face."[49] Stein was utterly amazed to meet the messenger sent by the wazir to notify the political officer at Tashkurgan of the party's approach. "The man had left Hunza on the morning of the 18th, and now [the 25th] was returning with the reply and a considerable load of merchandise which he was bringing back as a private venture. As an illustration of the marching powers of the men of Hunza, this feat deserves record. The distance from Hunza to the Kilik [Pass] is about eighty-one miles," and of the route's nature Stein had had ample experience. "In addition to this and half the return journey, the man had covered twice the route along the Taghdumbash Pamir to and from Tashkurgan, a distance of at least eighty miles each way." Two hundred and eighty miles of brutal terrain in one week! Such prowess explained how Kanjutis could strike at a distance and, because of the "rapidity of their movements, usually with impunity."

At the next hamlet, Khuḍabad, Stein was again in Hunza territory, the Wakhi villages behind him. What delighted him there was the cosmopolitan quality of his camp, the "strange variety of tongues which could be heard."[50] He enumerated the linguistic riches: with Sadak Akhun, he spoke Turki, with his Wakhi guides, Persian; the coolies spoke partly in Wakhi, partly in Burisheski, the Hunza language; the Dard dialect was used by an orderly whose services had been lent by Captain Manners Smith; Ram Singh and his Rajput cook conversed in Hindustani. "Notwithstanding this diversity of tongues things arranged themselves easily, for everybody seemed to know something at least of another's language." Having listed what he heard, he added that if Sadak Akhun had not replaced his Kashmiri servant: "I should have had an opportunity to keep up my Kashmiri also."

The next day's march to Misgar, he had been warned, would be the worst part of the route. By starting before dawn while the river was still low enough to ford, he avoided a long detour and a perilous crossing on a rope bridge. Then the going reached a climax of "scramble up precipitous faces of slatey rocks . . . with still more trying descents to the riverbed";[51] slower still was the progress along rafiks clinging to cliffs hundreds of feet above the river. But the previous five days had toughened him, and he felt fresh when he emerged from the rocky gorge to an open valley. Met there by levies from Misgar, he was conducted to their village whose smiling green fields refreshed his eyes. An uncultivated spot in their midst, just large enough for his tent, gave him the feeling that he was once again "camping on a green sward." The view across an open valley was a tonic: "Far away to the north-west I even beheld the snowy ridge which clearly belonged to the watershed towards the Oxus. I felt at last that the Pamir was near."[52] Here he discharged the "hardy hillmen who had carried our impedimenta over such trying ground without the slightest damage." Beyond, the route was open to baggage animals at all seasons.

At Topkhana, marked by a ruined watchtower, Stein was met by a "jolly looking Sarikoli, whose appearance and outfit at once showed he came from Chinese territory." He was one of the soldiers stationed at the Mintaka Pass, and he "assured me that the expected yaks and ponies were already waiting for me, and tried to make himself as useful on the rest of the march as if he belonged to my following of Hunza levies." A four-hour march the next day brought them to Shirin Maidan, the "Milky Plain," close to the foot of the Kilik Pass. At an elevation of 12,000 feet, "when the sun passed behind light clouds at noon and a fresh breeze blew down the pass, it was bitterly cold and I was glad to get into my fur coat as soon as the baggage arrived."[53] There he discharged his Guhyal coolies and Hunza levies. The party shrank to the Sarikoli headman and his seven relatives who had brought the yaks to carry the baggage onwards." They were cheerful to look at and to talk to. They understood Turki quite well and were more communicative. In their midst I felt that I had passed out of India." It was 28 June.

On slow-moving, sure-footed yaks the party ascended the open, steep valley leading to the Kilik Pass. In that world of rugged, snow-covered peaks was one, higher than the others, which "seemed to mark the point where the drainage of the areas of the Oxus, Indus and Yarkand Rivers meet."[54] But where, exactly where on the flat plain crowning the top of the Kilik, was that point? At a spot that looked likely, Stein stopped "to boil the water for the hypsometer. It proved a troublesome business in the bitterly cold wind, and by the time I got the reading which gave the height as circ. 15,800 feet, it began to snow." To the north, the bleak Russian Pamirs stretched away, shrouded in clouds. Wind and elevation, clouds and snow intensified the cold; the draggingly slow two-hour ride down to the flat of the Taghdumbash was decidedly uncomfortable.

Stein might have said adieu to India, but he had not passed beyond the influence the British wielded from there. "An imposing cavalcade met me as I approached the place where my camp was to be pitched."[55] At its head was the Indian political representative stationed at Tashkurgan, "the chief place of the mountain tract known as Sarikol,"[56] a place historically and archaeologically interesting. With him rode the Sarikoli governors of the several districts. The political representative, sent from Kashgar by Macartney, was an attractive, energetic young man who "introduced himself as an old pupil of the Oriental College at Lahore."

It was here that the surveying was to begin. Conditions were exceptionally favorable: "the surrounding ridges, all snow-capped, stood out with perfect clearness against the blue sky." Stein and Ram Singh, the surveyor, climbed a 16,800-foot spur which was to serve as the topographical station. The yaks, carrying men and instruments steadily up steep grassy slopes, fields of snow, and shingly beds of rock, charmed Stein. He suggested "that by a judicious use of the yak the difficulties which the high elevations offer to mountaineering in these regions could be reduced for the initial stage."[57] In this, his first experience, Stein was enthusiastic about the yak.

His camp was pitched so close to the Wakhjir Pass, the watershed separating the drainage systems of the Oxus and the Yarkand Rivers, that he "could not resist the temptation to visit it during the two days required by the Surveyor's work." His "topographical conscience" insisted that he visit "the glacier which Lord Curzon first demonstrated to be the true source of the Oxus." He was equally eager to try out his photo-theodolite in Central Asia. When, after a sudden snowstorm, he started for the pass, an easy yak ride of an hour and a half brought him to the watershed clearly marked "by the divergent directions of the small streams which drained the melting snow."[58] He climbed another eight hundred feet to photograph the impressive assemblage of snowy peaks and glaciers where the mighty Oxus began. He felt "a strange and joyful sensation [to stand] at the eastern threshold of that distant region, including Bactria and the Upper Oxus Valley which has had a special fascination for me ever since I was a boy."[59] In those trickles of water he could see "all the interests of ancient Iran cluster in one form or another around the banks of the great stream. Since the earliest times it has brought fertility and culture to the regions which it waters. . . . [He] found it hard to leave this desolate scene."

With Ram Singh's return, the party started down the wide grassy valley, a "novel sensation after the weeks spent in narrow gorges and amidst snow-covered heights to ride along these broad, smiling slopes gently descending from the foot of the mountains."[60] On that high, clean, spacious world, wherever the soil was touched by water, flowers and herbs covered the ground and scented the air—it attracted small groups of Wakhis and Khirgiz who pastured their flocks there in the summer. The sight of a troop of ponies enjoying their freedom and rich grazing reminded Stein of his boyhood and the bright summer days he had spent on the Hungarian plain. There was a crescendo to his many-layered sense of pleasure; at the end of an eighteen-mile march he had his first sight of "the glistening mass of a great snow dome": the Muztagh-Ata, the "Father of Icy Mountains," "which I had so long wished to behold." After Muztagh-Ata came Tashkurgan, rising atop distant cliffs. Its walls did "not hide any imposing structures or special comforts. Yet they marked the completion of a considerable part of my journey and my entry upon the ground which was to occupy my researches."[61] That night he was luxuriously lodged in the splendid yurt Major F. E. Younghusband had used when probing the Pamirs to assay Russian influence. But Stein's greatest satisfaction was the welcome news that the "Chinese Amban [the district officer] of the Sarikol region raised no objection to my proceeding westward of Muztagh-Ata."[62]

Tashkurgan, the "Stone Tower," had been identified by Sir Henry Rawlinson with the place mentioned by Ptolemy, the Alexandrian astronomer and mathematician (ca. A.D. 150). Ptolemy's information had come from Marinus of Tyre, the adventurous trader, who spoke of it "as the emporium on the extreme western frontier of Serike, i.e., the Central Chinese Dominions." It was the entrepôt to which men from the West— Romans, Parthians, Indians, and Kushans—came to meet the Chinese and their middlemen (both classed as Seres) who brought silk; silk soft, sheer, and lustrous, the luxury craved by the wealthy of Imperial Rome who paid for it with the gold they had amassed by conquest. Tashkurgan remained a convenient place for trade on the ancient, once-important route connecting Central Asia with both the West and the Far East. "From Tashkurgan one road lies open to Kashgar [on the northern rim of the Taklamakan] and another to Khotan [on its southern rim] and thus to both the great avenues which lead from Turkestan into the interior of China."[63] The town "rests on a great rocky crag and is backed by the river," Hsüantsang had recorded; Stein could discern the "ruined town within which the modern Chinese fort is built," but he knew that the town itself, racked by numerous earthquakes and repeatedly rebuilt after periods of neglect, defied "any distinct criterion of age."[64]

To identify and verify the "old remains, such as that of a ruined stupa," Stein felt it necessary to make an exact survey of the site. To make the survey properly "required some diplomatic caution, as the Chinese commandant or his subordinates might have easily mistaken its object." Then and there, Stein behaved as he had not before: always forthright, he dissimulated; always open and responsive, he was secretive. His behavior was dictated by the caution urged on him by the Oriental College graduate whom Macartney had sent. With Ram Singh, Stein went over the site in an apparently casual fashion. He waited "with the surveying until the hours after midday when the whole garrison is wont to take its siesta. When the work continued beyond the safe period, the clever diplomatist went to see the Amban and so skilfully occupied his attention with various representations concerning my journey that he and his underlings had no time to grow suspicious about the work around their stronghold." Thus Stein describes his tactics. What he could not say at that time was that the maneuver was directed not against the Chinese but against the Russians

—Macartney would spell out their machinations—who, for their own reasons might have chosen to interpret his act as "spying" and used this charge as a stick with which to beat the Chinese.

A new trouble was weakening the already vulnerable Chinese: at Tashkurgan Stein received a letter sent on by the Gilgit political officer informing him of a telegraphic message that flashed the news of the Boxer uprising in Peking. His answer to the warning, date 7 July 1900, was reassuring. "It appears very unlikely that disturbing news from Peking should spread to this outlying province, the connection of which with China is so difficult." Appreciative of the government of India's concern for his safety, he showed his concern for the Indian government's political representative at Kashgar: "The fact of a consignment of revolvers and ammunition having been entrusted to me . . . for delivery to Mr. Macartney is a further reason for my completing my journey to Kashgar.[65]

His journey to Kashgar, pursued until then with vigor and an unflagging attention to tempo—how many hours to do how many miles? —assumes after Tashkurgan, a change of pace and purpose. For thirteen days, from 10 July when he left Tashkurgan, until 23 July when he began the journey through the Gez Defile to reach Kashgar, he was occupied with a private project: to survey the Muztagh-Ata massif. The effort seems to have been dictated less by his passion for topography than by his desire to accomplish a feat thrice denied Sven Hedin just six years before. For three days the party rode in an alpine world of "broad rolling plains and of low-looking ranges that fringe them,"[66] passing small settlements whose grazing yaks, furnishing a bounty of milk and cream, were the only reminder of the high elevation. From afar, the "Father of Icy Mountains" announced its frigid presence: Stein put on his fur coat again. An icy rain drenched them as they crossed the 14,000-foot Ulugh-Rabat, the High Station Pass, "a slight depression in a broad transverse ridge connecting the Muztagh-Ata *massif*" with the "eastern brim of the Russian Pamirs."[67]

That night and the next day the rain continued: it was "by necessity a day of repose."[68] But the weather did not prevent Stein from riding out to the Little Karakul Lake by whose shore Sven Hedin had camped. Clouds blotted out the view of the "huge masses looking over the Roof of the World with Russia, India and China round their bases."[69] For 15 July, Stein wrote in his diary: "It rained and snowed through the whole night, and mist and grey drizzling rain covered the little of the valley I could see when I got up. There was nothing for it but to sit in my tent and write up notes and letters that were to go down to Tashkurgan to catch the next Dak for India and Europe."[70] The clouds lifted a little in the afternoon and showed fresh snow a few hundred feet above the valley. Stein rode over to visit the yurt of the Kirghiz chief who had accompanied him from Tashkurgan.

"In the middle of the yurt a big cauldron ('Kazan') of milk was boiling over the fire. One of the Beg's wives, no longer young but of a pleasing expression and cleanly dressed, was attending to the fire of dwarf juniper ('Tereskin'). While the dish was getting ready, I had time to look about. . . . The wicker-work sides and the spherical top of the yurt are covered with coloured felts, which are held in position by broad bands of neatly-embroidered wool. All round the foot of the circular wall lie bundles of felt rugs and bags of spare clothes. . . . A screen of reeds with woolen thread worked in delicate colours and bold but pleasing pattern separated a little segment of the Yürt apparently reserved for the lady of the house, who again and again dived into it, to return with cups and other more precious implements. The floor all around, except in the centre where the fire blazed, was covered with felts and thick rugs made of yak's hair; for my special accommodation a gay-coloured Andijan carpet was spread on one side. The warm milk, which was offered by the presiding matron, tasted sweet and rich."[71] As a gesture of hospitality, the beg presented Stein with a big sheep, "distinguished for toughness as for size,"[72] and was assured that his gift would be returned by more than its equivalent before Stein left the valley.

Night brought a change in the weather: the rain stopped, and the clouds, still draped around the peaks, were being shifted by gusts of icy wind. Stein moved his tent to the hill he chose for work with his photo-theodolite. The pictures would supplement and check Ram Singh's plane-table work. Chilled to the bone, he waited until the clouds lifted to show glacier after glacier flowing down from the white wall of ice-crowned peaks, "the worthy rivals of Muztagh-Ata."[73] Inadvertently, he broke one of the crosshairs, replacing it "thanks to the ample supply of delicate hair which Mrs. W.'s kindness had provided in Gilgit." It was a day to his liking: that night in his tent, warmed by the cakes of compressed fuel, he contentedly wrote "till midnight readying mail that was to carry my news to distant friends."

"On the 17th I awoke to a glorously clear morning. Without a speck of cloud or mist the gigantic mass of Muztagh-Ata towered above my camp." So began Stein's account of a five-day attempt to reach the summit of the 24,321-foot giant. Sven Hedin had described how yaks had carried him over ground almost free from snow to over 20,000 feet. Stein, through a small telescope, studied the ridge along which Hedin had advanced; he estimated that the snowline appeared to be a good 3,000 feet lower, an estimate confirmed by the Kirghiz who had guided Hedin. The snow, newly deposited and treacherous, raised the odds against Stein's success. Almost, it seems, it added to his determination.

The yaks whose wonderful surefootedness Stein had admired began to show a "sluggishness of temper"[74] which made them "a trying mode of locomotion." At 16,500 feet they panted for air struggling up the steep snow-covered slopes. Progress grew slower: "more and more frequently we had to dismount and drag the stubborn animals out of the deep snow-drifts into which they had plunged."[75] Another four-hundred feet and the Kirghiz yakman succumbed to the violent headache and nausea of mountain sickness; at 19,000 feet, the Punyali, a "skilled mountaineer . . . fell out and received permission to descend."[76] Stein felt no symptoms of mountain sickness and with the two Hunza companions plodded steadily on. "The snow became deeper and the mist that settled on the peaks above us showed clearly that a further ascent would offer no chance of a close survey of the summits. . . I fixed upon the buttress of the ridge just above me as the final object of the climb. By 2:30 P.M. I had reached its top and settled down by the side of the precipitous rock wall descending to the glacier." He had reached—he took the measurement—exactly 20,000 feet: "Our bodily condition would have allowed of a further climb, though I, as well as my Hunza followers, felt the effect of our six hours' ascent through the snow."

"The multiplicity of the ranges over which my gaze traveled was the best demonstration of the height we had obtained."[77] Below him lay the entire breadth of the Pamirs, "a seemingly endless succession of valleys and ranges . . . the vastness of the "Roof of the World"—Marco Polo's term. His companions' old freebooting spirit was aroused as they looked over "rich grazing lands, greater than the dwellers of the narrow valleys of Hunza could ever imagine."[78] Returning to the base camp, Stein discharged his Hunza men, happy to be home-bound and well pleased with their pay. By them he sent the return gift for the sheep presented by the Kirghiz beg. At the camp was a troop of ponymen with ten ponies sent by Macartney, who had thoughtfully anticipated the need for fresh transport. They also brought letters and papers and "little parcels, the result of orders I sent to Lahore six weeks before from Gilgit, after the first experience had shown me the *lacunae* of my equipment."[79] They also assured Stein he could proceed to Kashgar via the Gez Defile which, though flooded, was possible.

Possible was an ambitious word. The Gez Defile connects the high Sarikol with the Taklamakan and partakes of the awesome nature of both regions. A labyrinthian stone corridor, it was formed by the river thick with the huge rubble thrown off by the glacier madly forcing its way down, water and rubble cutting deeper and deeper. "Great serrated coulisses of rugged rock, several thousand feet in height, descend from the main mountain spurs on both sides. Along the face of one of them the road is carried by a gallery, a true Rafik. . . . It was getting dark between the mountain walls when this awkward part of the route had been passed."[80] From the Taklamakan end, heavy yellowish clouds, the end of a dust storm, blotted out the sky. The next morning Stein saw his first example of Chinese engineering: where mighty cliffs constricted the river to about forty-five feet, a wooden bridge, six feet wide with sturdy railings and painted a bright yellow, spanned the abyss. That night another dust storm brought with it the desert's oppressive heat. Where the corridor ended was a Kirghiz village. From there they took the circuitous trail over the Nine Passes since the path along the flooded riverbed was impassable. At the summit of the second pass, the snowy peaks were still visible while "a succession of serrated ridges in the foreground looked like lines of petrified waves."[81] A permanent spring at the head of a dreary, miserable valley was "the only one in this maze of mountains . . . and after the long hot climb men and beasts were equally grateful for its blessing."

A double march on 28 July was to bring them to Tashmalik, the first oasis at the end of the Gez Defile. The last two passes were steep: the first so steep that "the ponies even without their loads had difficulty scrambling up,"[82] while the second forced them to make "a last great detour into the wild, barren waste of conglomerate hills"—made the long day longer.

The point where he struck the plain at 6:30 in the evening was a wasteland. Keeping by the side of the river, he soon came to a canal and then to "fertile irrigated fields. Men were still working and in all directions rills of water betokening by its muddy color its glacier sources made a picture of life "doubly impressive after the stony wilderness."[83] Stein pushed on ahead to contact the beg to whom he had sent word requesting fresh ponies. Finding the beg's house was not easy. He enlisted the help of a farmer riding a lively donkey. "We passed miles that seemed endless to me between fields and gardens and little houses. Yet the Beg's place was ever ahead."[84] And when at last they reached the house, they were told the beg was in Kashgar. There was nothing for Stein to do but retrace his steps; his exhausted pony almost broke down before reaching the camp. "It was long after ten o'clock before I managed to get a 'wash' and close to midnight before I could sit down to a well-earned dinner." Yet the next morning "we were up as soon as day broke. The vicinity of Kashgar was an irresistible attraction to hurry on."

Stein started for Kashgar, fifty miles away, taking only the essential baggage loaded on a couple of ponies, leaving the heavy cargo to follow. "I knew that at the other end of my long march a hospitable roof was awaiting me."[85] Crossing the Gez River took time. The water running through half a dozen branches, "was 4–5 feet deep in most of them and the flow so rapid that it required careful guiding of the animals by special men stationed to assist at the fords." Riding along lanes shaded by willows and poplars and passing fields rich with ripening fruits and vegetables, Stein took precious minutes to have a modest meal with apples and plums bought at a wayside stall: "the first fruit I had tasted for months."[86] Then came a sample of the desert—a long ride across a sterile, sandy plain. After a few hours the desert yielded to the cultivated lands of another hamlet, and then another, and another, as Stein rode "among green avenues of poplars, mulberry-apricot, and other fruit trees. The mud walls of the houses with their bright yellow and brown looked singularly neat in this setting of gardens and orchards. . . . I scarcely saw any grown-up person that was not occupied in some way."

Evening was falling as they came out on the main road, "as dusty as it was broad. High lumbering carts, or 'Arabas' dragged along by little ponies and droves of donkeys kept up a continuous cloud."[87] Stein was eager to end the journey—it had been a long day and the day before had been long and hard. The rice fields stretched endlessly on. The ponies were tired when they reached the Kizil-su (Red River) which flows past Kashgar. Once across, Sadak Akhun guided him "along dusty suburban lanes where women with quaint caps of imposing height sat in groups enjoying a chat in the twilight."[88] Then, suddenly, out of the darkness the city walls appeared, massive and imposing. The city gates were closed. Sadak took a turn to the left, and on a short poplar-lined avenue the light of a lantern marked the outer gate of Chini-bagh, the Macartney residence. "As I descended from the spacious court to the terraced garden I found myself welcomed in the heartiest fashion by Mr. and Mrs. Macartney."

7

Stein's friendship with the Macartneys has the air of inevitability about it. Far more than a relationship between two men in the service of the government of India who were thrown together in Chinese Turkestan, it was a meeting of temperaments, and, deeper than that, of emotional orientation. In both, unassuming exteriors belied a tenacity of purpose tempered by a deep-seated courtesy—qualities which won for Stein the expedition that brought him to Kashgar and for Macartney an influence which, though long devoid of power, played a vital role in the remote but strategic post where British, Russian, and Chinese interests were responding to political maneuverings.

George Macartney (1867–1945) was by any measure an unusual man. Child of a most improbable marriage, a Scottish father and an upper-class Chinese mother brought together by the violent finale of the Taiping rebellion, his formative years were extraordinary. Until he was ten he lived in Nanking, growing up in a Chinese home. This familiar setting had a token continuation in London, where he accompanied his father who had been appointed by the Chinese to be secretary and interpreter in their British legation. But London was a world without a heart: his mother had remained behind and died when he was twelve. The "family" with whom he spent his lonely schoolboy vacations in England was that of his father's boyhood friend, James Borland. As a university student, his unorthodox upbringing continued: he matriculated at a French university (perhaps because his father had married a Frenchwoman). Languages came easily to him: in addition to Chinese and English, he was fluent in French, had a certain ease with German and Russian, and later became competent in Persian, Hindustani, and Turki.

In 1887, the same year as Stein, he too sailed for India. His deepest wish, to serve in a consular post in China, was denied him. Instead he was assigned to be the Chinese interpreter to the Burma commission, a lesser service of the government of India. His biographer, Sir Clarmont Skrine, himself a member of the Indian Civil Service, who held the Kashgar post four years after Macartney had left, summarizes the differences that set Macartney apart from other young men in his profession. Drawn from the "cream of the public schools and universities,"[1] they "went to the East imbued with a sense of national superiority and convinced of their ability and therefore of their right to rule." Products of upper-class Victorian society, their "beliefs in social and cultural superiority turned into assumptions of racial superiority and of the divine right of the English people to rule over the coloured races." This social Darwinism solidified into the rigid code of the Anglo-Indian world: "any stepping over the lines of caste and convention was not treated kindly."[2] Macartney had absorbed that world's eleventh commandment: Thou shalt not marry a native woman; throughout his life he never mentioned his mother—not even, it is said, to his own children.

The Kashgar post that Macartney held for twenty-eight years symbolizes his situation in life. He lived inside China's westernmost border in a region different from Nanking but one that nevertheless could give him the feeling that he had come home; while serving as Britain's representative there he preserved his filial tie to his father's land. For him Kashgar was the place and the post that validated his parents' marriage.

Macartney first went there in 1890 as the Chinese interpreter for Captain (later Sir) Francis Younghusband (1863–1942), already famous for his overland journey from Peking to India and for his knowledge of the Pamirs. Curzon knew and admired the way he combined man-of-action and mystic. Younghusband's visit to Kashgar was part of the "Great Game," the decades-long British-Russian confrontation for control of the high passes leading into India's Northwest Frontier. The British watched the Russians extend their empire a thousand miles southward from Central Siberia as "the independent khanates of Central Asia were successively swallowed up. Tashkent capitulated in 1865, Samarkand three years later. Khiva was taken in June, 1873," a bare five months after the Russians solemnly promised the British that "nothing of the sort would happen."[3] Again, after giving "assurances that Russia would not take Merv, . . . Russia took Merv [in 1884] announcing that the inhabitants had asked to be annexed." If it seemed that Russia's diplomatic right hand did not know what her military left hand was doing, it was clear that prearranged incidents and prevarications were used to justify the Russian army's advance.

The Russian consul-general in Kashgar (his term began in 1882) was Petrovsky, a brilliant, arrogant, charming, and devious man who for the next twenty-one years planned and plotted, bullied and bluffed, threatened and coerced the Chinese officials. His objective: to create the situation that would pry Sinkiang's westernmost oases loose from China and so give Russia control of the strategic passes leading to India's back door. Britain wanted the approaches to the passes to remain under the sovereignty of Afghanistan. Britain's imperial defenses, Balfour had stated, could be summed up in one word: Afghanistan.

To see what the Russians were up to, Younghusband was sent to Kashgar; to keep watch on them Macartney, who had earned Younghusband's esteem during the difficult year they spent together, stayed on. At twenty-four he became "the sole representative of British interests" in Sinkiang.[4] Yet for nineteen years his title "Special Assistant to the Resident in Kashmir for Chinese Affairs" had no official status: the Chinese tolerated his presence, Petrovsky ignored or harassed him.

Macartney's position and isolation can be felt in the answer of the assistant secretary to the government of India, Foreign Department, Calcutta. Is there "a British Consul or Agent in Kashgar and is he accessible by telegraph?" The reply: "There is no British Consul at Kashgar, but Mr. George Macartney, Special Agent to the Resident in Kashmir for Chinese Affairs, is stationed there. If it were of any urgency to communicate with him, it is possible to get a telegram through in three or four days, via Pekin; but in this case we have to telegraph to Her Majesty's Minister at Pekin, who sends the wire over the Chinese wire. This is very seldom done and is not always safe. Ordinarily we should send instructions to Mr. Macartney through the Resident in Kashmir, who would wire it to Gilgit, whence the message would go on to Kashgar by dak runners, arriving in about a fortnight."[5]

Until 1909, when he was named consul-general, he functioned by virtue of an oft-tested integrity and tact whose wellspring was his immense sensitivity to deep-seated cultural attitudes—the European contempt for the Chinese and the Chinese contempt for European barbarians. Accustomed since boyhood to the quicksands of xenophobia, he knew how to keep his equanimity, his equilibrium, and his sanity.

Did Stein appreciate Macartney's delicate situation until he saw it for himself? Stein might have thought the caution suggested by the Oriental College graduate at Tashkurgan was directed against Chinese hostility;

from Macartney he learned it was the Russians and not the Boxers who could make trouble. The arms he was bringing were inconsiderable: a few rifles and revolvers for Macartney's hunting trips and a modest present for the *tao-tai,* the highest local Chinese civil administrator.

Now under Macartney's roof, Stein could profit from his host's personal influence with the Chinese officials with whose etiquette Macartney was familiar. As Lockwood Kipling had introduced Stein to Greco-Buddhist iconography just so Macartney interpreted and instructed him in the niceties of Chinese officialdom. Certainly for Macartney, Stein's presence at Chini-bagh was a welcome event. Outside the Russian consulate—and there was a two-year period when Petrovsky did not talk to Macartney— the Europeans in Kashgar could be numbered on the fingers of one hand: two married Swedish missionaries with their wives and Father Hendricks, a Dutch Catholic priest, a kind of spiritual brother to Csoma de Körös. Dismissed by his society, the priest had wandered in China, Mongolia, and Siberia before settling in Kashgar where he lived on charitable scraps. A superb linguist, a fascinating conversationalist, learned in astronomy and geology, "he was always cheerful, hopeful, kind-hearted and enthusiastic."[6] Though he never made a convert, "he said mass by himself every day on an altar made from a packing-case." Macartney invited him to stay at Chini-Bagh after Petrovsky had connived to have him put out of his mean little house. Until Macartney was married in 1898, the priest was his closest companion.

Catherine Macartney was a daughter of James Borland. Despite a proper ladylike reticence, her qualities of courage and stamina, warmth and devotion can be discerned. She was, she wrote, "the most timid, unenterprising girl in the world. I had hardly been outside the limits of my own sheltered home, and a big family of brothers and sisters, had never had a desire whatever to see the world, and certainly had no qualifications for a pioneer's life, beyond being able to make a cake."[7] She commanded Stein's admiration. "I cannot praise my hostess's hospitality and consideration highly enough," he wrote to Hetty from Chini-bagh, 30 August 1920: "How this young woman, married less than a year and a half, manages to run a house in Kashgar that is more English than most of those of Anglo-Indians, is difficult to explain. Trained servants are absolutely unavailable here, and furniture and so forth must be made under the direct supervision of the person ordering it."

Chini-bagh would become almost a second home to Stein. Perched on the edge of a bank above a broad stream, it commanded a panorama of fields and gardens surrounded by low hills and, after a rain had washed

the air of its light dust-haze, the distant, mighty ice-crowned peaks revealed in all their majesty. Macartney had been permitted to buy an old,
tumbledown, mud-brick garden house which he transformed into a residence that "offered all the comforts of an English home while its spacious
out buildings and "compound' had all the advantages of an Indian bungalow."[8] Stein arrived there on 30 July 1900. "After two months of inces-
sant mountain travel I felt the need of bodily rest."[9] With his friends he
found peace and leisure as well as the help and guidance to plan for the
work ahead. He learned that his coming created new tensions; he would
have to prove beyond suspicion that he was not the spy Petrovsky claimed
but a scholar who would reveal Sinkiang's past in the glorious pre-
Islamic days of the Han and the T'ang.

Stein's preparations were governed by two necessities: on the one hand, it
was essential to carry supplies for work to be done in extremes of climate
and terrain, with each item packed so as to be readily available. On the
other hand, it was essential to limit the baggage in order to move rapidly.
He estimated he would need a caravan of twelve ponies and eight camels.
The care taken in their selection was justified by the result: "notwithstanding the fatigues and hardships . . . of travels which covered an
aggregate of more than 3000 miles, none of the animals I bought in
Kashgar ever broke down."[10] Then he had to engage a staff. His Punjabi
servant, who had been wilted by the rapid mountain marches, was replaced by a hardy half-Kashmiri, half-Yarkand man who had had long
experience handling ponies on the Karakorum route and also had "the
indispensable rudiments in the art of looking after a 'Sahib's' kit." Niaz
Akhun, a Chinese-speaking Kashgari, of "imposing appearance and of
manners to match"[11] doubled as Chinese interpreter and pony attendant.
Of Niaz, Stein would say in the rhetoric of a written reference, "as an
interpreter he served me honestly, . . . he had small personal failings, such
as his inordinate addiction to opium and gambling and his strong inclina-
tion to qualified looting." Poor Niaz, the Taklamakan proved his undo-
ing. Two men skilled in the difficult art of camel management came, as it
were with the camels: Roza Akhun and Hassan Akhun. Young, adventurous, curious to see new places and sample unusual experiences, they
stood ready to overcome the traditional fear of desert journeys. "They
made up for their cheerfulness and steady conduct on the march through
the sand-wastes by an irrepressibly pugnacious disposition wherever the
varied temptations of a Bazar were near." With Ram Singh and his
servant, these four formed the permanent core of Stein's expedition.

Equipment was repaired and new items supplied. "Ever since we emerged from the gorges of Hunza, yaks and ponies seemed to have vied with one another . . . in knocking and rubbing their loads against every rock passed."[12] Then there was the need to augment their water-carrying capacity: the two tanks brought from Calcutta were insufficient for the size of the party that would be traveling and working in the desert. Four more were improvised out of much-battered tins of kerosene imported from Russia.

Stein still had long and pleasant hours to spend "refreshing by systematic study my knowledge of the ancient accounts . . . such as the Chinese Historical Annals, the narratives of old Chinese pilgrims and of the earliest European travellers."[13] Chini-Bagh was ideal for this: in addition to Macartney, whose fund of knowledge was wide-ranging and acute, his Chinese secretary, "a literatus thoroughly versed in his classics," explained details of Chinese lore. "As I listened to his vivacious explanations, which Mr. Macartney kindly translated, I could not help thinking of my dear old Kashmirian Pandit Govind Kaul, and the converse I used to hold with him in Sanskrit during the long years of common scholarly labour. Bitterly I regretted . . . my ignorance of Chinese." In addition Stein spent an hour or two daily "in the study of Turki texts with grave Mullah Abdul Kasim, a shining academic light of the chief Madrash of Kashgar."[14] But before and after, in the early afternoon hours when the summer heat was greatest, Stein worked in an improvised dark room developing the photographs he had taken.

Punctuating his tasks and studies, Stein paid formal visits on the Chinese officials to acquaint them "with the purpose of my intended explorations and to secure their goodwill."[15] Without this Stein would be crippled. "In the course of these visits, followed as they were by 'return calls' and other less formal interviews, I was introduced to a rudimentary knowledge of the 'form' and manners which Chinese etiquette considers essential for polite intercourse. . . . Every act and formality . . . thus acquired due significance, until in the end when visiting strange 'Yamens' [official residences] far away from my Kashgar friends, I found a comfortable assurance in the rigid uniformity of these observances."

Stein wanted the tao-tai, the provincial governor, to "issue clear instructions to the Amban [District Officer] of Khotan"[16] to furnish him the necessary assistance to secure transport, supplies, labor, and the freedom to move about, to excavate, and to make surveys. This was not easily or quickly obtained in the face of Petrovsky's hostile attitude. It took many interviews and a lengthy correspondence to obtain what he wanted.

With Macartney translating, Stein explained the "historical connection of ancient Indian culture and Buddhist religion with Central Asia." His most telling argument rested on the tao-tai's familiarity with Hsüan-tsang: "All educated Chinese officials seem to have read or heard legendary accounts" of that pilgrim's visit to the Buddhist kingdoms of the "Western Countries." Because he never appealed in vain to the memory of Hsüan-tsang, he felt justified in claiming that "saintly traveller as my special patron."

At Chini-Bagh, the "little oasis of Anglo-Indian civilization,"[17] the worst of the Turkestan summer heat passed pleasantly in work, in study, and in preparing for the tours ahead. Plenty was on the land. Apricot, peach, and plum trees bent low by their abundance made it a time for picnic parties. Kashgar families rode out to their orchards to feast on fruits; to take their ease and visit. Father Hendricks, "an equally frequent visitor at Chini-bagh, the Russian Consulate, the Swedish Mission and the Chinese Yamen"[18] brought his harvest of impartially gathered news and gossip." In the cool of the evening, Stein would ride out to trace the remains of Kashgar's Buddhist past. Surveying the badly decayed mound of a once large and splendid stupa, Stein noticed that its base was fifteen feet below the level of the nearby fields. "I had here the first indication of that remarkable rise in the general ground level, mainly through silt deposits, which my subsequent observations on the site of the ancient capital of Khotan clearly demonstrated."

Stein left nothing to chance, especially in an expedition different from anything he had previously done. Refreshed by five weeks of rest, on 4 September he had a "shake-down" trip, "a kind of experimental mobilization of my caravan."[19] Even on this trial tour, a discovery hinted at the surprises in store for him. Near a fairly well-preserved stupa, whose dimensions and proportions resembled Buddhist monuments known to him from the Afghan border, was another ruin called the "Pigeon House." A square enclosure, the inner sides of its walls lined with little niches and its interior strewn thickly with human bones, it "curiously recalled a 'Columbarium.' "[20] This, a subterranean sepulcher, had niches in its wall similar to those used in ancient Rome to hold cinerary urns. Since neither Buddhist nor Muslim "custom would allow of such disposal of human remains," Stein had his first indication "of the times when Kashgar held a considerable population of Nestorian Christians." The voices from the distant past so long silent were beginning to be heard.

The short excursion quickly showed up weak spots: some of the camels had sores from ill-fitting saddles, some of the boxes needed stronger packing crates. Repairs were speedily made; on 11 September Stein bade farewell to his hosts and started for Khotan by way of Yarkand.

Like a man determined to take a swim in an unknown sea—first testing the water temperature, then standing knee-deep in the breakers to gauge their power and the pull of the undertow, and slowly, cautiously entering, feeling for where the land falls away and he must swim—so Stein readied himself for the Taklamakan.

Away from the main caravan road, he chose the route leading to a famous pilgrimage shrine in the desert. Approaching it gave him his first taste of crossing sand dunes of ever-increasing height. Tiresome for the men and tiring for the ponies, it was almost dark before they arrived. The shrine, a collection of huts and rest houses that accommodated the local custodians and visiting pilgrims, was threatened by a thirty-five-foot sand dune. Nearby were the ruins of a former settlement that had been "overwhelmed at a previous date by the advancing dunes and left bare again when the latter had passed in their gradual movement."[21] The water from the shrine's well had a strong brackish taste, and neither "filtering nor the lavish use of 'Sparklets' could make it palatable."

Leaving the shrine, they rejoined the main road, celebrating that night with a feast of the grapes and peaches to be had almost for the asking. A twenty-four-mile march across an arid, gravel-covered grey waste, its desolation unrelieved by a single tree or shrub, then, in sharp contrast, eighteen miles through the cultivated fields of the sprawling Yarkand oasis, brought Stein to the city itself. Situated at a point "where the routes to India, Afghanistan and to the north meet,"[22] Yarkand was the commercial center of Chinese Turkestan. The resident Indian traders, dressed for the occasion, had come out to welcome a *sahib*. Escorted by a clattering cavalcade, Stein was led to a summer palace which Macartney had engaged for him.

Stein had already learned how powerful Macartney's name was with the Hindus established in the oases between Kashgar and Yarkand: moneylenders from Shikarpur practiced their traditional Punjab usurious operations in Central Asia (an eighteenth-century traveler met them as far away as Samarkand). Their large numbers and their exorbitant interest rates could "only imply the progressive indebtedness of the cultivators."[23] Business, they told Stein, was brisk. "To protect the interests of this class is a task which the representative of the Indian Government cannot afford to neglect, however unenviable it may often be. So I was not surprised that my welcomers were loud in their praises of Mr. Macartney."

Their spokesman assured Stein he would have no difficulty in having his checks cashed. Unfortunately, most of their cash was already on its way to India, and their high rate for cashing checks convinced Stein to send to Kashgar for the gold and silver coins he would require. Awaiting them, he extended his stay at Yarkand.

Stein did avail himself of their skill in money transactions: preparing for the desert was simple when compared with finding his way through the currency "jungle." Chinese currency with its

> "Sers" or "Tels" "Miskals" and "Fens" arranged on a plain decimal system, would be convenient . . . but its simplicity is of little avail in this outlying province . . . which stubbornly clings to its time-honoured reckonings in "Tangas" and "Puls." Each of the little square Chinese coppers known in Turkestan as "Dachin" is reckoned in Kashgar as equal to two Puls and twenty-five of them make up a Tanga. The Khotan Tanga is worth twice as much as the Kashgar Tanga. Coins representing this local unit of value do not exist. All sums have to be converted into Miskals, the smallest available coins, at the ratio of eight Tangas to five Miskals. . . . But the exchange rate between silver and copper is not stable and the silver Miskal was just then considerably above the value of forty copper pieces. . . . So after successfully converting Tangas into the legal coin, a varying discount has to be calculated before payment can be effected. It only adds to these monetary complications that prices of articles imported from Russia are reckoned in "Soms" (Roubles), which in the form of gold pieces of five or ten Roubles circulate widely through the markets of Turkestan, while the heavier Chinese silver "Yambus," of horse-shoe shape and varying weight, have a discount of their own. . . . For the well-trained arithmetical faculties of the Hindu trader these tangled relations offer no difficulty.[24]

At Yarkand Stein was appalled to learn that two camels and two ponies had developed bad sores. After dressing them, he ordered a week's rest and easy grazing. Angry that the condition had been kept from him until it had become serious, Stein took the lesson to heart. "Thereafter inspections of the animals were held almost daily, and those responsible learned that the hire of transport to replace those rendered temporarily unfit would be recovered from their own pay."[25] A fortnight passed as Stein waited for the money to arrive and the animals to heal. The days passed quickly. The city had large, long-established colonies of Kashmiris, Gilgitis, Hunzas, and peoples from other parts of the frontier; a steady stream of visitors called on the sahib in his spacious quarters—the "carpet of my presence" usually had picturesque callers.

Liu-Darin, the amban of Yarkand, had shown administrative qualities: near the city a sandy waste had been brought under cultivation by a newly opened canal that carried water from the Yarkand River. Though sand dunes still retained the scrub covering, they were surrounded by carefully terraced fields thick with ripening wheat. The effort had been planned and supervised by the amban. "The labour had been wholly unpaid . . . [but] the cultivators now were glad to occupy the ground they had reclaimed and thus to reap the direct benefit of their labours."[26] As soon as the amban returned from a tour, Stein paid his formal call. An elderly man, gentle, intelligent, responsive, the amban impressed Stein with his manners and looks. His visit was returned the following day, and, through a "not over-intelligent interpreter,"[27] Stein showed him Hsüan-tsang's book and explained that he was "searching for the sacred sites which the great pilgrim had visited about Khotan, and for the remains of the old settlements overwhelmed by the desert." The courtesies continued: Stein was guest at a three-hour dinner and, when the banquet was over, photographed his host. Liu-Darin "installed himself on a raised chair of office, with his little daughter and son by his knees, and some implements of western culture, in the shape of sundry clocks, etc. on a small table close by. . . . My sitters kept as quiet as if they were sculptured. Then we parted in all friendship."[28]

The next district, Karghalik, lay on the other side of the Yarkand River which drains "the whole mountain region between Muztagh-Ata and the Karakorum range."[29] In summer it filled a mile-wide bed; now it flowed through three channels each of which had to be crossed by ferry. Because the ferry boats could not take laden animals, it was necessary to unpack and refit three times—a lengthy, time-consuming process. All that time "the traffic of laden ponies and donkeys was sufficient to fill two or three ferry boats at each crossing as quick as they could be worked." The next river, fed by the Kunlun Mountains, was easily crossed on a good bridge "built, according to the Chinese and Turki inscription at its head, some twenty-five years ago and [which] measures fully 250 steps."[30] Late in the afternoon of 28 September Stein entered Karghalik, a thriving town located "where a much-frequented route to the Karakorum Passes joins the great road connecting Khotan with Yarkand."[31] His tent was put up on a large meadow with "beautiful old walnut trees that carried me back in recollection to many a pretty village in Kashmir. . . . My people found quarters in a cottage close by and the ponies excellent grazing; everybody was satisfied."

The amban, Chang-Darin, informed by the tao-tai of Kashgar, knew the reason for Stein's visit. At the first formal visit, he presented Stein with a sheep for the men and fodder for the animals, and in the formula of gift exchange, he was given Russian sweets, sardine tins, and highly scented German soap bought in Kashgar for this purpose. Returning from the yamen, Stein stopped briefly in the bazaar to shop for *paipaks,* the felt socks for which the region was famous; on reaching his camp, he found the amban awaiting him. He showed Chang-Darin the Chinese glossaries in Julien's translation of Hsüan-tsang and the plates of antiquities in Hoernle's publications. A man of quick, lively intelligence, the amban was excited by the pictures of ancient Chinese coins and fragments of Chinese manuscripts. He bombarded Stein with questions which were falteringly answered through an inadequate interpreter. The next morning a messenger from Kashgar brought the money—bags of Chinese silver coins and a small packet of gold rubles—and letters from home. Answering his mail and summarizing for the Calcutta bureaucrats the "currency complications in my 'Monthly Cash Accounts' "[32] kept Stein busy until 2 October when he left Karghalik by the Khotan Gate.

Within a mile the road entered a barren stretch "marked all along by wooden posts erected at short intervals—no useless precaution considering how easy it would be for the traveller to lose his way at night or during a sand-storm."[33] Caravansaries offered water and shelter to travelers, and post stations provided horses and postmen carrying official correspondence. Though the hills were faintly visible, no canal or watercourse descended from them; the dreary route was marked by the carcasses and skeletons of animals that perished there: the same waterless, empty waste as when Hsüan-tsang followed it on his return to China, the same as when Marco Polo was making his way to Cathay: "Nothing has changed here in the method and means of travel . . . since the Buddhist religion and the elements of Indian as well as of classical culture travelled to the land of the Sinae."[34] The oasis of Guma marked the end of the waterless stretch.

Guma also marked the start of Stein's serious antiquarian inquiries: tracking down the manuscripts and blockprints written in unknown characters which, since 1895, had been reaching Western collections. "Islam Akhun, the Khotan 'treasure-seeker,' from whom most of these strange texts were acquired, had in statements recorded at Kashgar by Mr. Macartney and subsequently reproduced in Dr. Hoernle's learned report on the Calcutta collection, specified a series of localities from which his finds were alleged to have been made. Most of these were described as old sites in the desert north of the caravan route between Guma and Khotan."[35] Stein was about to begin one of his stated objectives: to authenticate these strange texts or prove them to be forgeries. Guma was the first such place where he could test Islam Akhun's statements by direct inquiry. None of the local begs had ever heard of the discovery of "old books." Of the string of locales given by Islam Akhun, two, close to Guma, were known to the begs. Kara-kul Mazar, the Shrine of the Black Lake, had been described by Akhun as a vast cemetery clustering around a saint's shrine; Stein found a little reed-covered saline pond held in a semicircle of sandhills atop one of which were poles hung, with votive rags, the traditional marking for a saint's tomb. The other, a little settlement reclaimed only fifteen years before from the desert, had small open-air paper factories, the pulp made from the bark of the mulberry trees. Of ruined sites and old books its inhabitants knew nothing.

Meticulously, Stein made visits to and took considerable time to examine every debris-strewn place that indicated the likelihood of a former site. At Kakshal he first experienced the "weird fascination in the almost complete decay and utter desolation of the scanty remains that marked once thickly inhabited settlements."[36] But the copper coins picked up at the site had been issued by an early Muslim ruler, thus giving an approximate date for the bits of pottery: "there can be no doubt that these coins have been washed out originally from the same debris layers to which the pottery belongs."[37] Another place, "a small hillock from which skulls and skeletons were protruding" had attracted Sven Hedin's notice; but the rag-coverings mixed in with the bones were of fairly recent manufacture.

At the very border of Khotan, in a sea of low sand dunes, Stein visited a shrine popularly known as the "Pigeons' Sanctuary." Thousands of birds, housed in wooden buildings and maintained by pilgrims' offerings and the proceeds of pious endowments, were "believed to be the offspring of a pair of doves which miraculously appeared from the heart of Imam Shakir Padshah,"[38] who died battling the infidel Buddhists of Khotan. Like all travelers, Stein fed the birds and, watching the fluttering scene, thought of Hsüan-tsang's account of a "local cult curiously similar at the western border of Khotan territory." In that version it was rats whom all wayfarers fed, in the belief that they had appeared when the king of Khotan, in despair, prayed for deliverance from the fierce Huns who were ravaging his country. In answer, the rat king led his army of rats, "as big as hedgehogs, their hair of a gold and silver color," Hsüan-tsang had written. Overnight, the rats ate the Huns' leather harnesses, making them immobile and helpless. Just as the holy pigeons replaced the roly rats in recalling a great victory, so pious Muslim legend replaced the pious Buddhist one. "It was the first striking instance of that tenacity of local worship which my subsequent researches showed for almost all sacred sites of Buddhist Khotan."[39]

At Zawa, just three miles beyond where the sandy billows ended at a low-lying marshy plain, Stein was welcomed to the soil of Khotan. From there the road, ankle-deep in dust, flanked by shade trees wearing their autumnal yellows and reds—a welcome relief from the khaki of the desert—was filled with the sound and movement of trade. Strings of donkeys carried mountainous loads of *zhubas,* the lambskin coats for which Khotan was famous; streams of people, coming and going, all mounted on ponies, donkeys, and bullocks, heralded an important trade center. Soon his caravan reached the wide Karakash, Black Jade, the second main river of Khotan—the river which in summer is three-quarters of a mile wide was now reduced to thirty yards and barely knee-deep. A little further they came to the Yangi-Darya, the New River: both conformed to the Chinese historical accounts. This was an indication to Stein that the ancient capital had been some distance off the road. Later he would visit its site.

The next morning, 13 October, escorted by the beg deputed by the amban to greet him, and by Badruddin Khan, head of the Afghan merchants settled in Khotan, Stein "rode around the bastioned walls of the great square fort that forms the 'New City' of the Chinese, and then through the outskirts of the 'Old City' "[40] to where he was to lodge. But the house, the residence of a rich merchant, was a "maze of little rooms all lit from the roof and badly deficient in ventilation," and its garden, "a wilderness of trees and bushes [offered] little room for a tent and still less for privacy." It was unappealing. Stein invoked the traditional hospitality of the region by which "one may invade the house of anyone, high or low, sure to find a courteous reception, whether the visit is expected or not."[41] It was thus Stein found the ideal campsite. Its owner, Akhun Beg, a portly old gentleman, received him like a guest and "when informed of the object of my search, offered me the use of his residence. I had disturbed him in the reading of the Turki version of Firdusi's Shahnama."[42] When Stein recited part of the epic in its original Persian, he was invited—not merely the customary polite formula—to pitch his tent on the lawn in the shade of a clump of trees.

Even before settling into his camp, Stein sent messages and gifts to the amban, and as soon as his camp was arranged, he paid his first visit. Stein found Pan-Darin a "quiet elderly man with features that seemed to betoken thoughtfulness." Gentle in gestures and speech, respected for his learning and piety, the amban conveyed his understanding of the purpose of Stein's visit as it was explained to him from Kashgar. Of Stein's desire to seek ancient settlements in the desert, Pan-Darin stressed the dangers and difficulties he would encounter and cautioned him against a too-ready acceptance of any statements made by the local people. As for Stein's wish to survey the sources of the Khotan River, Pan-Darin remarked that "the routes were very bad, implying hardships and risks, and that beyond the valleys of the Karanghutagh lay the uplands of Tibet where Chinese authority ceased." He then offered Stein all the help within his power. In every way, Pan-Darin was to be a tower of strength. Without "his ever-ready assistance," Stein could gratefully say, "neither the explorations in the desert nor the survey work in the mountains which preceded it could have been accomplished."[43]

In his ambitious program—finding and examining ancient desert sites, surveying the headwaters of the Khotan River, and discovering the truth about the suspicious documents—Stein worked within a tight schedule. That he had the help of Pan-Darin enabled Stein to move with speed and purpose, but that he accomplished his plan reveals qualities that made him unique.

Gold washing and jade digging had long attracted rich Khotanese businessmen; treasure hunting, on the other hand, was a time-honored occupation for the poor. "A small fraternity of quasi-professional treasure-seekers had learned in their periodic visits to ancient sites to pay attention also to antiquities as secondary proceeds."[44] When questioned, the men were secretive, vague, unreliable, and Stein soon realized he would lose time and effort were he to start without evidence known to have come from specific sites. Badruddin Khan made himself responsible for sending out prospecting parties, but it would take a month before they returned, a month Stein would use for surveying.

While preparing for his mountain survey, Stein, who had not inquired about "old books," was offered one. He had already become suspicious when none had appeared, since so many "old books" had in recent years been sold from Khotan. The book, ten ragged leaves of birchbark covered with an unknown script, was brought by a Russian Armenian. "I saw at once that the birchbark leaves had never received the treatment which ancient Bhurja [birchbark] manuscripts, well known to me from Kashmir, invariably show. Nor had the forger attempted to reproduce the special ink which is needed for writing on birch-bark. So when I applied the 'water-test,' the touch of a wet finger sufficed to take away the queer 'unknown characters.' "[45] This obvious forgery resembled some of the block-prints in Hoernle's Calcutta collection, and, in casual conversation, Stein learned there was a connection between the man who had sold the mau-script to the Armenian and Islam Akhun. Islam Akhun, always Islam Akhun. The fact that he had not come forward when Badruddin Khan sent out the treasure-seekers strengthened Stein's suspicions. He did noth-ing, said nothing that might alert Islam Akhun; he counted on silence and time to allay Islam's apprehensions. .

Seeking to map the headwaters of the Yurungkash (White Jade) branch of the Khotan River was the kind of strenuous exercise that had an intel-lectual validation in Stein's military training and his early love of old topographies. In this he would not be competing with Sven Hedin.

The region where the Kunlun and the Himalayas approach one another is a cosmic epic of high ice-capped mountains, vast glaciers, and deep, narrow gorges cut out of naked rock. It is a forbidding world where small, isolated communities eke out a meager living and care for animals brought there for summer pasturage. Though the region was still unsurveyed, it was known that about 400 A.D. Fa-Hsien had gone across that region from Khotan to Ladak, Kashmir's easternmost district; that in 1865 a Mr. John-son had made his way in the opposite direction—Ladak to Khotan; and that in 1898 Captain Deasy, again working from the Ladak end, had "suc-ceeded in reaching the sources of the [Yurungkash] at an elevation of over 16,000 feet but was prevented from following the river down-wards."[46] Here, then, was a geographical puzzle and, thanks to Fa-Hsien, also an antiquarian problem that appealed to Stein. It would occupy him from 17 October to 15 November.

To Akhun Beg, his host, Stein gave "a five Rouble gold piece pre-sented in a little steel purse,"[47] and to Badruddin Khan's care he confided the baggage he would not need. Ten ponies transported surveying instru-ments and a month's food supply. With an eye to future work, Stein made an overnight stop at the last village before the stony desert rises gently to the foot of the mountains. From an ancient site near the Yurungkash, vil-lagers brought him "old coins, beads and a few small seals, one showing the figure of a Cupid."[48] Not far away were the burrows and heaps of pebbles that indicated jade was still being sought. Jade "which has been so highly prized in China and to which the river owes its name, 'White Jade,' is still an important product. . . . I thought of the distant lands to which it has carried the name of Khotan."

His immediate goal was Karanghutagh, the last inhabited hamlet at the northern foot of the Kunlun Peak. Terraces of gravel and coarse sand, steplike, ascended toward the mountains and, except for a distant view of

Khotan far below, the landscape was one of utter desolation. The last of the giant steps brought them to a pass which offered a "full view of the outer ranges through which the Yurungkash flows in a tortuous gorge." Stein's words sound a rhapsody of glory and joy when, looking toward Tibet from a 14,000-foot ridge, he beheld a "panorama more impressive than any I had enjoyed since I stood on the slope of Muztagh-Ata. To the east there rose the great Kunlun Peak with its fantastic ridges separated by glittering glaciers. . . . By its side the main branch of the Yurungkash could clearly be made out as it cuts through a series of stupendous spurs. . . . The deep-cut valleys and serrated ridges . . . presented a most striking contrast to the flat, worn-down features of the plateaus behind us."[49]

The ridge also made an ideal survey station. While Ram Singh worked at his plane table, Stein took a full circle of the scene with his photo-theodolite. Before they had finished their work, the afternoon was half over, and they still had to reach the valley of the Karanghutagh. The local guides, impatient and anxious, spoke of the bad stretch ahead. The first two miles of descent were easy. Darkness was beginning as they came to where the trail plunged down a series of precipitous, rocky gorges. Even in broad daylight it would have been impossible to ride down the steep zigzags to the river, three thousand feet below. The horses had to be dragged. Loose stones increased the hazard; the thick dust kicked up was smothering. For an hour and a half they slid and scrambled, man and pony choking, enveloped in a dust cloud. Thus they came to where a stream flowing out of the Karaghutagh valley joined the main branch of the Yurungkash. Night had fallen when they crossed the river "by a rickety bridge consisting of three badly joined beams laid over a chasm some seventy feet wide. The foam of the river tossing deep down in the narrow bed of rocks could be made out even in the darkness. In daylight, and in a less tired condition, the crossing might have affected one's nerves more. . . . Karanghutagh means 'Mountain of blinding darkness.'[50]

A grim valley clutched by an inferno of barren rock held a village—some farm houses huddled together and small fields of barley—that was the winter station for the men who herded the yaks and sheep of Khotan merchants. The occasional visit by a merchant was their sole communication with the outside; occasionally, too, their number was increased by criminals exiled from Khotan. Having reached his first goal, Stein hired surefooted yaks for the dangerous trip to the headwaters of the Yurungkash.

He summarized his tour in a letter to Macartney written on 9 Novem-

ber when encamped at the Ulughat (High) Pass on his way back to Khotan. "I got as far as man and beast can go—to the point where the main
branch of the Yurunkgash issues from an absolutely impenetrable gorge.
. . . Then following a route hitherto quite unexplored, we got into the
drainage of the Karakash [Khotan's other principal river parallel to the
Yurungkash; the two finally join deep in the Taklamakan]. About four
high passes had to be crossed, and luck and some little instinct brought us
above each of them to splendid survey stations. The tracks were bad
enough and the valleys and slopes are certainly more barren and forbidding than anything I have so far seen. . . . The weather has been wonderfully fine, or else the route would have proved impracticable. But the cold
was great.

"Here at the last pass towards Khotan, I got to my delight, a most extensive panoramic view which has enabled us to connect the whole ground
surveyed by triangulation with the Indian survey. The position of Khotan
can now be fixed with certainty. It is strange that nobody got to this side
before—only three marches from Khotan! I am camping here for a day
or two to complete the work, though want of water and the elevation
[over 12,000 feet] make it a little trying. Without the Amban's help we
could not have got over the ground at all. The few scattered Taghliks
[hillmen] are a shy and obstinate race, almost as difficult to manage as
their yaks. But the Amban's orders were evidently as straight as I could
wish, and I could have it my way as far as men and beasts were concerned.
I wish I could show my obligation to the Amban in some useful way."[51]
Camped on the Ulughat Pass, shivering with cold, with a single cup of
tea as his liquid ration, Stein was assuring his friend that the Taklamakan
would hold no terrors for him.

A passage in *Sand-buried Ruins* cannot mask a personal revelation. At
night, perched atop a mountain at the edge of the Western world—never
before had he been so far away from family, friends, and the familiar—
he saw the mirage of a lighted metropolis; alone—to his men he was a
stranger—he proved himself the complete homebody by joining and participating in the approaching festive season: by letters. He tells how, coming out of his tent, he saw the full moon, rising as though from the ocean,
float above the dust-haze that hid the desert. He watched it climb high:
"it seemed as if I were looking at the lights of a vast city below me in the
endless plain. Could it really be that terrible desert where there was no
life and no hope of human existence? I knew that I should never see it
again in this alluring splendour. Its appearance haunted me as I sat shivering in my tent, busy with a long-delayed mail that was to carry to distant

friends my Christmas greetings."[52] There is the sound of an adieu before plunging into that "terrible desert."

Back in Khotan he pitched his tent as before, warmly welcomed by "dear old Akhun Beg, my former host,"[53] and, while the men and ponies rested after the month of hard travel, Stein studied the evidence brought back by the treasure hunters sent out by Badruddin Khan. Such were the circumstances under which Stein met Turdi. An old, experienced treasure hunter, Turdi was a simple man of noble instincts; a villager from the Yurungkash district, he knew the western Taklamakan as well as he knew his village. From Dandan-Uiliq, the "Houses with Ivory," the furthest of the sites locally known, he had brought back pieces of fresco inscribed with characters written in Indian Brahmi, fragments of stucco reliefs of Buddhist iconography, and "a small but undoubtedly genuine piece of a paper document in cursive Central-Asian Brahmi." His samples seemed to indicate Sven Hedin's "ancient City of Taklamakan," which he had passed when crossing from Tawakkel, a village on the northern end of Khotan, to the Keriya River.

When Pan-Darin returned Stein's visit he was shown Turdi's finds. The amban was deeply moved that Stein's faith in Hsüan-tsang was vindicated: truly the great pilgrim had guided him to recover Buddhist shrines long-buried, long-forgotten. To ascertain whether Hedin's site was Dandan-Uiliq, the amban sent word to the beg of Tawakkel to produce forthwith Hedin's two guides. Brought by the beg himself, Ahmad Merghen and Kasim Akhun were questioned by Turdi: indeed the two sites were one and the same. Dandan-Uiliq would be Stein's first goal in the desert.

So far, so good. But a riddle must have teased Stein: How was it possible that an early site like Dandan-Uiliq promised an abundance of evidence where none existed for Khotan's ancient capital? Hsüan-tsang had described it as being rich in lavishly decorated Buddhist shrines and had located it precisely. How could so splendid a center have disappeared without leaving tangible, identifiable traces? Furthermore, Stein knew from Rémusat's *Histoire de la Ville de Khotan* that the capital's earlier name was Yü-t'ien (from *yü*, the Chinese word for jade). This precious mineral had been found exclusively in the valleys of the two branches of the Khotan River, the Karakash and the Yurungkash. Rémusat had collected and translated some of the vast lore devoted to jade which, from Shang times on, had ritual significance. A jade disc was once the symbol of heaven, the source of all light; jade amulets magically protected the living against harm and the dead against corruption—hence the fascinating princely jade suits discovered by archaeologists. Ancient Taoist writers equated its nine precious qualities with human virtues. A finely textured stone of great hardness, jade could not be worked by any metal known to the ancients; it was patiently, lovingly formed and finished by artists skillfully using abrasive sands. At Yü-t'ien, now known as Yotkan, Stein reread Rémusat's volume, enjoying "this earliest contribution to the literature on Khotan near the very pits which furnish the precious stone so learnedly discussed in it."[54]

On his way to nearby Yotkan, Stein passed beds of rubble where jade had been found; but the trenches crisscrossing the gravel plain were partially filled with sand, sure evidence of having been long abandoned. Soon came pits in the process of being worked: at a depth of ten feet were the layers of rubble washed down long ago from the mountains and then being searched for jade. Finds of great value still occurred, attracting merchants with capital and a taste for gambling. The diggers, recruited from the poorest field hands, worked for food, clothing, and a few pennies with no share in the profits. However haphazard the chances, profits continued: a Kashgar merchant told Stein that in three years he had cleared "a hundred Yambus of silver (say Rupees 13,000) at an expense of some thirty."[55]

At Yotkan, the "lesson of the yars," as it might be called, gave Stein the answer to the riddle. A *yar* is a fairly shallow ravine whose steep dirt banks are formed when flood water slices through the fine yellowish-grey loam called "loess." A yar, which his pony had negotiated easily, had ambushed a camel: it had slipped, fallen into the water, and been extricated after much time and ingenuity. Stein's "archaeological conscience"[56] was aroused by what the camel's clumsiness revealed. Where earlier travelers had remarked on the frightful ravages wrought by the flood water, the yar showed Stein that buried under the fertile fields of the Yotkan cultivators lay the remains of the ancient capital he sought. Old villagers gave him a timetable for the yar's formation: a bare thirty years before, the small canal bringing water from the Karakash began digging deeper into the soft loess, changing itself into a yar.

From a small marshy depression formed when several yars came together, "the villagers accidently came across little bits of gold amidst old pottery and other petty debris." The latter were thrown away; but the presence of gold sent a shiver of excited activity through the settlement. So rich were the finds, word of them came to the governor's attention. He sent large parties of diggers, first compensating the owners of the fields for the soil washed away in panning for the gold. Later owners and diggers shared the profits equally. Gradually, the search for treasure went beyond gold as fragments of ornamented pottery, coins, and incised stones were found to have value. The gold was mostly in tiny flakes of gold leaf—the villagers were reticent about having found coins, ornaments, or figurines. Such larger pieces, Stein supposed, were quickly melted down. (On his second visit he was able to buy a tiny gold monkey of superb workmanship that had been found during that year's washings.)

Why gold leaf? Stein's answer was that probably the statues and even parts of the buildings had been richly overlaid with it. The stratum (from five to eight feet deep on the south side, thirteen to fourteen on the north) which yielded the gold leaf also held fragments of pottery, bones of animals, and a plentiful number of copper coins, ranging from "the bilingual pieces of the indigenous rulers, showing Chinese characters as well as early Indian legends in Kharoshthi, struck about the commencement of our era, to the square-holed issues of the T'ang dynasty (618–907 A.D.)"[57]

This treasure-filled stratum lay under a layer of alluvium, lighter in color and from nine to twenty feet deep. Though he searched and surveyed, nowhere could he find traces of buildings: their sun-dried bricks had returned to the soil from which they had been made. He had already noticed that all over Turkestan the rivers, muddy and fast-flowing, carried enormous amounts of disintegrated soil from the mountains and that "the accumulation of silt over the fields on which the earth thus suspended is ultimately deposited, must be comparatively rapid."[58] Again and again the process observed at Yotkan was confirmed: the main roads throughout the oasis were considerably lower than the cultivated fields they passed; so too the old cemeteries showed a ground level below that of the neighboring fields.

Stein was satisfied that Yü-t'ien's ancient capital lay buried under the fields of Yotkan: its location agreed with the "topological indications furnished by the early Chinese Annals,"[59] while the most prominent Buddhist shrines mentioned by Hsüan-tsang had, as elsewhere, become translated into places sacred to Muslim miracles.

November was ending; winter was setting in rapidly. He was ready to begin his first expedition into the Taklamakan.

8

"The morning of the 7th December, a misty and bitterly cold day, saw our start for the winter campaign in the desert. My goal was Dandan-Uiliq, the ancient site I had decided upon for my first explorations."[1] There, preserved by the dry sands of the Taklamakan, Stein would find the proof that had eluded him at Yotkan where centuries of water and silt had erased ancient Khotan. Clear proof of his "discovery" needed more than coins and bits of stucco reliefs and golden figurines, more than Muslim shrines, inheritors of Buddhist sanctity.

Most noticeable in this first venture deep into the Taklamakan is Stein's air of complete confidence. He put his trust in Turdi to guide him in the desert and his reliance in the amban of Khotan to secure labor and transport as needed: thus was mirrored Stein's respect for men in all walks of life who responded to his antiquarian passion. Over and above such presumptions was his faith in the science of topography, which permitted him to arrange a rendezvous to meet Ram Singh on the Keriya River, a place neither had ever seen or visited. They had parted near Yotkan— Ram Singh to finish his survey of the Kunlun area before a reunion weeks later in a distant spot.

To reduce the actual desert travel "with its inseparable privations for men and animals," Stein chose the slightly longer route via Tawakkel. Following the Khotan River, a dreary monotonous succession of sand dunes on one side and reed-covered strips on the other, they were met late in the afternoon of the third day by a torchlit parade led by the beg of Tawakkel and escorted along the irrigation canal to the oasis. Stein could measure the tempo of the transformation wrought by water extended into the desert: built only sixty years before, the canal had created a prosperous settlement of a thousand households. He also noticed that the silt deposit had raised the fields almost a foot above the road.

His two-day stay at this last village where workmen could be hired and supplies obtained before entering the desert set the pattern for subsequent tours. He recruited thirty of the fittest men and ordered food rations for four weeks. The men were loathe to go—they dreaded the sandstorms, the vicious cold, and the demons that lived in the desert; but they dared not disobey the amban's strict orders and they were reassured by Ahmad and Kasim, who had traveled there with Sven Hedin. But most persuasive was Stein's wage scale. He paid twice the amount given unskilled workers and gave each recruit a liberal cash advance. He held the village headman responsible for seeing that each man was supplied with warm clothes and his ration of food. Each man also brought his *ketman,* the ubiquitous hoe that proved its worth excavating in the sand. Among the recruits, Stein chose one who knew tailoring, another skilled in leather work, and a young farmhand who had learned how to write Turki—his penmanship was wildly original but legible. Ponies, too dependent on fodder and water, were replaced with donkeys to carry the food and supplies. The camels were to subsist on a half-pound daily dose of rapeseed oil—"this evil smelling liquid . . . proved wonderfully effective in keeping up their stamina during the trying desert marches."[2]

A blood-chilling experience at Tawakkel gives a measure of Stein's stoicism. As the last place he "could pass in comparative comfort," he asked the local barber's help in "getting rid of a troublesome tooth." The scene is vivid. "This worthy first vainly tortured me with a forceps of the most primitive type, then grew nervous, and finally prayed hard to be spared further efforts. Perhaps—and Stein's explanation rings true—"he had lost confidence in his hands and instrument, since I had insisted on seeing them thoroughly cleaned with soap and water previous to the intended operation." Even this excruciating fiasco did not delay Stein's departure; he left, as planned on 12 December. Escorted out of the oasis by the beg, whose two attendants carried his falcons "as a sign of his dignity,"[3] Stein thanked the beg for his hospitality with a gift of some Russian ten-ruble gold pieces and at the same time insured the beg's goodwill in keeping communications open with the desert party.

An advance group under Kasim's guidance had orders to dig wells at suitable camping places and then to continue past Danda-Uiliq to the Keriya River to await Ram Singh.

The first night they camped at the river's edge so that before entering the sands the animals could have all they wanted to drink. Marching in the drift-sand where at each step the feet of men and animals sank into the soft surface was tiring, and to save the heavily laden camels from over-exertion the pace was slow. A day's march was rarely more than ten miles. "At first the tamarisk and 'Kumush' [a coarse grass] was plentiful . . . [but] it grew rare in the course of our second march, while the wild pop-lars, or 'Toghraks,' disappeared altogether as living trees. Luckily amidst the bare dunes there rose at intervals small conical hillocks thickly cov-ered with tamarisk scrub, the decayed roots of which supplied excellent fuel."[4] Close to these hillocks, hollows ten or fifteen feet deep, scooped out by the wind, promised the nearest way to subsoil water. In those Ka-sim's party had dug for water, reaching it at five to seven feet down, but it was bitter, almost undrinkable. As they moved farther away from the river, it became sweeter.

The bitter cold—at night the thermometer fell to 0° and even −10° Fahrenheit—iced the damp soil in the water holes. During the march, though the temperature stayed at the freezing point, the wind was still. "I could enjoy without discomfort the delightfully pure air of the desert and its repose which nothing living disturbs."[5] But notwithstanding, the cold posed problems: how, when asleep, to protect the head and still breathe? Pulling his fur coat over his head Stein breathed through the sleeve. He had to give up his habit of reading or writing at night because not even his Stormont-Murphy Arctic stove kept the ink from freezing. But his worst "domestic problem" was caused by "the tooth I had vainly endeavored to get rid of at Tawakkel [that] continued to cause trouble, and the neuralgic pains it gave me were never more exquisite than at night."[6]

The medicine in his kit needed a little water;[7] to get the water required melting the lump of ice in his aluminum cup over his candle. "The min-utes which passed until I had secured . . . the little quantity of liquid, were enough to benumb hands and fingers." After this his toothache is not mentioned again.

Stein's referring to these as "domestic problems" distinguishes per-sonal annoyances—and how limited they were!—from situations affect-ing the expedition. Thus, when on the fourth day, they met two messen-gers sent by Kasim to report that he had failed to locate the ruins, Stein turned to Turdi, who had mentioned casually that he thought they were traveling too far to the north. Whether because Turdi had approached Dandan-Uiliq from the Tawakkel side only once before, or whether from a "feeling of professional etiquette or pride," he had respected Kasim's guidance and kept silent. Now he questioned the two men closely: a few questions, a few answers, and Turdi knew the point Kasim had reached. He sent the men back with detailed instructions on how to proceed. "Old Turdi with the instinct bred by the roamings of some thirty years and perhaps also inherited—his father had followed the fortunes of a treasure-seeker's life before him—found his bearings even where the dead uniformity of the sand-dunes seemed to offer no possible landmark."

Guided by Turdi, they skirted the high dunes and reached a "belt of ground where dead trees were seen emerging from the heavy sand."[8] Like skeletons, stark and bleached, they stood, and Turdi named trees which had been planted, the *Terek* or poplar, the willow, and others: unmistakable proof that they had reached an area of ancient cultivation. Six days after leaving Tawakkel, "I found myself amidst the ruined houses which mark the site of Dandan-Uiliq."

In an area of low dunes, a mile and a half from north to south and three-quarters of a mile wide, were small, isolated clusters, the remains of buildings "modest in size but of manifest antiquity." Stumps of walls, layers of plaster on their timber framework, were occasionally exposed, but even where the houses were buried in sand their outlines could be discerned by the wooden posts sticking out above the dunes. All exposed structural remains "showed signs of having been 'explored' by treasure-seekers, and the marks of damage done . . . were only too evident." A quick inspection showed the remains of Buddhas and Bodhisattvas on the badly injured walls of some of the larger rooms, the ruins of places of worship. "Peculiarities in the style of the frescoes seemed to mark the last centuries preceding the introduction of Islam as the probable date when these shrines and the settlement to which they belonged had been deserted." Chinese copper coins, "picked up under my eyes from the debris-strewn ground near the buildings"[9] remnants of the Kai-yuen period, A.D. 713–741, confirmed his dating of the frescoes' iconography.

In all the times that Turdi had visited these desolate surroundings, the dunes had changed but little, and he recognized where he had previously searched. Only his scanty resources and the difficulty of carrying sufficient supplies had made a lengthy stay impossible, and all but the most exposed parts had been preserved. Anything deeply buried had not been touched. These—and Stein was grateful for "Turdi's excellent memory and topographical instinct which enabled him readily to indicate their position—were the places Stein chose to excavate. He located the camp-site near the main ruins to be explored: it enabled Stein to keep the men "at work as long as possible every day"[10] and minimized their tiring walks through the drift sands. The dead trees from what had been an orchard provided a plentiful supply of wood fairly close at hand. Once the living arrangements were settled and the baggage was unloaded, Stein had Ahmad take the camels to the Keriya River, a three-day march to the east, where they could regain their strength by grazing on the luxuriant beds of reeds. The donkeys were returned to Tawakkel. The next morning, 19 December, he began work.

Dandan-Uiliq was the classroom where Stein learned the grammar of the ancient sand-buried shrines and houses: their typical ground plans, construction, and ornamentation, their art, and something of their cultic practices. He also used it as a laboratory in which to find the techniques best suited to excavating ruins covered by sands as fluid as water, which, like water, trickled in almost as fast as the diggers bailed it out. He had no precedents to guide him, no labor force already trained in the cautions, objectives, and methods of archaeology. Here, too, he was without the notices left by Hsüan-tsang and Marco Polo: he was, when the work started, utterly on his own. He felt his way from what was easy to what was difficult, from what he knew he would find to discoveries he had not dared to anticipate. His approach was both cautious and experimental.

He began with a small square building which Turdi knew as a *but-khana,* a temple of idols. As yet undisturbed and a mere two or three feet under the sand, its excavation "provided useful, preliminary knowledge" of what would be a standard arrangement for a shrine: a square inner cella surrounded by a square outer wall. Between was a corridor which, according to Indian custom, allowed for the worshipper's ceremonial circumambulation. The walls of the corridor were covered with frescoes and painted low reliefs of larger-than-life-size images of Buddhas in the orthodox attitudes of teaching, with raised hand or seated in meditation. Though only the lowest portions of the walls remained, Stein could recognize "their unmistakable affinity to that style of Buddhist sculpture in India which developed under classical influences."[11] His first clearing gave Stein the information he needed "to direct the systematic excavation of structures more deeply buried," and also some "hundred and fifty pieces of stucco reliefs fit for transport." Having mastered the easy lesson, he next turned to some small buildings lying six to eight feet under sand, and "was able to gauge correctly their construction and character, though only the broken and bleached ends of posts were visible above the sand." As he surmised, the shrines were uniform in form and decoration.

Of a colossal painted-stucco statue in this second cella (a Buddha?)

only its feet, thirteen inches long, remained on an elaborately molded base; the rest—even the wooden framework that had supported the heavy image—had crumbled away. In each of the corners a stucco figure stood on a lotus-shaped pedestal. Both sides of the corridor were lavishly decorated with giant images of Buddhist saints, their feet supported by small divinities above a frieze of lotuses floating in water. Stein made a complete photographic record of the work as it progressed and of all the finds, identifying and labeling the specific place where each was recovered. He was able to remove and pack for safe transport a seated Buddha with an inscription beneath it. He was especially pleased to find that the robes of the Buddhas and Bodhisattvas were in the orthodox reddish-brown worn by Indian monks. The cella's outer wall had bands of small seated Buddhas and "rows of youth riding on horses or camels, each holding a cup in his outstretched right hand, while above the riders a bird, perhaps meant for a falcon, is swooping down on this offering."[12] This novel scene portrayed what Stein thought was a local sacred legend.

Elated by this art of a vanished world, he was also touched to find the broom used "by the last attendants . . . to keep the sacred space clear of the invading dust and sand,"[13] the same sand which "had been the means of preserving them in almost perfect condition."[14]

Repeatedly, Stein found Indian elements mixed with Central Asian scenes. Votive offerings, thin panels of wood placed against the main image within the cella, had figures from Buddhist mythology; there was even one of Ganesha, the popular elephant-headed Indian god. Similarly, a stucco statue of a man wearing the armor and dress of the artist's time and region—a knee-length coat of mail and the wide-top leather boots still worn locally—showed his feet firmly planted on the writhing body of a vanquished demon. The demon's thick hair, "elaborately worked spiral tufts, strongly reminded me of the treatment of the hair in many a sculpture of Graeco-Buddhist type familiar to me from the Lahore Museum."[15] Despite its Eastern garb, the figure represented a *yakshas,* one of the spirit-guardians of the gate popular in Buddhist mythology. One fresco was auspicious: it showed a teacher holding a closed *pothi*—a Sanskrit book whose leaves, strung together so as to keep them in order, were contained between boards of thin wood. (In Chinese Turkestan, the circular hole through which the string is passed is on the left side). Stein dared to hope that he would find such ancient manuscripts.

Next Stein chose to excavate a structure whose arrangement of wooden posts indicated a dwelling, while the bleached trunks of dead fruit-trees suggested a garden or orchard. The digging disclosed walls of what was clearly the bottom floor of a house. A "small scrap of paper showing a few Brahmi characters was found in the loose sand which filled the building."[16] To encourage the workers struggling to keep the sand from pouring back in, he offered a reward for the first man lucky enough to find a real manuscript. An hour later a shout announced "writing!" Stein himself freed the delicate paper, wiped it clean, and held "a perfectly preserved oblong leaf of paper,"[17] once part of a pothi. The page he held had "six lines of beautifully clear writing which . . . show Brahmi characters of the so-called Gupta type, but in a non-Indian language."

In quick succession other pages of Brahmi manuscripts appeared, some singly, some in little sets. Their paper, size, and writing indicated they belonged to three distinct pothis; all were Sanskrit texts of Buddhist scriptures. It became apparent that they had fallen from an upper story as the basement was filling up with sand. Stein reasoned that the sooner the fragments had fallen to the safety of the sand-covered basement, the more extensive the manuscripts would be. He watched with mounting eagerness as the men struggled to clear the floor. Soon two or more packets of leaves were found. Paleographic evidence gave a dating some time before the seventh century A.D. Stein had stumbled on what had been the library of a Buddhist monastery. The basement into which the manuscripts had fallen had been its kitchen.

Even as Stein was absorbed in the emerging manuscript fragments, a distant gunshot was heard. Turdi, closely watching the excavations by his side, immediately recognized that it announced Ram Singh's approach from the Keriya River. An hour later the surveyor appeared with his faithful cook. It was the afternoon of 23 December: "Considering the distances covered and the various incidents for which it was impossible to make proper allowance in our respective programmes, the rendezvous I had arranged for has been kept most punctually."[18] That same night, examining Ram Singh's work, Stein compared the position it indicated for Dandan-Uiliq with his own fixing of the site as being the easiest way to check the accuracy of their respective surveys. He was content when their independent surveys varied by half a mile in longitude and less than a mile in latitude. "If it is taken into account that since . . . Khotan, Ram Singh has brought his survey over approximately 500 miles (on which for the last 130 miles or so no intersections could be obtained owing to the absence of all prominent landmarks) while my own marches over 120 miles lay almost wholly through desert, this slight difference represents in reality a very striking agreement."[19]

Ram Singh, at all times reticent, had little to tell of the cold and scarcities he had endured surveying in the Kunlun mountains. Such conditions he understood and, for his work, willingly endured. But the desert: ah! that triggered what for him was a short outburst about the "weird desolation of the desert, the total absence of life of any kind among the high waves of sand he had crossed since leaving the banks of the frozen river."[20] Even his response to the Indian style of the painting and sculpture unearthed could not overcome the "uncanny feelings which these strange surroundings, pregnant with death and solitude, has roused" in the otherwise phlegmatic surveyor. He and his cook Jasvant Singh "complained bitterly of the water which our single brackish well yielded." The latter soon began showing signs of incipient scurvy, which Stein controlled by giving him lime juice brought from Gilgit.

A new, but not unexpected, element was added when a piece of a lacquered and painted wooden bowl gave evidence of the presence of Chinese at Dandan-Uiliq. Soon, confirming "a conjecture as to the dwellers of this room, the first finds of Chinese documents rewarded my search."[21] The first was a long slat of tamarisk wood with Chinese symbols written in vertical lines on both sides; next, almost stuck to the floor, a piece of thin water-lined paper folded into a narrow roll contained a petition for the recovery of a donkey leased out and not returned. The date it bore corresponded to 781 A.D.; the locality "may variously [be] read as Li-sieh, Lieh-sieh or Li-tsa." Other documents mentioned the name of the monastery: "hu-kuo, literally Country-protecting."[22] However mundane the contents, they indicated that the superintending priests of the monkish establishment were Chinese whereas the population that supported them was not. The triviality of the affairs recorded in the Chinese papers was the reason why they were left behind when the settlement was abandoned at the end of the eighth century. The light, flimsy papers, dated between 782 and 787 A.D., lying close to the floor proved "that the sand must have entered the room very soon after these petty records had been strewn about . . . for the little paper rolls could not have resisted very long the force of the storms that pass over the country each successive spring and autumn."[23]

In all Stein cleared and examined fourteen buildings scattered amid the dunes. A detailed, careful survey helped him reconstruct the life that had once flourished there. On 2 January 1901, he sent a brief message to the government of India. "The last weeks spent at this old site in the desert have brought interesting results in the form of ancient Sanskrit, Turki and Chinese Manuscripts, Buddhist paintings, etc. recovered from sand-buried ruins. I am keeping well though the cold is semi-Arctic just now and hope to do useful work elsewhere too."[24] He had no time to explain that the full meaning of his finds would come later with study and other excavations.

Writing thirty years later, Stein related Dandan-Uiliq to subsequent studies, placing this first site in a larger world. He did not change his opinion about the decoration, wall painting, and reliefs: they were "unmistakably derived from that Graeco-Buddhist art which flourished during the early centuries after Christ in the extreme northwest of India."[25] Though "the remains of decorative art in Buddhist shrines of distant Khotan are far removed in time . . . yet they reflect quite as clearly the impress of Hellenistic style."

The painted votive panels (later carefully cleaned at the British Museum) added other relationships. One, which Stein recognized when he first saw it, depicted the rat-headed divinity mentioned by Hsüan-tsang, the king of the sacred rats who saved the kingdom of Khotan. Another illustrated another story told by the Chinese pilgrim of a Chinese princess who brought to Khotan the first silkworm seeds from China, which had jealously prohibited their export. The clever lady who had hidden them in her headdress was subsequently deified in her adopted country for having founded their flourishing industry. The full interpretation waited on a careful matching of iconography, and legend. (Of the introduction of the silkworm to Europe, "Procopius says that certain Indian monks offered to bring it from 'Serindia'; Theophanes relates that the eggs of the moth were smuggled in a hollow cane by a Persian from the country of Seres. Hudson resolves the discrepancy by supposing the monks to have been Persian Nestorians . . . and that Serindia was . . . Khotan or Kashgar. Most writers have overlooked what seems to be the most interesting part of the problem, namely, who were the people who knew enough of the technology of silk growing, reeling, twisting, etc. to transplant the industry successfully from . . . Sinkiang to Syria and Lydia?")[26]

Other fields yielded only slowly to a full understanding and interpretation. There were the leaves of paper manuscripts and the little packages of folia which Stein recognized at once as being written in an early Indian Brahmi script but in an unknown language. Though the shape, the arrangement of the manuscripts, and their script were Indian in origin, the language was that spoken by the Khotan population. Old Khotanese was discovered to be Iranian, closely related to that spoken in ancient Bactria and elsewhere on the Middle Oxus during the early centuries of our era. At that early date, Buddhism had penetrated into that part of Eastern

Iran by way of the present Afghanistan. Armed with this linguistic insight, Stein felt it was clear that Indian Buddhism had absorbed Iranian elements on its way to the Tarim Basin.

This accretion of cultural elements was exemplified in a votive panel that showed "a powerful male, wholly Persian in physical appearance and style of dress, yet obviously intended for a Buddhist divinity."[27] His long, ruddy face, his heavy black beard are utterly alien to any sacred Buddhist figure, as is the martial look conveyed by his full, curving mustache and black bushy eyebrows. Above his head is a high golden crown like that worn by the Sasanian king of kings. "The body, narrow-waisted in keeping with the traditional Persian type of manly beauty, is dressed in a brocaded coat. Below this are shown the feet and legs encased in high black boot-tops. From the waist is suspended a short curving sword. From the neck descends a curling scarf which winds around the arm just as seen on Bodhisattva figures of Central Asia. As is so often among these figures, the divinity is shown with four arms. Of the emblems which three of them carry, two can be clearly recognized, and both of these, a drinking cup and a spearhead, are unmistakably secular."

To add to the mélange, the picture on the panel's other side is a thoroughly Indian type of three-headed demon. "The dark-blue flesh of the body, nude but for a tiger skin descending from the waist, the two bulls *couchant* below the crossed legs and the emblems carried in the four hands all suggest affinity to some Tantric divinity of India. The subject and style of this picture seemed so far removed from the Persian Bodhisattva . . . that to find any connection between the two was very puzzling." Fifteen years later Stein found the explanation.

Returning from his Third Central Asian Expedition (see pp. 389 ff), Stein visited Sistan, "a link between Western and Eastern Iran" that had attracted him "ever since the Old Iranian studies of my youth."[28] Famous as the home of Rustam, the hero of Persia's national epic, Sistan was, before Timur devastated it, known for its fertility. Buddhism had flourished there: Baladhuri, a ninth-century Arab historian, related how the conquering Muslims had destroyed "the great idol, Al-Budd,"[29] a statue of the Buddha. On 4 January 1916, a bare month after he had started his archaeological tour of this barren region, Stein wrote jubilantly: "I had the lucky chance of starting work with the discovery of a Buddhist sanctuary which yielded interesting frescoes, the first brought to light in Iran."[30]

As he probed in an imposing site, his sharp eyes discovered a large mural secreted behind a wall. It depicted "a scene of homage and offerings presented to a youthful male of martial bearing, seated in a dignified pose.

The upraised right arm carried a curving mace surmounted by an ox head,"[31] Rustam's *gurz* (mace), his recognized attribute. There, too, was the three-headed demon. The mural celebrated the homage Rustam received from one of the demonic adversaries he had overcome and forced into submission to his king. The late seventh-century Sistan fresco shows "the close connection in time with the remains of the Dandan-Uiliq shrines"[32] and suggests that it was the deified Iranian hero who was introduced into the local pantheon of Khotan. Language and legend, iconography and history: "The Chinese historical Annals tell that Chinese control over the Tarim Basin under the T'ang dynasty" finally came to an end about 791.[33] "The collapse of Chinese authority and the successful Tibetan invasion" spelled serious trouble for an outlying oasis like Dandan-Uiliq, "dependent on an irrigation system which in turn depended on a stable, watchful administration."

It would take time and wider tours to establish the great arc of continuity between Eastern Iran and the Tarim Basin.

"It was with mixed feelings that I said farewell to the silent sand-dunes amidst which I had worked for the last three weeks. They had yielded up enough to answer most of the questions which arise about the strange ruins which they have helped to preserve. . . . I had grown almost fond of their simple scenery. Dandan-Uiliq was to lapse once more into that solitude which for a thousand years had probably never been disturbed so long as during my visit. . . . The recollection of this fascinating site will ever suggest the bracing air and the unsullied peace and purity of the wintry desert."[34]

The long view remembers Dandan-Uiliq for much more: there, for the first time, a trained scholar with archaeological discipline had ventured into the Taklamakan to stay and dig and find the evidence that he had hypothesized. Stein had made a beginning in bringing the Tarim Basin's complex pre-Muslim past into the modern world of scholarship.

Between closing the work at Dandan-Uiliq (3 January) and beginning new excavations at Niya (28 January), there was an interlude. Stein supervised the careful packing of his fragile finds, paid and dismissed some of the Tawakkel workers, and, before starting for the Keriya River, decided to make a slight detour to investigate ruins Turdi had visited nine years before—Rawak, "High Mansion." Behind a long, high dune he found two badly decayed mounds, the remains of small stupas that showed the scars of repeated searches for treasure. From the scattered small debris Stein picked up a fragment of unusually hard stucco "on which the practised eye of Turdi at once discovered traces of a thin gold-layer"[35]; a few old Chinese coins "without legend, as issued under the Han Dynasty" indicated that Rawak was probably deserted long before Dandan-Uiliq. Next he started for the Keriya River.

After two days of marching across mighty waves of sand, the party reached easier ground where at a mere six feet down they found water, salty but welcome. The camels had a good drink for, laden as they were with the cases filled with precious finds and after the strenuous going over the sand ranges, they were tired. There is a hint of fatigue when Stein writes: "Finally from the top of the last huge ridge of sand the dark lines of trees fringing the Keriya Darya [River] came into view. Four miles more . . . and passing a last low bank of sand, I suddenly saw the glittering ice of the river before me."[36] In no time at all they found the ponies the Keriya amban had sent, a clear sign that Stein already had the cooperation of the amban whose "territory extends over some five degrees of latitude."[37] Stein was especially gratified and reassured: it was to prevent such official goodwill from the amban that Petrovsky had fumed and threatened.

From Keriya he wrote Hetty, 13 January 1901: "It was very pleasant to see vegetation again even in hibernation; I rode along the frozen river for four days and regretted not having any skates for the glittering ice. Six miles before the town I was given a solemn reception by various Begs and on my arrival I found a roomy house readied for me." Stein was settling into an inner room lit by the usual small square opening in the ceiling, made cozy by felts spread around and warmed by a blazing fire, when the amban's interpreter arrived bringing greetings and gifts, firewood, fodder, and food. "I have never been received so solemnly and so cordially by any Yamen. From the Amban on down, everyone [including the scarlet-attired executioner] wore their formal dress and our conversation despite the ineptness of the interpreter, was like one between old acquaintances." The amban, impressive-looking in his elegant Chinese silk dress, was in his forties; his face was that of a man of culture and of good-nature. When Stein spoke of his wish to visit an ancient site north of Niya mentioned by Hsüan-tsang (always the magic name), the amban promised to issue orders for whatever was needed. Relating the interview to Hetty, Stein added: "In my dealings with the Chinese, I feel the same interest in archaeology as I did in India with the Pandits."

During his stay at Keriya, Stein was busy with paper work, since, as he wrote with high innocence, "Dandan-Uiliq was no place for clerical work,"[38] he put his field notes in order, answered his voluminous mail, sent brief official reports, and even wrote a summary of his discoveries for the Royal Asiatic Society's *Journal*. Because Keriya, unlike Khotan, was fairly new, it had no established treasure-seeking fraternity, and when he started northward (18 January) for the famous shrine of Imam Jafar Sadik, it was in search of an unknown, untested site.

The oasis of Niya, a series of hamlets and villages strung out along the river, was probably the Ni-jang that Hsüan-tsang had visited, describing it as the place where "the king of Khotan makes the guard of his eastern frontier."[39] It remained the easternmost oasis in the Khotan district until the nineteenth century when Keriya became a separate administrative unit. The only antiquity Stein was shown at Niya was a huge pottery jar, nearly three feet in diameter, found some years before at the site he was to visit. And then, quite accidentally, Hassan, his camelman, sharp-witted and friendly, who knew his way around the bazaars, met a villager who had two inscribed wooden tablets. Stein was electrified when shown them: they had Kharoshthi writing "of a type which closely agreed with that prevailing during the period of Kushana rule in the first centuries of our era."[40]

Their owner had picked them up on the road to the shrine; their original finder, however, was a young villager, Ibrahim, who when searching for treasure the previous year had dug them out of a ruined house in the desert. To him these wooden tablets were without value—some he had thrown away on his way back, some he had given his children to play with. Until that moment Kharoshthi writing had been found in Central Asia only on the earliest Khotan coins (first and second centuries A.D.) and on the unique fragments of a birchbark codex acquired in 1892 by the ill-fated French traveler Dutreuil de Rhins. Stein immediately hired Ibrahim as a guide. Anticipation shortened the three days' march to the shrine of Imam Jafar Sadik, the jumping-off place for work in the desert.

The shrine itself was picturesque, but to Stein its main attraction was its little lake made by damming up the river before it disappeared. The lake, the nearest source for water, could easily supply their needs. It was for such a contingency that Stein had brought the tanks. The two from Calcutta had withstood the strain when their contents froze, but those fashioned out of tin cans in Kashgar leaked badly. Faced with the problem of having a regular and adequate water supply for forty to fifty people, Stein devised a brilliantly simple solution: utilizing the below-zero cold, he had cakes of ice cut from the river and transported in sacks and nets.

Where the river had once flowed, they followed a path through a fantastic jungle of dead trees and vigorous tamarisk growth to a small clearing. Stein greeted telltale pieces of broken pottery, signs of ancient habitations; soon came a fence of thickly packed rushes enclosing the fruit trees and planted poplars of an ancient farm; then two buildings constructed like those of Dandan-Uiliq, but larger, more elaborate, built more solidly. Their greater antiquity was signaled by pieces of wood carved in the same motifs as those in Greco-Buddhist shrines. He pitched camp in a location central to the scattered ruins to be explored. Wondering whether Ibrahim's story was true, wondering whether other documents would be forthcoming, Stein spent his first night at Niya.

Still prey to apprehension and distrust, Stein followed Ibrahim to the ruin from which he had fished the tablets. On the way he picked up three that had been discarded and noticed that, in the year since they had fallen down, the covering of drift sand was too slight to shield the writing against the sun. Clearing the rooms Ibrahim pointed out, the workers found more than a hundred tablets so well preserved that even as they were handed to him, Stein could make out their use and outward arrangement. Most were wedge-shaped, seven to fifteen inches long, originally fastened together in pairs. The writing was only on their inner surfaces; their outer surfaces bore a clay seal and served as an envlope. The brief entry near the seal could have been for classification, or for the name of the sender or recipient. The ink on the inner surfaces of those tablets still remaining in pairs was as fresh as if it had been just written. Only a very few were illegible. The first day's result, carefully packed and labeled, equaled "the aggregate of all the materials previously available for the study of Kharoshthi, whether in or outside India."[41]

His jubilation was tempered by questions. Did the short introductory formula mean they were copies of the same text—a prayer, an extract from Buddhist scriptures? Did the seals on the tablets indicate secular matters— letters, contracts, official documents? If the latter, Stein knew their historical importance "would be increased beyond all proportion."[42] He also knew from past experiences in deciphering Kharoshthi stone inscriptions how difficult the task was, a difficulty compounded by the tablets' cursive writing and the uncertainties about the language itself. In his tent, wrapped in furs, he made certain, before the intense cold drove him to his bed, that the language was an early Prakrit of the extreme Northwest of India. Within a few days he had mastered the opening formula: "His Highness the Maharaja writes (thus)," and he knew that the documents contained official orders. Here was the stuff to give fresh and unexpected insights into historical studies, studies so far enveloped in darkness. Convinced of the importance and extent of the site, Stein sent to the shrine to recruit additional workers.

One wing of that building had a small antechamber leading to a room twenty-six feet square that had a raised platform of plaster on three sides and gave the impression that it had served as the office of a local official. The protecting layer of sand was a mere two feet, but it had preserved the three score tablets found on the platform, some in small, closely packed heaps, just as the last occupants had left them. The tablets were of many shapes and sizes. The wedge-shaped kind were outnumbered by oblong wooden boards, some thirty inches long. Most of the latter "showed plainly by the irregular arrangement of writing in small columns often ending with numerical figures; by the appearance of various handwritings, erasures, etc., that they did not contain texts or even connected communications, but in all probability memoranda, accounts, drafts and other casual records."[43]

The appearance of wooden tablets without a single scrap of paper was in itself significant. Paper, invented in China and used there by A.D. 105, had not yet spread to Eastern Turkestan. Wooden tablets are "mentioned in very early Indian texts, particularly Buddhist ones; and it is easy to realize that their use recommended itself in a country like Turkestan which produces neither palm-leaves nor birch-bark, the other ancient writing materials known in India."[44] The fact that the Prakrit documents were the "first specimens ever discovered of Indian records on wood," was immensely gratifying.

Again and again Stein had occasion to notice the desert's destructive as well as protective power. As he struggled to clear down to the floor, a northeast wind blew, light yet sufficiently strong "to drive before it a light spray of sand" which, passing over the tablets, "threatened to efface the pencil figures which I wrote with half-numbed fingers on the often soft wood of their surface to mark the succession of the finds." The capricious wind, undermining the ground on which the buildings stood, toppled them leaving only the heavy timbers. "Ultimately the wood, rendered brittle by long exposure, breaks up into splinters which the winds easily scatter, and only potsherds and small fragments of stone or metal remain to indicate the place of ancient habitations."[45]

From one room that had a scant six inches of sand, Stein retrieved a mousetrap and boot last as well as tablets on which were narrow, closely written lists of names—account items, obviously records from an office. The extent of the record-keeping could be measured, Stein thought, by the size of one oblong tablet seven and a half feet long and four inches wide. (Unfortunately, the writing had been bleached out.) In trying to reconstruct Niya as it had been, Stein made floor plans of the houses and cattle sheds. In one dwelling all that remained was its ice pit: two poplar trunks placed side by side had lifted the ice above the ground while a thick covering of leaves insulated it from the summer's heat—a method still practiced by the well-to-do villagers. The walls of another large hall had a design of large floral scrolls and elegant festoons of budding lotuses. Its four, handsomely carved 40-foot beams still were in place, supported by deep sand. Beams, frescoes, and the remains of embers on a hearth: nothing else, the last inhabitants had stripped the hall bare.

Stein was ahead of his time in not being satisfied with discovering only ancient documents and art: he wanted to know everything about this community. Only two years before Augustus Pitt-Rivers, who with Flinders Petrie is credited with creating a revolution in archaeology, had written: "Common things are of more importance than particular things, because they are more prevalent."[46] For Stein the emphasis was not wholly on the prevalence of common things but on their relevance. In this conviction he made surveys and precise layouts of residences, shrines, offices and halls, kitchens and ice pits; he noted the arrangement and kinds of trees in gardens and arbors; he collected furnishings and artifacts—whatever might be useful in constructing the daily life of a vanished world. Stein did not articulate the concept that archaeology is a historical discipline, the history of culture, but the concept animated his work.

Here and there they found oddments; snippets of cloth, felts, and even a piece of a rug worked in complex geometrical figures; pieces of ivory, a few objects of kitchen equipment, a bow and shield of wood, spindles, and even "a stout walking stick of applewood that I found to come in very handy."[47] The excellence of the ancient woodworking skill distinguished two elaborately carved chairs. Their motifs were the same as Gandharan and synchronized with the date of the Kharoshthi script and with the Chinese coins of the second Han dynasty. The meagerness of everyday objects made it apparent that the last inhabitants had cleared away "everything possessing intrinsic value or still fit for use."[48] Walking on what once had been a little country lane, identifying the withered fruit trees—the peach, plum, apricot, and mulberry—in orchards once rich with ripe fruits, Stein was possessed of a "strange feeling, obliterating almost the sense of time."[49]

Suddenly, a mindless spasm threatened to destroy the life and purpose of his search. Involved were three men of the permanent staff hired in Kashgar: Niaz Akhun, the flamboyant Chinese interpreter, Hassan

Akhun, one of the young camelmen, and Sadak Akhun, Stein's cook, whose culinary competence had been acquired at Macartney's agency.

The trouble began with Niaz. Having assessed Niaz's talent for avoiding useful work and his inability to adjust to the desert regime, Stein had left him amply provided for at the shrine to look after the ponies. He hoped that, with no one to fight or bully, Niaz would not cause trouble. But he underrated Niaz. The men bringing the ice reported that Niaz was demanding food from the shrine's officials; the next convoy told of Niaz's "amorous demands" on the women, demands "in excess of what the hospitality and easy morals of those parts would tolerate."[50] The third convoy brought a fervent petition that the shrine be relieved of Niaz. Then Stein acted: the shrine's goodwill was essential to his party's well being. He ordered Niaz to join the camp at once, leaving care of the ponies to a man sent for that purpose.

Stein tells the story. "The sinner arrived, weary with the two days' tramp through the sand to which he was little accustomed, yet in his genuine dejection acting the part of injured innocence. According to his story he was the victim of a wicked conspiracy. . . . In order to make his appeal for justice still more impressive, he had donned over his comfortable coat, white rags to indicate mourning." He had just received news of his mother's death in distant Aksu—but could not explain how the news had come. Overcome by shame at being recalled, he offered to commit suicide. But he grew calmer and seemed ready to accept the privations of desert life.

The next day the storm broke.

Having brooded, Niaz blamed Hassan and accused him of having secreted a gold ring picked up near a ruin. This was a violation of Stein's strict order that all chance finds must be reported to permit Stein to purchase them at a fair price. When Hassan came with the next ice transport, he readily handed over the ring and took the reward. (The ring was brass, not gold.) This angered Hassan. A fight started. Maddened by his disgrace, intent on vengeance, Niaz pulled out a knife. Stein separated the two using the antique walking stick to thrash both. Just at the point where Stein had succeeded with Ram Singh's help, Sadak, wildly excited, rushed in brandishing the sword that he proudly carried as a symbol of having served as guard at the British agency. Stein thought he had run amuck: his conduct had grown increasingly strange from *charas,* a drug to which he was addicted. Violently, he protested his misery and his wish to avenge it on the renegade Niaz; finally he allowed himself to be disarmed and led away.

While Stein was paying attention to Sadaz, Niaz made a desperate attempt to strangle himself. The "strange look [on his face] by the time we succeeded in loosening the convulsive grip of his hands"[51] and his utter exhaustion convinced Stein that this was no playacting. To prevent further violence, Niaz was kept away from the other Muslims: Ram Singh and Jasvant offered to share their campfire with him and keep watch over him. Hassan was sentenced to a number of lashes which Ibrahim "administered with an arm practised in such functions and which had a very salutary effect upon the young offender." Sadak remained a cause for worry with his fits of depression and his wild talk of running away at night. Poor Sadak: fear of getting lost prevented his escape. (When, finally, Stein brought Sadak back to Kashgar, he felt he could compliment himself.) Sadak, on his side, attributed the trouble he had caused "to the 'Jins' or evil spirits of the sand-buried ruins who had gained possession of his mind during the long nights in the desert."[52]

Quiet returned.

Work continued. As ruin after ruin was excavated, Stein realized that Niya had been cleared of everything its inhabitants thought worth taking. Somewhere they must have had rubbish heaps where they threw things no longer wanted. Trying to find them in that world of drifting sand seemed as impossible as finding a needle in a haystack. On one of his walks he noticed a ruin, in no way special, which had some bleached tablets lying exposed. On a chance he started digging. Was it intuition or luck? Within half an hour thirty inscribed tablets were found, including two novelties: one, a sliver of wood with Chinese characters, the other a fragment of well-prepared leather with Kharoshthi writing giving a date. All this was welcome, but nothing about the half-broken walls suggested Stein had found a Golconda. Continuing to excavate this rich mine, Stein realized he had inadvertently stumbled on a rubbish heap, "layer upon layer of wooden tablets mixed up with refuse of all kinds."[53] Hundreds of wooden documents were bedded down in a four-foot-high consolidated mass of refuse—bits of broken pottery, straw, scraps of felt, cloth, and leather, and other less savory refuse—which had escaped the destructive power of sun and wind. Enveloped in dust raised by a brisk wind from the rubbish, Stein marked and tabulated each tablet, "for three long working days I had to inhale the odors of this antique dirt and litter, still pungent after so many centuries"[54] and "swallow in liberal doses antique microbes luckily now dead."[55]

From the rubbish heap came two dozen complete Kharoshthi documents written on smooth, carefully prepared sheepskin—the first examples of writing on leather by people of Indian language and culture; the usual religious scruples did not prevent these pious Buddhists from using animal skins. It was from the great number of wedge-shaped wooden tablets that Stein understood how this form of communication worked. The tablets came in pairs, exactly fitted to each other; one end was squared off, the other tapered to a point where close to the tip a string hole was drilled through both pieces. The message always placed on the smooth obverse of the lower tablet was protected by the upper, or envelope, tablet. A hemp string passed through the holes in the pointed ends was drawn taut by means of grooves cut into the seal socket located at the square ends. When the socket was filled with clay, it covered the string, and when the sender's seal was stamped into the still wet clay, the tablets were firmly fixed and identified. The message could not be read without breaking the seal and cutting the string: "unauthorizd inspection of the communication was absolutely guarded against."[56]

On subsequent expeditions Stein would find this wooden stationery in sites far to the east, clear proof that "those ingenious devices originated in China."[57] Wooden stationery became obsolete with the invention of paper; (Stein was to find the earliest specimen dated A.D. 105), but paper had not yet reached Niya, though the place was not abandoned until the second half of the third century A.D.

Just as the wooden tablets originated in China, so the figures on the clay seal-impressions, still found intact on a number of letters, had their origins in the classical art of the West. One had a "Pallas Athene, with aegis and thunder-bolt, treated in an archaic fashion,"[58] while an Eros standing, an Eros seated, and a Heracles were in the style of "Hellenistic or Roman work of the first centuries of our era." Most happily, one covering tablet had two seals side by side: the one with Chinese lapidary characters belonged to the Chinese official, the other, a portrait head, was unmistakably cut after a Western model. Stein wondered, as he had earlier when looking at the engraved stones found in the debris of Yotkan, whether the classical seals had been engraved in Khotan or imported from Gandhara. Some day he would get his answer.

For the moment he was content. He had secured tangible proof that the classical art that had established itself on the Northwest Frontier of India had made its way into Chinese Turkestan. The extraordinary items recovered from the Niya rubbish heap, however small, were eloquent witnesses to the meeting of influences of the Far East and Far West. Their existence seemed to echo "the old local tradition, recorded by Hsüan-tsang, . . . that the territory of Khotan was conquered and colonized about two centuries before our era by Indian immigrants from Takshasila, the Taxila of the Greeks, in the extreme northwestern corner of the Punjab."[59] The evidence from other ruins holding other rubbish heaps, made Stein certain that the end of the third century A.D. was the terminal date for Niya and other similar settlements in the Khotan territory, settlements that were doomed when Chinese authority was withdrawn from those parts.

Stein closed his work at Niya having examined every ruin traceable under the sand, and on 13 February he started back to the shrine of Imam Jafar Sadik. Saying adieu was hard; "Where will it be next that I can walk amidst poplars and fruit trees planted when the Caesars still ruled in Rome and the knowledge of Greek writing had barely vanished on the Indus?"[60] He felt that somehow he had not exhausted the riches of Niya, a feeling transformed into fact when, on taking a different route back to the shrine, the party stumbled on a group of ruins previously hidden from sight by dunes of formidable height. Stein solemnly promised himself that his "farewell at this time should not be final."[61]

Niya was the second of the sites whose documents, significant in quantity and variety, Stein recovered from their protective coverings of sand and rubbish and made available to scholars. They were to be the materials for studies in many disciplines. His, he knew, was but the first step in the research that would continue far into the future and in places far from the Taklamakan. His role, as he saw it, was to provide substantial nourishment for generations of scholars. At Yotkan and at Dandan-Uiliq he had secured the proof for the "discovery" projected in his proposal. Niya, too; but Niya would become the raison d'être for future archaeological exploration. The riches of Niya increased Stein's appetite to explore other sites from the Imam Jafar Sadik base: he had been told of other ancient ruins still further to the east along the Endere stream. Already dust storms were signaling that the deliciously clear days were numbered; he wanted in the short time remaining to spread his net as far as possible.

9

It is easy in reading Stein to overlook an element that characterizes this initial venture into the Taklamakan: he imposed a schedule for work where the amount to be done was unknown, at sites of questionable existence and location, and in a region where communications moved at the pace of a camel or pony or the measured walk of the dak runners. There was no telephone to coordinate last-minute plans. It was a world whose work patterns were dictated by the sun's movements and the seasons' changes and not in obedience to the petty tyranny of the clock. And yet Stein and Ram Singh kept their several rendezvous without fail and punctually. That he could do the same with locally recruited workers dispatched by local officials seems almost miraculous. Stein always called it a "welcome surprise" when he received a sack of mail somewhere in the vastness of the Taklamakan. It happened again and again.

Back at the shrine of Imam Sadik, he was lodged in the hospice maintained for pilgrims where "for one brief afternoon [he] enjoyed the cheerful warmth of a fireplace and indulged in that long-desired luxury, a thorough 'tub.' "[1] His letters sent from there began the long journey back to Europe and India with the first notice of his exciting discoveries.

Rather than spend the time required by the long, roundabout road via the Niya oasis, Stein decided to cut across the desert straight east. No such approach existed, he was told, but persistent questioning finally won the admission that a shepherd had visited flocks grazing on the Yartungaz stream, which flowed between the Niya and the Endere Rivers. Soon one of the shrine's devotees, who was considered half-crazy, claimed he had visited those ruins. With both men as guides, Stein started eastward. Two days later, 17 February, after tiring marches across 150-foot sand dunes, they reached the icebound Yartungaz stream where a little oasis gave rest and water to the weary animals. A handsome, older man who welcomed Stein was proud to show that he knew a little Persian; he was the son of a pilgrim who had come to Imam Jafir from distant Badakhshan, that mountainous, eastward-pointing finger of Afghan territory along the Oxus headwaters, and had received a grant of land in that forlorn little desert outpost.

The Yartungaz, the old man explained, had recently shifted so that the main channel now flowed four miles west of the colony. Anticipating greater displacement and also preparing for the possibility that the old irrigation channel might fail completely, the cultivators had started fresh fields in the newer delta. Such vagaries of the desert streams were of interest to Stein: what was happening in the present helped one to understand what had happened in the past. The Endere was also shifting westward. Some time before, the river had moved eastward, and the little colony that had formed along what was then the "new river" was returning to its fields along the old channel. These constant changes in the rivers' final surrender to the desert sands would account for the series of dry depressions the party had crossed since leaving Imam Jafir.

From the lonely shepherd station, "where the black goat sat," they marched across low dunes to the ruin. Stein recognized it as a decayed stupa, rifled long ago; the houses said to have existed there were indicated by the familiar rows of wooden posts sticking up out of the sand. At Endere, "the high brick walls of some large buildings, and the remains of a massive rampart encircling the ruins, presented a novel feature."[2] And just then, as they reached the site, *mirabile dictu* (as Stein might have said), he greeted the arrival of a new contingent of twenty workmen. "Considering the great distance, some 120 miles, from which the men had been brought, and the difficulty of communicating with them over wholly uninhabited ground, I felt not a little pleased at this well-managed concentration which enabled me to start work at once."

Digging in one corner revealed the two concentric squares of the Dandan-Uiliq model, and within an hour pieces of sculptured stucco, also reminiscent of Dandan-Uiliq, came to light. Then came parts of paper manuscripts, a Sanskrit text in cursive Brahmi script, but again in the non-Indian language. On the cella's other side were Tibetan Buddhist documents, written on a peculiar paper easily recognized by its toughness and yellowish color. (Subsequent study claimed them as the "oldest so far known of that script and language.")[3] Tibetan graffiti were scratched on the stucco walls; amidst them was a Chinese one giving a date corresponding either to A.D. 719 or A.D. 791. Nothing in the buildings within the ramparts shed light on the Endere site. Were the ruins "the quarters of a well-to-do monastic establishment which found it advisable to protect itself by walls and ramparts? Or did they mark a fortified frontier-post which sheltered also a Buddhist temple?"[4] The decoding of the nature of the ruins was overshadowed by a more vexing problem. The always reliable Hsüan-tsang had passed no settlement on his desert march from Niya eastward to Charchan; but he did mention hearing about some *abandoned ruins* whose location corresponded exactly with Endere's. That was A.D. 645. How was Stein to reconcile his patron saints ruins with the sure evidence of the Tibetan presence there more than a century later? The Chinese Annals told of a Tibetan invasion in the second half of the eighth century that had destroyed the Chinese authority. How, then, to explain the Chinese graffiti with its date?

Stein was unhappy to leave such a mystery unsolved. Endere, like Niya, was filed away to be pursued on his next expedition: next time he would stay longer. This thought had to console him when he left the ruins, 26 February, for a rapid march back to the oasis of Niya.

The warm sunshine of the last few days was beginning to free the Endere from its icy covering; crossing it was hazardous—the heavily loaded camels were led across one by one. Turning southward through a jungle of wild poplars and a plain of coarse grass, they could see the prominent peaks of the Kunlun Mountains, sixty to eighty miles away. At midday they vanished behind a curtain of dust lashed by a strong north wind. So they reached the tracks connecting Keriya with Charchan, "once the great line of communication to China, but now a lonely desert track with practically no traffic."[5] Camp that night was a miserable one: scanty fuel, a scanty supply of brackish water, and a wait until midnight for the camels to appear. "Dinner was accordingly an affair of the small hours of the morning." On the way back to the Yartungaz stream the party met Stein's pony-man from Khotan who was "bringing mails and sadly wanted articles from the stores" left there. Receiving home mail was always a holiday; this time it was doubly celebrated because Stein had "time to sit down by the roadside and pore peacefully over its contents. . . . A look over the *Weekly Times,* nearly three months old, put me again in contact with the affairs of the far-off West and East."

Skirting the desert en route to the Niya River, the road passed spring-fed ponds, marshes, and reed-covered lagoons. On the last day of February, "after a ride over much boggy ground, I once again entered the little oasis. In the twilight it seemed like a return to civilization."[6] Since the middle of January, when he left the oasis for the shrine of Imam Jafir Sadik, he had, in addition to his excavations, completed a large oval loop of over 300 miles. Leaving the slow-moving caravan to follow, Stein pushed on ahead, covering the eighty miles to Keriya in two days. He needed time to write reports and letters, rearrange the baggage, discard winter clothing, enlist new workers, and secure additional supplies.

The day after his arrival, the amban, who had been away on a "little tour of criminal investigation," returned. Stein immediately called on him to thank him for his assistance and, from the stores forwarded from Khotan, to present him with the best of his tinned goods. His visit was returned immediately. "I was able to satisfy his curiosity about my finds with specimens of ancient tablets, etc. With the historical sense which all educated Chinese seem to possess, he at once rightly surmised that the use of wood . . . indicated a period corresponding to that when split bamboos were employed in China prior to the invention of paper." The amban took a certain pleasure in repeating the wild rumors that were the talk of the bazaars as far away as Khotan: that Stein had found coffers heaped with gold. Together they smiled that the bazaar-mind had a single definition for treasure and could not understand or appreciate what Stein was seeking and what he had found: theirs was a mutual confidence and respect.

Karadong, the black hillocks 150 miles down the Keriya River and then ten miles or so into the desert, was the next objective. Turdi knew the place; he considered the ruins unimportant, but, because Sven Hedin had called it an "ancient city," Stein felt it his duty to examine the site. Not to lose too much time in covering so great a distance, they pushed on by forced marches. By 10 March they were at the place on the Keriya where they had first struck it on the overland march from Dandan-Uiliq. Then it had been a sheet of glassy ice; now a muddy current rolled by. Spring had come.

They made their way over vast fields of *kumush,* the coarse yellow grass that looked like ripened grain, and came upon a small colony of shepherds. Here Stein found two guides. As they were approaching the site Stein had his taste of a real *buran.* First the sand merely peppered the air; then the mounting force of the wind made it impossible to see even a hundred yards ahead. "With the sand driving into my face and accumulating under the eyelashes in spite of the goggles, it was difficult to see much of the route. . . . The sand dunes rose in height. After plodding among them for another hour, our guides declared we were near Karadong."[7] While they went ahead to locate the ruins, the party sat in the lee of a high tamarisk-covered cone watching "how the sand was driven in a thick spray over the crestline of the dunes, just as if they were storm-tossed waves."[8]

Karadong was different. Immediately noticeable was the complete absence of stumps of white poplars cultivated everywhere for use in building; there were only the same wild poplars as grew in the riverine jungle —dead. Pottery fragments were rare and limited in area. "To talk of an 'ancient city' here would imply more imagination than an archaeologist need care to take credit for."[9] What Stein saw was a badly damaged quadrangle, 235 feet square, on whose mud ramparts were timber-built rooms; within were the ruins of a structure. What purpose had Karadong served, situated "in the narrow belt of forest land between the desert and the river"? Its location and plan suggested that it might once have been a frontier post or a roadside resthouse. But why there? Stein was reminded that a Moghul historian had suggested that, even as late as the sixteenth century, the Keriya River had reached the Tarim, thus connecting Khotan directly with the important oasis of Kucha on the Taklamakan's northern rim. Karadong would have been about halfway between.

The principal discovery after two days of hard digging was "a little store of remarkably well-preserved cereals [including] a couple of pounds of 'Tarigh,' a kind of pulse still cultivated around Keriya. . . . I had a small quantity of the 'Tarigh' boiled and found the ancient porridge made itself useful for glueing envelopes."[10] A second visitation of a buran made the inside of the tent as unpleasant as the atmosphere outside. Stein felt "heartily glad when by the evening of the 17th March our work at this desolate spot was concluded."

The next day he received mail sent on from Keriya. A note from home and a "communication from Mr. Macartney based on Russian intelligence, informed me of the death of our Queen-Empress. I could see that my two Indian followers to whom I communicated the news, understood, and in their own way shared the deep emotion which filled me."

"The season of Burans had now fully set, and the gales that were blowing daily [were] from different quarters and of varying degrees of violence."[11] The month from 18 March to 19 April was, despite its thorough and careful scheduling, one of frantic activity carried on in conditions of mounting hardship. To the sand set in motion, "was added the trying sensation of glare and heat all through the daytime. The sun beat down with remarkable intensity through the yellowish dust-haze and the reflection of its rays by every glittering particle made the heat appear greater than it really was." With the setting of the sun the thermometer plunged: the men suffered from agues and fevers. "Luckily the chills I caught freely could be kept in check by liberal doses of quinine until my work at those fascinating ruins was done."

Southward from Karadong, Stein went in search of the site of Pi-mo. Hsüan-tsang had visited the town on his way from Khotan to Niya, and, most likely, it was the settlement that Marco Polo had also stopped at and called Pien. The Keriya amban had suggested it might be in the desert beyond the oasis of Gulakhma, some thirty miles west of Keriya on the Khotan road. Going there Stein saw a peculiarity of the Keriya district—hundreds of square miles covered by kumush. In the absence of any rainfall, these savannahs could only have been nourished by subsoil water percolating from the district Kunlun. A later traveler thought the oases between Khotan and Keriya "owe their existence to this water, which appears in frequent springs after an underground passage of forty or fifty miles beneath the belt of absolutely barren desert between the main road and the villages at the foot of the Kunlun."[12]

So far, the shepherds who had guided Stein at various times had been competent. But the two sent by the beg of Gulakhma to guide the party to what Stein hoped would be Pi-mo looked unusually stupid. He discovered too late that they were also too timid to admit they had not the slightest idea of where he wanted to go. The search for Pi-mo was led by the offspring of stupidity mated to timidity. First the party got mired down in an extensive "boggy marsh, treacherously covered with light sand,"[13] from which they extricated their animals with difficulty. Next a day was spent plodding "through a maze of tamarisk-covered sand-cones"[14] from which happily they emerged on the cultivated ground of a village, recently settled by people from Domoko, the main oasis on the Khotan-Keriya road. There Turdi, who had been on mail duty, rejoined the party; Stein hoped that he would be able to organize whatever local knowledge the guides had. For safety's sake he had the water tanks filled.

Then followed a three-day odyssey through overgrown fields crisscrossed by dry irrigation canals belonging to the abandoned homesteads of "Old Domoko." Again Stein saw how whole communities had shifted within the memory of living men and that "such shifts of the cultivated land, backwards and forwards, had recurred repeatedly."[15] Unlike Dandan-Uiliq and Niya, "Old Domoko" had no sand to cover and protect its deserted villages. From there the luckless guides led them aimlessly out into the desert, and, when weariness of men and animals forced them to pitch camp, one of the shepherds deserted during the night. The other, under Turdi's watchful eye and tutelage, led them to Uzan-tati, the Distant tati, a large area covered with potsherds and small debris. Destroyed by the erosion of the wind and generations of treasure seekers, it made any attempt at excavation useless. But its location coincided with the location Hsüan-tsang had given for Pi-mo, and a Chinese copper coin of the Southern Sung dynasty (1127–1278) which Stein picked up gave him the proof that it had been occupied up to Marco Polo's time.

Stein had a further demonstration—if he had needed one—of "the serious difficulties which must always attend a search for scanty ruins hidden among deceptive sand dunes if made without adequate guidance."[16] Looking for another site known to be close to Uzun-tati, the party spent two days of exhausting marching and countermarching over twenty-five miles of deep sand to find the site a bare three miles from where they had started. The fierce heat and glare aggravated the hardships, and the men suffered cruelly from thirst barely alleviated by the limited water carried in the tanks. Everyone was relieved when they reached the oasis of Gulakhma (29 March). This oasis, with the neighboring one of Domoko, "is undoubtedly the modern representative of the Pi-mo oasis."[17] The "young green of cultivated fields and orchards"[18] was a clear signal that Stein dare not permit his caravan to remain there even for the short rest its members deserved. The main party under Ram Singh started by easy stages for Khotan; Stein himself, traveling as lightly as he could, "hurried back to Keriya to bid farewell personally to its kindly Amban."

Spring, delicate but invincible, held the land. Among the presents for the Keriya amban Stein included personal souvenirs. On his farewell visit he commended the excellent services of Ibrahim (different from Ibrahim, the Niya villager) whom the amban had sent to Imam Jafir with fresh ponies. Stein hoped the amban would laud him publicly and promise him a "comfortable berth and good emoluments. It was already well known at Keriya that the Amban of Khotan, on my recommendation, had provided Islam Beg, for similar good services in the Khotan district, with a fat Begship at Karakash, and the Amban of Keriya might well feel encouraged to follow the lead of his pious colleague."

His gratitude expressed, his farewells said, Stein returned by forced marches to Khotan. At the oasis of Chira, his campsite was "in a terraced orchard, where the white blossoms of the plum-tree covered the ground like fresh snow, while the air was scented with their perfume."[19] After a sweet repose, a forty-mile ride across a dreary plain of sand and pebbles was made longer and drearier by the strong dust storm that blew all day.

On the fourth day, with Turdi guiding him, he made a cursory examination of a ruined site: the remains of a stupa, it showed no feature of special interest. At the edge of the Yurungkash district, "Islam Beg, with the emblem of his new dignity,"[20] Badruddin Khan, and a group of local begs welcomed him to Khotan territory. From there he sent Ibrahim, the always trusty emissary, back to Keriya "with a liberal reward in glittering roubles for himself and an ample supply of specially desired medicines for the Amban."[21]

The next morning the caravan was on the march. It was to make a wide detour: in the brief time remaining for work, Stein wanted to see Ak-sipil, the White Walls, in the desert north and east of Khotan. Treasure seekers had recommended its ruins. After miles of hard going over dunes remarkably steep and high, they came to ground where, despite erosion, traces of ancient fields could be made out. Erosion and treasure seekers had done their work: segments of the rampart and parapet of an ancient fort were all that remained. Whatever had once been contained within the wall had disappeared, but from the debris Stein picked up Han-period coins, seals, and small stucco reliefs of an unusual hardness whose modeling, far superior to that at Dandan-Uiliq and Endere, recalled to Stein "the best products of Graeco-Buddhist sculpture in Gandhara."[22]

Although the ruins at Ak-sipil were disappointing, Stein felt it his duty to visit another Rawak, the "High Mansion," distinct from the shrine examined near Dandan-Uiliq. Turdi had casually mentioned an "old house." Neither his words nor his manner of speaking prepared Stein for so stunning a climax to his archaeological season. Its importance cried out even as he approached the large stupa and saw, enormous stucco heads, strewn around where treasure seekers had left them. Much of its quadrangle lay buried, but one exposed corner was sufficient to indicate the shrine's massive proportions. Rawak was an exciting, glorious discovery.

Everything favored him: within two miles of the stupa a well provided ample water, an ideal place for the workers' camp. He sent back an urgent request for additional men. From 11 April, when he started the excavations, until 19 April, when he headed back to Khotan, work went on under the irritating sand-drizzle of the burans. The clearing away of seven feet of sand disclosed the quadrangle's inner walls against which stood a row of gigantic stucco statues. While waiting for the reinforcements to shift great masses of sand, Stein's dozen workmen cleared enough for a preliminary survey. Rawak's measurements were impressive: the stupa court was 164 feet long and 143 feet wide; its enclosing wall, three feet thick, had been more than eleven feet high; the stupa rose in two stories to its dome and "bold projections on each face, originally supporting well-proportioned flights of stairs"[23] gave it the "shape of a symmetrically developed cross." From pieces of plaster adhering to the molding at its base, Stein imagined it as it had once been when sheathed in a thick coat of white stucco. Copper coins of the Han period, deposited as a votive offering, were witness of Rawak's venerable age.

Not the shrine's age, nor its size, not even its unique stairways mounting majestically from ground level to the dome were Rawak's greatest splendor; it was the presence of colossal stucco statues which lined the walls. Bits of paint still adhering to the draperies of the Buddhas and Bodhisattvas showed they had once been divinities and saints, while elaborate halos formed patterns of holiness. The excavation was difficult, treacherous. Moisture from the subsoil water had rotted the statutes' supports which, when the protecting sand was removed, threatened to collapse. As soon as the lowest portions of the statue had been cleared and photographed, they were immediately covered up to secure them. In some instances the men held the upper portions with stout ropes for the few moments required while their pictures were taken. Despite the extreme care taken, some damage occurred. Stein now understood that when the wind swept the sand away from the heads, they had fallen; he found no evidence "of wilfull destruction by human agency"[24]—no devout Muslim had smashed the human likeness—which led him to believe that "this great shrine was already long deserted and the ruins of its court covered up by the time Islam finally annexed Khotan."

In that week Stein briefly uncovered 91 large statues on 300 feet of the portions of the walls cleared. "The position of all statues was carefully shown on the ground plan and a detailed description of every piece of sculpture, with exact measurements, duly recorded. In addition I obtained a complete series of photographs of whatever sculptural work appeared on the excavated wall faces." To do this he spent every day from dawn until nightfall in the trenches "in atmospheric conditions trying alike to eyes, throat and lungs."[25]

In *Ancient Khotan*, from the mass of material collected, Stein later analyzed the sculptural art of ancient Khotan. Its style and arrangement adhered closely to that of the Buddhist shrines of the Peshawar Valley. And yet there were subtle, regional idioms, as for instance in the two figures of the Guardians of the Gates. At Rawak, these quasi-secular figures which Indian convention places at the entrances of all assemblies, whether real or mythical, wore Khotanese dress, including the wide, bulging trousers tucked in capacious boot tops that still showed their dark red color. What, perhaps was most satisfying to him, was that at Rawak he was still on the tracks of Hsüan-tsang: the "remains of gold-leaf stuck originally in small, square patches to the left knee of a colossal image"[26] tallied with his patron saint's description of a giant miracle-working Buddha at Pi-mo: "Those who have any disease, according to the part affected, cover the corresponding place on the statue with gold-leaf and forthwith they are healed."

Rawak had to be returned to the deep sleep it had so long enjoyed. It would have been an act of vandalism to try to move the statues; they would have crumbled into meaningless piles of stucco. "All that could be done in the case of these large sculptures was to bury them again safely in the sand . . . and to trust that they would rest undisturbed under their protecting cover—until that time, still distant it seems, when Khotan shall have its own local museum."[27] Small reliefs and pieces already detached were brought away. Stein could congratulate himself that the care and labor he had put into packing these friable objects brought them to London, transported by camels, ponies, railway, and steamer without any serious damage. The trenches were carefully filled in again. "It was a melancholy duty to perform, strangely reminding me of a true burial."[28]

There was nothing to detain him at a barren, debris-strewn site, the last reported to be near Khotan; he hurried back to Khotan. Though suffering from a severe cold, he concluded his endless details there, managing to rest at the same time.

In the exchange of visits with the elderly, scholarly amban, Stein relates an incident that both touched and impressed him deeply. "After looking at the antiquities arranged for his inspection, Pan-Darin's sharp questions about the relative age, the meaning and characters of the assorted ancient documents, prompted Stein to show him Bühler's *Indian Paleography* and from its plates explain how writing could date with reasonable accuracy the various ancient documents. Pan-Darin's reply was to write down Chinese characters to show how they had been modified at various periods. At that point Western scholarship and Chinese learning spoke the same language. However, when Pan-Darin repeatedly asked "why all these old records [were] being carried away to the Far West, Stein was at a loss to answer. Was he unable because of having a "not overintelligent interpreter"?[29] Or was it an inability to explain to a man schooled in non-Western learning—be it Pan-Darin or Govind Kaul—the paramountcy of Western scholarship? He must have been relieved when Pan-Darin turned to precise questions: What would he have to show the governor-general of Sinkiang who "had been so inquisitive" about Stein's excavations and who, the amban pointedly mentioned, would want a report? To satisfy the governor-general's curiosity, Stein promised to send from Kashgar photographs of the several types of manuscripts. " 'But they should be in duplicate,' was the cautious reply of my learned friend . . . [who] seemed eager to retain for himself some samples of the strange records which the desert had yielded up after so many centuries."

Stein included photographs when, from London, he sent his *Preliminary Report,* feeling that, of the copies presented to Chinese officials, the one given to Pan-Darin was "at least duly appreciated."[30]

During his eight-day stay at Khotan, Stein completed his remaining task: discovering the forger, or in Stein's words, conducting "a semi-antiquarian, semi-judicial inquiry." In the great number of manuscripts Stein now possessed—in Kharoshthi, Indian Brahmi, Khotanese (Central Asian Brahmi), Tibetan, and Chinese—not a single one was in "unknown characters." Furthermore, not a single yield-site, such as Guma, had provided a document as had been claimed. The finger pointed to Islam Akhun, the Khotanese treasure seeker, as the one "to whom it was possible to trace most of the manuscripts that had been purchased on behalf of the Indian Government during the years 1895–98." Islam had already been found guilty—and punished—for passing himself off as an agent of Macartney to blackmail an innocent hillman by threatening to accuse him of owning men captured by Hunza raiders and sold into slavery. Obviously, Macartney had briefed Stein about such nonliterary activities; Stein was eager to have a "personal examination of that enterprising individual whose productions had engaged so much learned attention in Europe."[31]

Pan-Darin, at Stein's request, acted promptly and discreetly. Islam was fetched from Chira where, when his "old books" fell under suspicion, he had established himself as a healer. The beg who brought him also had a strange assortment of papers seized with Islam, including "sheets of artificially discoloured paper, covered with the same elaborate formulas in 'unknown characters' that had appeared in the last batch of 'ancient block-prints' which had been sold in Kashgar." Questioning so versatile an individual was a delicate matter. Islam readily admitted his guilt in the impersonation case; he was contrite at having falsely obtained money from Badruddin Khan (he had forged a note claiming it was Captain Deasy's); but he stoutly maintained his innocence of have anything to do with "old books." He had merely acted as an agent for others—one, alas, had fled to Yarkand, another to Aksu, and a third had escaped by dying—

who told him they had found the manuscripts in the desert. "When he found how much such 'old books' were appreciated by the Europeans, he asked those persons to find more. This they did, whereupon he took their finds to Kashgar, and so forth. Now he lamented he was left alone to bear the onus of the fraud—if such it was."[32] Two lengthy cross-examinations failed to shake him: he vehemently denied having seen or visited any of the alleged find-places, or desert sites. Stein did not challenge his denial.

Later Islam Akhun, momentarily off·guard, in the presence of witnesses, repeated his disclaimer. Stein promptly produced Hoernle's *Report* with the statements Islam had made before Macartney concerning "his alleged journeys and discoveries in the Taklamakan."[33] Silence: in quick succession Islam was startled, confused, and proud that the lies he had served up years before, garnished with topographical details, "had received the honor of permanent record in a scientific report to Government," that Stein could list every "old book" he had sold in Kashgar, on what occasions and what he had told about each discovery. First Islam confessed only to having seen manuscripts being written, but gradually, bit by bit, he gave more details, and finally, when Stein assured him that he would not be punished, he told the whole story.

Islam Akhun, a man of native intelligence and quick wit, a small dealer in coins, seals, and similar antiquities collected from Khotan villages, learned from Afghan traders that the sahibs from India valued manuscripts—even such scraps as men like Turdi brought back from Dandan-Uiliq. As a businessman he was interested; as a man averse to the hardships and uncertainties of treasure hunting, it was pleasanter to manufacture manuscripts. In this innovative venture he had partners: one specialized in supplying the Russian market; he himself dealt with the British and other collectors. For a while he tried imitating the writing of genuine manuscripts. When he realized that his European buyers could read neither the genuine nor the imitations, he concluded that such painstaking copying was excessive. Business was good. He sold everything. To satisfy the growing market, he started a little factory turning out the "unknown scripts." Quite soon even the free-form handwriting seemed too slow, too cumbersome, and to increase his stock the enterprising businessman turned to woodblocks—woodblock printing was done extensively in Sinkiang. Business was profitable; Stein could judge how profitable since he knew that about forty-five "block-prints of imposing dimensions, fully described and illustrated," appeared in Hoernle's *Report*. He thought Islam Akhun would have been very proud to know that many of his documents, bound in fine morocco, were in great libraries, even in the Manuscript Department of the British Museum, albeit the forgery section.

Finding Stein an interested listener, Islam Akhun went on to explain his production methods. Khotan, as a center of the region's paper industry, was an ideal location: he had access to paper of any size and kind. The sheets were first dyed to give them a stained look; then, when printed, they were smoked over the fireplace to achieve an antique look (hence the scorching on some books sent to Calcutta); after this they were bound (crudely and unsuitably in imitation of European books); and finally they were liberally sprinkled with fine sand (sure proof that they had been long buried). When the full story was out, both Stein and Islam Akhun knew that the business was finished. Nor could it be revived. Stein asked Pan-Darin to be lenient. It was punishment enough that, for defrauding Baruddin Khan, Islam Akhun was sentenced to wear the cangue (portable pillory) for a long period.

At the end of the questioning, Stein jestingly told Islam Akhun that he "thought him too clever a man to be allowed to remain in Khotan among such ignorant people."[34] The spirit of the forger was cowed but not out: he suggested that Stein take him to Europe. Jest or no jest, Islam Akhun was, after all, a man of ability and enterprise trapped in a world that offered little scope for his talents. For Stein he remained an unforgettable figure.

The time for farewells had come. Its prelude was the departure of the heavy baggage under the care of Ram Singh. Three days later, 27 April, Stein said farewell to Pan-Darin; farewell to the bazaars of Khotan where he bought mementoes for family and friends; farewell to local acquaintances to whom he gave medicines for real or future needs; farewell to those who had served him: the yamen attendants who had guarded his camp, villagers who had collected information or antiquities, headmen who had secured supplies—to each for services large or small, with words of thanks, Stein gave the expected "tip," the proper amount of gold or silver, the customary acknowledgment for services rendered, favors done. Farewell, too, to the nearby height of Utlughat-Dawan where their last triangulation station had been, and beyond, farewell to the mighty Kunlun Mountains whose icy ridges looked down from a distance.

Delaying their farewells, three men rode with Stein on his roundabout way out of the Khotan oasis: Turdi, Niaz Akhun, and Badruddin Khan, "who claimed the privilege of keeping me company to the very border of the oasis."[35] Farewell to Yotkan where Stein collected "samples of soil from the different strata which contain the ancient deposits and the silt that buried them" and additional ancient coins, seals, terracottas obtained from owners who had not come forward on his previous visits. Farewell. Their outward path led to the still unvisited village of Karakash at the northwestern edge of the oasis. As they rode, the Afghan trader explained the system which allots one day each week to each of the seven main bazaars of the oasis. As the distances are not large and the sequence is conveniently arranged, traders go from market to market. A steady stream of ponies loaded with goods and owners crowded the road. "Badruddin Khan knew them all well, goods, ponies and men and had much to tell of their financial fortunes and personal characters."[36]

Farewell to his final duty at Khotan: the visit to Kara-dobe, the Black Mound—scattered debris and a decayed mound where once a stupa might have been. The long detour with its abrupt changes of farm fields, sandy jungle, high dunes, and marsh gave Stein, as though a parting favor, a miniature example of every type of Khotan scenery: Farewell. Outside Zawa, the last village on the Khotan territory, they reached the main road where Stein's long journey westward all the way to Europe began. The moment was emotionally charged: could he, would he, Stein wondered, return to so rewarding a field and again find such faithful helpers? Parting from persons and places was painful. There, where the road home started, he said farewell to Turdi and Niaz Akhun.

"Turdi, my honest old guide, whose experience and local sense never failed me in the desert"[37] was rewarded with more treasure, cash, than all his wanderings in the Taklamakan had brought him—an unlooked-for return he had never dreamed of. And Turdi, on Stein's warm recommendation, had been appointed by the amban to serve as steward for irrigation for his native village, a pleasing prospect, since Turdi was beginning to feel he might be too old for the desert. Unashamedly, Turdi wept at the leave-taking; and Stein saw how genuine were the tears that trickled over the "weather-beaten face of the old treasure-seeker." Farewell.

The parting from Niaz Akhun was different. The Chinese interpreter had, it appeared, "fallen into a matrimonial entanglement with a captivating damsel of easy virtue, and had decided to remain, against the emphatic warnings of the old Amban, who plainly told him that, as a confirmed gambler and without chance of employment, he would soon be starving." He had obtained a mail-order divorce from his Kashgar wife: for a few tangas a Khotan mullah had provided the necessary document. With divorce so easy, Stein found "the organization of Turkestan family life . . . rather puzzling." To him, Stein's farewell sounded a "good-bye."

Badruddin Khan went beyond Zawa to the desert's edge, to the lonely shelter where Stein had spent his first night on Khotan soil. Farewell, farewell.

Stein went by way of the "Pigeon's Sanctuary." "When I had passed here nearly seven months before, there was little to give me assurance that I should ever see the hopes fulfilled that had drawn me to this distant land. But now my task was done and I could rejoice in the thought that my labours had been rewarded far beyond those long-cherished hopes."[38] Mindful of the tradition of sanctity inherent in that place—the Buddhists' sacred rats transmogrified by the Muslims into sacred birds—Stein gave a liberal offering of corn, even though, as he said, "for success I had not prayed, but only worked."

Six rapid marches brought him to Yarkand where he caught up with his caravan. His short stay there coincided with a heavy two-day rain that caused widespread distress when the mud-brick walls of many houses collapsed. But the rain washed and cooled the air, making his three-day, 140-mile ride to Kashgar delightful. "The morning of May 12th, a brilliantly clear day and full of the sensation of spring,"[39] saw him welcomed back at the Macartney's home. What tales he had to tell, what rich results to show. In the telling and showing Stein repeatedly acknowledged that it was Macartney's influence and solicitude that had made the whole venture possible. At Chini-Bagh, Stein could get some rest while attending to the arrangements for ending his expedition and returning to India.

Before he had left Kashmir, Stein had received permission to return via Europe by arguing that thirty days would be saved by returning to India by way of Russia and London. Then, as the rich returns of his work mounted in quantity and significance, he requested two months' additional time in England. A very long letter to the secretary to the government of India (dated from the Niya site!) had described his successes and pointed out their relevance to Indologists, especially to Hoernle's current research. He enumerated the reasons why he should publish a resumé of his finds as soon as possible, a *Preliminary Report*. He anticipated any questions as to why he had not done this before. "To prepare such a report in the course of my tour has proved impossible by practical experience. Apart from the fact that hundreds of photographs and plans to which reference would be needed, are not readily available for use before reproduction, the complete want of leisure is an unsurmountable obstacle. In illustration of this I may mention that between the 10th January and 28th February, I had to cover over 500 miles by marches over desert ground where arrangements for transport and supplies required constant attention, and that during this period the 22 days spent in actual excavation work were practically the only halts I could allow."[40]

What was he asking for? Additional "privilege leave"; that is, leave with pay. He had already been allowed some six or seven weeks, but, he said, "I feel that after the fatigues of a year's travel under trying conditions I shall, in the interest of my health, require this short period for rest and shall not be justified in devoting it wholly to work on the report. . . . I believe that a period of *six weeks* added from the date my deputation ends . . . would suffice for this purpose." He bolstered up his arguments: "In support of my request, it may be mentioned that the expected savings [Kashgar to London, about 600 rupees; London to Bombay, 700; as against the estimated 2,000 rupees for the return via Kashmir] will fully cover the extra costs arising eventually from an extension of the officiating arrangements." The government graciously put him "on special duty for six weeks from date of arrival in London."[41]

With another government department, Stein carried on still another correspondence for a passport to travel through Russia enroute to England. Without having previously secured a passport, his plans to visit his brother and deposit his finds at the British Museum would become impossible. While at Dandan-Uiliq he had received a "certificate of identity" in lieu of a passport: the government was uncertain whether or not Stein was a naturalized British subject. His answer, written from the same site, explains the status of his nationality. "When I received my first appointment under the Punjab University, I made up my mind to become a British subject. I accordingly applied in Budapest, my native place, for dismissal from Hungarian citizenship. It took nearly two years, until I obtained in 1889 the formal papers conveying this dismissal. I ought then to have applied for naturalization in India. But as no question ever arose as to my nationality, the matter dropped from my attention. Having relinquished my Hungarian citizenship, and being settled in India with an appointment made under an incorporated body and, later on, under Goverment, I have naturally looked upon myself as a British subject. I regret that I did not complete the formal steps necessary for the purpose, but being now reminded of this obligation, I shall do whatever may be needful to obtain naturalization as soon as I return to India."[42]

It is hard to find a single other instance of where a matter (and one far from trivial) dropped from Stein's attention. Rather, it might well be that, during those years when Stein was hoping for an appointment in a European university, he was waiting to see where he might settle and planned to take out citizenship there. The frustration he had felt for so long gave no urgency to his taking the needful steps to become a British subject. But now, finally doing what he had so long wanted to do, the matter would not drop from his attention.

Answers to these various questions were awaiting him at Kashgar: he *could* return via Europe; he *would* find his "certificate of identity" honored by the Russians. It was to arrange for his transit that Stein met Petrovsky. During Stein's earlier stay at Kashgar, the Russian consul-general had avoided him, pleading illness. Now he responded to Stein's presence. Stein, always grateful for efforts made in his behalf, wrote that Petrovsky who "has devoted a great deal of scholarly zeal to the history and antiquities of the country, did all in his power to ensure the safe transit of my archaeological finds to England, and to secure for me the friendly assistance of the authorities in Russian Turkestan."[43] When twelve large boxes, carefully repacked, were presented at the Russian consulate, they were gently examined and received their seals with the imperial eagle, seals which were left undisturbed until Stein unpacked them in the British Museum.

Stein began the "demobilization" of the caravan. Ram Singh, his trusted lieutenant and faithful companion—and Stein spelled out his exceptional qualities as a surveyor and his fine qualities as a man—with his devoted Jasvant Singh, were to retrace the overland route back to India. With them went Yolchi-beg, the little fox terrier. "I confess I felt the separation from the devoted comrade of my travels, until we joyfully met again one November night on a Punjab railway platform."[44]

He also had to sell the camels and ponies—"a not insignificant portion of the grant allotted for the expenses of my journey was invested in them" —and was concerned to recover as much as possible of their price. The ponies sold practically without any loss, while . . . our eight camels realized not less than three-fourths of the purchase price." Stein felt justly proud that after "all the hard marching and exposure of our winter campaign in the desert the whole of the transport had been safely brought back in a condition which allowed of its sale with such small loss."[45]

With Macartney, Stein made several visits to the Yamen to thank the kindly taotai for his effective and invaluable assistance. The amiable old administrator "did not deny the genuine interest and goodwill with which he had followed my work; but he politely insisted on attributing all the sympathy and support I had enjoyed from him and his Ambans to the benediction of my patron saint 'Tang-Seng.' "[46]

And finally, a long farewell to the Macartneys.

On 29 May, exactly one year after he had started from Srinagar, Stein began the journey from Kashgar to Osh, where he was to start his train trip to Europe. (Caravans reckoned the distance at eighteen marches; Stein, eager to save time, covered it in eight "by keeping in saddle or on foot from early morning until nightfall.")[47] Eight sturdy ponies were hired to carry his precious cargo, his tent, camp equipment, and personal baggage. In addition to the ponymen, his party consisted only of Sadak Akhun, who "removed from the evil spirits of the desert (*recte,* the temptation to take too large doses of 'Charas') had become again a fairly sober character." The Kizil-su River system, which the route followed to the Russian frontier, was in high flood. As the path crossed and recrossed the rushing stream, great care and good luck kept the boxes from getting a bath. After five anxious days, they reached Irkeshtam, the border: behind lay China and the river's valley of barren rock and detritus, ahead a world of grassy slopes and Russia.

Stein wrote Mrs. Allen on 12 June 1901: "I had a fairly arduous but safe journey, crossing the Alai Plateau and the watersheds between the Tarim, Oxus and Jaxartes (with *you* classical names need no excuse) in a succession of mild snow-storms. Two days ago I reached the railway and with it practically Europe. . . . I write these lines in my compartment at a little station where a break of the line caused by heavy rains threatens to detain us for about 24 hours! At Samarkand I shall make a short halt and then travel through as quick as connections will allow. My brother's place does not lie far from my direct route and as I am just now my immediate authority, I shall grant myself a few days 'casual leave' before I begin work again at London. Government have agreed to let me have six weeks on special duty [relieved from regular duty to do particular task but still paid regular salary] at London to arrange my collections, etc. which is certainly a needed concession. After that I am to take two odd months privilege leave [vacation] which is due me. So you see my time will be by no means ample.

"I am longing to see you both again. . . . How sad Lahore will be without you two. Perhaps, it is better now that I am not to be posted there permanently. I know I could never be as happy there as I was while you and Publius offered me so cheerful a home."[48]

Stein's First Central Asian Expedition ended when in London, on 2 July 1901, he had "the satisfaction of depositing the antiquities unearthed from the desert sands in the British Museum as a safe, temporary resting place. Neither they nor my eight hundred odd photographic negatives on glass had suffered by the long journey."[49] His *Sand-buried Ruins,* finished some two years later, closes on a restless note: "The busy weeks spent in the basement rooms . . . of the British Museum [unpacking and labeling] . . . seemed to me a time of immurement for the sake of science. How often have I not, then and since, wished myself back in the freedom and peace of the desert."[50] With that brief, too brief, excursion into Sinkiang, Stein had proved its importance to the world of scholarship and his own skill and energy in digging where no European had dug before. He had arrived.

Intermission

1901–6

10

Stein's expedition was also a journey into himself. He had had long hours to think on the arduous marches and was aware of his new orientation. He could tell it to Ernst and Hetty, translating it into terms they understood. 1 December 1900: "My Calcutta experience shows me clearly how difficult it would be for me to have to adjust permanently to the conditions in the 'Vaterstadt' from which I have become estranged through my life and work in a totally different environment. Not only would it force me to give up my whole way of life, but also the chance to achieve independence. In India I can look forward to increasing my capital over the next eight to twenty years so as to secure a sum which would permit me a modest but free existence; in Budapest that would be impossible." Now, he could admit freely: "The thought of being close to the family draws me to Europe more than anything else."

He further defined his emotional disengagement from the European scene in declining Hetty's invitation to join the family on their summer vacation at the seashore. 13 January 1901: "I must admit I am not attracted to the sea and especially not to summer beach resorts. You must know I find solitude most enjoyable—Mohand Marg and the Taklamakan are my witnesses." Whatever Hetty's idea of solitude was, it was not Stein's. Neither on his alp nor in the desert was he ever alone: attendants were on hand when he wanted a change from his concentration on work. Chatting with them was explained (and excused) as an exercise in speaking Kashmiri, Sanskrit, Persian, Wakhi, or Turki. To Stein, solitude meant avoiding situations where he was expected to engage in polite, useless chitchat with persons to whom India was utterly alien. Purdah was not solitude. During his happy years at Mayo Lodge, after long hours of work, Stein enjoyed delightful times. Whether the mood was serious or light, the company of the Allens, the Arnolds, the Andrewses, bound together by their shared life in Lahore, was refreshing and stimulating. And necessary.

Ernst had been the sole confidant of his hopes, stratagems, frustrations, and successes; now, while in England, Stein found another in his friend Percy Allen, his Publius. Ernst, in whom brother and father were coalesced, spoke to him with the voice of Europe; Publius and Madam, singly and together, recently domiciled in Oxford, had an Indian-based experience—they shared a whole vocabulary of persons, places, and way of life. There was also a deeper rapprochement: even as Stein had headed for Khotan, the Allens began their lifelong quest to collect and edit the vast body of Erasmus letters; they could match Stein with discovery for exciting discovery.

As Stein was preparing his Second Expedition, he confided his first "Personal Narrative" to the Allens. Ernst had died. 3 November 1905: "I have made bold to entrust them [22 pages] to your friendly care. I cannot afford time for a full diary. So if you will put these pages aside they might help me someday to revive all the impressions I had." He explained his need. 5 November 1905: "There is no one else now whose unwearied interest I could rely on with such assurance. You know that the pages which passed into print under your care [Sand-buried Ruins] had first travelled home from Turkestan with letters for my brother. I wonder whether *you* would allow me to continue the practice when I shall be moving once more across the mountains. I should be most grateful knowing my personal record safe from accidents. . . . If you agree I hope you will let me order a little deed-box which would reduce the trouble of keeping the loose leaves for me after my return."

Why not Hetty? A strong, true, loving tie would keep Stein devoted to her and his niece and nephew when, after Ernst's death, she retreated into silence and inconsolable grief. Stein might be annoyed by her behavior, but he was neither alienated nor intimidated by it. Regularly, he bombarded her with insistences that she continue Ernst's weekly letters, demanding family news, imposing his constancy on her consciousness. It was Stein's acceptance of Hetty's collapse that necessitated his turning to the Allens.

Stein's arrival in London with his dozen cases was a triumphant homecoming. He had proved himself—not only to himself but to government officials and Orientalists. Mountains and deserts had yield to his efforts, and on the human level he had demonstrated that something called leadership. Commanding a succession of untrained, motley crews in an alien land, he had bound them to his work, his tempo, and his style. But he was not yet out of the woods. In the following years, as proposal after proposal came to naught, Stein would understand that his very success invited the bureaucratic reflex—to bring him to heel. This bitter realization came slowly.

At first, to the cordial reception given him in London was added the pleasant prospect opened by a letter from Colonel Deane, now the resident in Kashmir, announcing Stein's appointment as inspector of education for the new border province. He would serve directly under his staunch friend.

Sand-buried Ruins was finished in 1903, the halfway mark between his First and Second Expeditions, and it sounds with the disenchantment of those years. The book begins with the remembrance that "two years and bulky files of correspondence . . . had secured what was needed—the freedom to move and the means requisite" and closes with his "time of immurement" in the basement of the British Museum longing for the "freedom and peace of the desert." Stabilizing his discontent was the certainty that only service in India offered a favorable future in which he could amass the capital to provide a modest income: and its sweet reward —freedom. Between the two expeditions Stein's restlessness can be felt in the feverish expenditure of time and energy he put into arguments and proposals for being sent to Afghanistan and Tibet. His first major expedition had left Stein with an insatiable thirst for more.

Arriving in London, Stein was still in the spell of the "wonderful days at Jaworzno as a beautiful dream of rest and comfort." He listed for Ernst what was expected of him in the meager time allowed: to make official calls and personal visits to "old patrons" and highly placed administrators now retired from India; to write the *Preliminary Report* whose importance he had stressed; to visit Dr. Hoernle "who had accepted the undeniable [and] wants to have his report about the deciphered forgeries destroyed."[1]

Only the briefest mention is made of Stein's Oxford visit as Hoernle's guest. Its outcome speaks well for men whose friendship rested on common antiquarian pursuits and devotion to truth. The letter he penned from his host's garden mentioned that Hoernle was "very interested in the specimens I brought back with me. Understandably, he is very deeply disappointed by Islam Akhun's forgeries, but to my satisfaction has recovered and I am spared a painful discussion."[2] For an Indologist of

Hoernle's caliber, the chagrin and the years lost trying to decipher the handful of forgeries could only be compensated for by Stein's bringing him a wealth of bona fide documents written in unknown characters. Sufficient in number, they enabled Hoernle to recreate Khotanese, the lost language of the ancient kingdom of Yü-t'ien, and finally to identify it as an early Iranian language written in an Indian script.

Next Stein turned to the task of making a brief inventory of his assorted finds. He enlisted the aid of Andrews, the Baron, Lockwood Kipling's successor, who had been his companion on a short tour and had introduced him to photography and had shopped for necessary items for the expedition. The Baron, artistically inclined, careful, reliable, and devoted (he had, in fact, a quasi-dependency on Stein) thus began his lifelong vocation as Stein's trusted assistant. When Stein had first enlisted him to do errands, he had paid him for his time—for time Stein regarded as a precious commodity. "I will be able to compensate him, thanks to the savings I had from my grant,"[3] Stein explained to his brother. Soon he was complaining: "I lose valuable time because of the many visits paid to my collections."[4]

As day by day the period of his deputation passed, the pressure of work mounted. He wrote to Ernst on 10 August that he was "more or less up-to-date with the preliminary arrangement of the collection. But writing the inventory and the provisional report in so brief a span of time and with constant interruptions became physically impossible. I long for a few days of rest. I am glad my deputation ends so that I can reach you on the evening of the 15th. By the end of the month I will have to be back [in London] for the cataloguing, photographing, reproductions, etc. . . . There is a possibility I can get an additional short deputation which would delay my return to the Punjab by 1 or 2 weeks. All this requires an unspeakable amount of red tape. I would much rather be in the desert or Pamirs than in the bustle and humidity of London." The government did grant him what he asked for.

On 15 November 1901 his transfer from the Bengal to the Punjab Educational Service became effective.[5] He arrived at Rawalpindi, his new headquarters, two days before, having spent "a pleasant and useful day at Mussoorie, where the Trigonometric Survey Office is located. My charts are being worked out and the gentlemen of the Survey are fully satisfied with the results. The triangulation as far as Kashgar and Khotan proved quite reliable and I had many compliments for my phototheodolires. They will do everything possible to work out the technical details of the charts."[6] At Rawalpindi he learned to his dismay that his district had been separated from the Punjab—"this means I lose the interesting region around Peshawar and Hazara." But there was a happy aspect. "No inspector of schools has been planned for the new province and whether or not one will be actually created, depends on Col. Deane. I will discuss it with him. If this succeeds, complete separation from the Punjab would make me my own master. I like Rawalpindi enormously: it is cooler than Lahore, the nights are almost cold; the mountains are quite near and the hilly terrain and many gardens give a most pleasing impression."

In his letter of 13 November 1901 Stein wrote Ernst about his visit to Deane after the festivities that marked the brilliant inauguration of the new province. He "gave me almost his entire Sunday and could discuss events of the past and plans for the future. The separation took place as a result of Lord Curzon's urging. The cost of the new province had to be limited—at least on paper. For political reasons Col. Deane feels the creation of a European inspectorship for the Educational Service is necessary. He wants me in this position because he is confident that I can exercise a real influence on the tribes beyond the border. Lord C. will be travelling through the new province in the spring and D. intends to further the whole matter directly with his patron."

Then, even before starting his duties, Stein raised the question of a deputation to return to London. Deane was a powerful ally. "The Hon. Mr. Arundel, Lord C's communication minister, was [Deane's] guest and since he too is interested in archaeology, Deane arranged with him to bring my case directly before Lord C. It will have to wait until mid-December when the viceroy returns from Burma to Calcutta. By then I hope to have copies of my *Preliminary Report* and D believes I will get six or seven months leave." Abruptly—was he referring to a subject he and Ernst had talked over?—he adds: "then Balkh can be discussed." This is the first mention of Balkh, shorthand for Afghanistan, which would remain for the rest of his life the region he wanted to explore.

As inspector of schools for the new province, Stein set his own style. On his first tour he crossed the Salt Range to the Jhelum River, hoping "to visit some interesting places en route. I am taking camels to carry the baggage . . . and am living again in a tent, much larger than the one that gave me such good service in the last years. I feel quite at home in it."[7] For several days he stayed at Murti where he had made his first excavation. Could he in 1890 have imagined how far in space and achievement that first attempt would take him? "Ketas, an ancient Hindu pilgrimage site that I first visited twelve years ago is nearby. I am pleased to be on the tracks of my Chinese Patron Saint and enjoy the same lovely landscape he praised 1200 years ago. There is considerable work, but it is preferable to Lahore since I can spend more time outdoors which has become almost a necessity for me. I have finished furnishing my equipment and living as I do I hope will save nice little sums."[8]

Two weeks before Christmas he received his copies of his *Preliminary Report*. Copies had been sent to everyone on his suggested list—officials and scholars in India, China, and Europe, various institutions and libraries. His own copies were used to further his requests—the first was for eight months' special duty starting in May. He also went to "Lahore for a few days. I intend putting Sir Mackworth Young [lieutenant Governor of the Punjab] into a favourable frame of mind about my deputation next spring. I will spend Christmas with you in spirit," he told Ernst, "but where in body I do not yet know." Early in January he would start a new tour: "it entails a great deal of work but I feel refreshed by the constant movement and exercise."[9]

For Christmas he received from Ernst and Hetty "a change-purse, soap and tooth-powder." True, all were small items, chosen for a wanderer living in a tent, chosen too for a message. The toothpowder was, of course, a thoughtful response to the horrendous extraction-*manqué* at the Tawakkel oasis; the soap spoke of a caring, bodily intimacy; the change purse remembered the one given by "dearest Papa" to the boy Aurel when, at Dresden, he had first left home. Family love, distance, death joined together when at the end of his letter to Ernst of 25 February 1902 he whispered: "The thought of your health never leaves my mind."

Ernst complimented Stein on his *Preliminary Report*—a source of delight to both—and in reply Stein sent him a letter he had received from Edouard Chavannes, "the outstanding Sinologue in the field of history. The cooperation he offers for my Chinese findings will be of the greatest value. Paris, close to London, would give me the chance to get his and Sénart's advice occasionally." The French reaction stood in sharp contrast to that of the Germans: "Schroeder naturally received a copy but has not been heard from along with the other German colleagues. Perhaps they do not know how to react to a purely philological achievement. It eases my mind to know that these gentlemen will have to deal with my manuscript findings for many years to come." He added a P.S. "To my great regret, without my being at fault, I missed Sven Hedin again. He telegraphed me the night before his arrival at Rawalpindi (although I had written him about my continuous inspection trips) and of course did not receive the news until two or three days later."[10]

He sent his *Preliminary Report* to high officials, using the occasion to enlist their help in his request for additional leave. It would be proper to focus all the data—diaries, detailed notes, plans, topographical surveys, and his extensive photographic record—on the "antiquities themselves before they can be incorporated in a final report."[11] To each he explained how impossible it was even to attempt such a Report "while carrying on my regular official duties as Inspector of Schools. These are heavy enough and for the greatest part of the year leave me no leisure whatsoever for continued scientific work." In London he had complained of the stream of visitors who came to see his collections; now, with pride, he listed the most distinguished: "Messers. Sénart and Barth (Members of the French Academy), Professor Sylvain Lévi (of the Collège de France), Professor Flinders Petrie and other well-known scholars."

His persistence was successful. 26 March 1902, to Ernst: "I received a kind letter from Mr. [Denzil] Ibbetson, Minister of the Interior, telling me confidentially that the dispatch containing my reasons for a deputation to England left London last week with recommendation *in toto*. The definite confirmation will come by cable. I hope to be able to leave Bombay on May 3rd." At Lahore he had a satisfactory interview with "Sir Charles Rivas, the new Governor who showed the liveliest interest in my findings and plans. From Lahore I went to Peshawar to spend the day with Col. Deane so as to inform myself about Lord C's expected visit at the end of April and also my transfer to the new province. I will have the chance to show Lord C a selection of my Turkestan finds which could be very useful in the future. The post which is to be created for the Frontier Province and Baluchistan, which would give me greater freedom, is to be discussed on the occasion of Lord C's visit. I know Deane will do his utmost. . . . I have been overworked for the last four months and am making very slow progress with my popular account of the journey." It was then he wrote: "I beg fate to allow you to get well and allow me to see you once again." It was his constant prayer.

A letter from Publius was waiting for Stein when the S.S. Arabia on her run from Bombay to Brindisi stopped at Aden. Replying to this "kindest of welcomes across the seas," Stein wrote, 10 May 1902, to Allen: "No friend could ever surpass you in eliminating space by thoughtful kindness. In writing to me at Aden, you and my brother seem to have counted on what you are pleased to call my good luck. But you could scarcely have known what a close shave my start in the *Arabia* would be. The Secretary of

State's sanction (a formal matter, after all) was delayed most strangely. If it had not been for Lord Curzon who deemed it fit to act personally as a *deus ex machina,* I might still be waiting at Rpindi.

"On April 25th I had my interview with H.E. [His Excellency] at Peshawar, arranged months beforehand. During the hour he gave me, I had to answer questions about my Turkestan work which showed his interest in the results. When Lord C heard that my deputation was still without final sanction, he at once offered to wire the Secretary of State. The Viceregal reminder had its effect within hours and while still at Peshawar I received intimations from H.E. that all was arranged.

"Two more achievements will make those Peshawar Durbar days memorable for me: I talked to Lord C about my old desire to visit that forbidden land, Afghan Turkestan, the ancient Bactria, and dig at Balkh. I found him beyond all expectation taken by the idea and ready to help me. The apprehensions of the Indian Foreign Office, the chief obstacle, seemed to dissolve like a mere cloud. On the contrary, when next day I called on Mr. Barnes, the Foreign Secretary, I found him already instructed by the Chief Divinity of the Indian Olymp and eager to get details of my plan so as to take action accordingly. After my work in England is finished, I hope to start for Herat. . . . It seems like the fulfillment of a dream I had since I was a boy, and only the Gods know whether it will come out right. . . . The Foreign Office is to find the money and I shall take care not to be stinted in the matter of time.

"During the same Peshawar days, Col. Deane secured preliminary approval of the scheme which is to give me charge of Education in the Frontier Province and Baluchistan and to bring me permanently under the Foreign Department. It will be easier to operate from that base hereafter. Both this matter and the plan of my Bactrian Journey must be kept quite *entre nous* for the present. . . .

"When I land at Brindisi, on the 14th, I shall go direct to my brother of whom I had for the last few weeks better news, and get to London about the 23rd inst. I have been asked by wire to read a paper before the Royal Geographical Society, an unexpected honour at this season." Long before reaching England, Stein had publicized some of his important results: the previous February he had sent an article about his findings near the shrine of Imam Jafir to the *Journal of the Asiatic Society.* This and word of his impending deputation to London had alerted the Royal Geographical Society.

Before closing his long letter, Stein took up the invitation Publius had extended. "I propose to spend only the 23rd and 24th in London and then come to Oxford if you think you would allow me to go into 'Purdah' for a week or a little longer under your hospitable roof. I could not dream of staying longer unless . . . Mayo Lodge arrangements could be revived." The Mayo Lodge arrangements can only be surmised: a financial arrangement openly arrived at and mutually satisfactory. Stein was always punctilious about paying his own way whether with "gifts," as he had in Turkestan, or with cash, as with Andrews. With such "arrangements," Stein started what was to become an integral part of his life. Quite soon, in all but kinship, he would become a member of the Allen family.

From his Oxford purdah, he composed a long letter to Barnes, the secretary of the Indian Foreign Department, incorporating a historical resumé as Barnes himself had suggested. Stein explained why the region of ancient Bactria, corresponding to the present Afghan Turkestan with its adjacent district of Badakhshan, had real significance for students of early Indian history and antiquities. "From Bactria, the traditional home of Zoroaster's religion and one of the oldest centres of civilization in Asia, there were derived those elements of unmistakably Iranian (Old Persian) origin which meet us in the earliest extant remains and records of North-Western India. After the invasion of Alexander the Great, Bactria, under Greek rulers [Seleucids], became the seat of a remarkable Hellenistic culture, which flourished for centuries in that distant part of Asia, and from there triumphantly entered the Indus Valley and the neighbouring parts of the Punjab. The fascinating remains of so-called Graeco-Buddhist art, preserved in the sculptures of ruined Buddhist monasteries, in the Greek coins and other relics of the North-West Frontier, are eloquent proofs of the far-reaching influence then exercised by Bactria on Indian culture. The greater tribes of Central Asian invaders, the Indo-Scythians, White Huns and Turks, whose dynasties successfully held the North-West of India during the long centuries following the disappearance of Greek rule, had all been established in Bactria before their Indian conquests. During that extensive period, Buddhist religion and culture had in the same territory found a permanent home from which the Buddhist propaganda, together with an art and civilization of distinctly Indian character, spread over the whole of Central Asia and into the Far-East."[12]

After this brilliant, compressed exposition, Stein repeated the arguments for having a trained Orientalist carry out systematic exploration in the hope of filling in "some of the most interesting chapters in the ancient history of India and of Asia generally. For such exploration the site of *Bactra,* the ancient capital offers an exceptionally promising field. Its general position is marked beyond all doubt by the ruins of Balkh, the

'Mother of Cities' of Eastern tradition. . . . [its] site has remained almost wholly deserted since the thorough destruction of the medieval town by the armies of Changiz Khan and Timur." Then he gave reasons for seeking to explore the valleys of Badakhshan, particularly "those draining the northern slopes of the Hindukush watershed. [They] are occupied by a population of Tajik hillmen who probably represent the direct descendants of the original Iranian population . . . and are of the greatest interest for the philologist." As a special qualification he added, "I had devoted considerable time and attention to the study of ancient Iranian languages (Zend, Pahlavi·) and their modern derivatives, and I am able to speak both Persian and Turki."

This statement, with the addition of personnel, costs, routes, etc., made its way speedily to the viceroy. Curzon did not wait for it. On 12 June, in his most viceregal manner he addressed a letter to His Highness Amir Sir Habibulla Khan. Two weeks later the amir's answer was made without delay as if to state that he did not even want to spend time considering such a request; it concluded: "I now write to say that I cannot in any way advise the said Doctor [Stein] to carry out his intentions. . . . to secure such unobtainable and non-existing objects of research. No doubt your Excellency will also dissuade him from putting his intentions into practice."

With the amir's thundering refusal, Stein's first attempt to visit Balkh ended.

There is another side to the episode, of which Stein was unaware. Furthering his "interests" at court, paying respectful calls on Sir Mackworth Young and Sir Charles Rivas, having an interview with Curzon, consulting with Deane, Stein dealt with men whose training or political acumen had won them their commanding positions. Addressing men at that level, he had bypassed the civil servants, those who under a succession of administrators carried on the business of empire, who, if they could not grant, could delay, raise objections, ask embarrassing questions. The men who manned the bureaucracy made their power felt obliquely. They had a natural antipathy for mavericks like Stein who would not stay put but maneuvered their way from department to department, were always requesting leave, strained the budget with their own projects, or made shortcuts through sacred avenues of protocol. Could this have been why his sanction "was delayed most strangely"? Yet these men were neither stupid nor vicious; they were technicians who had a low threshold of tolerance for anyone who disturbed their routine or disregarded their informed reactions.

A Foreign Department expert had predicted to Barnes that "no matter in what way we address the Amir, he will not agree to the proposal." He based his opinion on "an incident reported in the Kabul diary received today. A man was charged with being a spy. A few letters in English, a pencil sketch and some pencils were found on his person. The Amir ordered him to be taken to the top of the Asmar Hill, there to make a sketch of the city, on completion of which he was to be rolled from the summit. This was how the man was put to death." Curzon had disregarded this warning.

That the foreign officer was proved right justified his pique. When such a man cannot harpoon a whale he sets his line for herring. Six months later a telegram notified Calcutta that the secretary of state for India proposed "to allow Dr. Stein's forthcoming book on the Pamirs to be reviewed by the Intelligence Branch of the War Office, as otherwise it will cause heavy loss to Dr. Stein.[13] What prompted this? Somewhere, somehow, some official in London (was it the man who had strangely delayed Stein's sanction?) decided that the forthcoming *Sand-buried Ruins* touched on the Pamirs, a sensitive area, cloaked in the wrappings of national security. The "Notes" dealing with the telegram were commented on by the same Foreign Department expert, who was still quivering at the Afghanistan rebuff and at Stein's ability to move wherever and do whatever he wanted.

After noting that Stein had "published a small book or pamphlet [the *Preliminary Report*] which has brought him much *kudos* from European savants," he scolds: "Dr. Stein as a Government servant of some years' standing, should be aware that he has no right whatever, without the authority of Government, to write a book for general publication conveying information which he acquired at Government expense. Dr. Stein was given special leave in England to enable him to write up the official account of his explorations. Did he utilize this in writing his book? . . . [The] telegram says that unless the work is reviewed . . . it will cause heavy loss to Dr. Stein. Surely this should read 'Dr. Stein will lose large prospective gains.' But whatever it means, I submit that this is Dr. Stein's own lookout." At the end the civil servant betrays his real dissatisfaction. "I do not think Dr. Stein deserves any special consideration. It will be remembered how he approached the Government of India directly and got us to move in the matter of his proposed exploration in Afghanistan, involving an absence of over a year from his educational duties, without having the courtesy to make any reference to the subject to the Punjab Government, under whom he is serving."

This took place in 1903 when civil servants, confined to routines like vestal virgins tending the sacred fire at the shrine dedicated to imperial beliefs and customs, viewed Stein's efforts to have the government finance his own scholarly research as behavior at best sinful, at worst ungentlemanly. The sign of his depravity was that he was not even a British subject!

During his deputation (May 1902–December 1903: his original leave of seven months had two extensions), Stein arranged the preliminaries necessary for the detailed report: *Ancient Khotan*. A master at scheduling, he had planned the *Preliminary Report* as the statement expected of a government official—a lean, factual resumé of where he went and what he did. *Sand-buried Ruins of Khotan,* his personal narrative, was to introduce a wide public to a region then little known but once the crossroads of the Eurasian landmass. These two accounts he could write himself. But *Ancient Khotan,* designed for the scholarly community, needed the best in scholarly excellence: it would be, as he saw it, a cooperative enterprise enlisting specialists to elucidate the diverse materials he had recovered and brought back to Europe. Fieldwork could still be a solo activity; but to decipher, translate, interpret, and interrelate those fields in which he lacked mastery—ah! that was different. *Ancient Khotan* is the orchestration of many scholarly specialists, each a virtuoso in his own field. At that time, and in archaeology, it was innovative. Today it would be called interdisciplinary.

Having decided that only so multipronged an approach could do full justice to the variety of his extraordinary finds, Stein had to involve the expert for each category, to furnish each with photographs of the materials to be studied, to outline the scope of the books, and to interweave the many contributions into a meaningful whole. How far and how well he cast his net can be seen in the book's Acknowledgements and Appendices.

First on his list of acknowledgements was Andrews. Stein thanked the Baron for his "unwearied assistance"[14] in the arrangement, the detailed descriptions, and illustrations of the hundreds of antiquities. The importance of this painstaking work was emphasized, since, because of "a decision of the Indian Government, the whole of the archaeological finds will now be distributed between the museums of Calcutta and Lahore and the British Museum." Inevitably, "future researches will depend mainly on the details furnished by the descriptive list and on the illustrations given in the plates." Stein's reliance on the Baron was fully justified: more than fifty years later, textile specialists from different countries compared scraps of Chinese silk found at Noin-Ula, a Siberian burial site in Outer

Mongolia, with those Stein had recovered from a Tarim Basin site, 1500 miles away. The scraps were like identical twins. The specialists lauded Andrews's "meticulous graphic description . . . of even the smallest detail . . . presented with technical precision and solicitude."[15] The scraps so analyzed substantiated the new concept of the role of the mounted nomads: more than merely feared marauders, they had been conveyors of artifacts between cultures distant and distinct.

To Hoernle, Stein turned over the documents "written in Brahmi, . . . [of] texts partly in Sanskrit, partly in two non-Indian languages."[16] Since Hoernle was engaged in presenting all previous Brahmi manuscript acquisitions and since the inclusion of Stein's texts and inscriptions would have necessitated a separate volume in *Ancient Khotan*, the happiest solution was for Stein to include Hoernle's analysis in his own volume and let the complete presentation of the documents be part of the latter's series of Brahmi publications.

Ancient Khotan has seven appendices:

A: *Chinese Documents from the Sites of Dandan-Uiliq, Niya and Endere*, translated and annotated by Edouard Chavannes, "the leading authority on all Chinese sources of information concerning the history and geography of Central Asia."[17]

B: *Tibetan Manuscripts and Sgraffiti Discovered at Endere*. Edited by L. D. Barnett, of the British Museum, and the Rev. A. H. Francke, Moravian Mission, Ladek.

C: *The Judaeo-Persian Document from Dandan-Uiliq*. Edited by D. S. Margoliouth.

D. *Inventory List of Coins Found or Purchased*. Prepared from notes of S. W. Bushell, a Sinologist, and E. J. Rapson, assistant keeper of the Coin Department, British Museum.

E. *Extracts from Tibetan Accounts of Khotan*. Communicated and annotated by F. W. Thomas; the "learned Librarian of the India Office."[18]

F. *Notes on Specimens of Ancient Stucco from Khotan Sites*. By A. H. Church, F.R.S.

G. *Notes on Sand and Loess Specimens from the Region of Khotan*. By L. de Lóczy.

In addition to the analysis of stuccos and soils, Stein also anticipated the reliance of contemporary archaeologists on investigations by natural scientists to clarify relevant factors in cultural studies: he requested J. Weisner, director of the Institute for Plant Physiology, to examine "the composition of the ancient paper materials represented."[18] At a time when race, not ethnicity, was the criterion for determining origins, Stein took anthropometric measurements which, interpreted by T. A. Joyce of the British Museum and secretary of the Anthropological Institute of Great Britain, formed the basis for his discussion "of the racial origin of the Khotan population."[19]

Getting *Ancient Khotan* off the ground necessitated several quick trips to Paris to renew personal contacts with Sénart, Foucher, Sylvain Lévi and others; to Hamburg (September 1902) to attend the Thirteenth International Congress of Orientalists. The congress's resolution extolling the importance and extraordinary riches of his work relieved Stein from the awkward necessity of having to use self-congratulatory words when pressing his requests to the government. Spaced with these foreign trips were pleasant visits to friends in England: a weekend spent at the seashore as the guest of Sir Mackworth Young, the former governor of the Punjab; a renewal of friendship with the Kiplings, who presented him with their son's newest novel, *Kim;* and a long-cherished brief outing through Devonshire with the Allens.

Emotionally, there was the devastation wrought by Ernst's death, to whose memory he dedicated *Sand-buried Ruins of Khotan.* His anguish at the dreaded finale sounds when he sent Hetty a copy on 25 June 1903: "How much pain lies in those years on top of all the work."

As the year (and his leave) was ending, political events in India spoke directly to Stein: the papers were filled with news and debates over Younghusband's mission to Tibet. Stein asked to be allowed to accompany the troops, reminding government that his first tour beyond British India's borders had been "with the Buner Field Force, during the campaign of 1897–98."[20] He listed his special qualifications—his recent work in the Kunlun Range on which Tibet borders—and argued an Indologist should accompany the expedition. "Tibet culture, religion and literature in the earliest attainable strata rest entirely on Indian foundations." He frankly admitted that his "knowledge of Tibetan, had so far, remained scanty and theoretical," a handicap easily remedied since the language was largely drawn from Sanskrit. His application went all the way to Younghusband who wired from Tibet: "I would not recommend Stein accompanying Mission at present. He might visit it later on."

"Might visit it later on." Those words had the power to reinfect Stein with hopes and plans. Another subtle, activating agent was his new title: starting in January 1904 he became inspector-general of education and archaeological surveyor, Northwest Frontier Province and Baluchistan. A cumbersome title but one to have far-reaching effects. How exactly was he to handle his dual functions? He was expected to "visit such sites as lay within easy reach of the route followed [when] on a prolonged tour of inspection."[21] Buoyed up by a title that recognized his archaeological accomplishments, within two months Stein returned to the Tibetan project this time emphasizing archaeological exploration of that area adjacent to "the northernmost edge of the Tibetan uplands, between the Lop-nor lake and the western extremity of the Kansu Province. This narrow tract separating Tibet from Mongolia has since ancient times been the main thoroughfare between China and Eastern Turkestan. I have reason to assume that the sand-buried sites of this tract would probably yield as rich a harvest as the ruined sites of the Taklamakan Desert."[22]

How much of Stein's strength was simply that he kept trying. Again his request was considered at the highest level. The gist of the official response he told to Publius, 6 April 1904: "A diplomatically smooth letter from Mr. Dane, the Foreign Secretary, informed me last week that my proposed deputation had been fully considered by the Viceroy; but as Col. Waddell, Capt. O'Connor and some other Political who had all made a special study of the Tibetan language, etc. were already with the Mission, they could not well let me go along." (Why, oh why had he raised the issue of the language?) Within two months, Stein had the immense satisfaction of telling Publius on 11 June 1904 that "I had a note from Younghusband, sending me a truly silly attempt at deciphering my Endere sgraffiti made by one of the 'experts' of the Mission. It had luckily been completely worked out by Francke, the Moravian-missionary scholar at Leh, and my friend Barnett."

He also told Publius that "the Viceroy had also taken note of my wish to explore towards the Jade Gate, etc., but had come to the conclusion that 'the time was not yet ripe' "—words Stein characteristically took as an encouragement. "I shall not abandon my efforts to pave my way to Lop Nor and the regions beyond. But it means evidently to begin the struggle afresh, as if I had never been to Khotan, and to face long years of waiting in wasteful routine work."

Summer was approaching. Stein wanted to use the vacation period to get on with the Detailed Report. For the hot months, he chose a mountain retreat: Colonel Deane gave him permission to camp on Battakundi in Kagan. Located at over 10,000 feet, his alpine kingdom was small. 11 June 1904: "I often feel as if I were on Mohand Marg again. But, alas, I miss my good old Pandit—and much else that has passed away from my reach forever." Spurred by his ardent desire to lead another expedition to Turkestan, he contemplated interesting an institution other than the government. "I intend to crystalize my plans on Lop Nor and the Jade Gate into a definite scheme for which I shall try to secure the support of the British Museum, the R. G. S. [Royal Geographical Society], and individual friends before submitting it to Simla. I do not wish to risk another refusal at that centre of intellectual sunshine." Stein's plan always had the support of his superior, Colonel Deane.

By 7 July he could outline his campaign for a second expedition to Andrews: "Col. Deane approves fully of my efforts, and I am endeavoring —at a heavy expense of correspondence—to secure the needful support from friends in Europe. I am sounding my way at Simla where luckily J. Wilson, an old Punjab friend, is now Revenue Secretary, but I am prepared for a long and difficult struggle. Everything depends on Lord Curzon's attitude, and I doubt whether anyone who knows my work will have a chance to talk to him on my behalf while he is at home [England]." In his letter to Wilson, Stein raised the specter of competition. "The German government which largely influenced by the success attending my journey had sent Prof. Grünwedel to *Turfan,* now, I understand, has allotted a fresh grant of £ 3500 for excavations in the same region. The Russian Government, even in its present difficulties [the Russo-Japanese War], is preparing an archaeological expedition to *Kucha* under Prof. d'Oldenburg a distinguished Indologist. My Khotan explorations were the first systematic archaeological enterprise in Eastern Turkestan."[23]

A stream of lengthy letters went out to secure friendly help, to inform, to alert persons of influence in India and England; such efforts were relegated to the late hours after a long day's work on the Report had satisfied him that he was making progress. His 7 July letter to Andrews continues: "I have cleared off most of the disquisitions into the pre-Muhammadan history of East Turkestan which were indispensable for the proper treatment of the antiquarian records brought to light. The description of the sites along the route to Khotan is complete, and the way for the rest would seem clear enough if only I could go straight on with the work for another six months. I shall try if possible to divest myself of the most irksome part of my inspection work in order to get some leisure during the winter. But this is by no means easy, since even so small a matter must go to the Foreign Office." "Irksome" was a misleading word for Stein to have used, for in the same letter he mentioned that a high school's magic lantern had no slides: "The Local Board sanctioned the magnificent sum of Rupees 20 (equal to £1 6s 8d) for the purpose." Would Andrews select slides illustrating geographical subjects? "I mean particularly 'rambles in English scenery, China, East or S. Africa, such as lecturers in English Schools are usually putting before their audience.?"

17 September 1904, to Allen: September's cold drove Stein into his tent, an "exact copy of the one I had with me in Turkestan (minus the serge lining!); so I can indulge more easily in the cherished illusion of writing to you from my camp in the desert. Will it be realized the plan from which grows the illusion?" He went on to tell Publius that the plan, incorporated in "a full application with all the detailed estimates which are to make the venture more palatable to the Secretariat-soul, started for Simla at the commencement of the week [14 September 1904]. . . . All this writing and pleading makes one feel only more than ever how petty a thing Government machinery is compared with personal independence or the money which secures it. Eleven years hence I may hope to have freedom. But shall Fate allow me to use it as I should now?" Was he soliloquizing or lamenting? Whichever it was, he knew that he must "try to forget the long years of labour spent in barren struggle—because there was no tide to lift one off a certain mud-bank."

His mood of self-pity passed. Before the cold drove him from his alpine retreat at Battakundi, he made it the scene of a momentous event. "I took my oath of allegiance before Douie on my own mountain fastness and was cheered by the warmth, personal and patriotic, with which he entered into the affair. Then we all clambered down by goat's track and I stayed once more for the night in one of their tents below my alp."[24]

On an alp, a setting of his own choosing, before friends who valued him, Stein carried through the final step of his naturalization, which had had its beginning when he was camped in the Taklamakan: henceforth he was not a man without a country—he was a British subject. "It is getting very cold up here and as snow threatens, I must descend as soon as mules are impressed."

11

Before the cold drove Stein down from Battakundi, he wrote to Allen, 12 October 1904: "[I] received a very encouraging letter from J. Wilson, now Revenue Secry to the G of I, whom I had privately addressed at length about my plan for a fresh journey to Khotan—Lop Nor—Shachou. He has always been a kind friend to me and he now promises unhesitatingly his best help for my scheme. So a formal application (12 pp. close foolscap!) with detailed estimates, etc., went up a few days ago to Col. Deane who had previously expressed his willingness to take up the matter. The result cannot be foreseen with certainty—except by the Gods who fitfully illuminate powerful brains." To Andrews he had given the briefest resumé of his application 23 September 1904: "I have asked this time for two years and Rs. 26,000—a bold demand which possibly may make an impression—or frighten. I must under all circumstances keep the plan strictly private in view of foreign competition and for other reasons. . . . You can imagine the burden of writing which it entails to get needful support. It seems scarcely possible to dispose of my Report Ms. until next May, unless I can get practically complete relief from my official work. I shall try to achieve this, but the hope is very slight."

With the plan for his Second Expedition launched—"how far it will reach only the Gods know, but somehow I feel I shall return to this charge again and again"[1]—the urgency of getting on with the Report did not stop Stein from accepting an invitation for an objective dear to his heart. "A lucky interval in the tribal wars permitted a visit to Mahaban, the mountain which had long attracted me as being the probable site of Alexander's Aornos." As previously he had allayed Ernst's fears for his dangerous sorties, so now he reassured Publius. "Pipon, who had exerted himself in the matter and upon whom Col. Deane had laid the political responsibility as far as tribal arrangements go, had asked me to accompany him as a companion. Of course I was ready enough to agree but I took care to explain that as I was going for a scientific task, I should have to claim full use of my time and could not arrange things in picnic fashion. We are to start on the 28th inst. [October] from Jhanda, the very place from which I looked up longingly to Mahaban in Jan. 1899. . . . The tribes have accepted full responsibility and will cater to us! Politically, it is proof that we are guests and hence sacrosanct." To make the tour sound less hazardous, he added: "I hope to prove to Pipon that my 'compressed foods' are also useful." On his way to Peshawar, where the week's tour was to begin, he inspected a few schools. Educational work is always accompanied by the adjective "tiring," whereas his strenuous assault on mountains is "invigorating," "refreshing."

Stein's scientific task was a picnic, as he made apparent in the account sent Publius, ostentiously dated, "Salar Gaduns' territory," 31 October 1904. "Four days ago we started on our tour to Mahaban, and though we shall not get back to British territory for two or three days, I cannot let the chance of a Dak (tribal this time) go by without sending you news. It must be brief; for after a twelve hours' climbing and surveying today over fairly stiff ground time does not suffice for a proper mail. . . . The heights of Mahaban successfully taken on the 29th, have furnished no ground for its time-honoured identification with Aornos. The tribes through whose valleys we passed made quite imposing efforts to ensure our safety . . . as their delightfully democratic anarchy permitted.

"We passed the first night at a little hamlet below the fir-covered southern slopes of Mahaban and climbed its long but disappointingly narrow ridges by midday of the next day . . . . the very confined space of the crest and the insignificant size of the ruins put Aornos out of the question unless we have to treat all the Alexander's historians' records as a most magnificent piece of Hellenistic fiction. It was a wonderful sight: the motley gathering of 400–500 tribesmen who camped around us on the top. . . . The Salar Gaduns whose territory we have entered now have given us a most cordial welcome and at some other sites lower down interesting ruins may be inspected tomorrow. Both Pipon and myself are justly elated by the success of his "tribal arrangements.'"

A telegram announcing his safe return was followed by a letter to Allen, 9 November 1904 relating his spectacular, scientific discovery. "I climbed Mount Banj and found on its steep slopes the ruins of a chief and long sought-for sanctuary of old Gandhara, the Stupas, etc., marking the place of Buddha's 'Body-offering.' [One of the Jataka stories tells how the Buddha, in a previous incarnation, threw himself from a cliff so that his body could feed a starving lioness and her cubs.] I have followed the track of my dear old patron Saint Hsüan-tsang through many lands, but never did I find it more accurate. Banj is rather a hard place to get to and the Maliks, whom my previous marching and climbing had somewhat worn out, did their best to keep me away from it. . . . Other interesting sites were also discovered though, of course, everywhere the sculptures had long ago been quarried by the poor hillmen for head-hunting Sahibs.

"Politically it has been a great success, this triumphant progress through successive tribal territories previously supposed to be inaccessible except for 'expeditions' which it was not thought worth while to send. Col. Deane, whom I saw after a 40 miles' ride . . . to catch the first train, was highly pleased with the success and hopes to let me repeat the experiment elsewhere. Pipon thinks the only difficulty will be that the tribesmen may be afraid of having to climb about with me."

Stein had hurried to Peshawar to attend the great durbar of his friend, Colonel Deane, appointed commissioner for the Northwest Frontier Province. "I found myself for the first time in my life established on a high and mighty dais, as (modest enough) part of the Local Government. . . . an afternoon party of a very official kind with crowds of picturesque Khans, Maliks, and so forth brought us all together again."

The long précis Stein had sent Wilson was forwarded routinely to those departments concerned with evaluating its worth. So it came to the director-general of the new Archaeological Survey of India, John Hubert Marshall, 1876–1958 (knighted in 1914). He was a mere twenty-six when named to that important position, a position designed and empowered to carry out a program of exploring, excavating, deciphering, and preserving the vast and varied material evidence of India's past. And for the British Raj, India included Pakistan and Burma.

Marshall, trained as a classicist, had served an apprenticeship at the British School in Athens and had worked briefly in Crete when Evans began uncovering the palace at Knossos. He had, as Sir Mortimer Wheeler would write, "the sketchiest training in the field-techniques of his day."[2] His two prolonged archaeological efforts were the careful uncovering of the many layers of the thousand-year occupancy of Taxila (500 B.C. to A.D. 500), and the excavating of the mighty prehistoric city of Mohenjo-daro (2300 B.C.). At the latter, by outmoded techniques he destroyed the evidence which might have established a timetable and "culture," even as, Wheeler said, he "dramatically flung a civilization at us which has taken its place in every subsequent history of the world." Whatever the faults in his fieldwork, of his time and training, his thirty-year tenure as director-general made the Archaeological Survey of India outstanding.

It is doubtful if the two men had met before Marshall was handed Stein's letter: Marshall reached India in the spring of 1902 as Stein was on his way to England. When they met first it was not face to face but will against will. In 1904, when Stein, in addition to his position as inspector for education for the Northwest Frontier Province and Baluchistan, was named archaeological surveyor for the same regions, Marshall, engaged in assessing the Archaeological Survey's personnel, noted: "Dr. Stein's publication [*Rajatarangini*] and recent work in Khotan proclaim him as a first-rate epigraphical scholar and explorer. I am not aware if his archaeological knowledge goes much beyond this."[3] So a classicist might have judged an Orientalist.

The deep commitment to Indian archaeology they shared was at first overshadowed by their differences in orientation and positions—Marshall, the administrator, Stein, the fieldworker. In that early period their views of Taxila might have been one measure of their differences. To Marshall it was then the easternmost reach of Greek expansion; to Stein it was the originating center for the spread of Buddhism eastward. Marshall's appointment required that he concentrate on India; Stein's philological training made him an ardent hunter of written words from the past. Or, perhaps, the differences came down to a simpler fact: Marshall, younger and untried, held the higher administrative post; Stein had made himself India's most famous archaeologist. Each had need of the other, a realization not immediately apparent.

Between the two there was a tension, clearly implied when Marshall drafted his reply to Stein's letter which Wilson, as was proper, had forwarded to the director-general. Marshall scratched out his first draft—"The project would indeed be an excellent one and you may rely on me to give my whole-hearted support when the time is ripe for laying it before Government"—and, changing his tone, wrote, "I have received a most interesting letter of yours in which you speak of your design to make another journey of exploration in Central Asia." Then—and the tone must have felt chilly to Stein—he continued: "In the meantime would you be good enough to let me know what your programme is for the remainder of the year 1904–05, and what exploration or other work you will be able to carry out *within the limits of British India*. There is a small excavation in Baluchistan which I am particularly anxious for you to carry out during the coming autumn. . . . The site is a mound a few marches south of Quetta at a little place called Nal." Why did Marshall's tone and attitude change. Why? And why, in 1904, Nal?

Stein's answer was businesslike. He accepted the assignment and planned to go to Nal after he returned from his Mahaban tour. But he also took the opportunity to add that, though Baluchistan was still practically *terra incognita*, "historical evidence [has shown] it to have belonged far more to Iranian than to the Indian sphere of culture. The fact only adds to my personal interest in that region, since my University studies were quite as much concerned with ancient Iran as with ancient India." Having given some of his credentials, he added that "the most important services I can render at present to Indian archaeology, are the early completion of my Detailed Report and the thoroughly effective preparation for my proposed journey."

The exchange of letters continued. Marshall hoped to be able to meet Stein at Nal, and in the exchange arranging their rendezvous (they did not meet) Stein decided to pierce the director-general's obtusity. "Now when my official connection with archaeology is at least partially established, it must appear, I believe, still easier to justify my employment on a special task the importance and urgency of which is recognized by all competent scholars. . . . I earnestly hope that no difficulties will be raised on this ocasion. If any weight is attached to scholarly opinion, the only question would be as to why so much time had to pass before I could continue those explorations." Stein, the veteran fighter, was sounding his first salvo.

After his enjoyable Mahaban tour with Pipon and between educational inspections, Stein hurried down to Nal. Located in Central Baluchistan where a series of valleys, corridorlike, connect the desert area to the west with the rugged hills fronting on the Indus to the east, Stein made "rapid preliminary surveys of such archaeologically interesting localities as came within"[5] easy reach. Glancing at that area rich in ancient mounds (one rising over 100 feet above the plain was encircled by cultural debris), neither Stein, nor Marshall nor any archaeologist then working in India could have appreciated that twenty years later Nal would speak eloquently of an as-yet undreamed-of past. Having acquiesced to the letter but not to the spirit of the director-general's request, Stein excused the brevity of his visit: "the demands of my main official duties [educational inspector] must at present restrict my archaeological activity on the North-West Frontier."

Stein had filled pages telling Publius about his Mahaban tour; of Baluchistan he wrote briefly 21 December 1904:: "I have rushed at great speed through a considerable part of the barren upland known as the Loralai District and have just returned from there to the railway which is to carry me back soon to the north." Christmas was four days off and Stein was looking forward to a "short interval of peace. . . . I enjoyed the long drives and rides through that maze of rocky gorges and desolate open valleys, notwithstanding the cold and wind." After telling of his immediate plans, he added a telling sentence showing that whether at Mahaban or Nal his thoughts were on his proposed expedition. "I have heard nothing fresh from Calcutta."

"Rejoice" thus read the cable sent by the Saint. One word. It was handed to Stein—he wrote Publius on 18 April 1905—"just as I was passing two mounds of ancient Peukelaotis [capital of the Yuzafzai, whom Alexander pacified on his march to Aornos] which my saintly patron, the great Pilgrim, has blessed by his mention. You may imagine the elated feelings with which I rode into Peshawar." Alexander the Great, Hsüan-tsang, and Central Asia—the triangle of Stein's innermost topography.

It was mid-April. Seven months had passed since Stein's offensive for a Second Expedition had begun, seven months while he wrote to friends to alert those strategically placed who could further his proposal. (His method, spelled out in advice to Andrews, was to keep in touch with well-placed friends" and talk "frankly to them of your wishes. The chance may come any day, but only if there is someone on the spot to catch it for you."[6]) Seven anxious months had passed while the bureaucracy processed his proposal. A cursory examination of the governmental procedure gives a glimpse of Stein's image as seen by the civil servants and the factors that operated for and against him. Their acerb comments and pointed questions were blunted. Did such men feel that by stating their reactions they could quash a human tornado intent on making its own path?

Few of the bureaucrats understood (nor did Stein himself) that his proof of the transmontane extension of ancient Indian culture was an instrument useful for transfrontier exploration. Was this the intent and meaning contained in the viceroy's cryptic: "the time was not yet ripe"? Younghusband's mission (the brainchild of Curzon's imperial mind), an incident in the "Great Game," had by then returned from Tibet. Was the time ripe to send out Stein's expedition to Lop–nor-Kansu-Shachou? Or, to put it most bluntly, was Stein a tool of jingoist plans? It is possible.

Did Stein knowingly go as such? The answer is no. Nothing in his most
private letters to Publius even hints at this.

The "Notes" in the Foreign Department's file contain the intramural
comments of the various departments considering Stein's proposal of 14
September 1904.[7] Calcutta's recommendation favoring Stein went to
London for official sanctioning 16 January 1905. Yet not until April
could Arnold cable his one joyous word, "Rejoice."

The comments of the Foreign Department: "As to the great value of
Stein's continuing his very interesting researches in Chinese Turkestan,
from a scientific point of view there can be no manner of doubt, and there
is no authority so competent as Dr. Stein to lay out a programme for their
extension. . . . He estimates the expenditure at Rupees 26,000 and it is impossible to cavil at the items which make up this figure. But this sum by no
means represents the total cost to Government. . . . We must include Dr.
Stein's salary for two years . . . bringing the total to half a lakh [50,000
rupees]. But this is not nearly the end: Dr. Stein wants to give up his
present work as quickly as possible [to prepare for the expedition] . . .
and that adds another Rupees 5,000. But we are not nearly ended, for his
sum is only to bring him back to Kashgar, and judging by the experience
of the last journey, he will want to make from there to Europe. . . . Then
will follow further applications, as on the last occasion, to remain on
deputation in London to arrange the proceeds of his tour, and applications
after his return to India for leisure to write up his reports.'

Having brought out the hidden costs in Stein's proposal, the official
next raised the issue of Stein's dual appointment "which though created
especially for him in July 1903, was not assumed until January 1904. . . .
I took the liberty of noting at the time my personal opinion that the
combination was not advisable in the interests of the public service, as
Dr. Stein was so keen on archaeology that I feared he would pay too much
attention to this branch of his duties to the detriment of education. . . .
Now, before he has held the post, specially created for him, for a year
he wishes to leave it again and as I have shown, for not less than three
years. The Secretary of State, in sanctioning the combination of the two
appointments, desired that the arrangement should be considered personal
to Dr. Stein. "Almost at the end of this sharp, quasi-hostile exegesis comes
a revealing sentence: *"But we are interested in trans-frontier exploration*
so this note may stand unless the Revenue and Agriculture Dept. prefer
that it should be withdrawn and that they take over the case" (italics
mine).

The Department of Revenue and Agriculture dealt with the cost figures: "The British Museum promises to purchase a part of the articles that may be discovered by Dr. Stein and to bear a substantial share of the cost. They also desire to be consulted in the matter. . . . The Secretary of State objected to the expenditure being met from Provincial funds [Bengal and Punjab]. It was also decided that the Lahore and Calcutta Museums should get their share of the collections [previously] made by Dr. Stein in return for the expenditure provided from Indian revenues. . . . I think we may safely say that the expenditure involved will not be less than Rupees 80,000."

The director-general of archaeology in India: after joining the others in making salaams to Dr. Stein's proposal, he discounted the threat implied by German and Russian competition: "Whatever collections which are made will be equally well cared for and made equally accessible to scholars in the Berlin or Hermitage Museum as they would be in the British. . . . It may be very well for museums to indulge in rivalry of this kind, but it should not influence us in deciding the present issue." Rather, Marshall trained his big guns on Stein's dual appointment, saying that during the year of its existence it had not proved a success: "As a matter of fact, he has accomplished practically nothing. . . . The excuse has always been that the preparation of his report on the first journey to Turkestan must come before all else, or that his administrative duties connected with Education allowed no time for Archaeology. . . . As regards the possibility of reductions in projected archaeological expenditure to meet the partial cost of this journey, . . . we should not, I venture to think, be justified in starving Indian work for the sake of trans-frontier exploration. . . .

"Dr. Stein's first journey should be completely published and the distribution of the antiquities between the British, Calcutta and Lahore Museums effected before he starts on his second. It cannot be disguised that the dangers attending the journey will be many. . . . Again, if he takes back to the British Museum another large collection, it is not unlikely that some confusion will result after a lapse of a few years."

The Finance Department contended at the time Stein's First Expedition was being considered that, if funds raised from Indian revenues were used, Indian museums should receive whatever antiques Stein recovered. This argument, seconded by Curzon, had soundly been made part of the sanction. For years it would be argued, debated, evaded—but all to no avail: the Finance Department's position with Curzon's endorsement would hold.

Referring to his earlier quid pro quo, the same official noted: "The expedition should, I fancy, be productive of very interesting results, but

there will be no special or direct gain to India (particularly as I understand
that we are not even to have the satisfaction of enriching Indian Museums) and we have full employment at home for all available funds. I
think therefore that we should limit our financial contribution to a *share*
of the cost . . . £3,000 would I think be very liberal and perhaps exces-
sively so in the interests of the Indian taxpaper."

Curzon: Again Curzon picked up this insistence that Indian museums
receive finds made with Indian-raised moneys. "Please see our dispatch
of 6 June 1901 [relating to the First Expedition], where we laid down
with the consent of the Secretary of State that the Calcutta and Lahore
Museums should have *first* claim to Dr. Stein's discoveries and that Dr.
Stein should divide his collection into three portions, subject to the above
condition."

As a result of this lengthy intramural discussion, the official recom-
mendation of the government of India in Calcutta went on 12 January
1905 to the secretary of state for India in London: it asked for sanction of
Stein's proposal for a Second Expedition. Three conditions were imposed:
that a limit of £3,000 be allocated from Indian revenues; that the Detailed
Report be finished and in print before Stein could start; and, that, if the
British Museum added the excess money required, the government of
India would grant it approximately two-fifths of the articles found.

Two weeks later, 25 January 1905, Stein wrote to Publius. "I can give
you satisfactory news about my main affair. On the 21st [Jan. 1905] I
received the letter of the G. I. [Government of India]. It relieved my
anxiety as to the fate of my proposal at Calcutta. They transferred the
issue to London. There, however, the ground was prepared already in
Oct. by Mr. Read and Sir. R. Douglas of the Brit. Museum who saw the
omnipotent Sir A. Godley at the I.O. [India Office] and secured his
sympathy for my plans. . . . A very reassuring letter from Maclagan (who
is now in the Rev. Dept., Calcutta) which I received two days ago, distinctly mentions that the 'G.I., rely for help on the Brit. Mus. authorities'
and 'hope from what they have heard' that this will be forthcoming. This
looks very much as if Lord C himself had been told about the plans when
he was still home; for I do not see anyone else could 'have heard' about
a matter so small in the eyes of the great. Is it not strange this elaborate
mutual encouragement in a matter so small? . . . I shall now propose
through Col. Deane that I be relieved from ordinary duties as early as
possible that I may have time to complete my Report & see it through the
press & to prepare detailed distribution proposals."

Stein's new proposals, logical and legitimate in his eyes, raised the subject of replacements, thereby throwing the spotlight on his dual appointment, an appointment "personal to Dr. Stein and . . . subject to reconsideration whenever it may be necessary to appoint a successor to him."[8] His leaving thus made two replacements mandatory and involved both the Home Department, governing education in that province, and Marshall's department. The former asked "What is to be done with Stein on his return from his approaching expedition to Chinese Turkestan?" Marshall, in the "Notes," described the present situation as highly unsatisfactory and the future "as a disaster if Stein resumed his dual appointment."

Stein told Marshall that he would relinquish his dual appointment if he were named archaeological explorer, explaining that though this was "a great improvement compared with the conditions amidst which I previously had to struggle it is far from giving me the scope in which I could turn my special qualifications to full advantage." Archaeological explorer: the post had the endorsement of that distinguished administrator, Sir Charles Rivas [lieutenant-governor of the Punjab], "who deemed it to be both feasible and desirable," and the *ex cathedra* blessing of Sénart, the eminent French Indologist, who considered Chinese Turkestan "un patrimoine de l'inde." Stein felt that if the idea of such a post must remain a dream, then financial consideration compelled him to "request adherence to the original terms of my appointment." Economic independence was a reality equally important.

Beneath a correct exchange, Marshall and Stein were continuing the struggle for power. In a letter of 29 July 1905 the latter could express his true feelings to Publius. "Marshall, writing to me very politely and with seeming deference, had suggested that in view of my absence for circ. 5 years, I might facilitate arrangts. about the Arch. Surveyorship by agreeing to have it filled independently and permanently! The argument was that the administrative labours of my main post made arch. work difficult of attainment for me, and this would be probably still more the case after my return when I should, perhaps, like to devote myself *solely* to educational duties! I thought it best to meet this ingenious and disingenuous attempt by taking the offensive quite frankly. Whether the result will be at least a partial success, only the Gods know. Col. Deane & O'Dwyer [revenue commissioner to the Northwest Province] thought that the case made out by me was overwhelming. . . . But things are queerly managed by the G.I. when they must depend on an advisor of M's knowledge and experience.

"It is a comfort to know that by 1909 I may hope to have the minimum private income I need, and small as it will be relatively, it will protect me against humiliations I should find it hard to put up with." "Humiliations," a word new to Stein's vocabulary, meant the wounds he might suffer in his struggle against Marshall; it revealed a new sense of powerlessness brought by a new antagonist whose authority lay outside that represented by men like Colonel Deane and O'Dwyer.

It is quite possible that Stein's anger against Marshall was a substitute for the despair in which Ernst's family was mired. Perched on his Kaghan alp, his pleasure in solitude was shattered by his remoteness. "I cannot get free of [concern for dear ones] when I think of the grief and always intense sorrow in which my poor sister-in-law spends her days. I wonder whether I mentioned to you," he asked Mrs. Allen, 12 August 1905, "that my nephew's transfer to my old school [in Dresden] proved a failure. My sister-in-law could not bear the separation and as the boy naturally felt miserable at first in strange surroundings (just as I did), she ultimately after barely three months' trial took him away. . . . He lost, alas, far too soon the firm guidance of his father."

On another front, his grief and pride were allayed by his victory over the "Gujar vandalism and official ignorance of half a century." He had watched the beauty and health of the alpine meadows being destroyed by the Gujars' sheep and goats and the denuding of the higher slopes. "The whole has been closed to grazing for three years and the awful scars on the mountain sides where the earth is sliding down at the slightest rain may yet heal. . . . The Gujars and their far more culpable feudal landlords, the degenerate set of Kaghan Saiyids, will remember this summer of judgment. Chaukidars [inspectors] have been appointed over all high Gujar forests—at their expense, of course, and the happy times of freedom to mangle, burn, ring and skin trees are gone."[9] The Saiyids, Stein learned, regarded his "residence in their valley as the 'will of God,' and my persistence in the matter of forest protection in much the same light."

The humiliation Marshall had caused him to feel was erased when, the end of August, he received word that his six months' special duty to be spent in Kashmir had been sanctioned. Of the two replacements who were to take over his duties as of October 1, he heard nothing, though his "reminders had gone out some time ago."[10] By October, his high alpine camp would be having a foretaste of winter; nevertheless he wanted to stay there and turn over his duties at Battakundi so as to go from there "straight across the mountains and the Kishanganga into Kashmir." It was not a whim. "By following a track which I partly 'explored' for historical purposes in 1892 (via the old sanctuary of Sarada, now Shardi), I

could reach the debouchure of the Sind in circ. 5 days. There I should like
to pitch my camp where it stood in May, 1900, in full view of Mohand
Marg, . . .

"I shall try to avoid Peshawar until next April when I wish to set out
from there for Swat, Dir, Chitral and the Pamirs! I think I told you of my
intended route through the Yarkhun valley up to the foot of the Darkot
Pass and then, if the Foreign Office lets me, up the headwaters of the Oxus
to the Wakhjir Pass and thus down into Tashkurgan. It would mean
passing through (uninhabited) Afghan territory for circ. 3 days. But if
the F.O. are afraid to ask for the Amir's permission, I have a route *in
petto* [confidential] which would allow me to dodge the last Afghan post
quite safely. It lies over the Kara-bort Pass and is practicable for ponies.
You see, I have not in vain used the 'Secret' Hindukush Gazeteer.

"The resignation of Lord C has, of course, touched me too. I had hoped
to thank him personally in Kashmir for what he had done to help me. The
manner in which his reign closes, is certainly regrettable though in keeping with a certain historical nemesis." Was Stein aware that most probably
it was Curzon's appetite for transfrontier activity, his "Forward policy"
imposed on the government that had made the Second Expedition possible? The ease, the speed in acceding to requests which would expedite
the journey, all reflected Curzon's power. With the viceroy's resignation,
an era had ended. During Curzon's brief rule Stein reached a plateau he
might never have attained without the luck to come to flower in so propitious a time.

Like an arrow let fly, at 4 P.M. on October 1 the date he had set, Stein
"marched off from Battakundi up the Kaghan Road",[11] hoping "to have
taken leave for good from all routine duties. For this liberation I shall
gratefully remember Battakundi." Due to his late start it was nightfall
before he caught up with his party, which had been sent on ahead. "The
last four or five miles' tramp over boulder-strewn slopes in absolute darkness was a somewhat trying experience. While I tumbled onwards, the
headache due to previous 'rushing' and inadequate rests passed away in
the cool night air, and when I safely got to my camp all official worries
seemed already to lie behind me a long way." His week's journey was
strenuous enough to satisfy him and each day's mileage brought him
nearer to his beloved valley. "On the morning of the 6th [October] I set
out for my final march. . . . It was a long one, some 34 miles and lasted
close to 14 hours, but its pleasures were great too. The mules went gaily
along the turfy paths, glad for the change from the awful tracks over

which they had passed so bravely. . . . Yesterday forenoon I passed once more under the Seven Bridges [Srinagar]." He pitched his tent near his old camp.

There "I gathered together my old entourage for a few hours. It was a pleasure to talk the language of the gods and though my interests have now moved far northward, I shall try to keep old Pandit Sahajabhatta by me when I occupy winter quarters at Gukpar. He had been with me and Govind Paul in the old days."[12]

Stein struggled to get his Detailed Report into print. His friends rallied to his aid. Publius proofread for him. Stein acknowledged his help in a letter of 23 December 1905: "I was truly amazed how thoroughly you had done the reading of all those dry pages and that you had gone to the length of verifying quotations! . . . You have noticed inconsistencies in Chinese transcription and other errors which had escaped me. . . . I feel great misgivings about their all being attended to unless corrected proofs are arranged for. Chavannes is to get a set but this will extend only to Chinese words. In this face of unexpected necessity I see no other quarter to appeal to but St. Thomas [Arnold] . . . [his] help would be most valuable because he could keep an eye on Muhammadan names and dispose of small queries that might arise in connection with them. . . . It seemed a special favour of fate that you should be in Oxford and in close touch with the Clarendon Press; but it is still hard that you should be made to suffer by all my wearisome labour."

Three days before 1905 ended, Stein received the telegram for which he had been waiting. The government of India notified "me that my proposals, written *after* Macartney's news about the German [expedition], had been sanctioned. So I can start even if the proofs are all not read and without waiting until all divinities on both sides of the Ocean have agreed to and formally sanctioned the proposals as to the distribution of finds. I wonder whether I could have secured these concessions if he who dictated these impossible conditions were still reigning. . . . So at last I can fix my starting time; before the middle of April I hope.[13]

Delays, irritations, distress vanished. "All practical preparations are well ahead. There is nothing left to order from Europe except little things that can go by parcel post. . . . Ram Singh's detached duty is sanctioned by everybody—from the Commander-in-Chief downwards. Is it not a wonderful machine that manages to get on with so many wheels!"

One by one the restrictions on his departure had been lifted. The way was clear for his Second Expedition. A year after the Saint had so ordered, Stein could rejoice.

3

The Second
Central Asian Expedition
20 April 1906–13 November 1908

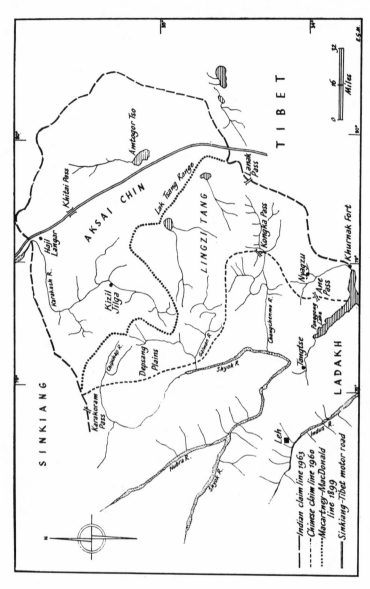

The Aksai Chin, the forbidding white stone desert south of the Kunlun. From Alastair Lamb, *The China-India Border* (London: Oxford University Press, © Royal Institute of International Affairs 1964), map 6.

12

The twenty-three days between 21 May and 13 June 1907 are perhaps the high point of Stein's achievement. It is almost as though his reputation rests on what he was able to do in that brief moment of his long career. These were, of course, the days he spent at Ch'ien-fo-tung, the "Caves of the Thousand Buddhas," near China's westernmost oasis, Tun-huang.

He had long known of the caves. Their location, their number, and their extraordinary frescoes and sculptures had been glowingly described by his friend Professor de Lóczy, head of the Hungarian Geological Survey, who, as a member of Count Széchenyi's expedition, had visited them in 1879. Twenty years later, C. E. Bonin of the French Diplomatic Service had mentioned them in a short published notice of his crossing of China to the west. It was to acquaint himself with these sacred grottoes—perhaps to acquire fine frescoes—that Stein's proposal had included travel in Kansu, China's northwestern frontier province. The fact that the existence of the caves was known to others cast German and French expeditions going to Chinese Turkestan at the same time as potential rivals. Their presence there made Stein complain endlessly at the precious time lost while the government was considering his proposal and made him see the conditions imposed by Curzon as prerequisites to his starting as heavy, unwarranted handicaps.

Was it the threat posed by competitors that gives an emotional unity to the Second Expedition? If he had handicaps, Stein also had advantages. No longer a tyro, he was a seasoned veteran of the Taklamakan, knowledgeable, prepared, sure enough of himself to take risks. He had tested himself against the brutal mountains and deserts and had found under the covering sands documents, religious relics, and artifacts of daily use in a forgotten past. For him, he thought, there could be no surprises. If he argued and listed reasons for his continued excavations and for extending his geographical explorations, within his heart he also longed to return to peoples and places where he had been a free agent doing what he wanted to do for reasons whose meaning satisfied his innermost being. And though the Second Expedition was phrased as a continuation and extension of his first, the introduction of a new element gives the second a unity, a dramatic unity. Entitled "the Secret of Tun-huang," the drama has four acts. Its first act, "The Race to Lou-lan," sets the stage, introduces the actors, and starts the action; the second, "The Treasure of Ch'ien-fo-tung," is an Aladdin-like episode whose jinn was Hsüan-tsang, Stein's patron saint; the third, "Two Daring Crossings," has a scenario of daring in mountains and desert; and the fourth, "Misadventure at High Altitudes," is a nightmare of near disaster; a final tableau might be called (with apologies to Shakespeare) "All's Well that Ends Well."

The play's second male role was taken by Chiang-ssŭ-yeh, the Chinese secretary who accompanied Stein. Macartney had picked him, judging him to be both clever and straight; Stein found him "not merely an excellent teacher and secretary but a devoted helpmate ever ready to face hardships for the sake of my scientific interests. His vivacity and inexhaustible flow of conversation lent attractions to the lessons I used to take in the saddle while doing long marches or else in camp when pitched early enough. Once I had mastered the very rudiments of colloquial Chinese, his ever-cheerful companionship became a great resource during long months of lonely travel and exertion. From the very first his unfailing care, good manners and tact assured me that I had not merely a faithful helper by my side but a gentle man and true comrade. With the historical sense innate in every educated Chinese, he took to archaeological work like a young duck to the water. With all his scholarly interests in matters of a dead past, he proved to have a keen eye also for things and people of this world. . . . What it took time to make sure of, and what always surprised me afresh, was the cheerful indifference and the physical toughness with which [the slight, wiry] Chiang could bear up against all privations and discomforts."[1]

So Stein described him, remembering the many months spent together and the secrets they shared about the treasure. How did they converse? Their first exchanges were in "the queer lingo which has grown up in Sinkiang by a constant process of clipping and transmogrification in Chinese mouths unable to pronounce the consonantal combinations of real

Turki or to use its elaborate inflectional system."[2] Yet on so wobbly and unlikely a linguistic bridge they quickly crossed over into friendship. When Stein floundered in the "eel-like perplexity of Chinese phonetics and the terrible snares of tonic accents so hard for unmusical ears to distinguish," Chiang's sympathetic understanding kept him afloat. At the critical time, Chiang would turn the key that unlocked the hidden treasure.

13

Act One: The Race to Lou-Lan

If Stein's secret hope was to visit the painted caves of Chi-ien-fo-tung, it had less urgency than his clearly stated purpose—to examine "the ruined sites north of Lop-nor, discovered by Hedin on his memorable journey of 1900,"[1] Stein wrote in his *Ruins of Desert Cathay*, the two-volume Personal Narrative of the expedition. Lop-nor, he was convinced, was a target at which the other expeditions would aim. To the handicaps imposed by Curzon—to complete *Ancient Khotan* and provide the distribution list of his finds—and his tardy start, he himself added another: to approach Lop-nor by the longer, southern route while his rivals used the quicker one along the northern oases of the Taklamakan. His unique advantage was having Macartney at Kashgar to observe and report on both the German and French parties.

The tension began even as he started in 1906. His letters to Publius relay the information he received. 6 January 1906: "Macartney on Nov. 30 [1905] reported that Lecoq and Bartus [of the German expedition] were still at Kashgar impatiently awaiting Grünwedel [head of the expedition] mysteriously delayed at Osh [the Russian railhead] for over a month—perhaps through troubles on the Russian railways. M reports jealousies between them and Grünwedel their chief; but even these may not help much. Gr. is a slow-moving man who wants to do things thoroughly. May he succeed in having his will and keeping his young Museum assistants to Turfan. But the true race will be with the Frenchmen—and how easy it would have been to avoid it if only this great Indian machine could move quicker."

Macartney kept close watch. (Whether out of patriotism—for expeditions like sports or military maneuvers have political implications—or friendship, it is impossible to say.) Stein passed Macartney's news on to Publius 20 January 1906: "It is a relief to know that 'the party' will stick to Kuchar [a northern oasis] and may the genius loci and Grünwedel's personal disposition keep it there until I have got to Lop-nor. It illustrates what I always held to be the drawback of parties. Lecoq & Bartus, Gr.'s ardent young myrmidons—and rivals—have been in Kashgar since early October & might have (no doubt to Macartney's great joy) spent Christmas there. Pelliot & the Frenchmen are to set out from France 'in the spring.' I am wicked enough to wish that the Russian route might continue to be barred even then. If they decide to go via India I may well hope to get ahead; for a hint to the Foreign Dept. would probably suffice to keep them to the Ladak route which is slow and difficult. My own plan is to keep council to myself & be on the ground before either Germans or Frenchmen know exactly of my start. It may be necessary to push on to Lop-nor during the summer, no pleasure indeed, but possible, & to raid desert sites as soon as the worst Burans are over."

At the same time Stein was trying frantically to get through the proofs of *Ancient Khotan*. Working against distance as well as time, he was aided by his three friends—Publius, the Baron, and the Saint. 10 February 1906, to Publius: "What am I to do about proofs that do not arrive before my start? I should like to read them enroute in spite of want of time." Question and suggestion went to Publius, the mainstay in the task. "I find occasionally little mistakes due to our friend Colton's editing. He turned e.g. honest Mirzā Haidar into a robber! But his care is great. Is debouchure really not in English nor French? I find it in Confidential Gazetteers on every page." The oft-repeated cry had been uttered in the previous 20 January letter: "But why could they not let me start last summer & spare me all these anxieties. I owe it solely, I think, to the 'Great Proconsul' [Curzon] who could not leave this humble affair to Wilson. If I now get in time to my goal it will be only through that good luck which is said to have more than once saved British interests in the face of great odds. But I am afraid I cannot claim such special providence."

As he wrote to Publius on 10 February 1906, Stein could feel satisfied that " the long letters of instructions I had to write about [my preparations] to the Baron, the Military Equipment Co., the Royal Geographic Society, etc., have achieved what is needful. But if you saw the bulk of the files, official and preparatory! I am going to keep a note of their weight before the start." In his letter of 17 February 1906, he indicated that he was still waiting word from Sir Louis Dane, secretary to the government of India, Foreign Department, who was trying to secure permission from the amir of Afghanistan for Stein to use the "short cut across the Pamirs. I do *not* hope that they will apply to the Amir and court a snub. All I hope for is for a discreet closing of the Foreign Office eyes. That is what Sir Harold Deane did." With his customary timetable for preparedness, Stein had begun assembling "the most necessary of my camp equipment, cameras and compressed stuffs."

A minor irritation was a ponderous concern raised by "that somewhat 'heavy' Major Godfrey who is now Political Agent for Dir, Swat & Chitral whose personality is not unknown to you both, [who] is making sapient objections to my early crossing of the passes. However, Col. Deane who knows them just as well as G does, and me far better, believes in the practicability of my programme & as luckily he is the Chief and not that 'meritorious' amateur archaeologist (Islam Akhun's firsthand dupe), I hope to have my way. But how much needless writing it takes."

By April the prelude to the start began. He reached Lahore on the 4th, "having successfully covered the 220 miles from Gupkar to Rpindi in 38 hours in spite of broken bridges and many land slips—to miss the night train by 15 minutes owing to a change in the timetable!" He allowed a four-day stopover for dental work: the trauma of Tekwakkel was, if possible, to be avoided. By chance (or was it Stein's good luck?) "the Viceroy is to be in Peshawar just about the same time when I shall say goodbye to India for a time, just as in 1902. But change there is." His objective in going to Peshawar was to get "the Viceroy to take a little interest in my 'Archæological Explorership' (of which the official gods have weakly 'deferred consideration' until my return!) and in Balkh, too, as a reserve scheme."

Lord Minto, whose great-grandfather had served illustriously as viceroy, had but recently replaced Curzon. Stein first saw the "new Lord of the Indies when the 'Administration' of the Frontier Province gathered at the Peshawar station for his official reception."[2] Busy as Deane was with the viceregal visit to his province, he still had time for a quiet talk with Stein. It was their last conversation. Sir Harold, "who had always been my truest friend and patron," a "born ruler of men whose strength of mind and body impressed the most turbulent tribesmen," died while Stein was in Turkestan.

At the official reception, almost on the eve of his departure, Stein was handed the amir's permission to pass through the Upper Wakhan, a permission regarded with much satisfaction by the government. Stein had been asked to communicate the number of men accompanying him. He wired Calcutta: "My party will comprise Surveyor Ramsingh, Naik [Corporal] Ramsingh, one Hindu follower [Jasvant Singh], two Turki, three Muhammadan followers; eight in all besides some local pony men needed for about 15 baggage animals."[3] The firman permitted Stein his choice of the route and commanded the "Frontier Officers of the God-granted Government of Afghanistan on the Chitral Fronter" to exert themselves and be attentive to the protection of Stein's party. Stein would learn how crucial would be the amir-given assistance.

From Peshawar he went to Abbottobad, where men and supplies were collected. "Muhammadju & another Turki safely joined me there after narrowly escaping being buried in an avalanche on the Burzil [Pass] which killed 7 of their companions. Surveyor Ram Singh and Jasvant Singh also joined me there; they are much the same as when I last saw them five years ago. Naik [Corporal] Ram Singh, of the Sappers & Miners [royal engineers], proves invaluable." The naik "brought a mass of small things made for me at the Workshops [of the First Sappers & Miners]: a set of carpenter's tools 'compressed' on the lightest scale"[4] which will enable him "to make strong crates, scaffolding, etc. Alas, it means ½ a pony more."[5]

Because his route passed through a potentially dangerous region, Stein asked the military for two carbines, three revolvers, and ammunition. "I should have them available for my use and that of my Indian assistant. I shall be visiting localities situated at a great distance from the ordinary trade routes or Chinese military posts. . . . The provision of some means of defence seems under the circumstances a reasonable precaution, particularly if it is taken into account that my work will necessarily give rise to the belief that I collect hidden treasure."[6]

At Chakdara, in Swat, first visited ten years ago, two telegrams awaited him. Macartney answered requests made in a February letter: "Can find Turki speaking Chinese Munshi [secretary]. Am doing utmost to purchase tengas at good rate, Sovereigns not current but gold roubles are easily exchanged in tengas. Cook understudy difficult to procure; am on lookout. Sadiq now in Chinese prison but if you want him can probably get him out. Have written to Taotai requesting Amban Sarikol be instructed to furnish twenty yaks to be ready on Chinese side, Wakhjir Pass about 8th May. . . . If you have not Afghan permission to pass through Wakhan premature arrival of yaks there may cause trouble." The second, Major Godfrey's, dealt with earlier arrangements. "After further information from Chitral will endeavour to get you and coolies over the Pass by end of last week in April provided weather fine as at present. If weather at all doubtful start should be delayed until clear."[7]

"I started yesterday from Chakdara," Stein wrote the Baron on 29 April. "Four marches through easy but not exactly too safe tracts have been covered. Tomorrow I expect to reach Dir and then comes the difficult bit of work, the crossing of the Lowari Pass (10,250) still deeply buried in snow and exposed to avalanches. However all precautions will be taken & the risks minimized. Beyond I shall find things easy in Chitral where a hospitable reception awaits me." The same day he had put the tribally doubtful zone behind, Stein reassured Publius, "I hope with these lines," he added, "the telegraphic message of our safe crossing [the Lowerai] will reach you. There is no doubt about it being an exceptionally bad season. But I shall use all humanly possible precautions and, of course, not dream of attempting the passage unless the weather is quite safe as it seems at present. . . . The snowy range towards Chitral is fully in view from here though some 36 miles of (good) road still separate us."

Crossing the Lowari is not a measure of Stein's bravery—that was reserved for other situations—but an example of his prudence, careful planning, and long experience on those glaciers and snowy peaks. The pass, separating Dir from Chitral, was the dividing height between Pathans and Dards, between a miserable, rain-soaked, narrow, prisonlike valley and an open boulevard leading toward Afghanistan. On 3 May, Stein moved his camp to the foot of the Lowari, above Mirga, "the last hamlet where Captain Knollys, the assistant political agent for Chitral, was caught early in December of the previous year by an avalanche and, though buried himself, by heroic exertions saved his party."[8] Stein's advance was cautious: "every load was lightened so as not to exceed forty pounds or less; two men were detached for carriage in turns." The carriers were divided into three groups: the first led by Stein, the other two by the two Ram Singhs; each detachment had four men who carried spades and ropes to rescue anyone in trouble; each group was to advance at fifteen-minute intervals to preclude a weight-starting movement in the treacherous snow slope. Finally, the crossing was to be done at night when the snow had a hard crust. At 1 A.M., while the moon still shone, carrying lanterns, Stein started with the first party. Fifteen minutes later, the naik "with military punctuality" followed; the third, after another fifteen minutes.

Almost immediately the "valley bottom completely disappeared under snow. . . . no sound of the streams flowing beneath this continuous slope of snow bridges reached me. Huge avalanches had swept down at intervals from the steep spurs and gorges on either side for the past months, leaving their tracks marked by moraine-like banks of hard snow. For the sake of the load-carrying men steps were cut in these banks by the spare men with me. The advance was very slow work. . . . From falls of fresh snow we were protected by the bitter cold of the night which had frozen the surface hard. When the first flush of the dawn showed over the spurs eastward, the narrow saddle of the pass came into sight, and a little before 5 A.M. my party gained the top."

The descent was, if anything, worse, "A real snow wall, some 80 to 100 feet high," reared up from a "remarkably steep slope where any heavy object once in motion would be swept irresistibly down. . . . The Dak runners' tracks descended abruptly in narrow zigzags, the steps trodden into the hard snow being often at three or four feet vertical intervals." Only when all the heavily loaded coolies, assisted by Chitralis sent by Captain Knollys, had safely negotiated the end of those spaced, icy steps, could Stein start down and escape the icy wind sweeping the summit.

From Drosh, the capital of Chitral, Stein announced the safe crossing to Publius by telegram. The friends had devised a code, joining thrift to swift communication. "Alburnum Lowari crossed comfortably singma hope Madame fralpari[?] greetings."[9] Nine words which said, "Arrived here after very good passage. Lowari crossed comfortably. Weather very fine. Hope Madam making good progress. Greetings." Madam was very much in Stein's thoughts: Publius had written that she was expecting a child and they had decided it was to be named Aurel or Aurelia. The child, alas, was stillborn.

The two-volume *Ruins of Desert Cathay: Personal Narrative of Explorations in Central Asia and Westernmost China,* tells in fascinating detail about the small mountain kingdoms through which Stein advanced. Here it is only possible to relate the major experiences while enroute to Kashgar by this different approach. From Chitral he sent Sir Louis Dane a letter thanking him for "smoothing my way to Turkestan. I ought to let you know how carefully the Afghan authorities in Wakhan are preparing to carry out the Amir's orders. . . . I arrived here after a safe crossing of the Lowari and received the cheerful intelligence that five Wakhis sent down from Sarhad [headquarters of the general commanding the Oxus Provinces of Afghanistan] to meet me had just come in. The men report that an Afghan colonel with about 100 men had reached Sarhad, that ample supplies and fifty (!) ponies had been collected to help me onwards, and that special orders had been received from Kabul to look after my 'safety' and comfort. All this sounds very promising for an easy progress to the Taghdumbash Pamir. I only wish the Afghan colonel will not doubt my identity—because I come with a modest following, as his predecessor in 1894 pretended to do in the case of Lord Curzon."[10]

Making his way through Chitral and neighboring Mastuj, Stein was not content to limit his inspection to sites Hsüan-tsang had mentioned, and went out of his way to visit the 15,400-foot Darkot Pass. His long day's climb, from 3 A.M. to 8 P.M., was inspired by the "record preserved in the T'ang Annals of the memorable exploit by which the Chinese General Kao Hsien-chih in 747 A.D. led his force over it for the successful invasion of Yasin and Gilgit"[11] to rout the Tibetans established there. As he retraced the general's steps, he found that again and again what should have been the crest proved to be but a shoulder "on the easy but seemingly never-ending slope. I began to understand the story of Kao Hsien-chih's crossing, the dismay and confusion of his Chinese troops when, brought face to face with the precipitous descent on the south side [a hamlet 6000 feet below], they realized to what height they had ascended."[12] Stein would have liked to raise a monument to the general who for the first and perhaps the last time led an army of 3,000 men across the Pamirs; but "there was nothing but snow and ice for many feet below us. But I could not refrain from writing a note on the spot to my friend M. Chavannes, the great French Sinologist, whose learning had first revived the story of that memorable expedition."[13]

Stein continued to work his way through the Pamir Knot, the name given to that area of rock and ice and snow where the Karakorum, Pamirs, and Hindukush meet. The next pass was the Baroghil saddle, "that remarkable depression of the Hindukush range where the watershed between the Indus and the Oxus drops to only 12,400 feet."[14] Soft snow and a brilliantly warm day caused the laden ponies to become hopelessly stuck after they had made a few hundred yards. Further advance with the exhausted ponies was impossible. Just then, when the prospect of the easiest route to the Oxus looked most gloomy, a lively young Wakhi arrived bringing news of help from the Afghans. How, Stein wondered, would this worn-out Chitral transport have fared had he not secured the amir's permission and been forced to take the route he had chosen as a secret desperate alternative.

The night's bitter cold froze the snow into a glittering surface. Starting at dawn, they reached the level plain of the saddle and started down. The first few miles were easy, but by 9 A.M. the animals were floundering in the deep snow, and "the help of the fifteen sturdy Wakhis who had met us at the saddle proved most welcome. It would have been quite impossible to get the animals, even unladen, through the snow-choked gorge into which the Baroghil drainage passes farther down." Out of the snow, on Afghan soil, Stein waited "while the baggage was being brought down in driblets" and then sat down to have a modest breakfast. Just then the arrival of two Afghan officers was announced. "Painfully aware as I was of my sadly neglected appearance, . . . and of the increased regard which, once beyond the furthermost limits of Indian authority, Oriental notions had a right to claim from me, I hastened to don *en plein air* my best travelling suit, brought down by forethought in a saddle bag."[15]

"I had a glorious ride down to the Oxus, felt back once more in my beloved Turkestan and had a grand reception from the Colonel [a strong, six-foot, sixty-year-old man], deputed to look after me [he had been patiently waiting for four weeks] and the 'Governor' of Wakhan," wrote Stein to Allen, 30 May 1906. "If I had been a 'General Sahib' they could not have done more for me. I had even a guard of honor to inspect, their presence in this barren valley for the last four weeks must have been a sad burden to the poor Wakhis. The Colonel is a dear old man full of Badakhshan which is home and which I am so eager to explore. . . . I have done my best to impress him with my hopes in that direction. Who knows whether after all Sarhad will not prove the gate to me for the middle Oxus. . . . I am fit and rapidly getting a fresh skin on my face which the sun on the Darkot had blistered. But that is a trifle."

Stein wrote a letter to His Majesty, Amir Habibulla Khan, Ghazi Bahadur, etc., king of Afghanistan, and sent it through the Foreign Office so that it could be accompanied by a copy of his *Sand-buried Ruins of Khotan*. There it caused a bureaucratic flurry: "Dr. Stein uses the expression 'Your Majesty' which we avoid. For this, among other reasons, it will perhaps be better not to send the letter formally to the Amir, though we may give a copy to the Envoy for transmission."[16] Lord Minto brushed aside such civil-servant niceties: "It does not appear to me to matter that Dr. Stein has addressed the Amir as 'Your Majesty.' Dr. Stein is not the Government of India, and it need not implicate us." There is no evidence that Stein ever received an answer to his letter.

A month after starting from Chakdara, Stein was at Sarikol. "I have been greeted as an old friend and enjoy thus being remembered at the very threshold of my Turkestan campaign. My next goal is Kashgar which I hope to reach about the 9th of June."[17] At 9:30 P.M., 8 June, "I arrived after a sixty miles' ride, . . . a long & hot ride of some 17 hours, partly in a dust-storm of true local colour. But I kept to my date and found my way to the open gate of Chini Bagh in spite of the darkness. I must have looked like a 'Taklamakanchi' of the true sort with all the dust I brought in."[18]

Added to the friendship and comforts offered by Chini-Bagh, was the joyous "advent of a new master, the British Baby. Eric, the Macartney's little son . . . running about the garden."[19] Memories of his own early childhood and "of days quite as sunny spent in gardens which then to youthful eyes seemed as vast a kingdom" are the measure of how much at home he felt under his friend's roof. Even the tensions he brought with him were allayed. As he wrote in the 9 June letter to Publius, "Macartney helps me most vigorously to push on with my preparations & thus to keep my start. . . . Up to the present my French rivals have not turned up. . . . the Germans are about Korla [an oasis in the north] and apparently undecided whether to go on to China—or India. M., of course, watches their plans and movements." A letter from Publius had brought sad news. "I wish," Stein replied, sensing Madam's despair, "that Mrs. Allen may gain strength daily and all care about her health be taken off your mind. But I rely with you on her courage and inexhaustible vigour."

In another letter to Publius, written ten days after his arrival, 19 June 1906, Stein could indicate how quickly he was getting his expedition organized. "With Macartney's help I have managed to complete all my preparations. I have got my eight camels, fine strapping creatures, together with Hassan Akhun, my versatile young camelman of 1900–01, who is eager once more to search for 'treasure.' Ponies, too, have been picked up with no small trouble, for prices have risen greatly at Kashgar and I must try to exercise economy. . . . Everybody is bent on spending his days at picnics in gardens in spite of the increased 'struggle for life.' Kashgar is still the happy old place where mere existence is no serious business. . . . My visits to all the Yamens and the Mandarins' return calls took up much time but revived happy reminiscences. . . . It was arranged easily that I can draw on the Khotan & Keriya Yamens whatever my financial needs are. . . .

"The Russian consul [Kolokoloff, who had replaced Petrovsky] has been very pleasant & attentive, and things are far smoother—on the surface—in that direction than they were in former times. With Pan-Darin, now holding a high post at Urumchi [the capital of Sinkiang], I exchanged telegraphic greetings in Chinese!" Every free minute Stein spent working on proofs of *Ancient Khotan*. Even in the shady garden, "Kashgar was getting decidedly hot."

Father Hendricks, who had been wasting away from a cancerlike illness, died in loneliness the day before Stein had set for his departure. "His last journey to the little Russian cemetery was a pathetic function for the sake of which I put off my start until late in the day [23 June]." In the priest's locked-up house, the body rested alone,"but the grizzly-haired Chinese shoemaker, the solitary convert whom the priest claimed, had faithfully kept watch on the house-top."[20] His two rooms were dim and dusty, and books, maps, pamphlets "mingled in utter confusion with household objects and implements used for his chief practical occupation, the making of Kashgar wine. There was a humble altar at which he used to say his solitary masses, and not far from it the open trap-door giving access to the roughly-cut cavity which served as wine-cellar and laboratory." Sturdy cossacks made the coffin. "The Russians behaved very well and 'some' coldness between the Consul and Macartney was dropped. So the old man acted his accustomed part even in death."[21]

Heat had settled on the land. "I did the desert march to the edge of the Yarkand oasis in comfort during the night, starting at 1 A.M." From there he returned "corrected proofs pp 441–456 without having had [Publius's] revision. Forgive that I asked the Press [Oxford University, Clarendon Press] to let you glance through those pages again. The completion of the Appendices: what a load will be off my mind!" He merely paused at Yarkand and pushed on to Kokyar, a village in the mountains, to spend the two weeks he had estimated "to finish off all the book in peace and coolness and at the same time have a rest, the first since my journey to Battakundi."[22] His retreat was well chosen. "I live & work in my little tent under the shade of ripening apricot trees."[23] The camels and ponies were sent to fatten on lush grazing ground—"they are off my budget for the time which is an advantage. By the 25th [July] I intend starting for Khotan by the route along the upper hills which will be cooler. . . . I should like to join [Ram Singh] about the middle of August for a little survey work up the Yurungkash."

On 5 August, "the day which brought me back to 'the Kingdom' will long live in my memory as one of the happiest I spent in Khotan,"[24] Stein was welcomed at its border. His letter to Publius, 8 August 1906 gives the details: "The new Amban, a very intelligent man & full of energy, seems bent on paying me every possible attention & the way in which I was conducted to the capital from the Pigeons' Shrine onwards would have satisfied even old Hsüan-tsang. The alacrity & evident attachment with which old friends & acquaintances have gathered around me, was even more pleasing. I quite felt coming back to my field again—and the knowledge of having worked to get to Khotan all these years makes me accept these favours of Fate with good conscience." His elderly friend, Akhun Beg, in whose garden he again pitched his tent to enjoy its peace and shelter, was about to set off for a pilgrimage and Stein worried that he would not be able to stand up to the rigors of travel. Would he, as they said their farewells, see his good friend again; would he return?

"Ram Singh joined me yesterday [7 August] after a very rough but successful passage of the Hindukush [Pass]. He has managed to fill up the survey of 1900 with the piece of terra incognita left in the great bend of the Karakash R. and our proposed work in the Karanghutagh Mts. will thus be considerably simplified. I am setting out for this after two days, many repairs being needed for Ram Singh's equipment which has suffered greatly over ice & rock. The Amban is offering all possible help; Islam Beg, my energetic Darogha though now a full-blown Beg, has eagerly volunteered for the trip. If . . . the weather is tolerably fair, I may hope to finish off the exploration of the Yurungkash, too, from the valleys surveyed six years ago.

"Would it be possible to secure a small number of reprints (say twenty each for the collaborators of my Appendices. They have, of course, no remuneration for their contribution . . . and would probably be glad to have them for reference." And then, the pleasant events described, the requests explained, Stein makes an emotional equation at the end of his letter. "How cheering a thought it is to me that five years' absence had changed so little for me at Khotan. It will almost be as long a time before we can be happily reunited again. May the changes affected by time seem small to us, too, when I can return once more to that haven of friendly sympathy and comforting kindness which you hold open for me."

On this expedition there was no need to spell out for the Allens and Andrewses the testing quality of mountains or desert; this had been explained before and had been made real by the photographs he showed them. This time his letters told of problems and successes, the new, the unexpected, and always the increasing discoveries of the past.

Returned from topographical work in the Kunlun, though it was early September and the heat, dust, and glare still ruled the desert, he hurried to get on with his excavations of desert sites. By 14 September he was back at the Rawak stupa. He writes to Andrews, 8 October 1906: "I found much of the sculptures I had cleared buried again under big dunes and the rest, alas, destroyed by treasure seekers. I steered south"—the nautical language so appropriate for the sea of sand—"to the Hangúya Tati [débris-strewn old site]. There from the débris of a temple with walls levelled down to a few feet already in old times, I recovered a mass of small terracotta sculptures most closely resembling the friable stucco decorations of the Rawak halos. You [Andrews] must have often wished with me that these had been worked in a harder material and here, a ruin of the identical period, furnished them in hundreds. Many, or perhaps most, had been gilt—definite confirmation of my conjecture as to the origin of the Yotkan gold. . . .

"From Hangúya I marched to Domoko near a site [Khadalik] not previously known to me. . . . Because there were no structural remains and there were traces of manifold burrowing, I ventured to hope but little. However the debris layers had only been scratched & though I worked with as many men as there was room for, it took three long working days before I had exhausted all the MSS. deposits which came to light here. Among them are Brahmi tablets, excellently preserved Chinese texts (one roll over 2½ feet long!) and Sanskrit texts of many kinds including birchbark. The date of these shrines is proved by style and conclusive evidence of coins to be the same as Dandan-Uiliq ones. . . . The pieces of stucco, frescoes, etc., which I carried away safely packed fill six large boxes. For nine days I was hard at work from daybreak until dusk and at the close of each day I felt like a chimney-sweep and too weary almost for all the writing of notes."

A serious problem had developed which made their stay at Keriya troubled as well as busy: it became necessary to purchase fresh camels. "To judge of camels' points requires . . . the inherited knowledge of a born camel-man. . . . I could rely on Hassan Akhun's honesty quite as much as the effect produced among the wily owners by his sharp tongue. My brave camels from Keriya . . . [did hold] out splendidly against all privations and hardships, and after nearly two years' travel were so fit and fine looking that, when at last I had to dispose of them, they realized over 70% profit—of course for the Government of India."[25]

The Khadalik site had been discovered by Mullah Khwaja. An elderly, respectable village official desperate to realize some money to pay off arrears in revenues, he had been told by a former guide who saw the "treasure" Stein valued that old manuscripts could realize cash—an improbable statement that proved to be true when Macartney paid him for the few he offered. Stein not only rewarded him when he acted as his guide but also promised to intercede on his behalf with the tao-tai at Keriya. The discovery of Khadalik, so unexpectedly rich in finds, made an auspicious beginning. Examining the strange artifacts found there, Stein wrote the Baron (8 October 1906) that "after all the study we have given such objects I feel as if I could date a ruin by a broom, bootlast or an old shoe."

Despite the goodly amount of relics found, Stein felt hurried, tense, impatient: he had received word that Pelliot's party had reached Kashgar.

10 October 1906, to Publius: "I could not help clearing the site completely, for being so near the oasis [of Domoko] it would have been swarming with seekers for 'Khats' [papers] as soon as I turned my back.

"I had to load myself with nearly Rs 10,000 worth of cash, since there are no places beyond as far as Sha-chou where the Chinese could arrange for payments. My demand has nearly exhausted the local Treasury reserves. . . . I am hoping to get off day after tommorrow [12 October] with an entirely fresh convoy. I cannot forego another visit to the Niya site, for I think that (new) ruins have emerged from the sands since those happy days of 1901. I shall do my utmost to finish with them quickly & push on to Charchan, 8 days' marching from Niya." Stein was keeping the promise made to himself to return to Niya.

The report to Publius was continued in letters of 29–31 October 1906: "It was delightful to be back once again under the dead poplars & fruit trees of circ. 250 A.D., to study at 30 newly-cleared houses the details of rural life in that age. Of course, the familiarity gained by my previous labours made it so much easier to observe and note them correctly." Niya is always wrapped in an elegiac tone. "At the end of the site where the dunes are absolutely bare, one really enjoyed the sensation of being in the sand ocean. The air was so clear that we managed to see the snowy range some 120 miles away to the south. I took fifty labourers [including carpenters and leatherworkers for on-the-spot repairs] to expedite work and had nothing to regret in spite of the additional trouble which the supply of water for so many has cost me. . . . The water service worked quite well. How glad I was to have the 25 Mussocks [goat skins to contain water]. The tanks would have sufficed only for some ten men, seeing that the camels could cover the 25 miles in two marches. . . .

"My greatest haul was made in the very house from which I had to turn back on the eve of my departure in 1901. I found there a great mass of office 'papers' left behind by the Hon. Cojhbo Sojaka, and discovered carefully hidden below the floor quite an archive of documents belonging to that worthy magistrate. Nearly three dozens of perfectly preserved deeds on double rectangular tablets, still with their seals & fastening intact. . . . There is an abundance of classical seals, some old friends among them."

As always, receipt of letters from Publius gave Stein renewed strength and quiet joy. It was midnight, 31 October, the end of a long day of preparations to leave Niya. "The autumn tints in the forest through which

I returned yesterday from my ruins & their dead world, were glorious. How I wished you could have been there to enjoy this feast for the eyes! But winter is close at hand." Before the cold—it was twenty below—froze the ink and sent him to his cot, Stein outlined sites yet to be explored between the Yatungaz and Endere Rivers and his intention of reaching Charchan by the third week in November. From there "I shall push on to Lop-nor as quickly as I can. Christmas ought to see me in the desert beyond the Lop lake. Pelliot's party was still at Kashgar at the 10th of October. But they, too, are bound for the Lop region and I ought to be there in time. If too much of interest has to be left undone, well, I shall have to contemplate a second campaign after my return from Sha-chou."

During six long desert marches, Stein read and reread Publius's letters: their "inexhaustible sympathy is my greatest comfort. Besides my work & its interest there is little else left, as you know, to encourage me.[26] He erased the desert with "fascinating visions of walks with you both through Italian towns." But his fantasy becomes clouded over; his words are iced with loneliness: "I wished you had Erasmian interests to draw you to Tübingen, a quaint old place to which dear memories make me cling, though all those who kept me there for three happy years of work have passed away. It seems hard at times to have lost practically all who taught and guided my work. But why complain of this when all who were dearest to me, have all been torn away."

Of the Endere ruins "which previously had not been reported to me by the distrustful shepherds of the neighboring jungle," Stein wrote that they had proved insignificant. "Yet they have yielded conclusive evidence that the site chosen for the 7–8th century fort had been indeed occupied until 4–5 centuries earlier. . . . Today I made the curious discovery that the rampart of the circular fort of the 8th century had in one place been erected over an ancient refuse layer which promptly revealed Kharoshthi records on leather! I rejoice at seeing old Hsüan-tsang's accuracy once more vindicated. He mentioned exactly in this position a deserted settlement of the Tukharas [Indo-Scythians known to him by their principal seat of power on the Oxus—Tukhara]. Now I can prove that when he passed through here the site had not yet been reoccupied by the later fort. . . . About five marches further east there are ruins which must be visited. It will depend on the work they offer when I reach Lop-nor.

"That I shall have to reckon with the French expedition is certain. But though Pelliot has publicly announced his intentions upon Lop-nor & Sha-chou, I need not abandon by own plans formed long before. . . . I fear my adherence to my original plans may alienate from me some of the friendly interest with which I have met so far in Paris. It seems rather hard to face this risk. I have certainly received more encouragement from Senart, Barth, etc., than from Orientalists elsewhere." In his long letter to Publius, Stein included one he had received from Deane telling him he had "influential friends." "I have so far found their voices inarticulate and do not expect to hear them raised with any emphasis in the future. If my present journey should attract interest at Whitehall, the G.I. may possibly have the courage of sanctioning my being employed again on archaeological exploration as a 'temporary measure.' " If not I shall be expected to go back to educational work."

In some of his letters to Publius, especially those written after a hard day's routine work—with just the grim getting on with the job and nothing unusual or important to fire the mind—there is less the sense of a communication than of a kind of thinking aloud. Not quite a diary, not quite a report, the past, the present, and the future tumble out in what might seem disorder. Such letters are Stein's venting of heart and purpose, an "interior monologue," made possible by Publius's "inexhaustible sympathy."

The excavations he had scheduled on his way to Charchan were finished, his laborers paid off, and even his considerable correspondence with Calcutta about the finances of the expedition completed. He expected to reach Charchan in six days, "just as old Hsüan-tsang had done. It was the same, silent uninhabited waste he describes between Niya and Charchan with the drift sand of the desert ever close at hand. The tracks of the wayfarers get effaced, and many among them lose their way. On every side there extends a vast space with nothing to go by; so travellers pile up the bones left behind to serve as road marks."[27] On so grim a road, Stein used the dreary "marches for peripatetic Chinese lessons."[28] He pushed forward as quickly as possible. On 3 December 1906 he wrote: "The Amban of this poor little district [Charklik] had received news of my coming from dear old Pan-Darin, now Taotai at Aksu. You will realize how grateful I am to my old Khotan friend and to the fortunate chance which put him in the right post just when I needed his support again. I please myself with the thought that Hsüan-tsang had arranged it in the heaven of the Arhats [Buddhist saints], and my Chinese secretary & mentor is indicting the right sort of learned epistle to inform Pan-Darin of this pious surmise."

The official command enabled Stein to secure thirty workers within a day, as well as the two Lop hunters who had been with Hedin and were to act as guides, and to be provided with supplies to last the party until it reached Sha-chou: all of Charklik was ransacked for food and fodder. "I shall make a depot at Abdal, the eastern-most inhabited place, so as to be ready for the rush to Sha-chou if the appearance of Pelliot's party should force me to hurry on. You will feel how these possibilities add to my cares. . . . It is reassuring to learn here how the long-abandoned ancient route to Sha-chou which Hsüan-tsang followed & Marco Polo too has been gradually opened up again with recent years. Traders now pass along every year, and I may hope to get my animals & baggage across the 20–25 marches without running too great risks of loss."

14–16 December 1906, Camp 121, Lop-nor Desert: Stein's mid-December letters to Publius were from the small fisherman's hamlet near where the Tarim River empties itself into the Lop marshes. "I marched to Abdal on the Tarim by a route which allowed me to visit an old site close to the Miran of the maps. I found there a much battered fort and guided by an intelligent Lop man's information, hit upon a number of rubbish-filled rooms." Within two days his large contingent of workers had cleared the greatest part of the rubbish accumulation. "They yielded between 300 & 400 Tibetan records on wood & paper, which must date from the 8th or early 9th century. . . . Under other conditions I should have been glad to do more but the desert journey called for haste. It is no easy task to take 45 men over about 90 miles of absolute waste [with] the impossibility of getting more than 27 camels . . . we dodged the difficulties by using donkeys for food & ice relays! Today after four marches we still have 10 camel-loads of ice & are safe, I trust, on that side." Stein was glad to have the two guides, "but, of course, the real guide is the compass."

Though Abdal was "a wretched hamlet composed of fishermen's reed huts,"[29] Stein was welcomed by "two Begs masquerading in official Chinese get-up" who made a brave attempt to boast of the profits locally made from fishing and grazing. He established his depot on the other side of the Tarim. To ferry the camels across the deep, rapidly flowing river, the Lopliks made workmanlike rafts by lashing five of their dugouts together. Looking at the river, Stein knew that he was seeing "all that remained of the united drainage which the great snowy-covered ranges of the Kunlun, the Pamirs and the T'ien-shan send down into this thirsty basin of Turkestan." At the Abdal depot he left the cases of finds made since Khotan and his entire reserve of uncoined silver. Left behind, too, were Tila Bai, his most reliable servant, and Chiang-ssŭ-yeh. Stein knew the latter would be unequal to the long, trying tramps across the dunes and that the supplies and ice he and his servant would need would have increased the baggage load. As he confided to Publius, he regretted what he would lose in companionship: "One longs for helpers really interested in the work. Ram Singh, the Sapper, is willing & cheerful, but his namesake, the Surveyor has heaviness of body & mind." His depot at Abdal was well chosen: Miran, to whose exciting excavation he would return, was close by.

The Lop ruins toward which (from his start in Chakdara) he had been hurrying were now but two days away. "It is an anxious thought, you can imagine, whether I shall not find the French there already. . . . We shall then have to find a *modus vivendi:* The springs from which we shall have to replenish our water supply, are too salt for drinking, but the ice is all right. A fortnight ago it would not have been available. . . . What a desolate wilderness, bearing everywhere the impress of death! On these salt-covered old lake beds I long for my rolling Khotan dunes and Dawans [ridges]!"

17 December 1906, Camp 124, Lop-nor site: "9 P.M. I have just safely reached this longed-for site with the whole column of men & camels. It was a tramp of over 100 miles from Abdal . . . over desiccated salt marshes, dunes, and for the last three days mainly over ground eroded by the winds into an interminable *hachure* of steep ravines, all lying across our track. For the last two days we had a fair experience of what this delectable region can offer in the way of winter. A NE wind travelling at the rate of 40–50 miles an hour when the temperature at night went down to 10°F. is not exactly pleasant, and early this morning it nearly carried off my tent. When I say early, I mean 4 A.M., the hour I usually had to get up for the last week in order to assure a timely start of my unwieldy caravan. But the effort has not been in vain. I found the whole site with its known ruins scattered over ten miles, clear of French or Germans and thus the 1000 mile race from Khotan (done exactly in three months incl. one spent in excavations) is won—for the present. . . .

"A short snowstorm has been a 'blessing' for the camels & the men, too, can get now their full potations of tea. Some 9 days' ice supply still remains! The sites have not been disturbed by treasure-seekers since Hedin was here and the 6 men he let loose at the ruins can scarcely have done much burrowing. A large number of interesting objects were picked up today from the route, while the march across the supposed old lake bed yielded implements of the stone age. We shall have to work hard, but the men are willing & cheered by the accuracy with which the ruins were struck across this forbidding desert. It is, of course, the merit of Hedin's remarkably exact survey. You should have seen the group of men, tired out by the long weary tramp when from a great distance I showed them the first Stupa. If they had read Xenophon they would have cried Thalatta, Thalatta!"

27 December 1906: On Christmas eve, "Turdi, the Khotan Dakchi, emerged from the maze of eroded trenches which surrounds one here" with a big dak. "Turdi had left me for Khotan on Nov. 15 at the Endere River, and how he managed to cover the 1300 odd miles to Khotan and back here in so short a time is still something of a puzzle to me. To your living presence as conveyed by these letters [August 23, 30, September 6, 13, 19], I owe a happy & peaceful Christmas.

"The results have proved worth all my efforts. The ruins which we have cleared in rapid succession have given up far more than from their numbers or state of preservation I might have been justified in expecting. Erosion has scooped out the ground with terrific effect, & the heaviest timber has been undermined. Small as this ancient settlement was, compared even with my Niya site, it lay on a great route & had seen much traffic pass through it. So every one of the dozen buildings or so which make up the main site, yielded documents in plenty, . . . One huge solidified rubbish-heap alone supplied over 200 records on wood and paper, Chinese and Kharoshthi, and of the latter enough has come to light in every ruin to prove conclusively that here too, as at the Niya site, the local language of the administration was an old Prakrit. To see Indian influence firmly established so far eastwards in the 3rd century A.D. is an important new fact and one which has wide historical bearing. . . . The conformity with my Niya finds is truly surprising too, in all art ware and industrial relics. Pure Gandharan style prevails in all wood carvings & relievos, and even if there were not the dates in the Chinese documents which Hedin carried off, it would be easy to prove that these ruins were deserted at the same period as those beyond the Niya River. Of variety there is not much left to wish for—a camel load or two (two, in fact, now that the packing is done) of architectural wood carvings, pieces of fine carpets, lacquered furniture, a lady's slipper delicately embroidered, bronze art ware, etc., make up quite a representative little collection. . . .

"Climatic conditions here are far more severe than about the Khotan desert. The bright calm winter days of the Niya site seem unknown here, & with even a slight wind 0° Fahr. is too cold for enjoyment. I am glad the men have stood the bitter nights well. But the Surveyor is laid up with rheumatic pains & can render little help. The Naik does his work with Sikh doggedness though he confides at times that no 'Sapper-Miner' had ever so much of it. Jasvant Singh, the little Rajput, is the hardiest & cheeriest of all. I expect to finish . . . by tomorrow [28 Dec.]. The camels have had a rest at a salt spring at the foot of the Kuruck-tagh, that terribly barren range of hills we clearly see from here. The water of its few springs has begun to freeze only since our snowstorm. It was very good we did not get here earlier, for the ice supply I had to take from them would have failed us. Even the camels refuse to drink the delectable liquid the springs yield; but they have found plenty of snow to drink. The main body is to return to Abdal through the desert by the 'road' we have marked with cairns & signposts. I myself with a small party intend to strike SW to the Tarim, to a site known as Merdek-Shahri."

That site, a small circular fort overgrown with reeds, had significance as having been built in Han times when the Chinese first gained control over the Tarim Basin. "I have heard or seen nothing of the French. I wish they may keep off sometime longer & give me the chance of clearing the Miran ruins completely before setting out for Sha-chou. The men & camels would also be better for a short rest. The ink is beginning to freeze in my fountain pen, though I have sacrificed an extra cake of compressed fuel to keep up the temperature in the tent for this long chat with you."

"Vale amicissime." Farewell, dearest friend (the dearest made permissible by Latin)—so Stein often closed his letters to Publius. With these words the year 1906 came to an end. Stein had won his race to be the first at Lou-lan, the city of Lop-nor. He could not have imagined that the discovery that lay ahead would almost obliterate his victory at Lou-lan.

Before Tun-huang—unreported, unexpected—lay the murals of Miran. Stein began the New Year marching steadily westward through an unexplored part of the desert toward the Tarim River. The entire route lay through a weird solitude "fascinating in its torpor which nothing living has disturbed for many centuries. After the first two days, dead trees and tamarisk grew very scarce,[30] so scarce they had to salvage every bit of "half-fossil wood" as they walked along. "It was only a small party I had to care for. Still, it was a relief to see them safely brought back to the river. Supplies were running low at the end, but as we luckily found a fisherman's hut just where we emerged from the desert, all troubles on that score passed. The line of lakes & lagoons looks bright with its glittering ice-sheet, . . . so clear as to show the poor half-benumbed fishes below."

Miran, one march south of Abdal, had yielded a "mass of Tibetan records from an old fort guarding the route to Sha-chou which Hsüan-tsang followed and also Marco Polo."[31] Earlier he had left off excavating to race to Lou-lan. Now he returned. To the Baron he described the riches of Miran. "What will interest you more than [the Tibetan records] are the dozens & dozens of lacquered leather scales from their [Tibetan] armours! All kinds of pieces are represented & you will probably be able to design a complete outfit. . . .

"All the time I had hoped that some of the Stupas & shrines near the fort would prove older. My hope has been justified by finds of a kind which would delight your eyes. From a much injured Vihara [monastery] I secured several fine stucco heads, as good or better than the Rawak ones. Some were colossal, unsafe for transport; others will, I hope, be duly laid at your feet. The same shrine also gave up a palm-leaf MS. fragment, circa 300 A.D., the first find of that kind of material I have made. But better things were yet to come. A few days ago I discovered on the walls of a circular passage, once vaulted & enclosing a small Stupa, frescoes of unexpected beauty. There was a dado formed of angels' heads, quite Graeco-Roman in style & with an Oriental tinge so slight that it might have been painted no further off than, say Alexandria. The whole conception & the expressions of the eyes, etc., are to my eye, Western. Fresco portions which had slid down from the upper wall were also found in large fragments. To remove these and the best of the angel panels has been a most delicate business. But with Naik Ram Singh's help it was done without serious damage. I have been busy all day packing these frescoes, as brittle as pastry. Luckily, the jungle is near enough to supply wood for boxes & reeds, etc., for stuffing. Remains of silk prayer flags with Kharoshthi inscriptions make it highly probable that the shrine was deserted already in the 3rd or 4th century. So I have got what I always longed for—specimens of pictorial work of the same period as the Niya ruins."

Like fireworks, his success lights up the letter. Yet, in the very next paragraph they fizzle out, due, perhaps to an ambiguity in his relationship with Andrews: though he needed him in his work, Stein was always the stronger, the more aggressive in attacking those bureaucrats who made his work possible. Andrews, for all his competence lacked the single-minded direction Stein shared with Publius. "I have photographed everything & noted colours, etc. But this does not prevent me from wishing you could see it all with your own eyes *in situ*—And yet, I am conscious it is better so." Had Andrews at some point suggested that he accompany Stein? As if to remind him, Stein always sends his greetings to the Baroness and their daughter Norah. Such attachments Stein himself had eschewed. "This is a trying climate for anyone & I should not like to see my friends exposed to its hardships. The Surveyor has been knocked out for some time, by rheumatism, and so are two of my Yarkand men. Nor could I offer you much care for food, etc. My Kashmiri cook, a hardy plant though not sweet to look at, is also on the sick list, & the Tibetan understudy would make you judge mildly of old Shib Ratan [their Mayo Lodge cook whom Andrews deplored?]. However, I am quite fit & hope to get rest on the long journey to Sha-chou when I shall not have to keep my eyes on everybody & my hands on everything." How Andrews read these lines is not known.

A week later (5 February) Stein was still at Miran, kept there by the "further discovery of fine frescoes . . . on the walls of another vaulted chapel enclosing a Stupa reduced on the outside to a shapeless debris mound. . . . Only portions of the main frieze & the dado below have escaped destruction, but these are fit to rank with the best Gandharan sculptures. The *putti* carrying a festoon & the medallion heads of feasting youths and girls between them look quite Graeco-Roman. What an illustration of the joy of life in this Buddhist sanctuary on the Chinese border." Stein described the same scene to Publius with a light, playful tone: "Where could these beautiful girls have got their diadems of roses from and where the youths their cups & goblets of wine? It seemed almost as strange as if magic were to create a desert with rolling sand dunes all about the Carleton—and a belated dinner party came out to wonder at it."[32] "Above appears a triumphal procession with a fine elephant, quadriga [chariot], prince on horseback, etc. more Indian in type but still full of classical borrowings. Kharoshthi inscriptions show that the frescoes must be about as old as the Niya records. Photographing was a trying business what with a gale, the cold & the dust. It is impossible to remove the larger fresco frieze, & even the cutting out of specimens from the dado seems a sacrilege."[33]

Miran, which Stein had thought to excavate in four or five days, kept him hard at work for eighteen—so dazzling, so unexpected its riches. From Abdal, Stein dispatched a convoy of six camel loads of antiquities to Kashgar; the manuscripts and documents on wood he kept with him—he would have them to show to the mandarins he would be meeting in China. He stayed at Abdal for eight days. "The mud hut of the 'Abdal Beg' was quite a cozy shelter when laid out with Khotan felts & I have got through a lot of writing, accounts, etc."[34] He thanked Publius for the Christmas gift of a copy of Wood's *Oxus*. "The 'Introduction' was the first work of Yule which I studied, already in my Tübingen days. . . . As one gets older early loves & interests revive. So my thoughts travel more to Eastern Iran & the Oxus."

Oxus: the very word was like a bell tolling his discontent. "There is little doubt that . . . I shall have to struggle to the very end for every year of freedom. Perhaps if I had been born a Frenchman in France or a German in Germany, I could have hoped for official employment on the right lines. But where research, etc., is left to 'private enterprise' the first need is for independence, i.e. money. . . . I cannot hide the fact that this treatment, undeserved as it is, gives me cause for bitter thoughts. I know it is due more to narrowness of horizon & indifference to scientific aims than anything else. But it discourages me that fate has withheld from me the gifts—or the patronage which alone can cope with such obstacles where dependence exists. Forgive this outpouring." Stein's "bitter thoughts" were not confined to Publius. They permeated his "Progress Report" which came to Marshall. The director-general's reaction is impatience: "It is difficult, indeed, to know what Dr. Stein wants. The Government of India has given him practically everything he has asked for, and he still goes on harping on his 'bitter experiences' and 'discouragement'! If he wished to stop the Government doing anything more for him, he could hardly choose a more successful method of effecting his purpose."[35]

"I am *quite* fit after all the roughing & glad to have things & personal kit put in decent order again here. The Surveyor is on his legs again & will, I hope be got safely along. All the same I sent a wire from Karashahr to the Survey for a substitute to join if possible during the summer. . . . My plans are difficult to fix in full detail until I see what the Buddhist caves and the deserted sites about Shachou can yield me."[36]

Sha-chou, "City of Sands," lay on the far side of the "Desert of Lop," as Marco Polo had called the dead lake bed. After twelve days of long and trying marches, they reached a little oasis consisting of reeds and five trees. During a day of rest there, Stein wrote to Publius, 5 March 1907, describing their seven-day March "along the edge of dried-up Lop-nor marshes, salt-encrusted & without vegetation; a drearier sight than any dunes. The few springs past which the route leads issue at the foot of the gravel slopes marking the shores of an ancient, big, lake basin. . . . I was glad when at last the dead lake bed contracted & the barren hills of the Kuruk-tagh range came in sight northward. Glad, too, that the heavily laden camels no longer had to trudge painfully over the hard-baked salt marsh. . . . I had plentiful occasion for interesting geographical observations; with every march it became clearer that we were moving along a depression which once, long ago, carried the waters of the Sha-chou, now lost in the Kharanor lake down to the Lop-nor basin."

"With 23–26 miles to be covered daily, there was plenty of time for conversational lessons with my honest Chiang. . . . He has told me many little secrets of the official machinery, of the chequered careers of proud Ambans & of their unholy profits. The 'New Dominions' are a sort of India for Chinese officials, where everybody knows everybody else—or his friends & antecedents at least. Often we have talked of Marco Polo who had described this old route so truthfully, & today, for the sake of practice, I gave an oral translation of the chapter on the Desert of Lop. I wonder whether Ser Marco ever thought of his book entertaining Chinese readers. . . . We see our route crossed by the tracks of wild camels and duly remember the spirits which make the traveller lose the true road.

"It is a big caravan I am moving along; for to carry food & supplies for the men and grain for our 15 ponies, a whole flock of donkeys, some 30, had to be added to the camels and, of course, the patient little beasts require to be fed too. It is rather a complex business to calculate supplies for such a journey. But we have moved along at a better rate than our Abdal friends had expected and about ⅔ of the distance is done. So there is no fear of a breakdown and we can give our beasts liberal rations. Some half a dozen donkeys had to be abandoned, no doubt animals which their owners had purposely underfed before starting—to earn compensation. But I took care that they should be left near to where there was water of sorts. So they may yet recover after a good rest. Here [at the oasis of Besh-toghrak: "Five Toghraks"] a larger detachment is to remain in charge of a young fellow who will have to make the best of his solitude— or the visit of goblins. Plenty of flour & a box of matches ought to keep him safe enough until the rest of the party march back from Sha-chou."

Did the "young fellow," marooned alone in the pitiful oasis, and in utter desolation, symbolize Stein's darkest feelings? "With long hours to dwell on things of the past, I mean those of one's own life, I realize only too well how much is drifting away in the current of time. It may be my fault or that of the distant tracks I have followed, far away from the associations of my old home. So it is great cheer to feel sure that there is a peaceful corner in Oxford where the effacing hand of time cannot affect my welcome. It is so different elsewhere. Were it not for scenes dear to me by congenial work & happy memories, I might fear to find myself a stranger in India, too. In Europe, apart from you & our common Lahore friends,

I have become it already." Publius was the haven, his weekly letters the lifeline holding Stein from being lost in the desert of his loneliness.

Even before he reached Sha-chou on 12 March, Stein was immensely gratified to identify an important sign of the past. The last six days' march to the edge of the Sha-chou oasis was "along ancient, desiccated lake basins reproducing most strikingly all the curious features of the Lop-nor, but on a small scale & consequently more easily studied & interpreted. There was the belt of sand-dunes deposited by a river that may have ceased to run before earliest dawn of history. Then a maze of strange, towering clay terraces sculptured out by erosion from an ancient lake bed. To the south & north of the old lake chain there extends hopeless desert in which one might for weeks search in vain even for a salty spring. Soon after we had gained the gravel plateau which edges this great depression, I came to my joy upon remains of old Chinese road towers, and on the third day after we had refreshed our beasts at a forlorn little brook, I succeeded in tracing the line of an ancient 'Chinese Wall' which traversed the desert for a great distance. [Stein had been first alerted to the existence of such a wall by a brief notice published in 1899 by C. E. Bonin, of the French Diplomatic Service, who attempted to cross from Tun-huang to Lop-nor.] He had written to Publius on 5 March 1907: "From Mullah of Abdal, the true pioneer of this route, what the observant old fellow had told me, gave me the hope I might come across the first Pao-t'ais, as he called the watch-towers.

"In many places it had almost been completely covered by drift sand; but the tamarisk layers which had been used to strengthen the *agger* cropped up so persistently that the eye caught the straight line as it stretched away for miles on the coarse sand. I thought of the Roman walls in Northumberland & elsewhere & saw how much easier it would have been for the local antiquarians to follow their traces if desiccation had not made all the surrounding ground a bare desert! Luck assisted us encouragingly. At the very first of the towers I surveyed more closely, we found with other rubbish a Chinese tablet which had been deposited in the earth wall. The writing points to an early date. The tracing of the towers popping up again & again on the horizon was quite a fascinating variation of survey work for three days. In addition I came across a large palace-like ruin, fortified posts & similar traces of ancient occupation of the marshy shores we were following. It all seemed a good augury for my first work on Chinese ground."[37]

14

Act Two: The Treasure of Ch'ien-fo-tung

Tun-huang, as the local Chinese call Sha-chou, using its ancient Han name, is undoubtedly the place most closely associated with Stein. There he found the treasure. What it was, where it was, and how he found it is an extraordinary story. Forgotten (or dismissed) is his patient excavation of the watchtowers of the Chinese limes, a discovery as stunning, if not as significant, as that of the hidden treasure at the Caves of the Thousand Buddhas. Stein has been called a "robber," a "bandit," etc., and by today's standards he can be judged to have violated China's own scholarly interest in her past by removing unique documents and relics to the West. But at the time he was there, he was not so considered: then archaeology considered scientific validation as sufficient reason for removing the documents and relics to the West. The case can be argued; but it must also be said that charges have not been leveled against Stein for what he recovered from sites deep in the desert such as Niya, Dandan-Uiliq, and Miran. Furthermore, after all the arguments are given a haunting question persists: if Stein had not taken what he did, would the treasure have survived as a collection or would it not have drifted (as a good many pieces did when some ten thousand manuscripts were being brought to Peking) piece by piece into dealers' hands and by such circuitous routes reached Western collections? Certainly, none of the items Stein brought back to the British Museum wandered into private hands or was cut up or lost. But whether Stein is condemned or excused, the events that occurred at Ch'ien-fo-tung are most interesting.

"Tun-huang welcomed us with an icy Buran, such as fortunately are rare in the region. . . . I managed to secure fairly comfortable quarters for my men in an old garden about a mile from the quaintly walled city."[1]

It was the 12th of March, and for the next four days Stein wrestled with his accounts. To him, the bookkeeping was an incubus; "its weight may be gauged from the fact that it meant not merely extracting all and sundry items, however small, from my general cash record into properly balanced 'Monthly Cash Accounts' in due official form . . . but also dividing all entries into transport and the like whether they were to be debitted against the Government grant, or the Survey of India's subsidy meant for the 'Survey Party' or, finally against my personal purse which would in due course recoup them from authorized 'Travelling and Halting Allowances.'"[2] Of the two discomforts, the buran seems to have been the less unpleasant.

On his customary evening walk, he noticed that "the people of Tunhuang are still as pious as in Marco's days; curious Buddhist temples with coloured woodwork & frescoes are scattered throughout the hamlet & the half-deserted big city enclosure. I was eager to pay my preliminary visit to the Grottos of the 'Thousand Buddhas' which had first attracted my eyes to this region."[3]

His first sight of Ch'ien-fo-tung was thrilling. "On the almost perpendicular conglomerate cliffs," as Stein described it, "a multitude of dark cavities, mostly small, was seen here honeycombing the somber rock faces in irregular tiers from the foot of the cliffs where the stream almost washed them, to the top of the precipice. . . . The whole strangely recalled fancy pictures of troglodyte dwellings of anchorites such as I remembered having seen long, long ago in early Italian paintings. . . . But the illusion did not last long. I recrossed the broad but thin ice sheet [of the stream] to the lowest point, where the rows of grottoes did not rise straight above the rubble bed but had a narrow strip of fertile alluvium in front of them and at once I noticed that fresco paintings covered the walls of all the grottoes. . . . The 'Caves of the Thousand Buddhas' were indeed tenanted not by Buddhist recluses, however holy, but by images of the Enlightened One himself. All this host of grottoes represented shrines. . . .

"The fine avenue of trees, apparently elms, which extended along the foot of the honeycombed cliffs, and the distant view of some dwellings farther up where the river bank widened, were evidence that the cave-temples had still their resident guardians. Yet there was no human being about to receive us, no guides to distract one's attention. In bewildering multitude and closeness the lines of the grottoes presented their faces, some high, some low, perched one above the other without any order or arrangement. In front of many were open verandah-like porches carved out of the soft rock with walls and ceilings bearing faded frescoes. Rough stairs cut into the cliff and still rougher wooden galleries served as approaches to the higher caves. . . .

"As I passed rapidly from one cella to another my eyes could scarcely take in more than the general type of frescoes and certain technical features of the stucco sculptures. The former, in composition and style, showed the closest affinity to the remains of Buddhist pictorial art transplanted from India to Eastern Turkestan, and already familiar from the ruined shrines I had excavated at Dandan-Uiliq and other old sites about Khotan. But in the representation of figures and faces the influence of Chinese taste made itself felt distinctly, and instead of the thin outlines and equally thin colouring there appeared often a perfect exuberance of strong but well-harmonized colours. Where deep blues and greens predominated there was something in the effect distinctly recalling Tibetan work. . . .

"Within the cella was ordinarily to be found a group of images occupying either an elevated platform or else placed in a kind of alcove facing the entrance. All the wall faces were covered with plaster bearing frescoes. . . . But whether the wall decoration showed pious compositions, or only that infinite multiplication of Bodhisattvas and saints in which Buddhist piety revels, all details in the drawing and grouping of the divine figures bore the impress of Indian models. . . . In the subject of the friezes and side panels, which often apparently reproduced scenes from the daily life of monks and other mundane worshippers; in the design of the rich floral borders, the Chinese artists seemed to have given free expression to their love for the ornate landscape backgrounds, graceful curves, and bold movement. But no local taste had presumed to transform the dignified serenity of the features, the simple yet expressive gestures, the graceful richness of folds with which classical art, as transplanted to the Indus, had endowed the bodily presence of Tathagata [Buddha] and his many epiphanies. . . .

"Of the sculptural remains it was more difficult to form a rapid impression; for much of this statuary in friable stucco had suffered badly through decay of its material, mere soft clay, and even more from the hands of iconoclasts and the zeal of pious restorers. In almost all the shrines I visited, a seated Buddha, sometimes of colossal proportions, was the presiding image; but by his side there appeared regularly groups of standing Bodhisattvas and divine attendants more or less numerous. I readily recognize representations of Dvarapalas, the celestial 'Guardians of the Quarters,' . . . But from the first I realized that prolonged study and competent priestly guidance would be needed. . . . It was pleasing to note the entire absence of those many-headed and many-armed monstrosities which the Mahayana Buddhism of the Far East shares with the later development of that cult in Tibet and the border mountains of Northern India."[4]

As Stein and Chiang were making this quick preliminary survey, they were joined by one of the resident Buddhist monks, a quiet young man who responded to their interest in the things of the past. He pointed out inscribed marble plaques that wealthy devotees had reverently placed there during the Sung and Yüan dynasties (from about A.D. 1127 to 1368). Excitedly, Chiang studied them; Stein, for his part, knew that most "had already been published by M. Chavannes from impressions brought back by M. Bonin."[5] His eyes were otherwise engaged. Stein was seeking to discover the recess in one of the cave temples where a huge hoard of old manuscripts was said to have been accidentally found. He had learned of the existence of the cache from Zahid Beg, a Muslim trader from Urumchi who had come to Tun-huang to escape his Turkish creditors. Less imaginative, less enterprising than the forger Islam Akhun, Zahid was on the lookout for any place that might be profitable. He was the type of man who amassed and dispensed information, true or false, real or rumored; to Stein, he was a welcome change from the Chinese who maintained a stolid, steadfast ignorance of anything pertaining to the past. What Stein felt to be their secretiveness might well have been their devotion to the sacred grottoes as part of their pious faith. No such qualms lulled Zahid's curiosity.

Nor Chiang's—as Stein found out when privately discussing the problem with him. Like a seasoned, skillful navigator who seeks to make harbor, Stein sought a local pilot. Such a person was necessary to gain "access to the find, and [help them] break down, if necessary, any priestly obstruction. I had told my devoted secretary what Indian experience had taught me of the diplomacy most likely to succeed with local priests usually as ignorant as they were greedy, and his ready comprehension had assured me that the methods suggested might be tried with advantage on Chinese soil too."

The execution of the plan was delayed because Wang, the Taoist priest in charge of the Buddhist sanctuary, "was away in the oasis [Tun-huang], apparently on a begging tour with his acolytes." While awaiting his return, Stein had a foretaste of the contents of the cache. The young monk, eager to show his appreciation of the visitor's interest in holy things, brought them a manuscript he had borrowed to add sanctity to the private chapel of his spiritual guide, a monk of Tibetan extraction. "It was a beautifully preserved roll of paper, about a foot high and perhaps fifteen yards long, which I unfolded with Chiang in front of the original hiding-place. The writing was, indeed, Chinese; but my learned secretary frankly acknowledged that to him the characters conveyed no sense whatsoever. Was this evidence of a non-Chinese language, or merely an indication of how utterly strange the phraseology of Chinese Buddhism is?" Stein reserved an answer until he could examine more of the hidden library.

"It was a novel experience to find these shrines, notwithstanding all apparent decay, still frequented as places of actual worship. . . . I reflected with some apprehension upon the difficulties which this continued sanctity of the site might raise against archaeological exploitation. . . . Only experience and time could show. Meanwhile I was glad enough to propitiate the young Buddhist priest with an appropriate offering. I always like to be liberal. . . . But, unlike the attitude usually taken up by my Indian Pandit friends on such occasions, when they could—vicariously—gain 'spiritual merit' for themselves, Chiang in his worldly wisdom advised moderation. A present too generous might arouse speculations about ulterior motives. Recognizing the soundness of his reasoning," Stein gave a piece of hacked silver equal to about three rupees; small as it was it brought a gleam "of satisfaction on the young monk's face."

This account as set forth in *Ruins of Desert Cathay* (pp. 28–31) is hardly that of a man who thought of himself as a "robber" or "bandit." Only a man committed to the primacy of Western learning could write so fully, so honestly of what had transpired. Nor is he more secretive in his letters to Publius and the Baron. There is no sense of guilt or of having been a party to sacrilege. How could he have foreseen what the attitude would be both in China and the West half a century later? Furthermore, he was reinforced in his thinking by Chiang: "I thought I could in his attitude detect something closely akin to that mingled regard for the cult and self-conscious pity for its ignorant representatives, which in the old days never allowed of easy relations between my learned associates in Kashmir and the local priests at pilgrimage places we used to visit together for archaeological purposes." With unabated candor, Stein mentioned that "news of the discovery [of the cached items] ultimately [had] reached provincial headquarters, and after specimens had been sent to faraway Lan-chou, orders were supposed to have come from the Viceroy to restore the whole to its original place of deposit." Since then, the matter had rested.

More than its proximity to the sacred caves had attracted Stein to Tun-huang: he was struck by its geographical position. Tun-huang was "near the point where the greatest old highroad of Asia from east to west is crossed by the direct route connecting Lhasa [Tibet] and, through it India, with Mongolia and the southern portions of Siberia.''⁶ It was Stein's awareness of the importance of this juncture—of the movement of peoples, the flow of traffic, the passage of ideas—that made him plot the ancient limes and sift through its watchtowers' rubbish piles. Geography was coequal with archaeology when he set his course for Tun-huang.

On arriving there, though the buran was blowing, he visited the yamen, preceding his call with gifts (including a much-appreciated piece of Liberty brocade). He dressed for the occasion in European finery: "black coat, sun-helmet and patent leather boots." Wang Ta-lao-ye, the newly appointed Amban, was a slender, middle-aged man with a lively, intelligent face. "In his combination of courtly manners, scholarly looks and lively talk [he] reminded me of dear [Pan-Darin]. . . . I instinctively felt that a kindly official providence had brought to Tun-huang just the right man to help me in my first work on these ancient Marches." When Stein referred to Hsüan-tsang he was not surprised that the widely read Amban was well acquainted with the great pilgrim's own memoirs of his journey to the western regions. Back in camp, Stein was getting his feet, half-frozen from the icy buran, into warm fur boots when the amban returned the visit. "From the mule-trunks close at hand I brought forth specimens of ancient Chinese records excavated at Niya and Lop-nor, reproductions of earlier finds, and anything else that might be relished by the Amban's antiquarian eyes. The effect was all I could wish for."⁷

Then, too, there was Lin Ta-jen, the military commandant. Because Tun-huang had once been a bulwark of the empire on an important frontier and was still a point of strategic importance, Lin had a status equal to the amban's. Stein also called on him. A jovial, burly, self-made man, he had come first to Turkestan as a corporal in the army that had reconquered the province from Yakub Beg; twenty years later, still a petty officer attached to the empress's bodyguard, during the tense times of the Boxer Rebellion, he had attracted imperial notice. He soon rose rapidly. "He did not lay claim to much education nor to any particular interest in things dead and buried but would chat away gaily about such haunts as Chiang and I knew. . . . Our prolonged visit to the district was evidently welcomed by him as a pleasant diversion, and the help of his myrmidons which he pressed upon us as a safeguard against the obstructive indolence and occasional turbulence of the Tun-huang people, soon proved useful in more than one way. . . . Luckily the two dignitaries were on excellent terms."⁸

At an elegant dinner party given jointly by the amban and the commandant, Stein was shown a "Chinese volume containing a kind of official gazetteer of the Tun-huang district and full of historical extracts. . . . I was able with Chiang's help to glean a good deal of interesting information about the history of the 'Thousand Buddhas' during T'ang times, as well as to ascertain how vague modern knowledge was in respect of the ancient routes leading westwards. With the ruined line of wall we had traced in the desert of Tun-huang, local Chinese scholarship had evidently never concerned itself. . . . When I thought of the physical conditions prevailing on the desert ground I was eager to search, I could scarcely blame Chinese confrères for not having cared to explore it before me. . . .

"In the course of our dinner talk, . . . I realized there were difficulties ahead such as on Turkestan soil no Amban had ever hinted at":[9] securing the dozen diggers Stein had mentioned as the minimum for his work. Stein spelled out the difficulty to Publius in letters of 3 and 9 April 1907: "To catch men for digging was no easy matter. The people of Tun-huang are few & delightfully lazy. The Amban & other officials are obliged to respect the autochthon *vis inertiae* & had a good deal of trouble on my account before we could start. But they did their best & what I am grateful for, did it cheerfully." Stein's immediate objective, to be carried out while winter weather made work in the desert possible, was the exploration of the ancient frontier wall. To get the project under way, Stein had to overcome the "deep-rooted secretiveness of the local population [that] effectively prevented any offer of guidance from the herdsmen or hunters who occasionally visit the nearer of the riverine jungles." Nor could the officials help—nothing about the wall was known to either one.

"The eight men whom the Ya-men attendants had managed to scrape together looked like the craziest crew I had ever led digging—so torpid and enfeebled by opium were they; but I was glad to have even them."[10] Thus, inauspiciously, late in March the search began. "After three marches northward we struck the marsh-lined depression of the Su-lo-ho River where it was impossible to find a passage. . . . Then I turned east in the sandy waste which lines the marshes & by the fourth day luckily hit the hoped-for line of the 'Great Wall.' You can imagine how pleased I was when from a high clay terrace I first sighted the watchtowers, or Pao-t'ais, as they are called along the main roads of the 'Middle Kingdom,' glittering far away in the setting sun. They proved to extend over a line fully fifteen miles long, and for a considerable portion of this distance the ancient border wall itself could still be traced subsequently. You must not imagine a big wall . . . only an *agger* of gravel & tamarisk layers which the desert winds have scoured unrelentingly. But I soon learned to respect these remains & the solid clay towers, 20–30' high which still guard their line.

"Near the very first tower I reached there was a little refuse heap & from it there soon emerged 'Khats' [documents]. That they were written on wood was promising enough, & better still when the debris of a small building near the next tower yielded tablets with dates. My good secretary could not fix the reign off-hand (Chinese chronology is quite exact but a little complex as befits such a nation of learning). So it was an exciting hunt in my 'Chinese Reader's Manual' which followed in the evening. I had all along put my faith in the antiquity of the wall, & yet felt elated when the first finds proved to be dated in a reign corresponding to 25–51 A.D. The following five days were spent in clearing from tower to tower the little refuse heaps left behind by the soldiers who had kept watch on the wall or the Dak men, etc., who travelled along it. The finds of wooden records, of course all Chinese, were ample, over a hundred in all & many give curious glimpses of daily life along this desolate border. The oldest tablet is dated 36 A.D., quite a respectable age for this region which the Chinese had conquered only a century or so earlier, I . . . believe that China at present knows no authentic document of equal antiquity, of course excepting inscriptions on stone or metal. . . .

"It is a dreary world here outside the oasis. . . . Luckily, within the oasis things are better." Publius had sent Stein his just-published volume of Erasmus's *Letters.* "Your Erasmus is by my side as I write. As usual I have been reading it while the tent was being pitched." Stein had returned to Tun-huang to secure fresh supplies and workers, and the ambans, eager to help him, paid an official visit "before I scarcely had time to wash the dust off my face." They were "duly impressed with the finds of documents dating from the 1st century A.D." Thanks to their help and attention he was able to set out again after just one day spent in the town.

"Nan-hu which we reached in two days proved a delightful change with its limpid springs & the peace of its two dozen holdings. The oasis was once far larger. . . . As in the case of most ruined sites still adjoined by cultivation, there was little scope for systematic excavations. To move dunes 30' to 50' high was not in my power. But I managed to collect ample evidence in coins, etc., to trace the story of the slow decay of this westernmost of all true Chinese colonies. It seemed as if fate had wished to show me in broad daylight how Dandan-Uiliq & others of my cherished Khotan 'sites' were gradually abandoned to the desert. . . . Two days marching across the bare gravel desert separates Nan-hu from the chain of marshes along the Lop-nor route & the ancient fortified line defending it. . . .

"We have been hard at work digging. My finds at the ruined posts of the wall east of the Sha-chou River had proved that the *limes* was as old as the 1st century A.D. The discoveries along this part of the line have surpassed the hopes then raised in extent & interest. Our previous mapping had acquainted me roughly with the line followed by the wall, the position of the watch-towers, etc. So no time was lost in seeking for the tasks of each day. The climatic conditions in this desert seem to favour preservation of all 'old things' to a degree unknown even in the Taklamakan. There is practically no drift sand. Yet inscribed tablets, rugs, utensils, etc., emerge absolutely uninjured from below only a few inches of fine gravel. Rain there seems to be none, & though the winds rarely cease, they do not scoop out & grind from want of abrading sand. So whatever the soldier once guarding this frontier had thrown away or left behind at their walled stations has remained untouched by time. . . .

"At the very first towers we found quarters for the men who had kept guard here with many curious relics. But 'Khats' are what attracts first attention. It was a cheerful surprise when the first dated one carried us back to 20 A.D. Chiang & myself take such finds more calmly now, since successive discoveries have shown that the wall with its posts must have been built at the first Chinese occupation of the 'gate to the Western Regions,' towards the close of the 2nd century B.C. There are plenty with exact dates from 96 B.C. down to the middle of the 1st century A.D. ["The abundant remains of elaborate calendars . . . written on tablets of a special size, over 14 inches long turned up. . . . The calendars to which these tablets belonged were issued for the years roughly corresponding to 63, 59, and 57 B.C."][11]

"Bitterly I regretted my own ignorance [of Chinese] as in spite of Chiang's untiring decipherment & explanations I must miss a mass of curious details. Still I gather enough to gain some idea of the lives which were lived here, on a border compared with which the dreariest bits of the Wazir or Afridi border [Northwest Frontier of India] would look like the very paradise. . . . At one post there was a neat epistle left by three worthies who had called to see the commandant '& had found no one to take their message.' At another, an imposing tablet gives notice that the officer in charge of this line is about to celebrate a big family feast & expects congratulatory offerings, etc. About the time of the usurper Wang Mang (9–25 A.D.) they seem to have had a lively time along this frontier with the Huns pressing from the north & things in China upside down. There are reports of troop movements, rapid changes at headquarters—& urgent reminders about starving detachments. . . . One seems to see the past brought to life again!

"I feel at times as I ride along the wall to examine new towers, as if I were going to inspect posts still held by the living. With the experience daily repeated of perishable things wonderfully preserved one risks gradually losing the true sense of time. Two thousand years seems so brief a span when the sweepings from the soldiers' huts still lie practically on the surface in front of the door, or when I see the huge stacks of reed-bundles used for repairing the wall still in situ near the posts, just like stacks of spare sleepers near a railway station. I love my prospecting rides in the evenings especially when the winds have cleared the sky. That is the time to see many things, the white brick towers glittering far away on the commanding ridges they usually occupy; the track within the wall trodden by the patrols of so many years as the slanting rays show it up on the grey gravel soil—and weak points along the marsh-edge where prowling Hun freebooters might have lurked for a rush.

"I feel strangely at home here along this desolate frontier—as if I had known it in a previous birth. Or is it, perhaps, only because I heard my beloved father tell so often of the Roman walls traversing parts of Southern Hungary? He had spent many a hot day in tracing their lines; but, alas, the day never came when he could show me what had puzzled & fascinated him. The people against whom they were built may, after all have been distant relatives of those Huns who had haunted these parts about the time of Christ. . . . Among many curious finds, the discovery of a packet of letters in what I take to be early Aramaic, is of particular interest. They came from a rubbish heap with Chinese records dated 1 A.D. Were they the papers of early Syrian traders to the Seres? [China.] Some are still tied neatly in silk or cloth envelopes. How did they get to a small post miles off the main road?—Indian fragments on silk have turned up also with Chinese documents of the 1st century A.D. and are reassuring me that after all I may have been guided by some right instinct when without any pretence to Chinese knowledge I fixed my eyes on this ancient gate of Cathay as a region worth exploring.

"Taking into account the work east of Tun-huang, I have traced every portion of the wall for over more than sixty miles. Apart from the watch towers [about 2½ miles apart] & the small quarters adjoining them I managed to locate two sectional head-stations. It is their rubbish heaps which have yielded the best harvest in documents. I have just now [28

April] surveyed a most imposing structure, some 500' long without its walled enclosure which puzzled me greatly when I first saw it. There is not a ruin above ground equally big & massive this side of Delhi or Persepolis. The relics are few in these huge halls for they proved to be an imperial Commissariat godown [warehouse] for the troops of this part of the frontier. From the office room of the humble Babu [clerk] sheltered in a corner of the courtyard we secured wooden indents, etc. which show that the magazine was already being used in 53 B.C. [Here were kept reserve supplies of silk and hemp cord for the crossbows, axes, hammers, and other implements.]

"The work delightfully combines historical & geographical interest. The line had been laid down very intelligently with a clear eye for topographical features. So it now supplies a most reliable gauge as to the changes which have taken place in the physical conditions, a much discussed topic nowadays. I have been able to ascertain again & again the earlier level of the marshes. The changes are not sensational. . . .

"On April 1 we were still shivering in our fur coats and now [3 weeks later] the heat and glare are already trying. I am grateful for the winds. . . . The mosquitoes are pretty active already & other insects abound along the marshes where we must camp for the sake of water; it will be time, a fortnight or so hence, to take leave of this desert. I intend to carry my survey & digging right through to where, according to a reconnaissance of Ram Singh, the line of towers ends at the edge of a great range of sand dunes. Thence we shall try & reach the foot of the Tsaidam mountains. I have seen them several times tower like a big ice-wall above the gravel slopes of the desert. . . . In these parts humanity is scarce & exacts high wages for a minimum of labour. I am always counting my men knowing they would like to decamp at the first chance. Luckily, some herdsmen on the route near Tun-huang respect authority & have brought back the first who tried to desert."[12] Stein ends his lengthy letter to Publius announcing, as usual his next project. "The second part of May I should like to devote to the 'Thousand Buddha Caves.' "

On May 15 a long, hot ride brought Stein back to the orchard in Tun-huang where he witnessed the annual pilgrimages to the Shrines. "The great fête, a sort of religious fair, was said to have drawn thither fully ten thousand of the pious Tun-huang people, and from the endless string of carts I saw a few days later returning with peasants and their gaily-decked women-folk, this estimate"[13] seemed highly plausible. It delayed his start for five days. The pilgrimage confirmed what Stein had been told, that however decaying the caves might appear, they were "real places of worship. . . . My archaelogical activity at them, as far as frescoes and sculptures were concerned would, by every consideration of prudence, have to be confined to the study of art relics by means of photography, drawing of plans, etc., in short, to such work as could not arouse popular resentment.[14]

Stein drew the distinction: frescoes and sculptures belonged to popular cultic practices; documents rightfully belonged to scholarship.

Stein's recognition of the ancient limes and the documents its watchtowers yielded was the extraordinary curtain-raiser to the drama enacted at the "Caves of the Thousand Buddhas." The libretto, as he relates it in *Ruins of Desert Cathay*,[15] presents the action which took place in the three weeks spent at the caves.

THE PLACE: Ch'ien-fo-tung. In addition to its caves, it had two dwellings, "unoccupied save for a fat jovial Tibetan Lama who had sought shelter here after long wanderings among the Mongols of the Mountains. In one of the courts my Indians found rooms to spread themselves in, and the Naik a convenient place to turn into a dark-room. In the other my Muhammadan followers secured shelter . . . while a hall possessed of a door and trellised windows, was reserved as a safe and discreet place of deposit for my collection of antiques. Chiang himself had a delightfully cool room at the very feet of a colossal seated Buddha reaching through three stories and with his innate sense of neatness promptly turned it into a quite cozy den with his camp rugs. My tent could be pitched under the shade of some fruit trees on the one little plot, grass-covered, offered on the narrow strip of cultivation extending in front of the caves for about half a mile."

THE CAST: Stein, Chiang, and Wang Tao-shih. Of himself Stein wrote: "It was useless to disguise the fact from myself: what had kept my heart buoyant for months, and was now drawing me back with the strength of a hidden magnet, were hopes of another and more substantial kind. Their goal was the great hidden deposit of ancient manuscripts which a Taoist monk had accidentally discovered about two years earlier while restoring one of the temples. . . . [and which he] still jealously guarded in the walled-up side chapel." Chiang, the devoted, sensitive secretary had warned Stein "to feel his way with prudence and studied slowness. . . . He had induced Wang Tao-shih, the monk, who had discovered and guarded the hidden cache, to postpone one of his regular tours to sell blessings and charms, and to collect outstanding temple subscriptions until Stein arrived. Wang Tao-shih, the Taoist monk: "As a poor shiftless mendicant he had come from his native province of Shan-hsi some eight years before my visit, settled down at the ruined temple caves, and then set about restoring one to what he conceived to have been its original glory. The mouth of the passage was blocked by drift sand from the silt deposit of the stream. . . . When I thought of all his efforts, the perseverance and the enthusiasm it must have cost this humble priest from afar to beg the money needed for the clearing out of the sand and the substantial reconstructions, . . . I could not help feeling something akin to respect for the queer little figure." The way in which he lived with his two humble acolytes "made it clear that he spent next to nothing on his person or private interests. Yet his list of charitable subscriptions and his accounts proudly produced . . . showed quite a reasonable total, laboriously collected in the course of these years and spent upon these labours of piety."

Stein's first description of Wang is almost like a stage direction, setting the scene. "A very queer person, extremely shy and nervous, with an occasional expression of cunning which was far from encouraging. It was clear from the first that he would be a difficult person to handle."

Everything hinged on Wang. After their first meeting, when he had welcomed Stein, the latter contented himself with photographing "one of the ruined temple grottoes near the great shrine restored by Wang. . . . I could not forgo a glance at the entrance passage from which the place of deposit was approached. On my former visit I had found the narrow opening of the recess locked with a rough wooden door; but now to my dismay it was completely walled up with brick-work. Was this a precaution to prevent the inquisitive barbarian from gaining even a glimpse of the manuscript treasures hidden within? I thought of the similar device" used by the Jain monks of Jesalmir "to keep Professor Bühler from access to their storehouse of ancient texts, and mentally prepared myself for a long and arduous siege."

Stein felt that it was essential for him to see the entire collection of manuscripts in their original place of deposit so that he could judge the quality and approximate date of the documents. To Chiang was delegated the necessary negotiations holding out the promise of a liberal donation for the main shrine. "It proved a very protracted affair. Chiang's tactful diplomacy seemed at first to be making better headway than I had ventured to hope for. Wang explained that the walling-up had been a precaution against the pilgrims' curiosity. . . . But evidently of a wary and suspicious mind, he would not allow himself to be coaxed into any promise to show the collection to us as a whole." What Chiang did learn during

the hours of diplomatic wrangling was that the great find of manuscripts
had been reported to a high official at Su-chou and through that channel
to the viceroy of Kansu. It seemed—Chiang was not certain—that "the
latter had given orders for the transmission of specimens and for the
safe-keeping of the whole collection." Had an official inventory been
made, it would have jeopardized the transaction.

Stein's apprehensions on that score were dispelled. Wang, a nervous,
talkative man, glad of a sympathetic ear, had mentioned that indeed "a
few rolls of Chinese texts, apparently Buddhist, had been forwarded to
the Viceregal Ya-men at Lan-chou [the provincial capital]. But their contents had not been made out there, or else they had failed to attract any
interest. Hence officialdom had rested satisfied with the rough statement
that the whole of the manuscripts would make up about seven carloads,
and evidently dismayed at the cost of transport, or even of close examination, had left the whole undisturbed in charge of the Tao-shih as
guardian of the temple." This lack of official interest must have confirmed
Stein's attitude that scholarship would be properly served by securing
the collection.

"To rely on the temptation of money alone as a means of overcoming
[Wang's] scruples was manifestly useless. So I thought it best to study
his case in personal contact." With Chiang in attendance, Stein called
formally on Wang and "asked to be shown over his restored cave-temple.
It was the pride and mainstay of his Tun-huang existence, and my request was fulfilled with alacrity. As he took me through the lofty ante-
chapel with its substantial woodwork, all new and lavishly gilt and
painted. . . . I could not help glancing to the right where an ugly patch of
unplastered brickwork still masked the door of the hidden chapel. This
was not the time to ask questions . . . but rather to display my interest in
what his zeal had accomplished. Pride not greed would prove the avenue
to access."

Wang's restoration, alas, had been only too thorough. Fortunately, he
had concentrated on the sculptures and had not touched the "tasteful and
remarkably well-preserved fresco decoration on the walls and ceiling."
The devotion of the "worthy Tao-shih . . . to this shrine and to the task
of religious merit which he had set himself in restoring it, was unmistakably genuine." Chiang had quickly taken the full measure of Wang's ignorance "of all that constitutes Chinese learning, and the very limited extent of his knowledge in general."

Stein's response to Wang was different. He felt that Wang was that
"curious mixture of pious zeal, naive ignorance, and astute tenacity of

purpose [that characterized] those early Buddhist pilgrims from China who, simple in mind but strong in faith—and superstition—once made their way to India in the face of formidable difficulties." As before, but in this context, Stein spoke of Hsüan-tsang and his own devotion to the saintly traveler. "Surrounded by these tokens of lingering Buddhist worship, I thought it appropriate to tell Wang Tao-shih . . . how I had followed [Hsüan-tsang's] footsteps from India for over ten thousand Li across inhospitable mountains and deserts; how in the course of this pilgrimage I had traced to its present ruins, however inaccessible, many a sanctuary he had piously visited and described. "Always truly eloquent when speaking of his Patron Saint, Stein's peroration became ever more eloquent in response to "the gleam of lively interest which I caught in the Tao-shih's eyes, otherwise so shy and fitful. . . . Very soon I felt sure that the Tao-shih, though poorly versed and indifferent to things Buddhist, was quite an ardent admirer in his own way of 'T'ang-sêng [thus Wang pronounced Hsüan-tsang], the great monk of the T'ang period,' as I am in another."

Wang matched Stein in devotion to Hsüan-tsang. In a newly built, spacious loggia of a nearby temple, he showed Stein and Chiang that "he had caused all its walls to be decorated by a local Tun-huang artist with a series of quaint but spirited frescoes representing characteristic scenes from the great pilgrim's adventures, those fantastic legends which have transformed Hsüan-tsang in popular belief throughout China into a sort of saintly Munchausen. . . . Here the holy pilgrim was seen snatched up to the clouds by a wicked demon and then restored again to his pious companions. . . . Elsewhere he was shown forcing a ferocious dragon which had swallowed his horse to restore it again, and so on. But the picture in which I displayed particular interest showed a theme curiously adapted to our own case, though it was not until later that I appealed again and again to the moral it pointed. There was T'ang-sêng, standing on the bank of a violent torrent, and beside him his faithful steed laden with big bundles of manuscripts. To help in ferrying across such a precious burden, a large turtle was seen swimming towards the pilgrim, . . . a reference to the twenty pony-loads of sacred books and relics which the historical traveller managed to carry away safely from India. But would the pious guardian read this obvious lesson aright . . . and acquire spiritual merit by letting me take back to the old home of Buddhism some of the ancient manuscripts which chance had placed in his keeping?" The question was implicit, not expressed. "Yet when I took my leave of the Tao-shih I felt instinctively that a new and more reliable link was being established between us."

Chiang remained alone with the Tao-shih "to urge an early loan of manuscript specimens. But the priest had again become nervous and postponed their delivery in a 'vague way' until later. There was nothing for me to do but wait."

How long? Stein had his answer that same night. "Chiang groped his way to my tent in silent elation with a bundle of Chinese rolls which Wang Tao-shih had just brought him in secret, carefully hidden under his flowing black robe, as the first of the promised 'specimens.' The rolls looked unmistakably old as regards writing and paper, and probably contained Buddhist canonical texts; but Chiang needed time to make sure of their character. . . . He turned up by daybreak, and with a face expressing both triumph and amazement, reported that these fine rolls of paper contained Chinese versions of certain 'Sutras' [Scriptures] from the Buddhist canon which the colophons declared to have been brought from India and translated by Hsüan-tsang himself." The heavens had spoken. Even the most hesitant must listen and heed the celestial fiat.

"Of Hsüan-tsang's authorship, Wang Tao-shih . . . could not possibly have had any inkling when he picked up that packet of 'specimens.' Chiang hastened to inform Wang and then as" quickly returned to tell Stein "that the portent could be trusted to work its spell. Some hours later he found the wall blocking the entrance to the recess of the temple removed, and on its being opened by the priest, [Chiang] caught a glimpse of a room crammed full to the roof with manuscript bundles." He hastened to inform Stein who could no longer restrain his impatience to see the cache for himself. "The day was cloudless and hot, and the 'soldiers' who had followed me about during the morning with my cameras, were now taking their siesta in sound sleep soothed by a good smoke of opium. So accompanied only by Chiang I went to the temple."

Still prey to "scruples and nervous apprehensions, but under the influence of that quasi-divine hint, Wang now summoned up the courage to open before me the rough door. . . . The sight of the small room disclosed was one to make my eyes open wide. Heaped up in layers, but without perfect order, there appeared in the dim light of the priest's little lamp a solid mass of manuscript bundles rising to a height of nearly ten feet, and filling, as subsequent measurement showed, close on 500 cubic feet. The area left clear within the room was just sufficient for two people to stand in. . . . In this 'black hole' no examination of the manuscripts would be possible." Even taking out the contents would be an arduous task, a task whose difficulties were heightened by Wang's fears that his patrons might get wind of it. Fortunately, in addition to the loggia, Wang had also restored a small, side room that "had a door and paper-covered windows. So here a convenient 'reading-room' was at hand for this strange old library where we were screened from any inquisitive eyes, even if an occasional worshipper dropped in to 'kotow' before the huge and ugly Buddha now set up in the temple."

Stein examined the wall that had screened the fabulous cache, because Wang explained that the entrance to that particular cave temple had been almost completely blocked by drift sand when he first settled there eight years before. From the condition of nearby caves and the low level of that particular one, Stein estimated that more than ten feet had accumulated. Laborers paid from "the pious donations at first coming driblet-like with lamentable slowness, our [mark Stein's 'our'] Tao-shih had taken two or three years to lay bare the whole broad passage, some forty feet deep." Subsequently, while replacing the old decayed stucco images occupying the dais of the cella, Wang had noticed "a small crack in the frescoed wall to the right of the passage. There appeared to be a recess behind the plastered surface instead of the solid conglomerate from which the cella and its approach are hewn; and on widening the opening he discovered the hidden chamber" with its deposit.

When, the question immediately posed itself, when was the cache concealed? An inscription on a large black marble slab (at some time it had been moved to a more accessible spot) dated A.D. 851, recorded imperial eulogies for a pilgrim who had returned from India and spent his remaining years translating the sacred texts he had brought back with him. The terminal date, however, remained unresolved until much later: none of the manuscripts "extended beyond the reign of the Emperor Chên Tsung (998–1022 A.D.)." This suggested that the walling-up was in response to the threat of a destructive invasion. "The conquest of Tun-huang by the rising power of the Hsi-hsia . . . took place between 1034 and 1037 A.D. The total absence of any manuscripts written in the special characters which the founder of the Hsi-hsia dynasty [Tanguts, a Tibetan people; their complicated script based on Chinese characters with Khitan modifications] adopted in 1036 A.D." tends to corroborate Stein's assumption.

Satisfied that almost a millennium had passed since the collection was hidden, Stein faced the problem of how to process so huge a collection in the presence of a guardian "swayed by his worldly fears and possible spiritual scruples." Again Stein was engaged in a race: would Wang be "moved to close down his shell before I had been able to extract any of the pearls?" If the need for "energy and speed" was great, equally imperative was the need to "display *insouciance* and calm assurance. Somehow we managed to meet the conflicting requirements. . . . But I confess, the strain and anxieties of the busy days which followed were great."

The first bundles to emerge "consisted of thick rolls of paper about one foot high . . . in excellent preservation, [they] showed in paper, arrangement, and other details unmistakable signs of great age. The joined strips of strongly made and remarkably tough and smooth yellowish paper, often ten yards or more long, were neatly rolled up after the fashion of Greek papyri, over small sticks of hard wood. . . . All showed signs of having been much read and handled; often the protecting outer folds, with the silk tape which had served for tying up the bundles, had got torn off. Where these covering folds were intact, it was easy for Chiang to read off the title of the Sutra, the chapter number, etc." But a quick evaluation of their contents was not possible: Chiang's studies had not included Buddhist scriptures, and Stein, though he knew them in their "original Indian garb, laboured under a fatal disadvantage—total ignorance of literary Chinese. . . . On one point Chiang's readings soon gave me assurance: the headings on the first bundles were all found to be different. So my apprehension of discovering here that inane repetition of a few identical texts in which modern Buddhism in Tibet and elsewhere revels, gradually vanished."

Chiang's effort to list the documents was soon swamped as Wang, grown courageous, dragged load after load from its hiding place. In the first rapid examination Chiang failed to discover colophons with exact dates for the Chinese rolls: their very length made complete unfolding too time-consuming. "So I had reason to feel doubly elated when, on the reverse of a Chinese roll, I first lighted on a text written in that cursive form of Indian Brahmi script made familiar by texts found in the Khotan region." Here was indisputable proof that the bulk of the manuscripts deposited went back to the time when Indian writing and some knowledge of Sanskrit still prevailed in Central-Asian Buddhism. With such evidence clearly showing the connection which once existed between these religious establishments and Buddhist learning as transplanted to the Tarim Basin, my hopes rose greatly for finds of direct importance to Indian and Western research." Again Stein states his overriding motivation.

As Stein handled documents more than a thousand years old and remembered how others of a similar age elsewhere had rotted due to subsoil water seepage, he rhapsodized about the conditions prevailing in the hidden cache. "Nowhere could I trace the slightest effect of moisture. And, in fact, what better place for preserving such relics could be imagined than a chamber carved in the live rock of these terribly barren hills, and hermetically shut off from what moisture, if any, the atmosphere of this desert valley ever contained? Not in the driest soil could relics of a ruined site have so completely escaped injury as they had here in a carefully selected rock chamber where, hidden behind a brick wall and protected by accumulated drift sand, these masses of manuscripts had lain undisturbed for centuries."

His gratitude for such special protection increased "when, on opening a large packet wrapped in a stout sheet of coloured canvas, I found it full of paintings on fine gauze-like silk and on linen, ex-votos of all kinds of silks and brocades, with a mass of miscellaneous fragments of painted papers and cloth materials. Most . . . were narrow pieces from two to three feet in length, and proved by their floating streamers and the triangular tops provided with strings for fastening, to have served as temple banners. Many were in excellent condition . . . and showed when unfurled beautifully painted figures of Buddhas and Bodhisattvas almost Indian in style, or else scenes from Buddhist legend. The silk . . . was almost invariably a transparent gauze of remarkable fineness . . . allowing a good deal of light to pass through—very important since these paintings to be properly seen would have to be hung across or near the porches through which the cellas of the temples receive their only lighting. For the same reason of transparency most of the banners appeared to have been painted on both sides. Some had undergone damage [while still used] in the temples as proved by the care with which rents had been repaired." There were also "silk paintings much larger in size . . . up to six feet or more. Closely and often carelessly folded up at the time of their deposition and much creased in consequence, . . . any attempt to open them out would have implied obvious risk to the material. . . . But by lifting a fold here and there, I could see that the scenes represented were almost as elaborate as the fresco panels on the walls."

Wang attached little value to such "beautiful relics of pictorial art in T'ang times. So I made bold to put aside rapidly 'for future inspection' the best of the pictures on silk, linen or paper, . . . more than a dozen from the first bundle alone. . . . Even among the fragments there were beautiful pieces, and every bit of silk would have its antiquarian and artistic value. . . . To remains of this kind the priest seemed indifferent.

The secret hope of diverting by their sacrifice my attention from the precious rolls of canonical texts, . . . made him now more assiduously grope for and hand out bundles of what he evidently classed under the head of miscellaneous rubbish. . . . In the very first large packet of this kind I discovered, mixed up with Chinese and Tibetan texts, a great heap of oblong leaves in . . . Central-Asian Brahmi. They proved . . . to belong to half-a-dozen different manuscripts of the Pothi shape, some in Sanskrit, some in one or the other of the 'unknown' languages used in Turkestan Buddhism. Several . . . were of large size, with leaves up to twenty-one inches in length, and some . . . proved to be complete. None of my previous finds in Brahmi script equalled them in this respect or in excellence of preservation. So that first day Chiang and myself worked without a break until quite late."

"At the day's end a big bundle of properly packed manuscripts and painted fabrics lay on one side of our "reading room" awaiting removal for what our diplomatic convention styled, 'closer examination.' "

Under the slogan of that euphemism, Stein bought time. "The great question was whether Wang Tao-shih would be willing to brave the risk of removal, and subsequently to fall in with the true interpretation of our proceeding. It would not have done to breathe to him the unholy words of sale and purchase; it was equally clear that any removal would have to be effected in strictest secrecy. So when we stepped outside the temple there was nothing in our hands or about or persons to arouse the slightest suspicion." Though physically spent and emotionally exhausted at the treasures he had seen and handled, Stein remained on the loggia, talking in front of the fresco showing the holy pilgrim returning with the sutras; again he spoke of Hsüan-tsang. Wang agreed that the scholar-saint had not permitted the discovery of the "precious remains of Buddhist lore" only to have them continue to languish in darkness. And then Stein hinted (a shadow of a hint) that Wang would receive an ample donation for the benefit of the shrine he was laboring to restore to its old glory. Chiang suggested that Stein retire and leave him with Wang to tackle the question how to secure quietly the manuscripts and paintings selected.

"It was late at night when I heard cautious footsteps. It was Chiang who had come to make sure that nobody was stirring about my tent. A little later he returned with a big bundle over his shoulders. It contained everything I had picked out during the day's work. The Tao-shih had summoned courage to fall in with my wishes, on the solemn condition that nobody besides us three was to get the slightest inkling of what was being transacted, and that as long as I kept on Chinese soil the origin of these 'finds' was not to be revealed to any living being." Wang's fear of being seen outside his temple precincts at night—so contrary to his routine —was stilled when Chiang took it upon himself to act as courier. "For seven nights more he came to my tent when everybody had gone to sleep, with the same precautions, his slight figure panting under the loads which grew each time heavier and heavier, and ultimately required carriage by installments."

This "curious digging," as Stein called it, differed from other excavations. Stratigraphy, for example, "From the first it was obvious that the contents of the hidden chapel must have been deposited in great confusion, and that any indications the original position of the bundles might have afforded at the time of discovery, had been completely effaced when the recess was cleared out, as the Tao-shih admitted, to search for valuables, and again later on for the purpose of removing the big inscribed [marble] slab from its west wall to the passage outside. It was mere chance, too, what bundles the Tao-shih would hand us out." In Stein's haste to secure as much as possible any systematic study was reserved for the future. On that first day during his hurried tremulous handling, "certain items helped me to form conclusions as to the whole cache of antiquarian treasures."

He instantly recognized "the special value of those bundles filled with miscellaneous texts, painted fabrics, ex-votos, papers of all sorts which had evidently been stored away as no longer needed for use." Their irregular shape and fastening set them apart from the uniform packets holding Chinese or Tibetan Buddhist texts. Their irregularities had caused Wang to arrange them atop the wall-like structure of "library bundles," and this favored his bringing them out in steady succession. Yet from such "mixed" bundles most of the manuscripts and detached leaves in Indian script and in the traditional pothi shapes emerged. "The most important of these was a remarkably well-preserved Sanskrit manuscript on palm leaves, some seventy in number, and no less than twenty inches long. The small but beautifully clear writing closely covering these leaves showed paleographical features which seemed to leave little doubt as to the manuscript going back to the third or fourth century A.D. at the latest." Even though it turned out to be a well-known text, its venerable age, "not surpassed by any Sanskrit manuscript then known" made it important. The material on which it was written conclusively proved it to have been done in India: "But who was the pilgrim who had brought it to the very confines of China?"

One gigantic roll over seventy feet long and a foot wide, its inner surface covered with Gupta-type Brahmi writing, had at one end a charming picture of two geese standing on lotuses and facing each other. The parts Stein was able to examine at that time, showed "only invocation prayers in corrupt Sanskrit of a kind familiar to Northern Buddhism;" subsequent examination has shown "that interspersed with these prayers, there are hundreds of lines with texts in that 'unknown' language which finds from sites about Khotan had first revealed to us, but without any key to its interpretation. This key was since found by Dr. Hoernle, with the help of two bilingual texts I brought away from the 'Thousand Buddhas.' "

The rolls had historical as well as philological value. "The monastic communities established at Tun-huang among a population mainly Chinese were, until a relatively late period, maintaining direct communication with their co-religionists in the Tarim Basin, from which Buddhism first reached China." The abundance of Tibetan pothis was evidence that Tun-huang had also received powerful impulses from the south. Easily recognized—Tibetan writing is horizontal whereas Chinese runs vertically —hundreds of them came out in disarray but were easily arranged according to their various sizes (they had the traditional holes for the string but were loose). Stein's cursory examination showed that the paper on which the Tibetan pothis were written looked older than that of the rolls "and in texture markedly different from that of the Chinese texts. So the conclusion suggested itself that the Pothis represented mainly imports from Tibet itself, while the rolls had been written by Tibetan monks established locally."

Chinese annals told of Tibet's conquest of much of Kansu about A.D. 759, a control that lasted a full century: the "period when Tibet was a great power in Asia holding in subjection vast tracts of Kansu and even Central Asia. It was by way of Tun-huang that the Tibetans about 766 A.D. onwards gradually overran the territories of Eastern Turkestan, and finally in 790 A.D. overwhelmed the Chinese garrisons which had long struggled to maintain the imperial protectorate in the distant lands north and south of the T'ien-shan. . . . This political connection . . . made it easy to understand why Tibetan Buddhism was so amply represented among the literary remains of the walled-up cave."

Stein was delighted when Uighur manuscripts appeared. The Uighurs, a Turkish tribe, established their kingdom about A.D. 860 along the northern oases, Turfan, Hami, and others, and by the tenth century they had reached as far as Kansu. Uighur manuscripts were like "small quarto volumes, being written on thin sheets of paper, folded and stitched after the fashion of Chinese books and [were] complete from cover to

cover. Chinese glosses and rubrics . . . clearly indicated Buddhist contents." Their script, derived from Syriac writing, was widely used for
Turki writings before the Turkish population of Central Asia was Islam-
ized; it also appeared on the reverse side of many Chinese rolls. Stein's
trained eye noticed that these were "distinctly less cursive and of a firmer
shape than the manuscript books and the specimens of Uighur texts I
knew otherwise." Only later, in Europe, did he learn that the language
was, in fact, Sogdian: "that Iranian dialect which Professor F. W. K.
Müller's brilliant researches [showed had] been used for early translations of Buddhist literature in what is now Samarkand and Bokhara. What
a large share this Iranian element must have had in the propagation of
Buddhism along the old 'Northern route [of the Tarim Basin].' "

Additional manuscripts gave Stein "further proof of the remarkable
polyglot aspect which Buddhist religious places must have presented in
these parts when [I] came upon fragments of texts in that earliest Turkish
script known as Runic Turkish [from its resemblance to the runic alpha-
bets of Northern Europe]. Until a few years ago this had only been known
from the famous inscription of a Turkish prince discovered on the Orkhon
River in Southern Siberia and first deciphered by Professor V. [ilhelm]
Thomsen."[16] Subsequently the texts, stories composed for divination,
were characterized by Professor Thomsen as "the most remarkable, comprehensive and best preserved of the rare relics that have come down of
the earliest Turkish literature." Even more unusual was a narrow, fifteen-
foot roll, beautifully written in a special form of Syriac script, an "early
Turkish version of the confession prayer of the Manichaeans."[17] Mani's
church, to which the young St. Augustine had belonged, traveled to China
during T'ang times. Manichaeism, "a formidable rival to Buddhist and
Christian propaganda alike throughout Central Asia, had its followers at
Tun-huang." Its presence there made Stein marvel at "this curious relic
of a race and language that have spread from the Yellow Sea to the
Adriatic" and stopped lightly "at that crossways of Asia that is Tun-
huang."

Decipherment and marveling were far in the future. "Five days of
strenuous work resulted in the extraction and rapid search of all 'miscellaneous' bundles likely to contain manuscripts of special interest. . . .
These had been put mostly on top when the Tao-shih had last stuffed his
treasure back into their 'black hole'. . . . There still rose against the walls
of the chamber that solid rampart of manuscript bundles." Clearing them
out, scanning them quickly proved troublesome "in more than one sense,
though discreet treatment and judiciously administered doses of silver
had so far succeeded in counteracting the Tao-shih's relapses into

timourous contrariness." Stein felt it important to clear out the whole chapel; Wang, fluttering at the increased risk of exposure, became "altogether refractory. However he had gradually been led from one concession to another, and we took good care not to leave him much time for reflection. So, at last with many a sigh and plaintive remonstrance," he undertook the moving. "By the evening of May 28th, the regular bundles of Chineses rolls, more than 1050 in all, and those containing Tibetan texts had been transferred to . . . the spacious main cella of the temple."

A hasty examination became possible because the bundles, several rolls tightly held in coarse linen covers, were open at the ends. The search was to find pothi leaves with Brahmi manuscripts, folded-up paintings, or other special documents, embedded in the bundles. "Such we picked out and put aside rapidly. . . . In view of the Tao-shih's growing reluctance, I had a gratifying reward for my insistence on this clearing." At the very bottom, used by Wang as shims to level the manuscript wall, Stein found a number of "exquisite silk paintings of all sizes, and some beautifully embroidered pieces." One, though badly crushed (happily restored at the British Museum), was a "magnificent embroidery picture, remarkable for design, colours and fineness of material, and showing a Buddha between Bodhisattvas in life size."

The climax of the second act came when, the recess emptied of its contents, Wang and Stein confronted one another. Two men radically different in culture and training, ambition and way of life; two wanderers who, for whatever the reason, shared a common regard for the caves where they had met. Darkly, submerged, there was a brief struggle of sorts. Stein has told fully, flatly what brought him to Tun-huang and why he wanted to mine the concealed treasure of documents—a desire intensified because the sacred grottoes, still a living place of pilgrimage, made it inadvisable (or wrong) to remove frescoes or sculptures.

But what of Wang? His motives and behavior come filtered through Stein. Suppositions, interpretations, explanations: Wang's version will never be known. Patriotic? no; religious? maybe. What—the questions can only be raised—what directed Wang to leave Shansi, his native province, a cool, fertile upland, to beg his way meal by chancy meal halfway across the vast breadth of China to that western oasis set in a desert where, but lately, Tungan rebels had decimated hamlets and desecrated shrines? What at Ch'ien-fo-tung had engaged his deepest emotions? What vision commanded him to restore the shattered statues and sand-logged caves? What had stirred within that unlettered man when he beheld manuscripts cached in the bricked-up recess? How are his actions to be heard, actions translated by Stein as "nervous," "frightened," "furtive," "sullen," "truculent," and "anxious"? Did he feel anguish or panic when the anchor to his flotsam existence was threatened? Did his moods speak a deep-felt anger at being deprived of his life's purpose to restore the caves and, when he stumbled on the precious deposit, to guard it? He was greedy not where greed means personal enrichment but only to retain his guardianship of the sacred caves. All these qualities Stein acknowledged. (Did he, perchance, unconsciously envy the directness and simplicity with which Wang carried out his life's work, he who repeatedly spoke of the economics of "freedom"?)

Stein explained the problems Wang's presence caused. It made all previous "digs" radically different from Tun-huang. He had to make a choice, and his ignorance of Chinese made it impossible for him "to estimate the philological value of those masses of Chinese canonical texts. . . . Still less was I able to select those texts which for one reason or another were possessed of antiquarian or literary interest." It was like fishing in a grab bag for prizes. To carry off all the manuscripts, or, as he phrased it, "to rescue the whole hoard," would require a procession of carts—as blatant an announcement as could be made—and could stir up religious resentment in Tun-huang, and perhaps even compromise his further work in Kansu.

"For two long days . . . discussions had to be carried out intermittently with a view to gain time while my examination of the miscellaneous bundles was proceeding. I managed to complete this by the second evening." The next day Stein was dismayed to find that overnight Wang had shifted almost all the bundles "to their gloomy prison of centuries. But the advantage we possessed by already holding loads of valuable manuscripts and antiques, and the Tao-shih's unmistakable wish to secure a substantial sum of money, led at last to what I had reason to claim as a substantial success in this diplomatic struggle. He agreed to let me have fifty well-preserved bundles of Chinese text rolls and five of Tibetan ones, besides all my previous selection from the miscellaneous bundles. For all these acquisitions four horse-shoes of silver, equal to about Rs.500, passed into the priest's hands." To transfer these from their recess to Stein's tent, a load too heavy for Chiang, Stein enlisted the help of two trusted men, Ibrahim Beg, a veteran of the First Expedition, and Tila Bai, his ponyman.

A week passed while Stein worked at the delicate packing. Wang, his nervousness increased by so prolonged an absence from his clients, went through the Tun-huang district on his seasonal begging tour. "He returned reassured that his secret had not been discovered and that his spiritual influence . . . had suffered no diminution. This encouraged him to part with another twenty rolls" against an appropriate donation for the temple. All in all, the manuscripts filled seven cases and the [more than 300] paintings and embroideries, five. "There was some little trouble about getting enough boxes without exciting suspicion at Tun-huang. Luckily, I had foreseen the chance and provided some 'empties' before-hand. The rest were secured in disguise and by discreet installments. . . . The good [mark Stein's adjective] Tao-shih now seemed to breathe freely again, and almost to recognize that I was performing a pious act in res-cuing for Western scholarship these relics of ancient Buddhist literature and art which local ignorance would allow to lie here neglected or to get lost in the end."

Two tableaux, very different, closed Stein's stay at the sacred caves. He had just come back from a short walk up the lonely rocky gorge to find huge bags of postal accumulations awaiting him. It provided "a feast such as my heart, thirsting for [Publius's] news since nearly four months had longed for. Turdi [the dak carrier] had covered the 1400 odd miles from Khotan (by the route through the mountains) in 39 days, no small per-formance, & his arrival in the evening was a delightful surprise. There were altogether some 170 letters to be opened."[18] With no time for a full P.N., he hinted at events and summarized his feelings. "It has been a time of great strain though apparently there was physical rest. I prefer *plein air,* even if it be in the desert, to caves & diplomatically discreet bur-rowings. For the results of the latter I must thank Fate—and Hsüan-tsang, & last but not least my indefatigable zealous Chiang."

The other event was equally unexpected: a visit from the friendly, refined, and learned amban. It was explained, Chiang learned, because "instructions received from the Lan-chou Viceroy, enjoined him to dis-suade me with all diplomatic politeness from any attempt at excavation. The idea apparently was that my archaeological activity would necessarily turn towards tombs, the only find-places of ancient remains known to Chinese collectors of antiques, and that the popular prejudice thus aroused might expose me to personal risks. . . . My prolonged stay at Tun-huang had evidently given alarm. But since [the amban] knew better, and could in all honesty point out the harmless nature of my work, which all lay 'in the Gobi' [desert], I could hope to avert polite obstruction. . . . Never-theless, I took care to send Chiang to town for a couple of days, and to assure myself through him that the report dispatched to the Viceroy put my case in the right light. For all such gentle guidance of bureaucratic wheels Chiang's help was invaluable."[19] The viceroy's uneasiness stemmed from another situation which to Stein seemed far more serious. Chiang had learned that there was "contention over a matter of revenue assessment between the magistrate's office and a section of the headstrong Tun-huang colonists, . . . [and that] enforcement of the judgement might provoke resistence among these *soi-disant* protectors of the Empire's marches."

On 13 June, having sampled generously of the hidden cache at the sacred grottoes, Stein was eager to turn to topographical work in the mountains. His caravan of loaded camels and ponies was augmented by five, large three-horse carts; it took the road to An-hsi. Wang watched the departure. His face again wore its air of shy but self-contained serenity. "We parted in fullest amity."

On 14 October, on his return to An-hsi, Stein wrote to Publius. "It was a great pleasure to receive back safe & sound my 17 cases of ancient MSS., etc. from the An-hsi Yamen where I deposited them in June. Have I told you already how during the local outburst which took place in Tun-huang about a month after my departure, the Amban's Yamen was looted and burned? . . . As the rising had a purely local origin with no trace of anti-foreign feelings, I did not hesitate to take up once more my sap [to draw off] at the hoard of the 'Thousand Buddhas.' The secret of our previous search had been well kept & I knew that this would inspire my timid priest custodian with fresh courage. A proposal for further pur-chases sent through a trusty messenger met with a promising response, but in order to avoid all suspicions, I had to remain away from the scene & entrust the execution to my ever-zealous secretary. He started off under cover of suitably imagined pretenses, while four camels were marched off by the desert route to rendezvous in secret near the caves. A week later, in the dead of night, they turned up again near my temple, . . . The 230 bundles of MSS. which the good [again, mark the adjective] Tao-shih had been induced to part with, contain approx. close on 3000 text rolls, mostly Chinese Sutras & Tibetan Buddhist works. But no real examination has been possible so far; Central-Asian translations are apt to turn up in this big haul. For the present the new acquisitions travel in huge bags, disguised as well as we could manage it. The secretary & trusty Ibrahim Beg must have had a busy night when these loads were first cleared out of the 'black hole' of the temple. The story how they dodged some in-opportune Tibetan visitors to the caves, returned to An-hsi by night marches to escape attention, etc. was quite romantic. For the present we must keep this *entre nous* . . . all which the 'Thousand Buddhas' yielded has cost the Govt. only some £130. The single Sanskrit Ms. on palm leaf with a few other 'old things' are worth this."

15

Act Three: Two Daring Crossings

Tun-huang is tied to Stein's name. Yet before and after he had breached the walled-up recess, he had found treasures at scattered sites—Dandan-Uiliq, Niya, Miran, Lou-lan, the limes watchtowers, and others—which lay lost and forgotten under the relentless sands of the Taklamakan in ruined stupas, rubbish heaps, and archives of communities dead and abandoned and distant from living villages. The hidden treasure of Tun-huang was just another episode in his explorations—certainly more splendid in its contents and quality—but not the be-all and end-all of his continued questing in Central Asia. Tun-huang had been a heady experience and he did not deny his luck, but it did not differ in kind from his elation at his very first recoveries of ancient remains, that initial proof of Central Asia's past. Surprise, though always present, had given way to an intensive, energetic assault on desert sites. As, in what might well have seemed a former incarnation, the *Rajatarangini* had guided him to reconstitute the ancient geography of Kashmir, so now in Chinese Turkestan he was again following in the footsteps of Hsüan-tsang and Marco Polo—and always led by trustworthy living treasure hunters.

The heat of summer gripped the land as he took the road to An-hsi, where the Tarim Basin's southern and northern routes met to form the main road to Kansu. In T'ang times it had been the headquarters for the governor-general of Turkestan; but it had not recovered from the recent Tungan devastation—its walls were crumbling and its single street straggled through a forlorn town. In the safekeeping of its yamen, Stein left the hard-won cargo with the faithful Ibrahim to guard it. On the way there "to my great satisfaction I managed to pick up again our ancient frontier wall quite close to the town." This wall, built in Han times, was to occupy his attention as he went eastward into the mountains.

His letter to Publius of 11 July 1907 has the tone of a schoolboy off on holiday as Stein describes the initial stages of his topographical work in the mountains. "My first move towards the hills took me to the broad valley of Ch'aio-tzu, half desert, half marshy steppe, where I found the remains of a relatively large walled town, . . . abandoned about 700 years ago. For excavation there was little scope but the ruins & the ground around them served as an ideal illustration of the physical changes which go on here [and] . . . have solved whatever had looked puzzling about the far older sites N[orth] of Lop-nor. . . .

"Crossing the second range of barren hills, I marched in 2 days to the Buddhist caves of Wang-fu-hsia [The Valley of the Myriad Buddhas], a pleasing & instructive *pendant* to the 'Thousand Buddhas.' There were some twenty well-preserved shrines with frescoes of the same time & style. There is a little monastic establishment at this lonely *Tirtha* [shrine]. . . . I had a friendly reception & what an airy court I dwelt in at the feet of a colossal Buddha some 80' high! There were no hidden MSS to exploit here. After two days spent in photographing, I moved on to the alpine plateau . . . at the foot of the main snowy range where the last six days have been spent surveying. It forms a huge triangular alluvium, rising from 7,000 to 10,000 ft [and] . . . the peaks to 19,000 to 21,000 as at present estimated. There is so little moisture about that the snow-line does not descend far. We got up without trouble to the main pass, circ. 13,000 ft. opening towards the desolate high plateaus of Tsaidam & got a fair share of their gentle summer breezes, cold & piercing enough for non-Tibetan mortals. . . .

"Today it got delightfully clear & I could see through the glasses even those horrid desert-hills beyond An-hsi. Tomorrow I hope to reach the hill oasis . . . from where I shall try to use the least-known hill track to Su-Chou where a Tao-tai is waiting with a lot of treasure (Rs. 20,000) at my credit! I intend to make a long tour in the central Nan-shan which has been so far but partially explored by Russian surveyors. Much depends on what transport we can secure. Camels would not do, & my own have just been sent down for a well-earned summer vacation."

His next entry in that long letter to Publius was "after a glorious march of some 25 miles, first in view of imposing glacier-clad peaks stretching away in a long chain to the south & then through a mountain-girt oasis which looked delightfully green in its setting of bare red & yellow hills. Since Chitral I had seen nothing like it in the way of *pretty* scenery. The sight of a big stream issuing from springs at the foot of these interminable arid slopes of gravel was also a pleasing change. Even at the base of the big snowy range one scarcely ever sees any running water—so thirsty is the soil & so rapid the evaporation. . . . The display of local ignorance about all routes but the high-road is also as great as usual & we may have to track the desired route to Chia-yü-kuan [Barrier of the Pleasant Valley; West Gate of the Great Wall] ourselves. Luckily the ground is easy to survey & the distance not more than 5–6 days' marching may cover."

A fortnight later he was camped at "The Spring of Wine, Su-chou, Kansu," having reached there by "the new route I had mapped out. The Chinese . . . professed the utmost ignorance. Well, in the end they took us by the route I had wanted to follow and when, after six days of pleasant hill marching, we emerged in the great desert valley which Chia-yü-kuan guards eastwards, a good deal of interesting new ground had been mapped which so far was a cartographical blank. You can imagine how glad I was to keep near the snows as long as possible. But I found compensation for their cool air when I reached the 'Great Wall' & could take again to archaeological rides.

"A long day's survey along the Chia-yü-kuan border sufficed to settle the question as to the date & purpose of the wall which the maps show running along the northern edge of Kansu. I was delighted to find it was nothing but a continuation of the *limes* which was my old friend in the desert west of Tun-huang. But in addition to this old wall there is another one, far later, crossing it at right angles which was meant *not* to protect the road into Turkestan, but, on the contrary, to block it. Of course, it dates from the time after the T'ang when China had abandoned all ambition of expanding westwards & had retired into its shell. Hitherto the two walls have been mixed up by travellers & Sinologists alike, & the result has been a curious antiquarian puzzle.

"The change in scenery within Chia-yü-kuan was striking. From a stony glacis into a large belt of cultivation which continues almost unbroken to the Yellow River. . . . I knew that I had crossed a great climatic & physical barrier. . . . I had a most cordial welcome from the 'Guardian of the Gate,' an amiable old Brigadier-General, & was escorted to Su-chou with a grand display of mounted men carrying quaint straw hats & banners of imposing size. At Su-chou I pitched my camp in a delightful spot outside the walls where an ancient spring & reed-filled basin is adjoined by temples all in ruin. . . . For two days the 'Spring of Wine' saw plentiful people, the return state visits by three dignitaries. By an excess of polite attention, they insisted on treating me to a big dinner at my own place. It was all done very readily. The largest temple hall was patched up with paper & glass panes within a day & there we sat feasting for nearly four hours. . . . Towards the close we came to discuss 'business' about my projected tour through the Nan-shan to Kan-chou; all kinds of absurd difficulties were raised about 'wild Tibetans,' impossible mountain tracks, etc. It was easy to demonstrate that my route would not take me into Tangut [Mongol] or Tibetan territory & that one Russian traveller at least had passed through several of those high valleys.

"It took days before the fears . . . had been sufficiently allayed for orders about the transport to be seriously issued. The dozen ponies or so I must hire are now collected & tomorrow I shall set out for the foot of the mountains. The question remains how far into the latter I shall get their timorous owners to move. They, too, dread the mountain tracks."

Increasingly Stein had become disenchanted with Ram Singh, his surveyor. Months before he had written to the Trigonometrical Survey telling them of the surveyor's rheumatism and asking that he be replaced by Lal Singh. At Su-chou he received a telegram announcing Lal Singh's arrival in Kashgar. "Ram Singh's rheumatism has disappeared for the time being, but not his bad temper & I could not have expected effective assistance from him next winter. It has cost much firmness & constant care to get the needed work done by him so far & and you can imagine"— he told Publius—"that this means much additional strain. Lal Singh, who did so splendidly with me at Mahaban is a man of much mightier disposition & greater natural intelligence." It had also been arranged that the new surveyor coming out from the west and Stein coming from the east would meet at Hami.

August was half over when he sent his next series of letters to Publius, written 18–24 August 1907: "It has been a strenuous journey this expedition through the high ranges of the Nan-shan, but as complete a success as far as I could have wished. As we made our way over these huge parallel mountain ramparts, reaching an average crestline of 18–20,000 ft., and across the broad plateau-like valleys which separate them, my eyes reveled daily in vistas which I knew would have fascinated you. . . . It had cost a great effort to secure the 16 ponies & mules which our much reduced baggage & supplies needed, for the Su-chou Mandarins feared all sorts of trouble & only gave their help under steady pressure. In reality, the Amban had no easy task and . . . had to come out & personally assist in the start to prevent wholesale desertion. Now after all these days steady marching in conditions which are often trying enough, . . . I begin to wonder how we ever got them so far. Perhaps, it is due to the effective way in which the first attempt at passive mutiny was checked or to the total absence of guides." Without guides the men, as Stein pointed out to them, were utterly dependent on him to get them back to Su-chou.

"We had only rough sketches of two Russian travelers to guide us over parts of our route. But the topographical features of this mountain system are so clear, & the vistas so extensive that it was easy to select the right lines for the survey work. Day after day I enjoyed the sight of glorious snowy ranges with glaciers far bigger than one could have expected in this quasi-Tibetan region. . . . The most impressive part has been the four days we marched up the huge basin from which the Su-lo-ho gathers its waters. A wonderful contrast between the alpine solitude—the snow had just finished melting in the valley & most of the ground was like a bog—& that familiar desert where all these waters are destined to find their end. Only the absence of all human life is common to the birthplace and the deathbed of the great river.

"Yet there are glorious pasture grounds in these valleys where thousands of Kirghiz could enjoy life & grow rich. I thought of the remnants of the Yüeh-chih, our Indo-Scythians of the N.W. Frontier, which Chinese Annals mention as having found a refuge in these mountains. Little wonder that all these big valleys are now tenanted by wild yaks & kulans [wild asses]. The Chinese seem to have a perfect horror of pastoral life, & on this narrow Kansu border strip, ever threatened by Huns, Turki & Mongols from the north, they had good reason to keep at least the mountain flank south clear of troublesome neighbors. Of the 'wild Tibetans' i.e. Tanguts [A Tibetan people, who had a "semi-oasis" way of life][1] who in a mild way of pilfering, try to maintain time-honoured traditions, we saw none. So there was no need to make a display of our official armament [which] I am sorry to say failed to secure us wild donkeys, sharp little beasts but aggravatingly fleet, or wild yaks. There was no time for patient stalking. But an over-curious wild sheep was duly bagged, & my meat supply for a week has come from one of its fore-legs. . . .

"The last two days have taken us through a very interesting region from the source of the Su-lo-ho (which once reached Lop-nor!) past those of the Su-chou River to the headwaters of the Da-tun which is a tributary of the Huang-ho [Yellow River] & thus sends its waters to the Pacific Ocean. I confess a geographical weakness for reaching at least the edge of the drainage area of the Pacific. The weather has been terribly rainy . . . and to get thoroughly drenched at a height of 13,000 ft. or so is not pleasant & it has been difficult to get fuel of sorts (yak dung) even for cooking."

Five days later (21 August) he could give an account of the end of that mountain tour. "When I wrote in that nameless, dismal gorge beyond the Huang-Ho watershed there were doubts about our immediate progress, as heavy clouds hung about the mountains. It was easy to get to the waters that flow into the Pacific, but far more difficult to find a way from them northward into the valley of the Kan-chou River. . . . There was no time systematically to reconnoitre for passes. The weather had been too bad even for topographical conjectures to guide one. The feeling that nothing would be gained by descending further in a deep-cut difficult valley, induced me to try for a passage from the very gorge into which that day's deluge had driven us. Luckily, it proved just practicable for the ponies up to the watershed, & a track human feet do not appear to have trodden for many years, successfully took us across it. . . . The look of our Chinese when through the mists they caught sight of the broad open valley beyond & felt assured of escape, was a thing to remember.

"Two days ago the northern main range of the Nan-shan was crossed. Our descent has been delightfully varied. Yesterday I reached the first slopes clothed with firs & junipers. . . . Just there, at about 10,000 ft elevation, the first Mongol camp was reached. Milk and sheep were forthcoming to refresh the worn-out men. So today's hard climbs over a succession of steep spurs were taken without grumbling. . . . Two days more ought to see us clear of the mountains. . . .

"Four long marches over steep passes & then over big rolling plateaus still quite alpine and . . . suddenly we emerged in the barren detritus-filled gorges which fringe the edge of the Nan-shan. At Liyuen a dear old Chinese Colonel offered a hearty reception & what supplies a half-decayed watch-tower could afford to our hungry Chinese & their beasts. It poured for nearly a day. . . . but with a year in parched deserts & mountains 'desiccated' still before me I find it hard to grumble at life-giving moisture.

"The fording of the main Kan-chou River, close to the town & now spreading itself over a bed about 1½ miles broad, was a long & trying affair even in the huge wheeled carts sent out for the purpose." Kan-chou, where he arrived 28 August, was his eastern limit. Welcoming Stein as he finally reached the other side was the Naik Lal Singh sent on from An-hsi with mail. Outside the town gates, the Kan-chou dignitaries in dozens had gathered. "It was quite dark when I could install myself in the temple which the Naik had requisitioned for our quarters—next morning it proved to be a regular repository for tenanted coffins awaiting shipment to the distant province of Hu-pei, a sort of residential club for corpses in exile. [In the morning] I managed to exchange the funeral club for a more cheerful temple & orchard." Stein stayed in Kan-chou for five days to give men and ponies a rest, to have repairs attended to, and pay the numerous official visits. After this glimpse of Chinese city life, he started back for An-hsi.

He explained why he took the main, direct road. "You know I do not like it, but it allowed me to test my assumption that the ancient *limes* I had tracked from the Tun-huang desert as far as An-hsi, once continued further east & joined the Great Wall near Su-chou. It was not easy to combine with the daily progress of my baggage train reconnaissances many miles away to the north. I was amply rewarded for the long fatiguing rides over bad ground & practically without local assistance by conclusive proofs that my conjecture was right. I tracked the remains of the *agger* and its watch-towers again & again. . . . The extension of the Great Wall right through to the end of the Su-lo-ho river course where I made my excavations in April & May can now be put down as a fact. It is a line over 300 miles. . . ."

He had commented to Publius in his letter of 14 October: "I was eager to get back to An-hsi for there was much work to be done there before the start for Hami & my second winter campaign close at hand." From June when he had left An-hsi until he returned there (25 September), he had mapped in detail 24,000 miles of mountainous terrain.

10 November 1907, to Publius: "I am marching steadily northwestwards through the dreary hills of the Peishan to Hami, a sort of half-way oasis between true Cathay & the 'New Dominions.' " He stayed there long enough to outfit his men for the winter. "You can imagine how glad my men were to find themselves again among Turks . . . and what a relief it was to be able to pay at last in coined silver instead of the never-ceasing weighing and cutting of silver lumps! I had come to look upon my scales as a refined instrument of mental torture even though Chiang handled it deftly & honestly. . . . I had a very attentive reception at Hami from the Amban, a sort of Resident, and the 'Wang,' a Muhammadan local chief whom the Chinese have found politic to keep in power. He squeezes his people far more than the most rapacious Amban would, & the result was a little riot some weeks before my arrival which, owing to the Wang's possessing a supply of Mauser rifles, ended quickly with a good deal of needless bloodshed."

From Hami along the big road that ran all the way to Kashgar, Stein
sent his heavy baggage—eight camel loads carried the manuscripts and
wooden records—while he, cutting corners, started for the Turfan district (2 November). "It is fully 250 miles to its first oasis where I write
this [Camp Pichan] but I managed to cover it in 9 days including one
spent on the survey of some ruins. . . . It is likely that the main ruins of
Turfan, all close to flourishing places, have been dug up by the Germans.
Unless less easily accessible places have escaped them, I shall move soon
westwards. *I am anxious to regain Khotan through the desert. . . .*"
(italics mine).

The long, dreary marches were spent in a kind of dream world as he
read of the Allens' search for the Erasmus letters. "I have read through
this grand *catena* of epistles more than once since that evening when they
brought me a glow of real comfort & contentment, & every time I thanked
Fate for having given me friends such as you are for letting me have a
share, however far away, in your joyful experiences. Everything you had
to tell me of your explorations over half the continent, was sure of my
liveliest interest."[2] Detail by detail (as his "dearest Papa" had taught
him) with emotions dredged out of his youthful years, Stein commented
on his friend's leisurely account. "I could scarcely believe my eyes when I
first saw your handwriting on the envelope bearing Hungarian stamps.
Your description . . . brought back happy memories of early youth when
I used to spend parts of my summer vacations with an uncle [Ignaz?]
(long since dead) owning land in the Fehér county. . . . Your account
revived impressions & sensations which had become dim & effaced by a
life of long, self-willed exile!" Strange, disturbing words from a man who
was doing what he most wanted to do and had, in the doing, but lately
reaped a scholarly bonanza at Tun-huang.

Boy and man were magically joined by Publius's wanderings in "surroundings familiar to my recollections. On my first visit to Paris, just before I came out to India, I too stayed at the Hotel St. James & Albany. I
too spent a glorious day walking over the Arlberg when returning to
Hungary after one of the university years at Tübingen. Hotel Bellevue is
closely associated with my Dresden remembrances, for my brother always
stayed there. Need I tell you how touched I am by the trouble you took to
see my old schola Sanctae Crucis? You must have thought the Gothic of
the buildings (circ. 1866 A.D.) plain & poor. But to my young eyes it
looked quite solemn & worthy of a school which has an unbroken record
ever since the early part of the 13th century. . . . To Breslau too it was easy
to follow you. I had spent bright autumn days there with my brother in

1893 & probably also visited the church to which Erasmian piety drew you."

Stein went on to describe the Turfan district, four hundred feet below sea level, where he spent sixteen days: "remarkably dry & quite exempt from drift sand & erosion, & the ruins on the slopes of the hills were quite safe from moisture. Thanks to a curious system of underground canals [*karez*] little or no land has been abandoned since Uighur times." He described the canals in *Ruins of Desert Cathay*. "The lines of these underground channels [are] marked on the bare clay surface by the little circular heaps of earth which the diggers had thrown up mole-like at the mouth of each successive well. Starting at ground level from the area to be irrigated, a low, narrow channel is tunnelled from well to well up the natural slope of the basin, but at a gradient less inclined than its surface. The wells were said to range to depths of over fifty feet; but the diggers are so expert that, working in parties of four or five men, they could complete an average Karez within half a year."[3]

The sites, he explained to Publius, "had already been exploited by the Germans in a peculiar fashion, and yet nowhere thoroughly. . . . Big temples, monasteries, etc., were dug into with the method of a scholarly treasure-seeker, barely explored with any approach to archaeological thoroughness. The places most likely to yield 'finds' had been reached by this system for which the German language supplies the expressive term of 'Raubbau' [to mine carelessly]." Should he or should he not work at Turfan? After looking over the ground and after much soul-searching, Stein felt the futility of trying to accomplish in a few weeks what "Grünwedel & Lecoq with their parties had failed to do during an aggregate of 3-4 years. . . .

"Good chance or a little calculation led me to the only known site of Turfan which had escaped German operations, & this I managed to clear with thoroughness. It was a group of small temples situated not far from the salt lake of the Turfan Basin. In the Khotan region we should scarcely have called the surrounding ground a 'desert,' the nearest cultivated area was only 6 miles or so off and the sand spreading around showed only baby dimensions. Still this sufficed to keep off the Germans. The excavations yielded some good fresco & stucco fragments & MSS, in Chinese, Tibetan, Uighur which suffice to settle the question of date." Stein approved of the Germans' choice of "the Turfan ruins, first surveyed by an expedition of the St. Petersburg Academy. The continuity of work assured them by intelligent support at Berlin was another great point in their favour. They certainly have laboured hard & I only wish they may have the will & money to make up for their defects of their burrowing method."

This last batch of mail had brought Stein the first reviews of *Ancient Khotan.* "One, in the 'Revue du Monde Musulman' rather surprises me. What makes my book figure *dans cette galère?* It touches upon so little that is Muhammadan. A wholly unexpected recognition [came] from the Munich Academy of Sciences which within a fortnight of publication had awarded to the work the rather substantial prize (M. 1500) from the 'Hardy Stiftung' [Hardy Bequest], intended, I believe, for the encouragement of Buddhistic researches. I must value this mark of approval particularly in view of the competition from Berlin."

From Turfan to Kara-shahr at the northeast corner of the Tarim Basin, a distance of two hundred miles, Stein marched in "8 days along the regular caravan route [which was] provided with stations, for the track over several passes in these barren outer ranges of the T'ien-shan was pretty bad & the days are now so short." It was 8 December, and the long account he began at Turfan was continued as he moved along to the nearby ruins of Ming-oi (East Turkish for "a thousand statues"). "The Germans had paid them some spasmodic visits; but to judge from the accounts of their people had left much untouched."

Ming-oi, at the foot of the Korla Mountains, consisted of a few shrines crowded along a barren ridge. He was appalled by what he saw of the Germans' work. "A sort of scraping. But what distressing traces they have left behind! Fine fragments of stucco sculpture flung outside or on the scrap heap; statues too big for transport left exposed to the weather & the tender mercies of wayfarers, etc. I cannot guess what made them dig here with quite an inadequate number of labourers, & still less this indifference to that fate of all that was left in situ. Was I really too sentimental or over-conscientious when amidst the physical difficulties of the true desert I spent time & labour to fill up again my excavations at ruins so little exposed to human destruction as those of Lop-nor & the Tun-huang desert?"

He could understand, if not condone, why the Germans seemed to have left the Ming-oi site like a hit-and-run accident. 28 December 1907, to Publius: "Day after day an icy mist lay over the plain & the nightly hoarfrost practically amounted to a light snow fall. . . . The yield was as ample as could be hoped for at ruins which had been systematically sacked & burned down by iconoclastic invaders (probably early Muhammadan) & then exposed for centuries to a climate much moister than that of the great Turkestan plains. . . . Yet pieces of excellent woodcarving, gilt relievo panelling, etc. had escaped here & there as if by a miracle & give a good idea of the splendour which once reigned in the adornment of these shrines. The most striking finds were among the hundreds of stucco sculptures, large & small, which turned up in the debris of the big shrines. Broken indeed, were most of them. But before they fell from the walls fire had turned the friable stucco into a sort of terracotta quite able to stand up to all the rain & snow which descends on the shore of Baghrash lake. So it was easy to pack & transport this wealth of well-modelled heads. Of course the burning had lost all the colouring; compensation for this was offered by a series of fine fresco panels just big enough to be cut out and carried off safely. It was a hard task: the plaster stuck so well to the wall that the brickwork on which they were had to be chiseled off before the panels could be secured. . . . I felt heartily glad when I could take leave of 'the Thousand Houses' (please note Turki grandiloquence!) with good conscience & move off for the newly traced ruins up in the valley of the Karashahr river." The relics, saved from devastation, climate, and German burrowings filled ten heavy cases.

To compensate for the Cimmerian gloom of Ming-oi, Korla, the new site, was both picturesque and dry: a cluster of temples nestling on rock ledges and ridges at the foot of a big barren mountain. "Christmas day and its eve were spent most enjoyably clearing the ruins. A quaint old Mongol had made his much begrimed home in a half-cleared cella & did the *honneurs* of the place; a fine spring spread itself under a glittering ice-sheet & provided a bright patch of vegetation in the midst of the barren, stony waste. The men I had brought with me from Ming-oi had learned to wield their 'ketmans' well, & Christmas evening saw our task completed. . . . Two days have sufficed to bring me from that snow-bound Karashahr valley with its queer mixture of Mongols, Tungans and Chinese down to Korla. . . . It was glorious as I rode down the defile, by which the green waters from Baghrash lake rush down into the Tarim Basin, to see once more the unlimited horizon of the real 'sand-ocean' before me."

With its typical arrangement of "scattered roomy houses & big orchards," Korla was a flourishing oasis. Stein's quarters "at a Beg's house, are quite the cleanest & most comfortable I ever entered since leaving my Yarkand 'Palace.' What cheers me most is the prospect of congenial work in the desert. Information, first vague & romantic, then confirmed by tongues first fear-bound, had reached me at Ming-oi; it spoke of an 'old town' accidentally discovered two years ago by a hunter. It lay two marches south of Korla midst dunes & dead jungle."

A few days after the New Year, Stein received his Christmas package:

Publius's letters from 6 September to 1 November and "Madam's most thoughtful and welcome present—the beautifully packed boxes of chocolates. They have come with wonderful precision, just when last year's supply was beginning to give out. How often I have thanked you at late hours of the night when dinner was still far off & a headache approaching as a reminder of bodily needs, for your incomparable forethought. I shall need such nourishing & tasteful 'iron reserves' for the next few months for quick marches are needed to make up for time & there are good reasons to hurry back to Khotan."[4]

Time had been lost chasing a mirage. The "old sites" seen at different times and verified by numerous hunters "proved a mere phantom. It was a queer experience; a psychological puzzle not easily explained by any ordinary reasoning. . . . Yet the result of the thorough search made in the desert which these 'guides' had indicated with some topographical sureness, proved fruitless. . . . Of course we had plenty of marching through the true desert such as I liked to see again, & geographically these days were quite fruitful. But that does not lessen my regret at the loss of 8 precious days. . . . My foolish guides were expecting me with my powerful arts to discover those forts & houses they had seen in their dreams & where they might dig at ease for all kinds of riches. It is an essential part of the folklore, . . . hence my experience will not be of much avail to dispel the illusions of those good people of Korla."

He reached Kucha on 17 January. "Japanese, Germans, French and Russians have been at work for the last couple of years here; I cannot expect much chance of work, but I am anxious to see the caves & other ruins for purposes of 'rapid instruction.' Comparing these ruins with those I had worked at will be useful, & if by any chance there is room for more excavations I could return from Khotan in the summer. . . . Of course, the necessity of reaching Khotan before the heat gets too great for work in the desert involves a great strain." Working with speed and certainty, Stein set in motion the organizing of men, supplies, and provisions for a trans-Taklamakan journey. While the preparations were going forward, he spent busy days "visiting all the old ruins & caves within reach. They are situated at the foot of hills 8–12 miles away & in different directions. The caves proved disappointing with their frescoes terribly mutilated by pious vandals—& collectors. . . . The ruined monasteries & temples had all suffered so badly through moisture that Japanese, Germans & Russians contented themselves with mere scrapings. But Pelliot last year made a thorough cleaning . . . and was rewarded by an important find of MSS. at least in one ruins. It was pleasing to see a piece of excavation so different from the Turfan burrowings." Stein had satisfied himself that he need have no lingering doubts about possible work left for him to do at Kucha.

He could now give full attention to his plan. As always, he expected his schedule to work. "I have wired Macartney to get Turdi's acolyte Roze Akhun with a posse of diggers, sent from Khotan down the Keriya River to meet me. I am glad to get right into the heart of the desert again before the spring is over."

The map gave Stein his plan—to take the "thieves" road (as old Turdi had called it) straight across the desert; the eight precious days lost chasing fantasies justified it; and a message from Badruddin that the scouts he had sent out had located a number of new, untouched sites furnished a valid reason. Yet under the arguments there is an inescapable feeling that Stein was again pitting himself against Sven Hedin, to equal, perhaps to better, the latter's record. Had he not put time and strength to reach the 20,000-foot mark on the Muztagh-Ata massif though snow conditions were visibly unfavorable? Had he not raced for Lou-lan, repeating his laments at having been delayed? Competitive he might be, but never rash. While he went one way from Korla to Kucha (covering the 140 miles in five days of marching), he sent Lal Singh by another, via Shahyar, south of Kucha, to gather information and secure guides. "Guides are badly needed" for the perilous journey, he had written Publius. He was reversing Hedin's journey of 1896: Hedin had followed the Keriya River to where it finally spluttered out and then kept due north to the broad ribbon of the Tarim. Stein knew that twelve years was a long time to gamble on the steadfastness of a wandering delta; it could now be four or five miles in either direction, a serious, difficult gap. It was the needle he would have to find in the huge haystack of the Taklamakan.

The journey of "close to two hundred miles over absolutely lifeless desert"[5] began on 31 January. "Safety depended on right steering & there were absolutely no landmarks by which to find in a bewildering delta the single riverbed which might ultimately bring us to water. Up to the 6th the navigation was relatively simple. We had to keep straight south as well as we could amidst an endless succession of dunes rising often to ridges of 150 feet or more. The high sand and the solitude greatly frightened our Shahyar men, tame people from near a tame high road. Well-digging did not succeed at two camps where, according to Hedin's itinerary, we ought to have got water." Then, since he was writing the Allens, he described how he "celebrated Madam's birthday in due spirit at one of the waterless camps & was cheered the next morning by getting a well sunk enroute though the ground looked most unpromising.

"The real cares began when we struck the northern edge of the forbidding delta-like region where the Keriya River has buried itself in sands since heaven knows what prehistorical ages. Dry river-beds, smothered often for miles & miles under dunes & high sand ranges between them, spread out here like a huge fan in absolutely bewildering sameness. The actual river died away some forty miles further south & the point was to hit the continuation of the bed carrying water for the time being. For Hedin coming from the south it was relatively easy to follow this bed into the desert. In our case it was a matter of chance: for over such ground it is very difficult to make sure of one's longitude & in spite of all the careful mapping both he & ourselves were 'bound' to be 'out' considerably. We were lucky enough to sink a well at the very edge of the delta, but when we tried to steer by this old bed it vanished the next day in a wilderness of dunes & dead forest. . . .

"Ancient beds emerged again & again as we marched south, our safest course, to disappear with equal suddenness. Climbing high sand ridges was of no use. As far as the eye could reach there was on all sides the same dismal view of sand, dead Toghrak trees & tamarisk scrub. During five days' march only once well-digging succeeded. The water found at 15 ft. depth was scanty enough, but it saved the ponies. The ice supply I had brought all the way from the Tarim would have sufficed for us humans for 5–6 days more. Rations of ice had to be carefully husbanded, & 3 large teacups of water per diem are a modest allowance for a hard marching man—as I could verify from experience. It felt warm enough in the sun at 80°F or so, but the nights are still bitterly cold.

"By the morning of the 12th we were still without any indication of the river though the tracks of wild camels had become exceedingly numerous, & I knew from astronomical observations that we had reached the latitude to which the river had descended 12 years ago [though it] might have diverged greatly since Hedin's time. The Shahyar men had grown rather desperate & on the morning of that day attempted to leave us in the hope of returning to the last well. However, the little mutiny was stopped & the men agreed to follow us a short march further to the south. Of course, they would never have got back alive to the Tarim. By pushing reconnaissances right through to the east and west I was sure of hitting the river. Luckily, there proved to be no need for these.

"On climbing to the top of a sand ridge, some 300 ft. high, enroute I sighted to my relief a narrow white streak which looked like ice. You can imagine the excitement of the men as we pushed towards it. My surprise, too, was great when it proved not a forlorn salty lake as I had feared, but the real river under an imposing broad ice-sheet. It was flowing through as barren a desert as any we had passed through; it became clear at once that the river had widely departed from the forest-lined bed in which Hedin saw it. To see the men, ponies and camels drink was a treat I shall not easily forget. The last had been without water for 13 days & the ponies for nearly four. Yet all the animals held out splendidly, especially the camels which had only dry leaves to feed upon for a few days & nothing before. Of course we never rode the ponies.

"The 13th was a delightful day of rest. The two long marches up the ice-bound river which followed, did not bring us to the hoped-for forest but were enjoyable enough otherwise. We saw a few birds & a deer, the first living creatures in over a fortnight. This morning after some miles along lake-like bends of the river, I heard sheep bleat in the distance & after an exciting chase the men caught—a frightened shepherd. Thus we met the first human being again & emerged upon known ground. We had reached close to Tonguz-baste, my northernmost point in 1901 as I thought we should in all likelihood." His letter to Publius, written from that camp, referred him to the map he had drawn "in my P.N. of that journey. It will show you how the various beds of the river spread out here like the fingers of a hand. Well, three years ago the river broke through here into a new bed, or rather an ancient one as shown by the ruins of Kara-dong which lie not far from it. Hedin's bed has completely dried up & if we had hit lower down we should have found plenty of forest but no water.

"Thus has ended this 'short cut' through the Taklamakan which I shall ever remember with pleasure in spite of its hardships. Lal Shingh has proved the very man for such work, ever cheerful & untiring. The thought how I should have had to drag his predecessor through, makes me shudder. I am quite fit & ready for work, now of the archeological sort."

4 March 1908, to Publius: "A week ago marching up the Keriya River, I met the long-expected batch of Khotan Taklamanchis & received from their leader Roze Akhun, old Turdi's acolyte, the first mail bag via Khotan." It was the happy outcome of the meeting arranged months before. The Keriya shepherds met me kindly enough, with a curious mixture of feelings. My first visit was, of course, well remembered & had removed the original 'junglee' shyness. The reception accorded was that which a

semi-barbarous tribe might give to a chief one had stood in awe of & even liked in a way, yet not altogther longed to see returning from another distant world. They knew nothing of the Khotanliks who had come to seek for an old site since my visit, & I suspected that for once they spoke the truth. After a day's rest I moved to Kara-dong, now by the river's latest vagary brought within four miles of water. In 1901 sand storms had prevented my searching for smaller scattered ruins & this was on my conscience. The weather was clear & sunny & I discovered some half a dozen homesteads in the neighboring dunes. . . . The finds were modest."

The party moved up the Keriya River by the route followed seven years before. "I met much-needed supplies from Keriya & the party sent from Khotan. They told me that none of Badruddin's treasure seekers knew anything of a site below Kara-dong. I decided to march straight to the first of the new sites prospected by Badruddin's men in the vicinity of the Domoko-Chira oases. . . . The tamarisk jungle which here covers the desert in wide belts was so deceptive that it took my treasure-seeking scouts three days before they had located the ruins at which I am now encamped [Yailik-bashi] & which they had previously discovered from the south. . . .

"It is extraordinary that any ruins should have survived, since the distance to the remains of 'Old Domoko,' abandoned about 1840, is less than five miles. The shifts of the cultivated area in this particular tract, backward & forward in the course of long periods are most curious & of considerable interest geographically. . . . The position of the ruins is so instructive that I do not grudge them a thorough search in spite of all the damage done by those who worked them long ago for timber, saltpetre etc." . . . Eight strenuous days were spent at this old site that had once formed part of Pi-mo, mentioned by Marco Polo. "We cleared altogether four shrines which, alas, had suffered much already in old times, with further finds of wooden records, etc. belonging to T'ang times. But the chief reward came on the last day when I excavated a little temple which luckily had got buried under a big tamarisk-covered sand-cone. It contained some fine frescoes of a size which could be safely removed. . . . My treasure-seeking scouts are kept on the move now. There is not much probability of any important ruins having escaped my former searches, but welcome supplementary finds may still be gathered from minor sites, & on ground such as this of Khotan, everything which helps to fill up a picture has value."

By the end of March, Stein had made a quick trip to Keriya, moved to a site called Uzun-tati, and searched it thoroughly. He wrote Publius on 26

March: "Its identity with Marco's Pien is now fully confirmed. . . . I am now going to some ruins northeast of Hangúya. Soon I hope to get a few days' peace for writing tasks at Khotan where old Akhun Beg is awaiting me with his garden quite ready for the short flush of spring. It will be nice to see blossoms & young leaves before starting back to desert sites."

He had the feeling of coming home when he entered Khotan, and its people's high regard for him was manifested in their reception: 31 March 1907: "A big posse that came out to welcome me took me out of my way to inspect the new canal which is to reconquer quite a goodly slice of the desert. The main objective was to get help for a correct alignment of the canal. So Lal Singh is to go out to do the levelling. I stipulated as my royalty an annual present of walnuts to be duly dispatched to Kashmir when the trees of the new colony will have begun to bear fruit."

As before, Stein camped in old Akhun Beg's quiet garden. "Nothing pleased me more than to find [him] returned from his pilgrimage in full vigour & more cheerful than ever. It was no small undertaking for a portly gentleman over 65, but he seemed to have enjoyed everything keenly, even the many days' railway journey from Andijan to Sevastopol. You would have appreciated the old Beg's graphic account of how he fared while tossed about on the Black Sea. His sufferings there seem to have relieved him of all sorts of bodily ailments, gout included, & he no longer asks for my pills. His accounts of the glories of Stambul make me long more than ever for a tour of the Bosphorus." The remembering makes Stein add: "Oh, could we but globe-trot together all about the classical shores."

His return to Khotan was the first step in his long-planned, daring homeward journey. His preparations had begun long before while still at Tun-huang, when he wrote on 30 March to the Baron asking him to shop for farewell presents. The gifts he ordered and for whom give an index of Stein's relationships to his Central Asian friends. "What trouble —& friendly institution—it must have cost you to hit off with such precision what I was looking for," he wrote Andrews enthusiastically. "The Liberty brocades [four, three-yard pieces of brocade] are a huge success. Chiang thought them fit for a viceroy & the first recipient, the kindly old Amban of Keriya, seemed quite overcome by such largesse & did not know how to limit his counterpresents. I am keeping a piece for my old friend, Pan-Darin, at Aksu & am certain he will use it for a state robe."

Stein had ordered three watches. "1. One good silver watch in case, not to exceed £2-10-0 in price, bearing inside the lid, the following inscription: 'Presented by Dr. M. A. Stein to Chiang-ssŭ-yeh as a token of sincere regard and in grateful remembrance of his devoted scholarly services during exploration in Chinese Turkestan, 1906–08.' 2. Another silver watch, about £1-10-0, for Tila Bai of Yarkand with an inscription; and 3. A similar watch, same price, for Badruddin Khan, with an inscription thanking him for his friendly services." In addition to these gifts, he wanted for himself a pair of strong, washable suede gloves, "suited for travel use in summer (the heat in this dry atmosphere had been blistering my hands lately when exposed too long) and a pince-nez—it is a "godsend, seeing that my last pair got hopelessly broken months ago." Of the watches Stein wrote, "They are real marvels of cheapness, considering their elegant look & finish & have made their recipients quite proud & happy. Take a thousand thanks for all care & be assured that I shall always think gratefully back to this proof of unfailing helpfulness." The coming years would show that these were no idle words.

The distinctions made in his gift requests mirror the distinctions made in this friendly relationships. For some there is the mark of true friendship, the equality that united despite differences in culture, language, and status. Thus Stein distinguished between Badruddin Khan, to whom he had entrusted his goods, and Turdi, the illiterate treasure hunter, master of the Taklamakan's topography, to whom he had entrusted his life—and for whom he had secured an honorable position in his community. Turdi was his friend; Badruddin Khan and Tila Bai of Yarkand and Ibrahim of Keriya were in another category—reliable, honest business associates. As with others elsewhere, Stein's friendships had subtle gradations, and he planned his farewell gifts with this yardstick.

Pan-Darin was a friend. Though Stein maintained the proper etiquette with some other officials, with the former amban of Khotan he had a friendship solidly based on the amban's response to Stein's scholarly activities despite Petrovsky's fierce agitation against him. An initial gratitude had developed into a mutual appreciation: it motivated Stein to crowd his final months' schedule with a long trip to pay a visit to Pan-Darin in Aksu. On this expedition the official's blessing had facilitated Stein's work—the yellow brocade was to express thanks. But, for their mutual interest in things historical, Stein took along specimens of documents dug from the limes's rubbish heaps, and to see Pan-Darin "handle & study those relics of Han times was a treat."[6] But Stein always had practical reasons for heeding the dictates of his affections; he wanted to secure permission for Lal Singh to survey the T'ien-shan between Aksu and Kashgar while Stein, taking another route would complete his circuit of the Taklamakan.

Most of all he hoped to secure Pan-Darin's interest "to obtain the chance of official employment which my honest Chiang has striven for since 25 years. . . . So a detailed report on his services & all he did to help me was drawn up for submission to the Governor-General in Urumchi. Pan-Darin thought well of Chiang's plan & took the trouble of thoroughly revising the contents and style of the letter; he also agreed to send it up to the fountainhead of official favours under his own seal & envelope which is apparently a support."[7]

"My stay at Aksu extended to five days and passed most pleasantly. There was so much to talk about during our daily visits & cheerful reminiscences of the old Khotan days. I brought him news of how the new canal built under his orders had brought verdure over a belt of former sandy waste nearly a day's march in extent,"[8] and "how cheerful it was to hear all of the cultivators who came there as helplessly poor folk getting to feel themselves regular 'Bais' [men of substance], sing his praises." Stein found it hard to say good-bye to Pan-Darin for good. "I indulged in a sort of sanguine self-deception by promising to return again when Pan will be Governor-General in Urumchi. But can an honest, scholarly Mandarin ever rise so high? Pan gave us [Chiang was present] a pleasant little *symposium* [using the term in its original sense of a mixture of eating, drinking, singing, and conversation]. Over much diverting talk I forgot how many hours it lasted."[9]

Of all the friends, Chiang-ssû-yeh was the closest. Not a master treasure hunter like Turdi, not an official like Pan-Darin, Chiang was a younger, refreshing Pandit Govind Kaul, gentle yet able to stand up to the rigors of field trips. When Stein sent the Allens a snapshot of Chiang, he wrote that "his family in Hunan which he has not seen for 17 years will think him rather thin. He is most attached to his son whom he left behind as a baby & yet is content if he gets news once a year. Does this not take one back to Marco's times? I wish I could spend a few months of quiet study with him on Mohand Marg—he would think nothing of making such a trip."[10] When writing an account of their trip to Aksu, on which Stein told Chiang of his intention of speaking on his behalf to Pan-Darin, he tried to convey to Publius Chiang's amusing gossipy anecdotes. 8 May 1908: "Chiang is never more fluent than when his talk turns on the quaint hierarchy he knows so well with all its glitter & foibles. Now that he hopes to gain admission to it even though he knows he cannot raise enough money to buy rank straight away at Peking, his stories are more entertaining than ever. His optimism is truly a thing to be envied. At first he can only expect an acting-appointment. . . . Yet men have made a lakh [100,000 rupees] or do during 6–7 months of this. Chiang is far too modest & considerate to aspire to so much loot & has voluntarily promised not to make quasi-lawful gains. Whether anything will ever come of his dreams, only heavens knows, but his enjoyment of them is none the less real."

Stein's trip to Aksu was a prolongation of his investigation of sites along the Khotan River—it was also a companion journey to his cross-desert southward march from the Tarim to the Khotan-Keriya district. Early in April he crossed "the strip of sterile land between the Khotan and Keriya Rivers. I found myself at a most unpromising-looking place where were the remains of a large temple completely buried under high sands."[11] Of the same early date as Rawak, with frescoes finer than any yet found in the Khotan region, it had experienced, like Rawak, the same disastrous deterioration due to subsoil water. One after another, the twelve-foot fresco figures collapsed when excavated, and all Stein could do was to photograph them. "Some smaller frescoes I managed to remove though the wall surface had scarcely more consistency than a moderately well-boiled plum-pudding, & are now being dried in the sun preliminary to further hardening treatment. . . . What is really important is the evidence of the date recovered in the fresco inscriptions."

Four days' march down the dry bed of the Khotan River brought the party to ruins atop a rock upthrust, Mazar-tagh, which had yielded Tibetan records on wood to treasure seekers. With new supplies, a fresh contingent of Tawakkel workers, and Kasim Akhun, who seven years before had guided him to Dandan-Uiliq, Stein reached the strange range that rose from the surrounding desert. The "big open bed" of the river was now entirely dry but for an occasional limpid pool. He described the scene to Publius, 19 April 1908: "The heat was gaining rapidly upon us. On reaching Mazar-tagh I found a small ruined fort set on the bold cliffs where this curious red sandstone ridge ends just above the river. From its precipitous height it looked like a true robber's stronghold. Though it had been burned out long ago, its refuse heaps kept us busy for 12 hours or so per diem. . . . It was another Miran with all the unspeakable dirt which these old Tibetans seem to have cultivated wherever they held posts. Buried in thick layers of straw & unsavory refuse, Tibetan tablets & papers, Chinese & old Khotanese records with even a few Pahlavi [Mid-

dle Persian] & Uighur pieces cropped up. The whole has risen to 900,
. . . torn pieces are frequent but so, too, are well-preserved records, passports, requisitions & the like. The coins and dated papers show that the Tibetans had held this place as a frontier guard station in the 8th cent. A.D. With Miran at the eastern end their domination of the Tarim Basin at that time is conclusively proved. This was a main route to the north. . . .

"Not a trace of the humblest vegetation is on the barren sandstone ridge which stretches away into the sea of sand for an unknown distance. From the river below not a particle of moisture can rise. . . . With a bright winter sun it might look quite pleasant, but now the pink rocks are already glowing with heat. In spite of heat & sandstorms work was easy; for camels & ponies there is young scrub in the narrow jungle belt of the riverbed, & for us all deliciously clear water in the little lagoon. To-morrow [20 April] my diggers return, well-rewarded, to Khotan. We march off to Aksu which I hope to reach by May 1st."

Except for discomfort from the heat, the march up the Khotan River to Aksu was, as compared with the perilous journey down to the Keriya, a carefree jaunt. Stein suffered from the "sun burning down fiercely from a relatively clear sky and with a glare sufficient to irritate one's face. The temperature did not fall below 97° in the shade and the contrast with the cool nights, due to rapid radiation, was great. Regard for the camels which had to be 12 hours or so on the road daily & then needed daylight for grazing, obliged us to start at 3 A.M. which meant rising at rather unearthly hours. Luckily a succession of duststorms brought relief and kept us for the last four days under the protection of a thick haze. The bed of the river, completely dried up at this season, made a delightfully level road—from one to over two miles broad in places. With my old guide Kasim there was no risk of losing the way amidst the many river branches even in a real dust fog. He knew where the spots with small pools might be looked for under the banks last washed by the current. The water was delightfully fresh & its plenty a perfect blessing after such hot, dusty days. It was strange to pass through the very centre of the Taklamakan & yet feel free from all the usual cares of desert marching. . . .

"Kasim shot a deer & some wild ducks. Today's march provided mild excitement. . . . Shortly after leaving the camp we came upon the perfectly fresh track of a tiger, running for miles along the route. It must have been a huge beast judging from the size of its footprint & by no means shy to keep so steadily to a well-beaten road. That it had thought of paying a visit to where our animals were last night seems likely enough, for all at once the men remembered how Dash, otherwise a sound sleeper, had set up a violent bark, & how restless the ponies were. Tigers on the Tarim are by no means rare & are said to relish donkeys and ponies."[12] As planned, Stein reached Aksu on 1 May.

His return from Aksu to Khotan went through Uch-Turfan, a route planned so that he could visit "the last bit of ground in Turkestan which Hsüan-tsang had trod & I had not previously seen. The vaguely rumored 'old town' proved to be a fantastically serrated high mountain group, fortunately in the very range I had to pass on the unsurveyed route which I wished to follow to Kelpin. . . . The chain of bold rock pinnacles rising like castle & towers to heights of 12–14,000 ft. was a wonderful sight. In the arid but delightfully picturesque valley the only water to be found is in natural rock cisterns. Kelpin was a "charming oasis hidden away between the outermost ranges of the T'ien-shan."[13]

The route Stein chose to follow through narrow gorges and ragged canyons cut into the rock wall by rivers long since dried up had once been the most direct way to Yarkand. "The first three days led through absolute desert, almost clear of drift sand. . . . After skirting for thirty miles the rugged outer hills where we filled our Mussucks [goatskin water containers] from a natural rock cistern, . . . my guides struck south & duly brought me to the promised old site, Maral-bashi. [The road had once split there, one fork leading to Kashgar, the other to Yarkand.] All structures of what was once a large & thickly populated settlement had decayed completely through erosion. I obtained plenty of coins to prove the date of its abandonment, traced the walls of the central fort, etc., and subsequently cleared up the questions about its water supply—The Kashgar river had extended much further east than it does now. The conclusive evidence I discovered of the ancient highroad from Aksu to Kashgar having crossed this now absolutely waterless belt, made its study quite exciting. . . . A series of small ranges cropping up out of the desert plain of which existing maps know nothing are unmistakably connected in geological origin with those curious Mazar-tagh hills far away on the Khotan River. To track the connection right through the Taklamakan would be quite a pleasant task—in the winter with plenty of ice & good camels."

Yarkand, some 140 miles away, was reached in five days, and a short halt was made there. In the busy round of formal receptions—Chiang had served at the Yarkand yamen for many years and was much sought after— Stein had little time to purchase the "old brass things, embroideries, etc. for which Yarkand is famous and is still a good market." He had supper with the Raquettes, the Swedish medical missionaries who had moved there from Kashgar. "It is almost two years since I had seen the last European lady."

"Seven marches all done during the night—with a good bit of the morning added, for distances were great—have brought me here. . . . A succession of dust storms made it rather difficult. Two nights were enlivened by our losing the track in the darkness & driving sand, with men & ponies going astray. One occasion when no light could be kept burning in the lantern, there was nothing to do in the howling Buran but to lie down on the Sai [hard gravel desert] & wait for the dawn."[14]

At Khotan, reached on 9 June, Stein had planned the hard work of cataloguing the "finds," packing them, and tinning the cases. "From the Amban down everyone is trying to help & make things easier. It is not in vain that I call Khotan my own home in the 'New Dominion' & that its people know me." The big mullah, his former landlord at Narbagh, "a pleasant & spacious summer residence of advancing decay," had died, and the palatial property was divided among a number of less well-to-do relatives. "So my faithful factotum, Badruddin Khan had to treat with several people including a formidable dowager. . . . That this is the season when silkworms spread themselves everywhere needing cool & dry accommodation, added to the complications." The needs of tenants, sericulture, and Stein were all satisfied, and he found himself installed in an airy pavilion in the garden.

An enormous task confronted him: each "find" had to be listed and wrapped, each of the many panels carefully "backed," each of the fragile silk paintings tenderly rolled in cotton batting. Chiang had to catalogue the text rolls and bundles of miscellaneous manuscripts, often unfolding rolls of thin paper thirty yards long to find the colophons. "Loads over the Karakorum must be kept down to 240 lbs. per animal & many boxes with stuccos weigh too much. Luckily all transport arrangements to Ladak are concluded with a Kirghiz Beg from Shahidulla [a camping ground on the northern side of the Karakoram] whose reminiscences go back to the days of Hayward & Shaw & Forsyth.[15] I luckily gave the Beg an advance payment & finally clinched the contract negotiations at Khotan. I also hope to get all that is needed for our survey expedition to the source of the Yurungkash & thence to the back of those Kharangutagh peaks which stopped us with their glaciers. It ought to be an interesting bit of work before we rejoin the heavy baggage column at the foot of the Karakorum."[16] Stein was waiting for the Naik Ram Singh to return: "The Naik would never have been more useful than now when there is so much work for my hands."

Soon Stein was impatiently waiting the naik's return. In less than two weeks he arrived—a tragic figure. As the naik was led along the garden path, Stein responded instantly to the man's helplessness. Here, before

him, was the naik who had been so marvelously adept in all the skills he
had learned, blind. What had happened? To Publius he told the story,
from its beginning.

23 June 1908: "The end of March I had dispatched Naik Ram Singh
to Charklik with instructions to remove from the largest of the Miran
shrines a series of fine fresco panels which in Febr. 1907 we had found
impossible to attempt. We had not yet learned how to cut out walls behind the frescoed surfaces & to provide for later glued surfaces. I gave
the Naik plenty of time, the best of my Turki men, faithful Ibrahim Beg.
The Naik himself showed some keenness—for knowing something of his
ways, I had promised him a special money reward for successful work.
While away in Aksu I heard nothing of him & was now looking out for
his return when he arrived in a state which caused me the greatest grief.

"He had started in fair health, it seemed, & had got as far as Charchan
when he felt severe pains in his head. They increased as he was marching along to Charklik & during some days halt there he lost sight in one
of his eyes. A few days later complete blindness overtook the poor fellow.
With characteristic Sikh tenacity he remained there in the hope of yet
being able to set to work & did not commence the long return journey
until two more weeks had passed. I searched vainly in my medical handbook, . . . and could do nothing but arrange to send the poor sufferer
(with what comforts we could provide) to [Dr. Raquette] the Swedish
medical missionary at Yarkand. R[am]S[ingh] seems to bear his afflic-
tion with remarkable courage & calmness, perhaps a compensation for a
certain heaviness of mind & disposition. But whether this did not make
him lose precious weeks on his return for the proper chance of treatment,
only the gods know. He himself seems confident of an early recovery.
But, alas, I know only too well how delusive such hope may be & feel
the full weight of this care. As there is no local trouble apparent in the
eyes, I suspect some brain affection is the cause." Was Stein remembering
cases he had heard Uncle Ignaz tell about?

On July 15 he continued the melancholy tale, needing to share his dismay at the report he received from Yarkand. "Mr. Raquette has recognized the poor Naik's disease as Glaucoma & if this diagnosis is correct
there can be no hope that the sufferer will ever regain his eyesight. . . .
The only hope would have been a timely operation on the iris before complete blindness had set in. Never with a word had the Naik referred to
symptoms such as accompany the disease, else I should have been able to
look in the right place of my medical handbook & realized the risk he
was running. I should have sent him back to Kashgar or Yarkand with-

out a day's loss for the proper treatment. That he should have been so far away from me at the time the disease developed so rapidly, was a particular piece of bad luck. It is useless to speculate whether or not there would have been time if he had instead been with me on the journey to Aksu. Mr. Raquette assures me emphatically that I have nothing to reproach myself with, that the disease might have come on just as well had the Naik remained in India & that nobody could expect me to recognize initial symptoms.

"Glaucoma, I am told by Mr. Raquette, attacks men under 50 or so only in extremely rare cases. He strongly recommended that the Naik should return as early as possible to India where he could be treated at least against the recurring pains in the head. He had arrangements in Yarkand for him to travel with a party of Hindu traders just now leaving for Leh." The thought of a blind man being led over those passes must have haunted Stein and he was dismayed at the realization that he would not be there to make certain that the naik was properly treated and watched over. "After much anxious consideration of all aspects, I have decided to let him go ahead, at least as far as Leh where there is a hospital. July & August is by far the best season for the Karakorum & as I myself could not start for Leh until September when early falls of snow at that elevation are by no means uncommon, there would have been a risk of exposing the sufferer needlessly if he were to wait to accompany my caravan.

"I need not tell you"—the words are as painful to read as they must have been for him to write—"that I have been doing my utmost to assure all possible care & attention for the poor fellow on his way. It has been no easy matter to arrange over this distance & my hands have been full with urgent requisitions for trustworthy men to make up his personal party.

"The thought of his future weighs heavily upon me. Of course he will be invalided, but the amount of pension is bound to be very modest. Luckily, he had some land and has spent practically nothing out of his pay with me (nearly 5 times as much as what he drew in his regiment) so that his accumulated savings make quite a handsome sum for a man in his position. I shall try every means to obtain a special gratuity for him from Govt., & of course am willing to give personally whatever help I reasonably can. Yet what does this all mean in view of such a tragic fate?"

Compassion for Ram Singh and sorrow at the news of Dr. Duka's death cast a shadow over Stein's final days at Khotan. "I lose in him [Dr. Duka] practically my oldest friend in England & a truly paternal one.

I fear I have told you little of all the kindness & sympathy with which he ever followed my doings. . . . I shall feel doubly lonely in London now that I can no longer turn to his pleasant house near Earl's Court as a peaceful haven after the day's rush."

However Stein grieved, he had to keep to his schedule, a schedule set by two mighty forces: the inexorable fact that he had to cross the Karakorum before early winter set in and the expiration of his leave as set by the government. By mid-July he could tell Publius that "the packing is all done, the whole of the cases with antiques making up 47 rather heavy mule loads. All the more valuable cases are tinned & I hope safe against water. Thirty of the cases are filled with MSS & other written records on wood, etc. Chiang has got through only ⅓ of the Ch'ien-fo-tung texts. It would be risky to take the convoy up the gorges to Shahidulla before 5–6 weeks hence. But there I hope to rejoin the convoy for the crossing of the Karakoram."

Before leaving Khotan Stein paid a visit to Yotkan, where the flooding river was flushing out gold and stucco relievos from the kingdom's ancient capital's stratum that lay buried under present-day fields. Though he was meeting with difficulties in preparing for the Polu expedition—the reluctance of men to accompany him into the high mountains south of the Taklamakan, the same reluctance he had encountered in his trip through the Nan-shan—he left as planned on 1 August. Again there were the farewells to be said. As for Chiang, so integral a part of his expedition days, he could take comfort that Macartney had appointed him to be the Chinese secretary to the British consulate at Kashgar. Unexpectedly, the most moving farewell to Stein was to honest Turdi, the dak man, who stood ready to take his mail for the last time. "I thought of how he had managed to find me that Christmas eve in the heart of the Lop-nor desert, and how another time I had sent him off from the foot of the Nan-shan for a weary ride of months. But whatever the occasion, there was nothing to read in his face but calm unconcern and a sort of canine devotion."[17]

Only such Turdi-like devotion would save Stein when he was stricken on the high, cold desert of the Kunlun.

16

Act Four: Misadventure at High Altitudes

The accident came without warning; almost at the very end of a spectacular tour. The tour had had a long inception. On his First Expedition he had tried to reach the sources of the Yurungkash River (White Jade River) and had been stopped by the Karanghutagh Mountains (Mountain of Blinding Darkness) through which the river cuts its way westward, racing in narrow, deep, impassable gorges. Since the frontal attack had failed, Stein planned a flanking movement: to approach "from the east where that wholly unexplored mountain region adjoins the extreme northwest of the high Tibetan table-land. Thence I proposed to make my way to the uppermost Karakash Valley along the unsurveyed southern slopes of the portion of the main Kunlun Range which feeds the Yurungkash with its chief glacier sources. Climate and utterly barren ground were sure to offer great obstacles in that inhospitable region."[1]

Stein's plan included every element that had great value for him. He would be in high mountain country; in high mountain desert country; in a vast, empty, unsurveyed region, the Aksai Chin. Historically, it was the direct route from Kashmir to Khotan which Moorcroft in the 1820s knew as a secret, Chinese-forbidden route and of which W. H. Johnson (1864–65) had drawn a map, parts of which were "incredibly inaccurate."[2] Stein knew that when Johnson visited Khotan its ruler was an old man, Haji Habibulla Khan, who, as Lamb tells his story, was "much alarmed at the chaos into which Eastern Turkestan had fallen of late, and terrified of conquest by Yacub Bey (and rightly so, in view of his nasty death at Yacub Bey's hands in the following year). Haji Habibulla, it seems, in 1864 and 1865, was also taking an interest in Aksai Chin as a means of access to the south for his threatened little kingdom. He had, it would appear, first explored the possibilities of a more easterly route, . . . but had been rebuffed by the hostility of Tibetan nomads. In the Aksai Chin region he constructed a number of stone shelters (*langar*), and one of these was Haji Langar on the Karakash."[3] In that unmapped no-man's land it was a point Stein would have to find.

The approach to the Aksai Chin took Stein eastward to Keriya and then into the mountains to Polu, the last village at the north foot of the Kunlun. Traveling light and fast, he reached Polu where he secured supplies: stores of flour, fodder, and a dozen sheep to furnish meat during their lengthy wanderings in the mountains. Obeying orders from the Keriya amban, Stein also was furnished with extra transport to carry these necessities to the nearest Tibetan plateau and twenty hillmen to guide the laden donkeys. ("The Yaks I had hoped to obtain for this auxiliary transport column proved quite unaccustomed to loads, and after careening wildly through the one long village lane had to be exchanged for more donkeys.")[4] His first letter to Publius was written on 16 August 1908, Camp Saghiz-kal, Aksai-chin plateau:

"Last evening we reached this high plateau, 15–16,000 ft. above the sea, after four somewhat trying marches from Polu. I should not have thought much of the continuous scrambles in narrow gorges half-filled by glacier-fed torrents; but to get all our baggage & supplies through safely was rather a business. There were continuous crossings & recrossings of the tossing stream which our little donkeys had trouble to negotiate at all times &, of course, could not ford at all when the day's flood had come down. The track whether winding amidst slippery boulders at the bottom of the gorge or along rocky slopes above was impressively bad. Somehow one ought to come to such 'routes' straight from tame civilized mountains to appreciate their difficulties fully.

"Well, we made our way up here safely in four days & could contemplate in a peaceful state of mind the particular precipices where Capt. Deasy [1898] had lost ponies, Yakdans [mule-trunks] full of silver (as Polu tradition has it), & alas, one of his men, too. But it cost a good deal of care & strain I must admit.

"Yesterday afternoon we emerged from the main rampart of the Kunlun range over a pass some 16,000 ft high on a wonderfully wide & barren upland, draining nowhere. With Lal Singh I climbed a steep ridge above the pass & enjoyed a glorious view both of the range behind & the great wall of snowy mountains which flanks the Yurungkash river sources. It was blowing hard from the north, an icy continuation of a Buran which must have been scouring the desert plains, but the dust clouds were kept outside the great mountain wall & the view kept delightfully clear. It was

a grand panorama I photographed; I wish the operation may have succeeded in spite of half-benumbed fingers. A big unsurveyed mountain
area lay before us westwards & to this our efforts will be devoted for the
next 10 days or so. Lal Singh is most keen, undismayed by any fatigue &
ever a cheerful companion. I wish I had had *him* from the start. Then
followed a ten miles' descent over easy slopes of absolutely barren gravel
to the little lake by which we are now encamped. The icy gale pursued
us all the way & I was glad when we got under shelter.

"There is some grass of sorts fringing the lake shore, & I enjoy the
sight of the poor hard-tried donkeys peacefully browsing or else lazily
stretching themselves in the sun. That hardy low scrub, Burtze, of the
Tibetan plateaus also abounds so the men have plentiful fuel. This will
make a good base for topographical work & for resting the animals before we start 3–4 days hence for the unexplored valley of the uppermost Khotan river. How we shall make our way to the south towards
the 'Soda plains' & the Karakash depends on the result of those surveys.
If only we had a dozen sturdy Hunza men to help on climbs & trained
yaks to carry baggage! As it is, glacier passes are quite out of the question for our transport.

"Well, this feeling of being a stranger almost everywhere outside my
own camp is not to interfere with the satisfaction I hope for from my
return 'home' & my leave. No information whatever has so far reached
me about my official proposals. I have done what I could about re-
mainders."

3 September 1908, Camp Ulugh-köl, Aksai-chin: "Terribly scanty as
time is now what with daily climbs & long marches, I feel I must write &
tell you, however briefly, of how the first part of our explorations has
been accomplished & the Purdah [veil] lifted from the true headwaters
of the Khotan river. When I sent you my last news . . . I had my doubts
of how far we should be able to penetrate into that region of deep glacier-
girt valleys for which Karanghutagh experiences had prepared me. Well,
fortune for once played me a good turn & brought a guide into our hands
such as I had no hope of among those wily Taghliks [hillmen]. It was a
Keriya hunter who had been after wild yaks & who when luckily caught
by an energetic Darogha confessed that he knew tracks which would allow
us to make an almost complete circuit of the uppermost Yurungkash region. Of course, he soon caught from our unwilling Polu people, the in-
fection of pretended complete ignorance; but it was too late & as the poor
shy fellow was evidently hard up & besides not clever at lying, he served
us all right in the end.

"Within two days he brought us from that desolate high plateau north of the great outer Kunlun range into a wild side valley of the Khotan R. where gold-pits long suspected but never seen by Europeans had been worked since ages. It was quite a romantic event our irruption into this terribly tortuous & gloomy gorge of Zailik where now a few dozens of gold-seekers try to exploit what is left of auriferous layers probably worked for many centuries. The poor wretches are practically the bond-slaves of the Amban's contractor & their earnings so precarious that they soon got quite 'tame' & willing to leave their dark holes for work as guides & coolies with us.

"It was a wonderful place this long rock-bound valley where all the conglomerate cliffs overlying the live slate rock are honeycombed with galleries often in the most inaccessible places. There is no saying for how long wretches had toiled here under all hardships of a semi-arctic & prac-tical slavery. Graves spread over every little bit of level ground which the gorge in its 12–13 miles course down to the Yurungkash affords, & many more of the victims have their rest in old exploited galleries where the mouth has been perfunctorily walled-up. During the 'old Khitai days' [Chinese] & those of the soi-disant liberator Yacub Beg, the mines were still worked by forced labour, at what expense of human life one does not care to calculate.

"For us the gloomy gorge with its break-neck paths served as a splen-did approach to commanding survey stations. But what a climb we had to reach those precipitous ridges from the valley bottom, 14,000 ft. or so above sea level! The views obtained after much toil were glorious beyond hope. They showed us practically the whole of the unknown icy range that lies east of the Muztagh of 1900, & a maze of deep-cut valleys & glacier crowned ridges; it is such as human eyes can rarely have seen before. Here it was cheering to think that human gaze might, perhaps, never have rested so long on them. For who among hunters or miners would climb such high points, exposed always to icy winds & beyond the scantiest vegetation which this terribly arid region affords? We spent trying, long hours at our 'hill-stations,' Lal Singh busy with his plane table & theodo-lite & I with photo-panoramas, while our poor porters were getting half-benumbed even while the sun shone. But our reward was ample, & the miners, too, who toiled up with the instruments, soon found such exer-tions well paid for—of course a pleasant change to their burrowing in darkness.

"They got to like it; in fact, so well that we had no difficulty in secur-ing the dozen coolies without which we could not have attempted to proceed up the glacier-sources of the river by the track our queer guide

Pasa had confessed to. We cut down our baggage to the minimum, leaving everything that our little donkeys could not carry. The ponies were sent back to the plateau where we have just rejoined them—& our modest camp comforts. For seven glorious days, rarely spoilt by real snow-storms, though the tail-end of a belated Monsoon as I take it, blew icy blasts daily with heavy cloud masses, we crossed high ridge after ridge, over passes all 17–18,000 ft high. It was hard toil to climb daily four–five thousand feet, even though the glaciers which once filled the valleys below those passes, had long dwindled away to small snowbeds. For survey & photo work, of course, adjoining heights had to be ascended. And here again fortune was kind; for Pasa's track took us quite close to some ideal 'stations,' peaks well-isolated & commanding magnificent panoramas.

"We had got quite close to the huge array of glaciers which cover the north face of the big watershed range towards the Tibetan upland, when at last the main Yurungkash valley became accessible for progress. Lower down it is quite as impracticable a gorge as I found it about Karanghutagh far away to the west. The grey glacier water of the river with the increasing cold had just fallen low enough to be forded in places; but for the last day or two some of the glacier-fed sidestreams tossing down amidst slippery boulders gave us quite enough trouble. Luckily none of the donkeys swept off their feet came to grief, & the drenching of much of our loads did not seriously matter. It was a wonderful gorge on which we finally emerged on the great basin, 15–16,000 ft. above the sea, where the main feeders of the Khotan river meet coming down from a perfect amphitheatre of ice-cold peaks, mostly from 21–23,000 ft. The passage of three miles of precipitous cliffs & shaly slopes took us nearly a day; but then none but wild yaks & a few hunters like queer, half-tame Pasa had ever traversed it. The basin beyond was grand beyond description with all the ice-streams around & geologically most interesting. There was evidence of the glaciers still in relatively recent times having covered many square miles of what is now bare sodden detritus plain. I thought of my sites in the desert & how the recession of all these ice-streams must have affected their fate.

"After another glorious day spent at a 'station' over 18,000 ft. high which showed us all our newly-won friends, peaks & glaciers, without names, I said today goodbye to that fascinating lonely mountain world where the Khotan river is born. Almost without crossing any watershed we passed into the region of flat Tibetan plateaus sending their drainage into isolated, big lakes. It all seems flat & tame, in spite of great elevations, compared with the rugged peaks & narrow deep gorges of the

Yurungkash headwaters. In spite of all hardships I longed to remain among these for a little longer. But time is getting short & the more trying part of our journey still lies before us.

"After following for a few days the Lanak-la route to Ladak (along which Deasy explored ten years ago) we shall have to strike across practically unknown plateaus & valleys to the sources of the Karakash, the second big Khotan river. The ground is sure to be as barren as any in Tibet can be. So I have taken all care to provide adequate fodder for our animals. Our ponies have had a few days' rest & our donkey train which so far has been a great success has been strengthened by fresh 'remounts' from the depot I had left here. So tomorrow we are to set out for our Tibetan tour. I can only hope it will prove as useful as the first part of the programme now behind us & be accomplished without too heavy losses."

Benumbed fingers, icy winds, glaciers, high elevations: perhaps Stein, too accustomed to discomfort, misjudged the power of the cold. In the six weeks between his two letters the accident happened and the rescue took place.

16 October 1908, Leh agency: "You will feel surprised by the great interval between the date above the one when I wrote the preceding pages. The account of my concluding journey will explain this. So let me continue in chronological order, pedantic as it may seem.

"On Sept. 4 we started for the valley of the Keriya R. sources, across a range that ought not to have been very difficult, but in the end proved trying in consequence of the wretched snowy weather which set in. Pasa, our queer guide who had first volunteered to come on to the Karakash & hunt for us, showed due appreciation of the prospects by bolting during the first night. Poor silly fellow, he never waited for the reward he had earned during the preceding weeks! The snow & slush made our descent to Baba Hatim on the Keriya R, quite a difficult task. Moisture in these mountains is so scanty that a few days' snow & rain will make an otherwise easy detritus gorge become a regular bed of sliding mud. So you will not wonder at all the trouble we had in dragging our little donkeys down in safety.

"From Baba Hatim, a bleak windswept reach of the uppermost Keriya (which you will find in any map of Tibet) we crossed in three hard marches the great upland basin which receives the glaciers feeding the Keriya R. The route had been followed by a couple of travellers, but no existing map prepared one for the gorgeous array of glaciers which we skirted. They all proved to clothe the east slope of the great range we had discovered circling the topmost Yurungkash basin. So mapping work pro-

ceeded most effectively & cleared up many details. The ground was everywhere so sodden that the strain on the animals at an elevation of 17,000–
17,500 ft. was greatly increased. Of vegetation there was none & what
with the cold & the recurring snowstorms, no amount of care could prevent the first losses of animals.

"By the evening of the 8th Sept. we first reached tolerably firm ground
again, the region of that hardy plant *Burtze* which serves as fuel all over
Tibet. Then we left the road to Ladak & struck eastwards into unsurveyed
ground. My object was to follow the big range which feeds the Yurungkash sources, all along the south face & thus to complete what has now
proved to be the real main chain of the Kunlun. Existing maps showed
here a fine blank with the inviting name of Aksai-chin, 'the white desert.'
Well it proved something very different from a desert, a grand succession of lakes & high basins with the big snowy range always in view to
the right; but the intervening ranges so easy that our poor hard-tried animals could cover big daily marches. First, for two days we followed the
north shore of a beautiful big lake which Welby had first seen in one
corner. In the valleys coming from the big Kunlun there was scanty grass
& fuel. But apart from wild donkeys & yaks the solitude was complete.
Yet the scenery of the rock-bound lake with its many fiords was delightful & quite fit to attract shoals of tourists!

"After a day lost owing to the faintheartedness of our pony & donkey-
men who managed to remain behind while we discovered a high but perfectly practicable pass to the west, we entered a wonderfully broad &
easy valley or rather a succession of plateaus opening a passage towards
the Karakash. It was such a comfort to have 50–60 miles of open ground
before us, in spite of the increasing barrenness. After two days all surface
water disappeared in spite of the big valleys running down from the
snows & we had to take to digging well in the enormous dry flood beds
just as if we were in the Taklamakan. By the fourth day we struck extensive dry lagoons covered with salt & approx. in the position where over
40 years ago pioneers of the Trigonometrical Survey had roughly located
a big lake. The old lake shore was still distinguishable but of water we
saw only distant streaks. It was a terribly dead & depressing landscape.
I now decided to steer to the N.W. in the hope of striking the route by
which Johnson, a plucky but very unreliable subordinate of the Survey
Dept. was believed to have made his way to the upper Karakash & thence
to Karanghutagh. For two long days (Sept. 16–17) we skirted at 16–
17,000 ft. elevation desolate basins, once filled with lakes & now showing
only dismal soda beds between vast alluvial fans of detritus.

"During the night we spent between these salty wastes, I had the great grief to lose my dear Badakhshi pony which had carried me all the way to the Nan-shan & back & had so bravely faced the desert. What he suc-cumbed to I failed to make out. He was equal to the hardest of fares & would cheerfully chew even ancient dead wood. It was some consolation that he suffered but for a short time & had for his last night every com-fort we could provide in that wilderness.

"Fodder for our animals (now reduced by nearly $\frac{1}{4}$) was getting very scanty when at last on the evening of Sept. 17 we struck old road marks, cairns & skeletons of ponies in a barren side valley which lay in the direc-tion of the hoped-for passage to the Karakash. Soon I had definite proof on a decayed shelter-hut that we were on the route which Haji Habibullah, the rebel-chief of Khotan, had tried to open during his short-lived reign from Ladak direct to Karanghutagh. It could scarcely have been used for more than 1–2 years, yet there were still the heaps of *Burtze* root left be-hind over forty years ago by the last travellers, etc. On Sept 18 a well-marked track, with cairns as fresh as if built yesterday, took us over high but easy spurs of the Karanghutagh Kunlun down to feeders of the Kara-kash, & all anxiety about our safely effecting the passage had vanished."

So far, so good. Stein had every right to feel relaxed. He had succeeded in finding where the Karakash breaks through the flanking range from the south. Late on the afternoon of 19 September, Satip-aldi Beg, the faith-ful old Kirghiz who had been waiting for a fortnight according to the arrangements made months before, rode into Stein's camp. He brought five sturdy Kirghiz yaks, a few camels, and badly needed supplies. He also brought good news: Tila Bai was with the convoy where he was supposed to be. Equally welcome was a letter for Stein from Capt. D. G. Oliver, British joint commissioner at Ladak (who had helped the blind Naik get safely through to Kashmir) promising all help for the convoy when it should have crossed the Karakoram. That night "for the first time after what seemed like long months, I could rest without anxious cares." Just two days later the accident happened.

Stein's letter from Leh, differs from his usual emotion-tinted accounts: it is candid, precise, intended to inform but also phrased to spare the Allens any added worry. The full story, in which he finally felt free to include the pain and anxiety, was told in his *Ruins of Desert Cathay:*

"The only exploratory task still remaining . . . was to trace Haji Habi-bullah's route up to the point where it crossed the main Kunlun range above the Karanghu-tagh, and at the same time determine the position of the Yangi Dawan [Pass]," a pass east of and easier than the high and difficult Karakoram Pass. "Two of the Kirghiz told us that about sixteen days before, . . . they had, of their own accord, reconnoitred the glacier above in search of the 'Yangi Dawan' they knew of by tradition. After a steep ascent along the rocks flanking the glacier . . . they had found the surface of the ice practicable and, with its gentle snow-covered slope, apparently affording a possible approach to the watershed; but snowy weather had prevented them from ascending to this. . . . The depression they had seen in the range was really the looked-for pass."

Even such reliable information was, to Stein, hearsay evidence. "So I determined to make the ascent on the morrow if the weather permitted." In addition to four of the Kirghiz, Stein was accompanied by Lal Singh and the latter's hardy follower, Musa. "Accustomed as these Kirghiz are to hunting yaks in glacier-filled valleys, they fully appreciated the use of roping I indicated as a necessary precaution against crevasses." Snow showers fell in the evening, "but the sky was perfectly clear when I rose before 4 A.M. and though a restless night, due to an attack of colic made me feel somewhat below par, I decided not to miss that rare chance for survey work. There was, in fact, only the choice between making the ascent that day or abandoning the attempt at the watershed altogether; for there was absolutely nothing for the yaks to eat, and after a second or third day's fast they would not have been equal to giving the help on the high snow-beds which the Kirghiz insisted upon if they were to carry up our instruments. . . .

"By 8 A.M. we had reached the point previously attained by the Kirghiz at an elevation of 18,000 feet. . . . The power of the sun from a speck-lessly clear sky and through the rarified air soon made itself felt, and the yaks, unable to push on in the softening snow, had to be abandoned. Roped together now to guard against crevasses of which we were made aware again and again by the leading man sinking in almost to his arm-pits, [we continued up the] snow-covered slope that had seemed uniform and relatively gentle. . . . The snow was now so soft that the leading man at each step sank in thigh-deep, and those behind him had to struggle from one snow-hole to another. Our ascent on the Darkot in May 1906 seemed easy by comparison. We had now to contend with such trouble in breathing as the much higher elevation caused. But the Kirghiz, . . . cheerfully responded to my exhortation to let us reach the crest-line, for which I promised a liberal reward. . . . The 'Yangi Dawan' lay else-where. . . .

"It was 3 P.M. and it had taken us over seven hours to cover the ap-proximate distance of four miles from where we first got on the glacier";

we stood close to the "brink of a snowy precipice falling away hundreds of feet to the névé fields of a big valley north. The panorama before me was overwhelmingly grand. . . . We were close on to 20,000 feet. . . . We thought we could recognize the peak K_1, triangulated long ago from the Ladak side with an accepted height of 21,750 feet. . . . But the view which most impressed me by its vastness extended to the south. . . . across the valley of Haji Langar to the great dead upland basins we had skirted. . . . It was likely that those [ranges] farthest away to the south send their drainage to the Indus. The world appeared to shrink strangely from a point where my eyes could, as it were, link the Taklamakan with the Indian Ocean. . . .

"Even now, when looking back from a distance of time and sad experience, I can understand why the mind's feeling of triumph at the successful completion of our task let me forget what I owed to the body. . . . there was too much work to be done. The plane-table was set up first and it took time before . . . we could definitely fix and check our position. . . . My own photographic work . . . took much time owing to the bitter cold and deep snow [which made] it very difficult to secure the stability and correct levelling of the camera requisite for a panoramic series. It was half-past four before this trying task was concluded, in a temperature of 16 degrees Farenheit below freezing-point with the sun still shining. I scarcely had time to eat a few mouthfuls of food before the Kirghiz insisted upon starting downwards.

"My mountain boots in the course of the ascent had got wet through and through, and during the long stay on the col with a rapidly sinking temperature they must have become frozen. But I felt no pain then in my feet and attributed the trouble I had in descending . . . to the preceding fatigue and the deep holes of the track through the snow to which we kept for safety's sake. . . . It was dark by the time we struggled down to . . . where the yaks were. No halt was now possible from fear of getting altogether benighted, and knowing that on the treacherous moraine slopes below, with their piled up boulders and thin ice-coating, progress on yaks would be safer than on foot, I followed the example of the Kirghiz and mounted. Alas, I forgot that my feet had no such protection as their felt moccasins would offer while drying.

"The yaks were as sure-footed as ever but terribly slow, and this part of the descent in the dark seemed endless. I tried to keep my feet in motion but felt too weary to realize their condition. Where even the yaks could not negotiate the jumbled rocks without our dismounting, I struggled along with difficulty. I felt painfully the want of sure grip in my

feet, but attributed it wrongly to the slippery surface instead of their benumbed state. At last, when we came to easier ground above camp and
the difficulty of walking continued, I began to realize the full risk of defective circulation in my feet. I hurried down as quickly as the yak would
carry me to where the camp fire promised warmth and comfort, hobbled
into my little tent, and at once removed boots and double socks. My toes
felt icy cold to the touch, and a rapid examination showed that they had
been severely injured by frost-bite.

"I immediately set about to restore circulation. . . . A rapid reference
to my medical manuals showed this was the safest course to persist in.
On my left foot the toes under this vigorous treatment gradually recovered some warmth, though I could see that the skin and the flesh
below was badly affected in places. But on the right foot the end joints of
all toes remained quite insensible. At last I had to seek warmth and rest
in bed. . . . Thus a day of hard-achieved success closed in suffering.

"Next morning, September 23rd, I found myself suffering severe pains
in my feet and quite unable to move. The serious results of my accident
and the urgency of surgical help were only too evident. . . . My mountaineering manual, in which the subject was discussed at some length,
plainly indicated that in such cases gangrene would set in, and recommended that 'the aid of an experienced surgeon should be sought at once.'
The advice was excellent but scarcely reassuring. For how could I secure
such aid in these inhospitable mountains—and meanwhile might not gangrene spread further? So all my thought and energy had now to be concentrated on a rapid journey to Ladak. For only one day could I halt
in that bleak camp under the frowning rock-walls to gather a little
strength. . . .

"The pains in my feet had increased, and the next day when the start
was made back to our main camp, I found that riding on a yak, owing
to the low position of the feet, caused cruel suffering. The Kirghiz, . . .
absolutely refused to lend a hand in carrying me on an improvised litter.
They were not accustomed to burdens, and the great elevation, no doubt,
made it a trying business. So all I could do was to get myself strapped on
the padded saddle of a camel, as soon as the going in the gorge became
sufficiently safe for the animal under such a load. The constant jerks and
swaying were most painful, and I shall not easily forget the sufferings of
that day.

"At [the main camp] I found Ibrahim Beg with our ponies and there
I managed to have my camp chair . . . made into a sort of litter resting
between two poles which were fastened to a pony in front and another

behind. It took no small effort to improvise sufficiently long poles out of the short pieces of bamboo which jointed served as our tent poles. Every mile or so the pieces lashed together would get loose or slide from the ponies' saddles, threatening to deposit me on the ground. But at least I could keep my feet high up on the felts and rugs made into a foot-rest, and luckily the going in the broad Karakash Valley was easy. I always felt grateful . . . when at the end of the march I could be laid on the firm ground. . . . Portash, where I had previously ordered Tila Bai to meet me with the heavy baggage . . . was reached by September 27th, and there I had the satisfaction to see again my heavy caravan of antiques. . . . Not a single case had suffered.

"For two days I was kept hard at work on my camp-bed settling the accounts. . . . I also made all needful arrangements for the further transport of my precious convoy of antiques, which was to be moved on camels across the Karakoram and to be transferred beyond to yaks hired from Ladak for the difficult marches near and across the Sasser Glacier. The responsible task of seeing the whole of those fifty camel loads carried safely over the highest trade route of the world I entrusted to Rai Lal Singh, whose scrupulous care and untiring devotion I could absolutely trust. One of Satip-aldi Beg's hardy Kirghiz had already a week before set out to carry news of my coming to Panimikh, the first Ladak village, and to summon yaks. . . . Another of those indefatigable despatch-riders had since followed with a letter reporting my mishap, and asking for medical help to be obtained, if possible, from the Moravian Mission at Leh.

"On September 30th I myself set out from the Karakash Valley with the lightest possible baggage on ponies and only my few personal servants, in order to reach Leh as rapidly as [possible]. . . . Ibrahim Beg [always with the qualifying adjectives "honest," "faithful," who had cared for the blind naik from Charklik to Khotan], to whom I said farewell here, had managed to hunt up from Kirghiz felt yurts some staffs which somewhat improved the arrangement of poles required for the carriage of my improvised litter between ponies. By two forced marches I got myself carried . . . to where we struck the Karakoram trade route. Then we followed the latter with its unending line of skeletons, sad witnesses of the constant succession of victims which the inclement physical conditions claim among the transport animals, and by October 3rd crossed the Karakoram Pass, 18,687 feet above sea, and with it the frontier between China and India. . . .

"On the following day, [where] the track among the rocks of the Murghe defile became so difficult it would have been quite impossible to

get my gimcrack litter through on ponies . . . [Just there], at the very
first impasse, my sorry little caravan was met by a band of Tibetan coolies.
Without this timely help which Captain D. G. Oliver . . . had provided,
I could never have got myself carried in my litter over the ground before
us—and I do not care to think now what sitting in a saddle would then
have meant for me. On October 7th I was taken over the glacier slopes
and moraines of the Sasser Pass, the patient and good-natured Ladaki
coolies doing their best to spare me painful tumbles on the ice and snow.
. . . At last by the evening of October 8th, when descending towards
Panimikh, the highest Ladak village on the Nubra River, I had the great
relief of being met by the Rev. S. Schmitt, in charge of the hospital of
the Moravian Mission at Leh.

"Though himself still suffering from the after-effects of a serious illness, he had with the kindest self-sacrifice hurried up by forced marches
across the high Khardong Pass in order to bring me help. . . . Trained as
a medical missionary at . . . Livingston College in London, and provided
with abundant surgical experience by his exacting but beneficent labours
at Leh, he recognized at the first examination that the toes of my right
foot had commenced to mortify and were more or less doomed. . . . The
injuries received by the toes of the left foot were far less serious and
would cause no permanent loss.

"Owing to my exhausted condition . . . [and] the fatigue of the four
marches, my kindly Samaritan decided to postpone the operation necessary on my right foot until we reached Leh. . . . I reached Leh on October
12th, having travelled nearly three hundred miles since my work closed
at the foot of the Yangi Dawan. Two days later Mr. Schmitt successfully
effected the operation on my right foot."[5]

Thus, on 16 October 1908, but two days after the operation, Stein wrote
a nine-page letter to Publius, telling him the details "in all truth &
[hoping] you will join me in taking a philosophical view of the whole
case. . . . The skin of the toes on the left foot, though partly destroyed,
is now healing rapidly. On the r[ight] foot two middle toes had to be
removed altogether, as well as the topmost joints of the rest. But Dr.
Schmitt assured me that the three toes left thus for the greater part intact
will be ample to assure my full power of walking & climbing. The operation did not cause much pain, but though the wounds are healing very
well, I shall scarcely be able to start for Srinagar before the 26–27th
inst. The heavy baggage has been brought down quite safely over the
passes; it starts for Kashmir tomorrow."

News of Stein's accident had been telegraphed to Calcutta where his good friend Colonel James Dunlop Smith, private secretary to the viceroy, kept Lord Minto informed. "I had a very sympathetic message from the Viceroy suggesting my meeting H[is]E[xcellency] at Agra by the end of Nov. Of my deputation I have heard nothing officially except that a despatch has gone to the India Office. I fear telegraphic reminders will be needed." And then, as always, ruminating on his future and responding gratefully to Publius's suggestion that he might be able to secure him a "Fellowship" at Oxford, Stein reminds his loyal friend that "a Fellow's stipend would suffice for a modest scholarly life such as I look forward to. But where would the money for travels & explorations come from? You know how little is to be got even by English Orientalists for work in the non-Biblical East. I think it well to keep my connection with the I[ndian] G[overnment] as long as they give me a chance & treat me fairly." His long-range plans were already taking shape: did the thought of gangrene poisoning effect his thoughts of the future? "While I serve [the Government] I can save some money & this not only increases future independence but gives the means of doing something permanent for the studies which I am attached to, after my death."

"I know you are interested for news of my bodily progress," Stein wrote on 16 November, a month after his long letter from Leh; he was back at Srinagar, back in his Happy Valley, Kashmir. "The wounds on the operated toes went on healing quite well while en route & the fresh air and 'movement' in my improved Dandy [palanquin] did me a great deal of good. The road was pretty bad for a number of marches, but all along the route transport, etc. was kept ready for me under the British Joint Commissioner's orders & efficient local myrmidons saved all trouble. So I could cover those 250 odd miles in exactly twelve days with ease.

"The valleys of Ladak were sombre & cold; still I saw plenty to give interest to the journey. Sunshine & brightness came only when I had descended to Baltal. I spent a night at Nunnar, just below Mohand Marg & received there some of my faithful retainers from Manygam. It is true, I had thought of a different return to my own alp, but even its sight from afar was cheering. Next day I made my way here, gliding in glorious sunshine over the Dal waters. . . . Dr. Neve, my old mountaineering friend, in charge of the big Church Mission Hospital, soon came to see me. He has great experience & reputation as a Surgeon. So I awaited his examination with eagerness. Laus Deo: its result was reassuring.

"He was evidently relieved to find the ball of the right foot (the one which was operated upon) quite unhurt & assured me that as the ball is the part which practically alone takes all the weight, I shall be able to walk all right even on hilly ground. What is left of three toes, especially of the big toe, will be quite enough to secure this. Compared with this great fact it seems of small import that Neve will not allow me to start attempts at walking until the tiny wound left on one toe only has completely healed also. This, he thinks, may take a fortnight, & as I do not care to go down to the plains until I can use my feet in some fashion, I must submit to his advice & stay here for that time.

"To complete this bulletin for the pathological details of which I trust your sympathy will find an excuse, I must mention that the wounds of the left foot where no bone was affected & where nothing was cut, have already healed. So massage is being alone needed now to give it strength & practice to resume its duties. Of course, it will take some time after the healing of the right foot until I shall be able to walk freely again. . . . The Residency quarters I am in, are better than any I saw in India. Electric light, art furniture & what not, are provided. My host is a young political of the most amiable & attentive disposition, looking after me as if I were a relative. I get down to meals in my comfortable camp chair & lead again a quiet but civilized existence. The two Turki servants who looked after me with such care on the journey, are still with me.

"I found at last a copy of 'Ancient Khotan' awaiting me. Its binding & 'get-up' generally please me beyond all expectation. Let me confess honestly that in my author's vanity it was a feast for my eyes. This too I owe to you and our common friends whose unsurpassed care & devoted attention watched over the book right to the end. I feel always, no gratitude can be a return for such sacrifices."

As November was ending, Stein grew restless. Dr. Neve cautioned him against leaving Srinagar until the last wound had healed properly; nor did he want to appear in Lahore on crutches. While still waiting obediently, patiently, he received the welcome news that his deputation had been approved. He relayed it to Publius on the 26th: "It [came] in a kind message from the Viceroy. The chance of an interview at Agra had, of course, been abandoned long before. But I thought it would be worth trying whether by going to Calcutta I could not secure some help towards obtaining the compassionate gratuity for that poor fellow, Naik Ram Singh, & for my own surveyors the mark of recognition they have so amply deserved. I entertain no illusions either way but I should at least like to satisfy my conscience. . . . I have tried to find out from Col. Dunlop

Smith whether the idea of a visit to the Bengal Olymp is encouraged. I should go there only for a few days & then hurry on to Bombay."

Stein was suddenly seen as a hero. As the knight-errant who had freed documents languishing in a "black hole," he was impressive; as the victor of an ambush set by a merciless cold, he was irresistible. This double victory assured that his requests were no longer ignored or postponed. Suddenly all doors were open to him; he had but to ask and that "great machine," the bureaucracy, listened. If heretofore his work happened to coincide with the interests of the government, now the government bent to facilitate his work. The panorama gained by his new position extended to the furthest reach of his hopes.

17

All's Well that Ends Well

Stein's long letter to Publius, begun at "Government House, Calcutta" on 22 December and concluded on 6 January aboard the *Osiris,* "in view of the Calabrian coast," is part of the tableau—"All's Well That Ends Well."

"My last evening at this gorgeous camping place gives me a quiet hour and I use it to begin a letter which will not be posted until Brindisi is near. . . . At Dehra Dun dear old Col. Burrard had for two days turned the Trigonometrical Survey headquarters into a sort of consulting-room for my maps. All points about their drawing, reproduction, etc., could be settled readily with the men in charge of the respective departments, & I got through a great deal of work which otherwise would have cost strings of letters. There will be about 130 sheets published, all on the scale of 4 miles to 1 inch & in the style of my map for Anc. Khotan as far as their resources permit. Dehra Dun is a delightful spot, full of tropical vegetation & yet so cool, & my only regret was that there remained no time for drives—& walks, alas, were out of the question.

"Col. Manifold, the civil surgeon, whose guest I was, took every care about my foot & assured me that the inevitable 'rush' at Calcutta would not matter though the healing of the sore surface on what is left of my big toe, could not commence until I got on bord [*sic*] & had rest. It seems that what had covered the wound by the time I left Kashmir was not throughout real skin, & as this gave way towards the end of my Lahore stay I shall have to start the process afresh but luckily only over a very small surface.

"The journey down here was quite pleasant & gave me for the first time a glimpse of the green plains of Oudh. A grand Viceregal carriage fetched me from the Station to this luxurious tent in the Garden of Govt. House which was to be my quarters for these few days. Col. Dunlop Smith . . . had carefully arranged everything that could help me in the objects of my visit. I saw H. E. & Lady Minto who were also kind enough to claim me as guest for the first time at lunch after my arrival & was asked a great deal by both about the journey, etc. Govt. House is a gorgeous enough place; but what with the distances to walk through its halls & corridors, I was glad that the departure of the Viceregal family party for their prolonged weekend left me quite free until today to enjoy the less exacting hospitality of the Dunlop-Smiths—& to push on my items of business.

"They had gathered together for one evening the great men of the Home Dept. to dinner with some other luminaries just to put me in touch with those who could help. So next day I found it easy to secure from Sir Harold Stuart, the Home Secy. what I wanted—a promise to insist upon the originally proposed length of deputation; ditto to do all they could for getting poor Naik Ram Singh provided with a generous pension. The Viceroy had already shown interest in the matter & only an official application on my part was needed. I earnestly hope the poor fellow will get all the Viceroy is empowered to grant in cases of exceptional merit in excess of the regular pension rules.

"Sir Harold also showed me the despatch which was just then going to the Secy of State about my 'explorership.' So they at last after 4 years had found courage for this too—For Lal Singh, the Surveyor, Col. Dunlop-Smith, backed up officially by the Surveyor-General, hopes to secure the amply deserved 'Rai Bahadur' by Jany. 1st. Of course, it took a little arranging that my official & D[demi] O[fficial] supporters should set about it in unison. How he can keep everything that needs his action going is a marvel to me.

"At the Finance Dept. Hailey has acted like a true *deus ex machina*. Without my making any formal proposals I found a supplementary grant which covers all expenses since my return, already sanctioned. In the same delightful fashion I am presented with my pay up to the day of sailing, last pay certificate, etc.—all things which would have cost otherwise a lot of urging &—walking. Hailey is indeed a friend who can help—& spoil me.

"Today I had a great field day, starting in the morning with the arrangement of a little exhibition of my finds at the Priv. Secretary's office. It was quite a spacious room. But I found that the selected specimens of pictures, MSS. tablets, etc., from a single one of my cases quite sufficed to cover all available tables. After lunch I had the satisfaction of showing the Viceroy & Lady Minto round this sample room of 'finds' & was kept at this task for nearly two hours. Lord Minto greatly appreciated the fine sketch map prepared at Dehra Dun showing all the ground we had surveyed & was evidently impressed by the excellent preservation & the great variety of antiques recovered. He afterwards told Dunlop-S[mith] that in spite of all he had read in my letters he had no idea of the 'proceeds' being so ample. It was very pleasant to act as showman & reconteur. . . . That H.E. [His Excellency] was really interested in my work & willing to help me thereafter I could see. So I hope his promised letter to Lord Morley [Secretary of State for India] on my behalf may do some good.

"If I tell you that I had among my visitors today also Mr. Butler, the Foreign Secretary to whom I was anxious to talk about Balkh & Seistan as future fields of exploration and about the complimentary letters which are due to my Chinese patrons & honest Chiang Ssŭ-yeh, you will realize that my Calcutta stay short as it is, has been fully utilized. Everybody whose help I need, has been anxious to save me trouble & time. So the necessity of walking has been restricted to a minimum.

"In accordance with Col. Manifold's suggestion I saw his cousin, Col. Pilgrim, the chief local surgeon. He was not altogether encouraging & explained that the formation of real skin was bound to be slow where the wound was at an extremity & the patient over forty. He, too, expects most benefit from the rest & invigorating air of the sea. So it is just as well that after getting that refractory wound cauterized tomorrow I can start straight off for Bombay. My tasks here are completed & I feel I have done all I could hope for the reward of my helpmates.

"A very pleasant surprise awaited me here: news from the Hon. Secy. of the Bombay Branch, R[oyal] Asiatic Society that the first award of a gold medal for Oriental research, newly founded there in honour of Sir John Campbell, was to fall to my lot. Of course, there will be no time for the meeting they proposed to hold for the public presentation. I am glad to escape this & quite pleased with the fact that the first acknowledgement my labours ever received in India should come from Bombay to which I have always felt attached. . . . In Bombay I found a very hospitable invitation from Haigh, the Collector of Customs & Hon. Secry of the Asiatic Society, and spent with him & his young wife a very pleasant Christmas evening. Mrs. Haigh is quite a professional archaeologist (of the Greek-Cretan persuasion) but in spite of this quite amusing. [Was this a sly dig at Marshall?] . . .

"The 'Salsette' which took me off to Aden, proved quite new & comfortable, more like a huge yacht than a 'ferry boat.' I had scarcely been on the sea when that wretched sore place of my toe which just before leaving Calcutta had undergone rather severe treatment with blue stone [vitriol: sulphate of copper] from Col. Pilgrim began to show real signs of healing. It has gone on steadily since until now sound skin has covered the whole down to a spot not bigger than a big pin head. This, too, I hope will get under cover during the few days of rest I shall take at Vienna. . . .

"I have decided to go to Vienna where there is practically no one to look up outside my sister-in-law's home & where I can make quite sure of giving full rest to my foot. After 4–5 days I shall run down to Budapest for another four days or so & face the unavoidable round of calls at old friends. On my way to Vienna I propose to take it quite easy & to stop a day in Venice where if the weather is fine I can manage to get plenty of fresh air & light on the canals without straining my foot in the least. My intention is to go to London via Paris where dear old M. Barth eagerly awaits my visit. Of course, a day is the utmost I can spend there. Then 12 hours or so in town just to see the Saint & the Baron, to put in an appearance at the India Office & Brit. Mus. & then hurry to that true haven of rest, your incomparable cottage.

"My big consignment of cases ought to reach London by the 9–10th, & will, I hope be received into safe deposit at the Brit. Mus. . . . How I am to start real arrangement without the Baron is still a mystery. I did my best at Calcutta to prepare the ground for a proposal which might lead to his permanent employment & found the Home Dept. apparently sympathetic. But it may take years to effect anything definite."

Landing in Italy, Stein satisfied old dreams, renewed old ties. 10 January 1909, to Publius: "In spite of the shortness of time & my eagerness to get back to my brother's family, I managed to snatch a bright day in the city of Marco. I felt as if I ought to bring to the home of the great traveller the latest news from Sa-chou and the desert of the Gobi! . . . A very comfortable journey of some 7 hours, performed in a sleeping car for the sake of an 'invalid's' rest, brought me here [Vienna] yesterday. I found my sister-in-law & her children quite well & rejoicing at this long-delayed reunion. The former had only to look at me to be relieved of the worrying fears she had continued to entertain on my account in spite of all written assurances."

From Stein's letters to Hetty and his references about her to the Allens, it is clear that she appeared to him as the quintessence of long-suffering motherly love, of womanly warmth, and goodness of heart. Perhaps that is how he expected his brother's bereaved widow to act and feel. It set her apart as "family" and reinforced his unswerving commitment to his brother's memory. Reading his letters to his sister-in-law—and he wrote regularly and at length—Hetty is revealed as a worrier in the grand style: in the attic of her nature were trunks crammed with worries, some large, some small, some old and threadbare, some new, some real, some imagined, all ready to be opened and used. Sentimental, wet with the tears of self-pity, disguised—they commanded Stein's devotion.

From Vienna, 10 January 1909, Stein continued reporting his homecoming journey to Publius. "I can report that it was readily acknowledged that I look quite fit—& I feel it, too, and that the wound has finally closed, I trust for good, at Venice. The doctor I asked to look at the foot certified that the skin covering it was quite healthy. There remains only the question already pointed out by Col. Pilgrim & Neve whether Schmitt in his anxiety to save my 'anatomy' had not left a little too much of the big toe, i.e. whether the end of the bone had an adequate protection of flesh & skin. There is a certain strain of the latter which might be due to that cause. Before I leave Vienna, the chief surgical luminary here is to be consulted. Of course, I should not think of a further operation however easy it is said to be unless there is a real advantage to be gained by it. Anyhow I should choose my own time for it. In the meantime I am taking it easy in the most conscientious fashion just as you both would have me, not moving out of this comfortable flat and scarcely getting out of easy chairs. . . .

"My nephew received 'Erasmus' among his Christmas presents & has become a most enthusiastic Erasmian, translating great portions in writing, etc. He is a devoted humanist. His sister takes delight in the Vedas & comparative philology without, I am glad to find, losing her youthful humour over those dry old affairs."

After Vienna, Stein went to Budapest and, on his way to England, stopped "for a day or two at Paris where dear old M. Barth & other friends expect me. By the 20th or 21st I may hope to reach London. Homecoming and honors were anticipated when at Brindisi, awaiting him were letters from friends and co-workers, one, especially enjoyable for Stein to read was from Miss Yule, "the first I had from her in years." It was to the memory of her father he had dedicated *Ancient Khotan*.

Stein's journey to the sources of the Khotan River, the sun-melted ice of the glaciered Kunlun Mountains that pass through defiles to nourish the oases, might be viewed as symbolic of his Calcutta visit and the results that flowed from his interviews with government "luminaries." To Sir Harold Stuart he presented the full, sad details of the Naik Ram Singh's blindness. He mentioned that the "Naik's Regimental pay was Rupees 16 a month, but that I was paying him . . . Rupees 33 per mensem and argued that "under certain articles of the Civil Service Regulations he was entitled to a pension of full pay, viz. Rupees 33 per mensem."[1] The following May, 1909, "Naik Ram Singh was granted a special pension of Rupees 16 a month. . . . But, it is now reported that Ram Singh died on 25th March 1909:" After much consultation and considering the special circumstances, the government generously granted a pension of 12 rupees per month to his "elder son for the support of the family." The usual pension to the families of those who died "while serving in action was Rupees 3–8."[2]

Stein also called on another luminary, Harcourt Butler, secretary of the Foreign Department, to ask that the government of India thank the Chinese officials for the help they had rendered him. It would have been impossible to "achieve the archaeological and geographical results which I have had the satisfaction of reporting to the Government of India in the Home Department, had the Chinese Magistrates of the districts . . . not given me most effective assistance in all matters connected with transport, supplies, labour for excavations, etc."[3] He listed the governor-general of Sinkiang, at Urumchi, the viceroy of Kansu at Lan-chou-fu, Pan-Darin, the taotai of Aksu, the chief district magistrate of Su-chou, and the magistrate of Tun-huang. "I feel sure that an appreciative mention of their services would prove to the advantage of those administrators and might indirectly be useful for future British travellers."

To the list of important administrators, Stein added Chiang-ssu-yieh, of whose services "I cannot speak too highly. . . . As he is now employed as "Chinese Munshi" [secretary] of the British Consulate at Kashgar, I think that this recognition might suitably be accorded in the form of a complimentary letter accompanied by a present which Captain Shuttleworth [replacing Macartney on home leave] could conveniently arrange for." Stein's suggestions were acted on speedily: within six months Captain Shuttleworth could report: "The gold watch was duly received and presented to the Ssu-yieh in a suitable manner. I invited the Russian Consul and all the Chinese Ambans of the Old and New Cities of Kashgar to a dinner and presented the watch to Ssu-yieh in their presence after dinner. The Ssu-yieh was very gratified that it was presented before [these officials]. . . . he was visibly affected when I presented the watch to him after making a short speech, and I can assure you he very deeply appreciates the honour."[4]

Stein's high-level conversations dealt with his future plans as well as his past performance. He was not ready to rest on his laurels. Inevitably, he took the occasion to reiterate his hope of excavating in the vicinity of Balkh "should an opportunity ever [be offered to speak] to the Amir on the subject." Butler added it to the notes he jotted down of their interview, but said "that he would not write on the matter as there were already papers on record in the Foreign Office and he had nothing new to add." Butler's notes also give the first mention of Stein's serious interest in the Sistan, the homeland of the father of Rustam, Persia's epic hero. "He [Stein] asked me whether I knew of any difficulties in the way of his undertaking archaeological work there. . . . I replied that speaking off-hand I knew of none. No action need be taken until he refers to the matter." That would wait.

After five years of separation, Stein was reunited with the Allens. He was too happy to notice that there were ominous clouds in the sky.

4

The Third
Central Asian Expedition
1913–16

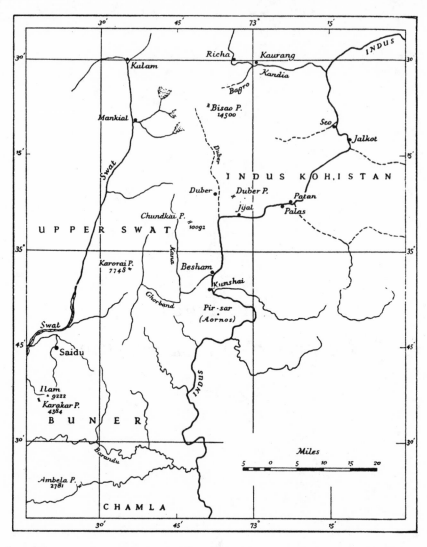

From Stein, "From Swat to the Gorges of the Indus," *Geographical Journal* 100, no. 2 (1942).

18

It is to be expected that some of the political and social changes which took place during the thirty years which span Stein's four Central Asian expeditions would have affected them. Larger configurations, political, patriotic, and personal, now one, now another, threw colored spotlights on his work. Stein, then, is the constant who shows up differently—hero or villain—as the colors change.

Politically, it was the Chinese Revolution that divided the style of the first two expeditions from the Third and Fourth. In 1911, six years before the Russian Revolution, the Ch'ing, or Manchu, dynasty, which had ruled China from 1644, was overthrown and the republic proclaimed. Thus, the First and Second Expeditions were carried out before that momentous event, when the centuries-old system was still operating. Stein's longest expedition, the Third, took place during the years immediately following the establishment of the republic, years when many patterns from the past lingered on—a confusing, unstable mixture, prelude to the bitter, long-drawn-out civil war that ended only in 1949 when the People's Republic of China emerged victorious. Then the seat of government was returned to Peking from Nanking, where it had been moved in 1926 when the struggle for power within China was postponed while both sides, under the Kuomintang, presented a united front against the Japanese. It was during this temporary lull, a spurious political stability, that Stein attempted his Fourth Expedition.

A very different kind of political involvement colored Stein's race against German, French, and Russian archaeological teams: the competition to which he responded was only in part personal; it was also tied to national, hence patriotic, emotions. As a naturalized British subject his allegiance was complete—the fact that his work, his friends, his way of life centered on India made him more *pukka,* more out-and-out British, than born and bred Englishmen. In part, then, it was for England that he ran the race.

His two earlier expeditions had been at the start of the twentieth century when the European overseas enterprise that began with Vasco da Gama's voyage to India had, for want of space, ended, when geographical exploration dwindled to prizes long withheld by ice-covered waters and lands. "Firsts" were chalked up that brought renown to individuals and satisfied national muscle-flexing. To the galaxy of polar heroes such as Nansen, Peary, Shackleton, Amundsen, Stein's name was added. Like the others, he had won his race and displayed stamina and heroism. But there the likenesses cease. The explorations of the others had taken them literally to the ends of the earth and, however scientifically significant, however delightful their encounters with Eskimos and penguins, they had advanced up dead-end streets. Stein, if he had not scored a "first," had mastered a formidable environment and retrieved from desert wastes just beyond Europe's backyard rich evidence of transcontinental contacts of peoples and ideas. His work had the double appeal of archaeology and exploration.

Stein's Third and Fourth Expeditions were affected by another, a wholly personal, attitude—his thoroughgoing acceptance of British imperialism with its rationalizations. The years spent in India, that brightest jewel in Imperial Britain's crown, had confirmed his belief that the subcontinent had benefitted from British rule; he subscribed wholeheartedly to the advantages accruing to India. To him, the presence of the British Baby was at once the symbol and measure of the advance of civilization. Civilization was, by definition, Western civilization—the golden cornucopia its science and technology had created. His belief was further confirmed by tours along the Northwest Frontier, where he saw how as one by one the tribesmen who threatened the Pax Britannica were brought to heel, tribal feuds were suppressed and the proper blessings of health, education, and trade moved forward.

Imperialism, one aspect of Western civilization, was deeply embedded in the status quo; its style was part of Stein's own experience: as, when young, he had enjoyed consideration and care from "patrons," so did he as sahib give in turn the same care and consideration to his Asian followers. He would have bristled had the word "inequality" been applied to him, and he could truthfully offer Pandit Govind Kaul, Pan-Darin, and Chiang-ssu-yieh as witnesses of his friendship and high regard. Inequality, he might have said, was the inescapable concomitant of illiteracy. Education would effect change, and change so effected was desirable. He would have agreed with Max Müller who had written, "India has been conquered once, but India must be conquered again, and that second conquest should be conquest by education."[1] Thus Stein thought in terms of a steady, uphill evolutionary pace—not revolutionary mutations. His thinking was colored by his unshakable belief in the primacy of Western scholarship. It justified everything.

There are quiet but noticeable differences in the tone of Stein's letters. Until he returned invalided and a hero from his Second Expedition, his letters to Publius are punctuated by his "bitter outpourings": complaints against the bureaucratic machinery, the *babudom,* which mindlessly forced him to educational administrative routines; humiliation when Marshall seemed to restrict his archaeological work; irritation and frustrations at delays. Beginning with the Third Expedition, a word, used lightly, gains in frequency and weight: "difficulties." It is not applied to the terrain, the men, the animals, or the organization of travel. On the First Expedition he brushed off the idea that the Boxer Rebellion posed "difficulties" in Sinkiang; he borrowed weapons from the government in case he had "difficulties" when crossing tribal territory; he noted his good fortune at having escaped "difficulties" by being at An-hsi when a mob stormed and looted the Tun-huang yamen. "Difficulties" was a word reserved for politically disturbing events—rebellions, hostile tribes, mobs.

Revolution to Stein was like a vicious buran that in passing obscures the sun and sets the desert sands in violent motion engulfing town and temple alike. A Christian by baptism, a British subject by choice, he was firmly rooted to the establishment. His whole being would have said "amen" to the idea Shakespeare gave to antiquity's great traveler, Ulysses:

> The heavens themselves, the planets and this centre
> Observe degree, priority and place
> Insisture, course, proportion, season form,
> Office and custom, in all line of order.

Troilus and Cressida, act I, scene iii

Is it not paradoxical that Stein, who had unearthed eloquent evidence of powers that had ruled and passed and of faiths that had perished, could not believe that the same might happen before his very eyes? The past was the past; his eyes would not see the same processes of social change even then at work. And, when political upheavals destroyed the empires of China, Russia, and Austro-Hungary, his eyes did not like what they saw.

Perhaps it was Stein's disinclination to accept the new that made him turn to the past. Back and back he went from Marco Polo and Hsüan-tsang, back to his first love Alexander the Great and still further to the new discoveries of the ancient Indus River civilization. To Stein the past offered not lessons but an escape.

"The Government find themselves required to provide for a white elephant in the shape of an educational officer, who had turned out to be a very distinguished Archaeologist."[2] That in a nutshell was the problem facing the bureaucrats. There was much mumbling but no grumbling, as point by point they moved Stein from the Education Department to the Archaeological Survey without his forfeiting his salary (1,250 rupees per month), tenure, and pension rights. One further stipulation Stein made: freedom from administrative routines, from the endless paperwork on which babudom waxed strong. That too was met. He was "attached to the office of the Director-General of Archaeology through whom his programmes of tours, reports, etc., will be submitted and who will generally control his work. He will not, however, be required to take any part in the administrative work of the Department."[3] One suggestion he had made was not allowed; he had proposed and argued that he be called "Archaeological Explorer." Marshall found the title too grandiloquent and too pointed to transfrontier tours—politically inexpedient. He suggested "Special Explorer" as an acknowledgment of Stein's unique qualifications; but even that was too fancy for the bureaucrats, and they named him another "superintendent"—a first among equals.

The tone of Marshall's comments in a long, intricate, interdepartmental exchange shows that by 1909 he and Stein had reached a stage of amicable equilibrium. Neither had an ounce of meanness in their nature. Each had learned to value the other's qualities and achievements; each, a tireless dedicated worker, honed his talents for the greater glory of Indian archaeology. This noble purpose united them and kept their alliance free from the slightest hint of opportunism. Their strengths were complementary. Marshall, a brilliant administrator, at once orderly and innovative, was battling to save his newborn department from being the pathetic victim of infant mortality; he shared the enthusiasm aroused by Stein's finds. Splendid and important results Marshall knew were "absolutely indispensible . . . to interest the public in our work and secure more adequate funds for it."[4] He never lost sight of this stark fact, and the sites he chose were chosen, as he candidly admitted, to secure "spectacular results." Thus as scholar and struggling administrator Marshall was drawn to Stein, whose adventurous discoveries and travels had won public interest.

For Stein, it was second nature to respond to all government officials willing to become enthusiastic about his historical world view. He welcomed the director-general as an ally. Their tie, warm but businesslike, never passed into friendship; but, once established, it lasted. While Marshall was director-general, Stein was practically a free lance, but he never abused the freedom so generously given to him. Both men regarded the honors awarded Stein as glory accruing ultimately to the Archaeological Survey. Stein's lectures, articles, books, and even the honors carried the Archaeological Survey's importance to thousands at home and abroad.

Stein received the Founder's Gold Medal from the Royal Geographical Society and honorary degrees from Oxford and Cambridge; the Belgian Academy elected him an honorary member and the Alpine Club a member. "I learn," he wrote Publius, 10 August 1909, "that among the twelve members with whom I am to share the distinction are Sir Joseph Hooker [founder of the Royal Botanic Gardens at Kew], Nansen [Arctic explorer, oceanographer, and humanitarian], Younghusband, Curzon, etc., names too big for such company as mine."

Most unexpected and gratifying was the official recognition: in June 1910 Stein was honored with a C.I.E., Companion of the Indian Empire. Dress and behavior were prescribed in his notification from the Central Chancery of the Orders of Knighthood:

> A companion to be invested by His Majesty will attend in Levée Dress, with Decorations and Medals.
>
> On being admitted into the Royal Presence he will advance opposite to the King and make the usual reverence by bowing, and will then kneel on the right knee.
>
> His Majesty will then invest him with the Insignia by affixing it to his left breast.
>
> The recipient will then raise his right arm horizontally and The King will then place his hand on the wrist of the recipient, who will raise it to his lips.
>
> He will then rise and retire from the Royal Presence with the like reverence that he made on entering.
>
> N.B. Those attending the Investiture will leave their Cocked Hats, Helmets, &c. in the Lower Hall. They should wear one glove on the left hand.
>
> Swords are *not* to be hooked up.

Never could he have imagined that the treasure he found at the Caves of the Thousand Buddhas would bring him into the Royal Presence, for the high, solemn ceremony of investiture. He described it to Publius in his letter of 10 July 1910: "It all went quite pleasantly, the hour's wait in the Royal antechambers being made entertaining by congenial company and the sight of grand uniforms. I saw old friends from India again & had a long, quiet talk with Marshall who also received his decoration. Everything was managed in grand style & without any fuss. . . . I may add that I felt quite comfortable in my black velvet dress,—but glad that it had not to be worn in the heat of Calcutta."

Budapest went wild when he visited the 'Vaterstadt.' "A 'rush' of luncheons, dinners, &—alas, suppers. . . . The first lecture was a great success. The theatre was crammed with the 'best people' & had been sold out for 10 days. The slides were excellently done, the acoustics of the place very good & I stood the 1¾ hours' talking without too great a strain. What amused me a good deal was the attitude of quasi-awe which assured perfect silence to the very end. . . . On Monday I repeat the lectures, then go to Vienna by the night express, hold my lecture there next evening & so on to Munich."[5] A week later he was back in Oxford resting "after 4 speeches in 50 hours & all the rest of it."[6] Tired but pleased, he returned to the depressing lack of movement at the British Museum. His working arrangements there were stalled: the paintings and artifacts, the thousands of documents so laboriously retrieved were still in their cases, homeless. Yet until his rich harvest was identified and photographed he could not provide scholars with samples and invite their aid. It was already late in the summer (1909) and precious months of his deputation-*cum*-leave were gone. (The term was from 26 December 1908 to the end of September 1911).

He escaped from the doldrums. He had long wanted to take a leisurely vacation with Hetty as his guest. The Tyrol suited them both; it was the first of many, planned for whenever he was in Europe. When it was over, Hetty returned to Vienna "which she dislikes so much yet which she has not the strength to abandon,"[7] and Stein went to Italy to find a place that suited his needs—he was organizing his survey notes so that the Trigonometrical Survey could begin preparing the sequence of maps. First he tried a hotel in the Lake Maggiore region; in February he moved to Ruta. Was it the Arnolds who had suggested it? They had spent a few days together, days filled with delightful walks, gay sightseeing, and talk that satisfied one of Stein's fondest dreams. Ruta, near Genoa, above the Riviera di Levanto, a coast untamed and picturesque, had a large hotel almost empty of tourists at that season, set in its own park and accessible to mountain walks offering sweeping panoramas.

An ideal "Marg" for concentrated work. But he was alone, alone as he had never felt when far away, and Publius's letters filled with Erasmian labors and Oxford chitchat permitted him to participate in his *amicissime*'s comings and goings and doings. Ruta was too near for such vicarious participation. "I feel increasingly the wrong of separating myself from the bodily presence of those whose thoughts are my dearest companions."[8] Publius and Madam had already sensed Stein's need and had arranged that, beginning in May, for the remaining fifteen months of his home leave Stein was to have a set of rooms at Merton College, Oxford. Gratefully, he thanked them for "the lordly retreat your care has prepared for me by the Isis [upper Thames]. At Merton I may hope to have peace & reasonable seclusion for steady work & yet see as much as I can of you both during hours of rest & exercise. It seems an ideal solution to a problem which had long weighed on me."

Stein's likes and dislikes, the rigid work habits so essential to a person having nothing but work into which to channel his energies, emerge clearly in the exchange of letters about the Merton rooms. It is also clear how well the Allens knew him. Thus Publius anticipated Stein's request for a floor plan: he did not need to be reminded of Stein's "penchant for exact localization."[9] With it, Stein could visualize that "These big rooms, worthy of a canon at least, occupy the finest quarter of the quad & give a view over that noble garden. . . . I should certainly like to use the southern big room as you suggest for work and B, the eastern one for my bedroom. C, the smaller one, would suit excellently for an occasional guest." Thinking in terms of "camp" arrangements, Stein preferred to rent "the few bigger articles of indispensable furniture—for I have always fought shy of encumbering myself with property of that sort and should not know what to do with it when my time at home is up."[10] He was, he knew, a nomad: "I am like the Kirghiz who cling to their tents while they live and think them the emblem of freedom."[11] He gave Madam carte blanche to do whatever she thought necessary; as a caution, however, he added that "a tear in my tent-fly, or a tear in the wall paper is a thing which must be mended."[12]

Having been finicky about minor matters, he came to the heart of the doubts that increasingly assailed him. 10 March 1910, to Publius: "The main thing for me is that I should be able to settle down in peace *as quickly as possible*—& take to that strict arrangement of my working times & domestic economy without which I cannot get ahead with all my tasks before me. My plan, often tried before, will be to keep the day up to my afternoon walk absolutely protected from interruptions, however attractive. I shall try to keep two hours daily for exercise before dinner & you know I never work after it. . . . Will it be possible to stick to this programme which suits me perfectly and to keep in *absolute Purdah* for the earlier & greater part of the day without coming into conflict with College notions & customs? I am sometime attacked by troublesome doubts on this point, yet it is a question of freedom to which I attach value for the sake of my work."

The "caves" was Stein's name for the basement rooms in the British Museum where the collections of the Second Expedition were finally housed. As before, Stein entrusted the unpacking, arranging, photographing, and the Descriptive List, the detailed technical summary of each item of art and artifact, to Andrews, the Baron. Because of his position at the Battersea Polytechnic Institute, the Baron could only work half-time; it was then that Miss F. M. G. Lorimer began her long association with Stein's work. She is the "R.A.," the Recording Angel, whose "zealous and painstaking work"[13] was gratefully acknowledged. Madam had recommended her as a valuable addition, and her sobriquet indicates that she quickly became part of the inner circle; she continued her role when she moved to India, and then, after many years, when she exchanged roles, she disappears from Stein's letters: she married. Through her full and precise notes, made in duplicate and sent weekly to Stein, he could follow the progress being made in the "caves." Neither Andrews nor the R.A. could handle problems of an administrative and scholarly nature. To deal with these Stein was fortunate in having the help and advice of Dr. Lionel B. Barnett, an eminent Sanskritist and Keeper of Oriental Books and Manuscripts at the British Museum, whose department adjoined the "caves." Stein's letters to Barnett dealing with the personal aspects of their professional activities show how deeply Stein was indebted to Barnett's learning, wisdom, and acumen.

In these correspondences the lines of authority and friendship, however liberally seasoned with tact and expressions of gratitude, are clearly stated. But the correspondence with the Baron offers a soliloquy (the Baron's letters are absent) depicting an emotional ambiguity; each man, to varying degrees and in different areas, was dependent on the other. Stein acknowledged this privately and in print; Andrews seems never to have admitted and almost to have resented his dependence. In interpreting the nonliterary finds of the First Expedition, the Baron proved how

justified Stein was in appraising his artistic sensitivity and technical understanding. These attributes plus his dependability and devotion made him
indispensable in handling the much greater collections of the Second and
Third Expeditions. Yet how reluctant he seems to exert himself to realize
his full potential. The Baron was fearful of risks, slow in taking the initiative, touchy. Big, and splendid like a Percheron, Stein wanted to re-
lease him from his brewery truck so he could display lightness and speed.
By comparison Stein was a wiry, canny steeplechase mount clearing the
hurdles, if not with style, with success.

While at Maggiore Stein began his long-distance supervision of the
ongoing handling of the collection. He wrote to Andrews on 24 October
1909: "Now that wall cases are available would it not be possible to start
the new assistant [Mr. J. P. Droop] on a series of greater archaeological
interest & variety? say miscell. objects from Yotkan, Khadalik, Niya or
L.A. [a Loulan site]? I think a certain amount of preliminary unpacking
would be advisable & would stimulate & help them [Droop and the
R. A.] to realize the variety & scope before them." There is a kind of
pedagogical restraint in Stein's suggesting that Andrews "guide Droop
and Miss L. by the light of your own experience & keenness & leave to
them as much of the drudgery as you possibly can & reserve yourself for
the really novel tasks, such as the silks and large frescoes. . . . I am very
glad for the interest of Colvin [Sir Sidney, then writing his article on art
for the 11th edition of the *Encyclopaedia Britannica*] & should not mind
your accepting the help of his Dept. for laying out the silks, provided it
is made quite clear that no exhibition can take place until after our collection has had its show as *a whole*. . . . I hope special care is taken that
all slips are initialled by the asst. responsible. Any slips with doubtful
points are to be kept apart & submitted to you for decision. A *very brief*
daily record by Droop and Miss L. as to progress made seems desirable for
control & would be 'stimulating.' "

The work went forward; the problems seemed to lessen and by the
early spring (1910) the more spectacular finds were ready to be exhibited
at the British Museum. Stein could then concentrate on the Baron's position, which permitted but half time at the "caves." 7 April 1910: "It is
a real relief for me to know that you are safely beyond that awful &
protracted 'rush' & toil at your Polytechnic. [Andrews was the Director
of Arts.] I need not waste time by telling you how I grieve for your
exertions in such a machine & what I think of a system which rules it.
I never doubted that your care was keeping up steady progress at the
B. Mus. Your discoveries about the main materials in the fabrics is very

instructive. Your slips on those examples will prove valuable. I think attention must be concentrated on getting all artistic remains (paintings, frescoes) into a condition when they can be safely handled and studied. Estimates for illustration plates will have to be made out early this summer & for them a preliminary survey of all materials must be made possible. . . . I wish the hope of setting you free could be brought nearer realization."

Two months later there is the first mention of handling the manuscripts. 10 June 1910: "Until an inventory list is made of all MSS. & those to be entrusted with detailed examination definitely selected, the issuing of photos can only cause complications. At present all Turki materials are reserved for [E. Dennison] Ross who will make a preliminary list & submit proposals for further study." That May, Stein moved into Merton College, and the next month "the 'private view' at the B. Mus. exhibition brought me into touch with some interesting and interested people. My ancient painting seemed to make a strong impression," Stein wrote the Allens who were in Switzerland hunting Erasmus letters. "The embroidery picture, M. Mizion, the conservateur [at the Louvre] thought we might well ask any price within the total cost of my expedition. I wished he would approach the Early Victorian lord [Morley?] with the suggestion of a 'deal' which might lighten his conscience."

From Paris, Foucher and Pelliot came to see the exhibit. 1 July 1910: "Foucher arrived & has been most enthusiastic over my poorly housed collection. He has already cast a good deal of illumination on pictures & frescoes. The Roman look of those from Miran was to him too a revelation. . . . Pelliot, too, has been hard at work on the Chinese MSS where he has made numerous interesting discoveries, quite enough to assure his readiness for taking up the systematic inventory. He is a splendid worker in his own Sinological line, but a bit too self-centered. . . . I had lunch with Marshall who came to look at our collection. He behaved in a very pleasant way & seemed full of appreciation." A week later, when they again met at the investiture, Foucher was still in London "having made more rapid progress with my pictures than he expected. Pelliot after rapidly examining the 10,000 odd Chinese MSS pieces has found plenty to repay further work; for he asked to be entrusted with the inventory for the relatively modest remuneration of £200. "My 'collaborator' arrangements have lately progressed well, but, of course, cost a good deal of correspondence. A recent list shows that the number of individual helpers has now risen to 18! Chavannes as usual is well ahead with very interesting results."

17 July 1910: "The safe backing of the friable clay panels costs infinite labour and only an artist's hands can perform it properly. Sir. A. Church, the R[oyal] A[cademy]'s chemical expert is to examine colours & backing with a view to safe preservation thereafter. Other tasks in 'les caves' have been advanced a good deal: the big embroidery picture after three months' labour is now ready for exhibition. Some beautiful silk paintings besides 70–80 banners are now flattened & available for study. The lines of collaboration for my team of 17–18 specialists are also getting straightened out. But I begin to dread their demands for separate volumes: both Chavannes & Pelliot must have their own, etc." The collection was moving ahead; the exhibit had attracted favorable responses; specialists were to be had for the asking; Stein had every right to feel pleased and optimistic.

With all this activity, gratifying and auspicious, Stein was at Merton intent on writing his Personal Narrative of the expedition, the two-volume *Ruins of Desert Cathay*. It was completed the beginning of November 1911, and by the end of that month he was on his way back to India.

19

The year 1912 began much like any other. On arriving at Peshawar Stein found that the man who had temporarily replaced him "was an excellent scholar & did splendidly in starting the new Museum here [one of the many Marshall created] & in his diggings. But he seems to have cared little for the practical aspects of camp life—the tents he provided the office are almost entirely of the type which requires an elephant or bullock cart to move it." . . . It remains for me to settle by a rapid series of tours where a serious piece of excavation work can be put in before the official year closes. Of course, I do not regret this."[1] Contentment breathes in his letters; a sense of assurance in his long-distance directing of the work steadily going on in the "caves." Delays, problems are taken in stride. With no slackening of pace, there is an ease about Stein: he was fifty and could look back and see what he had accomplished in his chosen field.

"It has been a thoroughly enjoyable experience to be on the move again & to see my 'camp' shaking down into proper trim."[2] His feet did not trouble him, and the anxiety that had haunted him was exorcized. "Moving along the Swat Canal I inspected again Takht-i-Bahi with its extensive Buddhist shrines which have been partially cleared of the debris left by old 'unauthorized' diggings and are being carefully conserved by the Military Works Dept. under 'our' direction. I had to get a temporary shelter over two well-preserved small stupa bases before the heavy rain comes on. . . . Next month I am to give my lecture for the Divisional Institute at Peshawar. It will be quite useful to put me into touch with the military bigwigs & to form fresh personal links."

The same letter was continued on his return from "the tribal territory south of the Swat range between the Malakand and Buner. It was one gap left in my knowledge of this border & a blank for the archaeologist too, though native reports pointed to numerous ruins. Bolton, the Political Agent for Swat, accompanied me right through the Palai Valley to the Shahkot Pass [which] I had only seen before from the north. . . . On the pass itself a wonderfully well-preserved ancient road surpasses all I have seen in the way of such 'Buddhist' engineering. . . . It is clear that even a preliminary survey will cost 2–3 days, so I was heartily glad that Bolton (whom I knew in the days of Col. Deane) agreed to arrange for the Jirga's [tribal council] permission to let me pitch camp on their soil." Another aspect of his tour was discussed in a letter to the Baron. 30 January 1912: "At Jamalgarhi I had the good luck to discover a Kharoshthi inscription on a slate slab of the flooring! How it escaped discovery before is difficult to explain. I have had a strenuous time, scrambling all day over rocks & boulders and generally getting back to camp in full darkness. But it has shown that I am still fit for these terribly stony hills and I feel very well. The only trouble is getting time for writing work. I expect . . . to settle down for excavation at a big site near Mardan until the end of March & to close up field work with a tour in Beluchistan during April."

The big site yielded an abundance of interesting Greco-Buddhist sculptures and relievos. "There is quite an *embarras de richesses,*" he wrote Publius, 27 February 1912, wondering what to do with all the specimens partially injured or in duplicate. "In a new field like Turkestan every bit of material has its immediate value. But Gandhara has yielded already so much—selection is the great need. By a happy concidence Takht-i-Bahi ruins (also within a half mile of my tent) have just now given up exquisite carvings from a small chapel on a steep rock slope completely buried under debris. They were still in their niches, undisturbed by later occupation. So the excitement of clearing recalled the Turkestan work. . . . I have 200 men now on my present mound & make rapid progress." Totally unexpected and doubly rewarding was his luck in finding "a four-armed Parvati in late Graeco-Buddhist style, a welcome proof of old Hindu tradition on this ground, too."[3]

He concluded his excavations on schedule and hurried to Peshawar to be present at the viceroy's state arrival, 1 April. "I saw Lord and Lady Hardinge perform the function with a charming combination of grand seigneur manners & diplomatic bonhomie. I was expecting my interview for next Saturday; instead I was asked to come to lunch & see H.E. afterwards. So I got a copy of *Desert Cathay,* specimen plates, etc., together in a hurry. The lunch was quite a small one & I was seated between Lord &

Lady Hardinge. . . . The interview which followed in the Viceroy's study—the very room where 6 years ago Lord Minto bade me good luck before my start—was most pleasant to me. Lord H. let me talk freely of my hopes & plans for B[alkh]. Sir H. McMahon [foreign secretary] had prepared the ground well. So, H. E. agreed that in connection with the presentation of my book a request for a visit to the Oxus, Badakhshan—and B[alkh] in the distance should be submitted by McM. to the Amir. Who can foresee the result?"[4]

This attempt to get to Afghanistan, which began with springtime hopes, ended in the fall, when Stein received the gist of the amir's reply. The foreign secretary wrote that "after stating that he [the Amir] takes a personal interest in archaeology and expressing gratification at the receipt of your *Ruins of Desert Cathay,* His Majesty says he must have the book translated into Persian in order that he may read and understand it, before he can decide whether or not to permit your visit to Afghanistan. The Amir adds that the work of translation will take a long time."[5]

Was it spring, was it a dazzling hope of Balkh that possessed Stein as he took the familiar road to Srinagar—and so to Mohand Marg? The two days' drive "gave me time for a quiet 'think' over things past and future. I reconstructed the chronology of all the journeys up & down that route I had done since 1888 (what a time it seems and yet how brief in a way!) and worked out itineraries for the future. Distant or near, who could say?"[6] At Srinagar, as the guest of the Frasers (the Kashmir resident), Stein was lodged "in what must be the best room in the place, a huge apartment that . . . is to serve for the Viceroy when he comes up in the autumn." Kashmir was his home of choice. There old friends greeted him: "Oliver from Ladak; the Neves; the Dunstervilles. . . . Old native retainers have already been entrusted with orders. But alas for the best of my Pandit friends my return comes too late. Still I have the relief to know his family is cared for—and that is something under Kashmir conditions."

An unusual creation of Stein had its beginning during his short stay with the Frasers—a Kashmir Art Institute; its director to be the Baron. Whether it was to test his idea or to secure an on-the-spot ally, he first mentioned the idea to Publius in a letter of 13 May 1912: "You can imagine my delight when during my first confabulation I discovered that Fraser was fully aware of the need for providing for the systematic guidance of Kashmir art industries. [By the end of the nineteenth century, the great Kashmir shawl industry had been battered to death by machine-made European imitations.] The idea of a technical art school seems to have been in the air for some little time, & I told F[raser] of Andrews' execptional fitness for creating and guiding such an institution. The impression of its feasibility & importance sank in visibly. At his request I had a long talk with Dr. Mittra, the Home Secretary of the State [Jammu and Kashmir] who prides himself on having been the first to perceive this need—however vaguely. Without pledging the Baron, I got that dignitary to agree that the Baron should be invited to prepare a working scheme. . . . They will no doubt take a very long time before emerging from the stage of mere talk."

Stein mobilized Fraser's desire to improve conditions in Kashmir, the resident's admiration for himself, and Mittra's awareness of the need for some direction in changes to be made to counterbalance the Baron's capacity for indecision and the home secretary's for procrastination. He undertook the delicate task of tilting the scales. "As regards the personal aspect of the matter I feel my responsibility keenly." He stated the problem to Publius. "Life in Kashmir is pleasant & the field for an artist like the Baron enticing. But, of course, infinite patience & tact are needed to face all the petty intrigues & annoyances. . . . I am doing my best to prepare the Baron for all needful caution before he decides. It would be a real relief to see him set free from the sordid surroundings of Battersea and his teaching slavery. The post will not be a bed of roses exactly, but there is infinite scope for congenial work, & life in Srinagar under the Resident's protecting shadow is after all easy."

Stein was equally forthright in stating the case as he saw it to the Baron. 6 May 1912: "I should advise you to make up your mind to take the Kashmir chance for good. The more I see of this glorious land the more I *pity* those who live & work in London whatever their pay, etc. For a pleasant existence in England one must have independence, plenty of money—or else tastes not too artistic or intellectual. Yours are! I may be prejudiced in my love for Kashmir & fredom. But if I were you I should rather aim at being able eventually to retire in Kashmir on say Rupees 600 per mensem, than at getting a post in London, however interesting, with the prospect of enforced retirement when one is 60 or so—and no pension worth thinking of. Excuse this frank statement of quite individual feelings. For me Kashmir will always seem the best base." Stein continued to press his argument in letters judiciously spaced. 29 June 1912: "I too am anxious—quite as eager as you are, dear friend—to maintain our collaboration to which I owe *so* much. When I thought of the chance that might open for you at Srinagar, I was selfish enough to remember that I am fully resolved to make Kashmir my working place for the future as far as it be in my power. You know how attached I am to the land and to my Marg in particular."

The collaboration which Stein valued was given greater urgency: "the arrangements with the British Museum will not be repeated. . . . There is a strong bias now at Simla in favour of Oriental research being encouraged in the interests of Indian politics (what a change from the attitude of indifference & worse we knew so well at Lahore!) and there is no desire whatever to take so rigidly an English institution into partnership. After Nov. 1915 I shall have personal independence practically within my reach. . . . The arrangement which would attract me most for later times implies a judicious blending: an Indian period of a year and a half with two Kashmir summers and autumn alternating with a Europe visit comprising two winters on classical soil. But all this must be rather vague at present since it does not take count of exploration chances which would make me quickly forget all such schematic ordering."

Which of the arguments finally moved Andrews to consider returning to India? Or was it his nature to move deliberately, cautiously, slowly? The Baron's initial No had over the months become a Maybe, trembling on the brink of a Yes. Stein cautioned him in a letter of 7 September 1912: "The only thought ever-present to my mind is whether you realize fully the peculiar conditions of service in a native state such as Kashmir. It presupposes constant exertion of patience and good temper, qualities you eminently possess. Mittra is easy going though vain and will probably give all support as long as his notion of his own profound knowledge is flattered. The Resident will soon realize the value of your work & do what he can to keep the field clear for it. But you must be prepared for many chances being lost through local inertia. On the other hand, you will scarcely be overworked: it is not an Indian fault."

Stein was on Mohánd Marg, from where his stream of letters went out. He had good reason for his euphoria: it was the summer when he felt fairly certain of getting to Afghanistan and was methodically having his gear made ready. After his long absence, he was given an especially warm welcome" by people more or less attached to my old domain. There came Rustam, the quaint old shepherd, now more withered than ever and no longer the proud leader of big flocks from which he used to spare, on occasion, a sheep or two for a Sahib officially reported as a bear! I was able to send the poor old fellow and his sick daughter for treatment at Neve's hospital. Then Hatim, the reciter & poet, whose songs & stories I recorded 15 years ago, turned up in due course. It was delightful to see him alive & hail; for Grierson who is to edit these texts, had many points to ask, and no better verification is possible than at the original fountain-head. . . . I shall not weary you with a list of all who came to renew their allegiance."[7]

As he wrote Publius, when 1912 was half over, "to my utter astonishment a heavy Dak bag brought me a letter from the Viceroy's hand announcing the K.C.I.E. [Knight Commander of the Indian Empire], with a bundle of congratulatory telegrams from Simla. I scarcely believed my eyes. How could I as a simple man of research foresee this more than generous recognition, the due otherwise only of men who have worked hard for the State. It seems in some ways an overwhelming attention almost."[8] And then, almost like his personal signature, he added: "All night my thoughts were turning to those dear ones who would have rejoiced so greatly in this acknowledgement of my efforts and who have all left me long ago." To Barnett, his good friend at the British Museum, he ascribed this official recognition "to the attention which Oriental research received in India since Sir Harcourt Butler in 1911 took charge of the Education Dept. in the Govt of India."[9]

It was indeed Stein's year of recognition. Twenty years earlier, when struggling with the *Rajatarangini,* he had pleaded, petitioned, and yearned for official consideration; ten years before, returning from his triumphant First Expedition, he had expected a slight nod of official approval. Now, almost inured to official disregard, he was in truth overwhelmed. The honor erased his sense of slight when the government had refused to give him the title of 'Archaeological Explorer'; it substituted another category, Sir Indian Archaeology. Knighthood was the clear sign that the Establishment had taken him unto itself.

His investiture took place on 12 October at Srinagar, where he stayed at the hotel run by M. Nedon, a Frenchman who had known him since he first visited Kashmir, as the green, untried, unknown registrar of Oriental College. Now his accommodations were the finest. Five days before the ceremony his black velvet court suit arrived; the Saint had used it at his investiture and thus endowed it with a "saintly blessing." The splendid occasion served Stein well. "The Maharaja's court was present which was a useful reminder to old native acquaintances of my existence. Lal Singh [now attached to the Kashmir Survey] had been duly invited & felt very proud of his Sahib."[10] Stein had a brief private interview with the viceroy. "I was able to thank him for all the kindness he had shown to me and [now that the Afghanistan trip had been indefinitely postponed by the amir] to tell him my plans for another Turkestan journey. He . . .

seemed fully to appreciate my arguments about the need for using the time while physical fitness lasts—and China remains open for archaeological exterprise."

A small but positive proof that he no longer need beg for favors was the permission readily granted him, by the commissioner for the Northwest Frontier Province, to apply the sanctioned leave already granted for his proposed explorations in Afghanistan for work on *Serindia*. Marshall acquiesced in the arrangement and asked only that he leave Kashmir for a fortnight to inspect the museum in Peshawar and the conservation work going on at Takht-i-Bahi. "As this will provide a change not unwelcome at the least pleasant season [January] and obviate any question being raised about a fresh deputation etc., I have readily agreed provided that I am officially instructed to concentrate on Serindia and protected against needless references official and D.O. [demiofficial]." The tetchy tone of his letter to Publius, 5 November 1912 speaks of the delays, the problems he was facing in writing *Serindia*. "I have reached the stage where the pains taken with *Desert Cathay* prove a most useful preparation." The earlier book sharpened his memory of events dulled by the passage of years.

But the essential difference between the two accounts of the Second Expedition remained: full as *Desert Cathay* was, it did not have the scholarly elucidation that *Serindia* was to present. Barnett, a tireless coadjutor, took the linguistic materials upon himself as the work in the "caves" went forward. To him Stein turned for help in dealing with the problems, pitfalls, and procrastinations that the specialists presented; this is what makes his letters to Barnett so informative. "I can scarcely tell you what a comfort it is to me to know dear Andrews and my collection are under your personal care as it were. . . . Your philological guidance is a great boon for them [Andrews and Miss Lorimer]. I hope Hoernle will do all that remains for the cataloguing of the C.A. ("South Turkestan") MSS. But when can we hope to see them analysed & published in the thorough way S[ylvain] Lévi is pursuing for Tokhari? As to Pelliot's catalogue I fear great delay. Luckily, this does not affect my Report directly."[11]

And again. "I cannot tell you how grateful I feel for all the help you are giving my collection. Miss Lorimer's full weekly reports show me what a support your watchful interest is to my hard-worked assistants. Slips are reaching me now steadily from Francke, Hoernle, [Louis de La Vallée] Poussin. Even from Ross half a dozen arrived by last mail. I have urged him to put essential facts about the Uighur MSS into a preliminary report for the J.R.A.S. [*Journal of the Royal Asiatic Society*] and should be very grateful if you could kindly help to secure this from him. You will understand my misgivings when he speaks of that Abhidharma text[12] & commentary offering work for a lifetime! This, of course, *entre nous*."

Some of the scholars had the temperament of prima donnas. "I am anxious to get some progress achieved about the Sogdian pieces. You are aware that photos of them have been with F. W. K. Müller [director of the Royal Ethnographic Museum, Berlin] for over two years without any tangible results in the shape of preliminary notices or inventory notes," Stein wrote Barnett on 29 December 1912. "In the meantime Gauthiot [Ecole des Hautes Etudes in Paris] who has been so helpful to me about the early Sogdian documents, has made constant progress with the Pelliot materials, and I should be very glad if he could take over the Tun-huang Sogdian rolls, too. He is on very good terms with Müller and would probably get him readily to agree to this devolution. M. has plenty to occupy him in Berlin. . . . Not knowing how things stand with Müller, I cannot approach Gauthiot direct without a great loss of time. Could you sound him out on my behalf? . . .

"The work presupposes sound Middle Iranian knowledge, and it is on this account that I do *not* care to fall in with a hint received through Miss L. and entrust [E. Dennison] Ross, ingenious & many-sided as he is, with this material, too. His strength is really elsewhere. Some 4–5 weeks ago I heard from Miss L. that Ross believed he found a date about 1350 A.D. in one of the Uighur texts and that he proposed to write to me *by the next mail* in detail. I have heard nothing from him. . . . The terminal date which Pelliot & myself on the strength of abundant dated Chinese documents, etc., assumed to be the early 11th century would have to be shifted down by three centuries. . . . If you could induce Ross to put me in possession of his observations instead of communicating news in scraps through Miss L. or via Berlin, I should feel very grateful indeed."

Ross's personal style as well as his carefree scholarship offended Stein's standard of scholarly behavior, a reaction Barnett seems to have shared. "I had a letter from Ross written while he was at Paris, full of interesting glimpses of all lectures, confabulations, etc., he was enjoying there, but without any of the hoped-for information about the dated Uighur MSS book. He does not seem to realize the necessity of clearing up the time discrepancy," Stein wrote on 2 February, confiding his annoyance to Barnett. "I fear my last letter must have reached you after R had flitted away again to the continent. His inventory slips are meagre and do not allow me to form an idea as to whether he thinks all Uighur MSS of approx. the same period. Could you get [Lionel] Giles, who is a sober & reliable worker, to look carefully at the paper used in the different Uighur MSS with a view to noticing its relation to that of the dated Chinese MSS."

Barnett, once-removed from the sources of Stein's irritations but inspired by his extraordinary finds, was a sympathetic helper. Stein thanked him for "the admirably worded 'diplomatic instrument' with which [Barnett] endeavoured to extract some materials from Müller. I fear, we must be prepared for delay with this most learned of men. . . . In case Gauthiot accepts the publication, it would be easy & safe to send a polite intimation to Müller in the sense you propose. M. will anyway have had the advantage of studying these texts for use with his own from Berlin."[13] These letters, the events they deal with and the men involved, give an idea of the effort, the vexations as well as quiet satisfactions, which were expended, ultimately to end as polite, generous acknowledgements when *Serindia* finally was published in 1921. Under the decorous, affable, scholarly patina were the scratches, mismatchings, and knotholes Stein confronted when putting together his Detailed Report.

Müller, Ross, Pelliot: the main targets in this exchange of which only Stein's letters remain. Ross is skittish: his slips "would look strikingly inadequate by the side of Hoernle's or Poussin's inventory. I suppose by the time this arrives he will have flitted away."[14] For a while Ross redeems himself: "It is a comfort to think that Ross has made himself so useful by bringing Müller's and Pelliot's notes. [Pelliot, overwhelmed by personal problems, had to bow out of the major task he had assumed for the Chinese texts.] I have not heard from [Ross] in reply to my last letter. I wish some fruits of his peripatetic researches may soon see the light of print."

Ten months after Stein had begun his campaign, Andrews agreed to come to Kashmir: "There is Andrews' Srinagar appointment to cheer me," he told Barnett. His coming "will be a great relief to me. I felt keenly how he suffered in that Battersea mill. Things move here terribly slowly, but at last the Resident seems to have got the 'Durbar' to face the necessity of providing the man who can run the machine already bought, as it were. A. would be able to do excellent work here."[15]

As spring (1913) came to Kashmir, Stein's worries were eased. "I am very pleased to know that Müller's paper is to appear in the J.R.A.S. This is an achievement which we evidently owe to Ross. So his meteoric flittings were of use here. For the rest of the Sogdian pieces we may be quite content with the slips you hoped to make up from M.'s materials.

It is encouraging, too, that Poussin has sent his transliterations of the Kharoshthi texts and that your systematic care will be able to evolve some useful paper out of them. What a pity, though, that our amiable and excellent friend does not seem to realize himself the advantages of orderly presentation! A letter Ross sent me enclosed a proof of his rejoinder to Legge [James Legge, translator of *The Travels of Fa-Hien*, 1886]. I am glad for both of these, but I cannot help wishing with you that 'he had not opened his mouth in the first instance.' . . . I wonder whether Ross will ever produce any of the labours on our Turki MSS, and whether if he does, their value will prove proportionate to the pains taken. *Qui vivra verra."*

By summer Stein felt the acute phase was behind him. He wrote Publius on 7 July: "By 1916 my learned silkworms will have all produced their finished threads which is by now no means the case—and possibly I shall feel then fitter for writing and better able to use a broad brush. But the main assurance lies in the fact that with my collection completely catalogued and a general account provided in *Desert Cathay,* which is exact enough for ordinary purposes, the results of my work of 1906–8 do not run the risk of being really lost." His lightness of tone, his quiet resignation stem from Stein's impending Third Expedition which was soon to begin. Thus he was far away in Turkestan when he wrote to Publius a year later, 27 June 1914: "Dr. Barnett, Keeper of Oriental MSS. & Books, has now been ousted from his official position as regards the collection & cannot fully support us. He was taken entirely by surprise at R[oss]'s appointment . . . An Orientalist's appointment without the only Head capable of judging qualifications, etc., being asked or informed. This makes me cautious of putting myself again into the meshes of that time-honoured godown [warehouse] of learned things & men." Ross's elevation over Barnett was, as Stein saw it, an ill omen; he was grateful that Marshall was planning a noble museum in New Delhi.

The timetable of Stein's Third Expedition points up his new position. No ifs or buts were laid down as preconditions, no haggling over costs, no negotiations for the British Museum's participation. His plan, long in his mind (and there is no reason to doubt his word) was automatically reactivated when the amir's refusal was received early in September 1912. A month later Stein called on Sir Henry McMahon, "who has to combine a Kashmir shooting holiday with the disposal of piles of papers all bearing the red label of 'Urgent.' I . . . found him as friendly as ever & genuinely interested in my new plans."[16] The plan, as he outlined it to

Publius was "to let my explorations in Chinese Turkestan be preceded by work in regions which have . . . the nearest approach to Balkh outside the Afghan borders. . . . To put it briefly, I wish to use the winter of 1913–14 for Seistan [eastern Iran], that miniature edition of the Tarim Basin, a region full of ruins & yet never surveyed by a professional archaeologist, then turn for spring and summer to the Russian valleys north of the Oxus . . . where I may hope to track Ptolemy's route of the Seris Silk caravans. Thus I could put in a most profitable nine months' work and yet always within postal reach—before organizing my real desert campaign for the winter of 1914–15."

Stein amplified the reasons behind his plan, sketching in for Publius the ancient history that validated his itinerary. "Seistan has, of course, a special nexus with Indian history as the seat of the Sakas who ruled the lower Indus, and the McM[ahon] Boundary Mission [1903–5] was greatly impressed by the abundance of its ancient remains. . . . McM. offered to give me all the help of his local experience & specially asked that I should come to see him for this purpose before the start. As to the Russian Pamir valleys, Darwaz, Roshan, Karategin, the outlook is equally attractive." That alpine region was the only portion of the ancient Bactrian Kingdom accessible for study, and its inhabitants were of the oldest Iranian stock. There he hoped to be able to identify a stretch of the silk road that in Roman times had brought that precious Chinese commodity to Balkh, a trade mentioned in Ptolemy's *Geography*. It was the kind of search that commanded Stein's enthusiasm.

Stein rearranged this plan as he wrote to Publius, 12 November, on receiving a letter from Macartney: "If the Afghanistan scheme has fallen through, and you are seriously thinking of coming here again, the sooner you start the better; for Lecoq is also contemplating another journey and would probably have been in Turkestan by now, had the German F.O. not been put off applying for his Chinese passport by the alarmist rumors circulated in the Russian press on the disturbances in Kashgar. As a matter of fact, never have the Chinese—those here at least—been more friendly towards the British than at the present moment and if you came, I am sure your old Amban friends would welcome you with open arms."

If Lecoq's intended journey was not news—Foucher had already mentioned it—it fired up his rivalry. "If I felt sure that L. would be aiming only at Turfan which the Germans know so well and where they have left so much behind after their former operations, I should not fear delay so much, for Turfan which still needs years of steady, tame digging within an oasis, is not quite the place in my line. But from Berlin . . .

they always go out hunting in packs, and though L. is not the man for the desert, he is likely enough to take out this time some young man with geographical ambitions who might well be guided by *Desert Cathay* to anticipate the few interesting tasks left for me. . . . So it seems after all as if it would be wiser to start straight for Kashgar instead of letting it be preceded by 6–9 months' work about Seistan and the Russian Pamirs. The latter would be in fact very convenient & attractive to descend to at the close of the Turkestan work, say in the summer of 1915, and even Seistan might very easily be tacked on then."

Stein's proposal was dated 23 November 1912. He asked permission and funding "for a journey of archaeological and geographical exploration in Central Asia and chiefly within Chinese Turkestan."[17] He itemized the sum of 44,500 rupees, reminded the government of the need for Chinese and Russian visas, and politely requested a quick reply—because "the Chinese Government [has not] as yet raised objections to foreign exploitation of ancient remains in the country. But it is impossible to foresee how long such favourable conditions will last." The bureaucrats hurried as best they could: on 17 May, C. Porter, secretary of the Department of Education, wired him: "Your proposals have been sanctioned. My congratulations."[18]

It should not go unnoticed that Stein himself raised the possible charge of "foreign exploitation" that might be leveled against him. He justified his archaeological work with the claim that Chinese Turkestan was a field "in which India may justly claim a predominant interest"[19] by virtue of the fact that "the spread of Buddhist religion and literature over Central Asia and into the Far East is the greatest achievement by which India has influenced the history of Asia in the past.

What he maintained was true, but it was his scholarly rationalization. Its motivation lay far deeper and was wholly personal. His digging at places like Dandan-Uiliq, Niya, Miran, and especially the documents retrieved from the refuse heaps of the Han limes (so near to China itself and yet unknown to the Chinese, so important in overland trade and yet ignored by the Official Annals) permitted Stein to feel that he had established his right to Chinese Turkestan. Having successfully mastered the conditions imposed by that vast and terrible terrain, Stein was deluded into thinking that he was indeed the master of the province itself. He saw himself as devoted solely to bringing to light its marvelously textured past, that culture variously composed of elements from the classical West, India, Persia, China, and Turki-speaking tribes. He was deluded into thinking that Chinese Turkestan was his personal preserve. This delusion

would be his hubris—his unhappy Fourth Expedition would make clear
the reality he escaped in his Third.

The Third Expedition lasted two years and eight months and covered
nearly 11,000 miles from "westernmost China across the whole Tarim
Basin [and back] to the uppermost Oxus and to Iran, from the Hindukush valleys in the south to Dzungaria and Inner Mongolia in the northeast."[20] His Detailed Report is the splendid and scholarly four-volume
Innermost Asia (two volumes of text with pictures, one of photographs,
and one of maps). Therein he set forth why he went seeking sites and
routes once alive with human activity and long since skeletonized by the
ravenous desiccation of the Taklamakan and the Sistan. His emphasis was
on geographical investigation; but the areas traversed were dictated by
archaeological interest—important corroboration of historical happenings.

For this expedition Stein wrote no "Personal Narrative" as he had before. Fortunately, the personal story is in his letters. Only a man who delighted in strenuous effort would have planned such a program, and only
a leader who had proved himself would have attracted the ablest and
toughest of those who had served under him before. Again he had Rai
Bahadur Lal Singh, " a veteran of indefatigable energy" and his servant,
the wiry Jasvant Singh; the "surveyor Mian Afraz-gul Khan, a Pathan
of the saintly Kaka-khel clan," who had worked with Stein on the Northwest Frontier in 1912; both Musa, his Yarkandi, and Tila Bai crossed the
mountains to start with him. Later, in Kashgar, with other trusted followers they would form the nucleus of his party. New this time, and
again on loan, was Naik Shams-uddin, of the sappers and miners, a
handy handyman.

Past mishaps had stimulated Stein to make a will while he was in
England. His lawyer was the grandson of Horace Hayman Wilson, the
first British professor of Sanskrit. It provided for the eight travel books
of his Second Expedition to be given to Publius in the event of his death
and, failing Publius, to Foster of the India Office. Thus they were to be
available for the completion of *Serindia*, for whose publication he felt
responsible.

A degree of danger might be expected in the route to Sarikol "via
Chilas, Darel, Yasin, the Darkot and Chillingi Passes. This would take me
over entirely new ground and along archaeologically interesting routes.
Darel visited by Chinese pilgrims is the main goal. Though recently
brought under political control in a way, it has never yet been visited by
a European. Of course, this is an added attraction."[21] Permission for this

came just ten days before his departure date. "Raja Pakhtun Wali [ruler of Darel and Tangir] had agreed to my visit on condition that no chief from the Agency Territory will accompany me. He is evidently anxious to prevent the appearance of official prying into his 'Raj'! I am quite glad to forego such local following and to be the chief's private guest, pure & simple. . . . He will be a most interesting specimen to study, the last successful usurper who built up a petty throne for himself with truly traditional methods. . . . His nephew is to escort me from Chilas."

The expedition began on 1 August 1913. The major part of the baggage, with two Indian assistants, started for Hunza; Stein, traveling as light as possible, took the road to Chilas—the two parties to rendezvous in Hunza. A large quantity of stores was left with Pandit Nilakanth, Pandit Govind Kaul's son, to be sent on to Kashgar by monthly parcels. For safety's sake he took a six-months' reserve. Stein, a virtuoso at long-distance, long-term exploration, was pleased with his arrangements. "I had the satisfaction to find that my estimate of baggage-weights, etc., made nearly six weeks ago proved surprisingly accurate. There was not one extra animal to be raised."[22]

The approach to Chilas was not easy. "The track became almost impossible even for Kashmir ponies, and two had a tumble over a precipitous slope. Luckily they recovered a footing in the thick jungle growth, and the loads though dropped down some 150 ft. escaped with minor injuries."[23] To forestall the needless worrying of his friends, he told them that "in order to get into training for Darel, I have walked all the way from Kashmir, except for the march of yesterday [10 August] which took me down to hot Chilas, only 3500 ft. above sea. . . . This morning I set out for [Pakhtun Wali's] dominion. A float down the Indus in flood, on a skin raft, was most exhilarating, 12–13 miles done in 50 minutes."

He wrote a series of letters to Publius, 24 August to 7 September 1913, from that region of the Great Himalaya Range where, to his surprise and delight, he found stands of grand deodar trees. "I have made my way through Darel and Tangir. . . . It was a *very* strenuous time but also most interesting and profitable. . . . There were five passes to be taken, between 10,000 and 16,000 ft., and several as stiff as I ever faced. . . . We reached the Raja's dominion where his nephew, Shah Alim, with an escort of delightfully rough & ready men-at-arms, select quondam outlaws, etc., but most pleasant to deal with & watchful, took charge of us. Darel must have looked delightfully civilized to my Chinese forerunners

[Fa-Hsien and Hsüan-tsang] as they passed down here from the barren rocky gorges north. Of ruined forts & sites there was a great abundance; just as in Swat it was quite impossible to visit them all.

"Adjacent to the commanding stronghold of the Valley, Pakhtun Wali had built his new castle and there he received me most hospitably. The human setting of his court when our first meeting took place, carried me back centuries into the good old days when kingdoms could be carved out all about the Himalayas and the Hindukush. We got on excellently even though P. W. could speak neither Hindustani nor Persian & after a long & confidential confabulation next morning I found it hard to say good-bye to this strong & crafty chief. He had no small struggle in establishing his rule after having come as a helpless exile, and now he may find it hard to turn to internal development. But he took very kindly to what friendly hints I could offer. Another day saw me busy among ruins in the lower part of the Valley: the site of the great Maitreya Buddha shrine is still worshipped in Muhammadan guise. Then followed a night's climb across the Hershat Pass into Tangir Valley. We saw the great bend of the Indus southward which no European eyes had yet beheld. It was only after 14 hours over awful rock slopes that the march ended at P. W.'s old stronghold. The heat below was great. Next day [20 August] we marched up the fine big Tangir Valley to the lovely forest region at its head. Here my escort saw need for business-like proceedings and the flanks were watched. But there was no incident. [Four years later, in the winter of 1917, Tangiri tribesmen, nursing old hatreds, murdered Pakhtun Wali "with an axe while he was watching the construction of a mosque. With his life the chieftainship also came to a sudden end"[24] and the old feuding between Darelis and Tangiris revived.]

"At the Sheobat Pass, about 14,000 ft. my jolly ruffian bodyguard handed me over safe & sound to the Levis of Gupis who crowned the Pass and showed little fraternal feeling for their old raiding visitors." Marching up the Yasin Valley, Stein was on the tracks of the old Chinese general, Kao Hsien-chih; it was a welcome change between weeks of strenuous climbing. He then crossed the Darkot Pass, remembered from his Second Expedition. "Several of my old Wakhi friends joined me & pleasant reminiscences could be thus revived. . . . By Sept. 1 we reached the head of the Ashkuman gorge at Chillingi, & a well-timed concentration of coolies from the far-off inhabited parts of that tract enabled me to effect the crossing to Hunza the next day. The 17,000 ft. Chillingi Pass, the only possible route that way, proved a stiff task. . . . It took us 8 hours to reach the top and four more to descend the big glacier on the Hunza side. But my feet stood this long climb in deep snow extremely well. . . . I feel quite happy & grateful for the successful execution of my programme. It has been 15 passes in all, between 10,000 and 17,000 ft. done exactly within five weeks. . . . Progress towards Kashgar will now be rapid & easy."

On the Taghdumbash, the high Pamir Valley, Stein renewed old friendships. When transport arrangements from the Sarikol side failed, former acquaintances among "the local people on the Chinese Pamirs proved very helpful & timely summons brought yaks, at least enough for the baggage. Snow fell daily between Sarikol and Kashgar; the same exceptionally wet weather which had pursued us from the start, was reported by the Kirghiz. At Tashkurgan old friends turned up on all sides. The pleasantest surprise was to find my Russian hosts of 1906, Captain and Madame Babushkin, who gave me the kindest welcome & made Sept. 12th a day of real rest. It scarcely seemed possible that seven years had passed by—all about the place was unchanged."[25]

Three days later, the big column divided: "I said goodbye to brave cheery Lal Singh who was to start for Charchan, Charklik, Lopnor & Tun-huang [!] Our rendezvous is planned for about January 15th. May he succeed in his tasks & keep fit. It was hard to part with this splendid companion. Afrazgul & the Sapper took the heavy baggage by the main route to Kashgar. I myself, with the second surveyor, set out to make my way down the Karatash valley, never before seen by a European, still less surveyed. . . . The gorges had become just practicable a week or so before and the fresh snow did not stop us. Difficult enough the route proved, quite equal to Hunza standards. . . . We were most lucky in securing a willing guide among the Kirghiz and some truly wonderful camels. I shall never forget the rock staircases & ledges over which they managed to scramble. Places where the ponies had to swim we found Rafiks à la Hunza to clamber along. It was a truly great pleasure when last evening [19 September] we emerged into the wider portion of the valley, & today's march brought us down to the fertile plain. It was delightful to return to familiar scenes by a new route, the last left to explore." On 21 September he was at Chini-Bagh, "The newly built Consulate provides a grand guest house and the Macartneys, creators of it all, with a spacious & comfortable residence."

Stein saw for himself the changes the revolution had made, changes Macartney had described in a letter. "The old governor had been replaced by a certain Yang Tsuang-hsu, a young officer about 30 years old, trained in Japan and formerly in Ili [a northern district in Sinkiang]. He it was who in Feby. 1912 started the revolution in that province by the murder of the Ili Tartar General. Yang arrived with some 500 of his own troops—a mixture of Chinese, Tungans and Turkis—all wearing Cossack overcoats. Yang himself parades in a Russian General's uniform and wears a mass of decorations, the founders of which are best known to himself. But he is most friendly. . . . Our old friend Chiang can't quite make up his mind as to the respective merits of the old and the new. His indecision is reflected in his head-dress. His queue has certainly gone; but now and then when a reactionary wave sweeps over the Chinese in Kashgar with murder in the air, he wishes he still had his appendage. One day he puts on an English cap and another day a Chinese hat, according to how he is influenced by the political weather. Today the English cap is in favour with him."[26]

Soon Stein could report his own news and rumors, a potpourri of facts and impressions. "Macartney is now the greatest power in the land and able to help in many ways. . . . I shall have to reckon with Lecoq's competition about Lopnor, though M. for a long time had expected him to return to Kashgar & thence go home. He had got dysentery at Kucha where he was clearing rock temples of their frescoes & where he still is. . . . He seems bent on getting those fine Miran frescoes which I discovered in 1907 & which poor Ram Singh's blindness prevented from being added to our collection."[27] Everything "is quiet and the private murders & squabbles among the Chinese which followed the so-called revolution have ceased. The only change is that the Ambans now wear European felt hats. . . . Dear old Chiang: his hearing has greatly improved & he does not need the expensive electrophone I provided. But for a long journey he would not be fit. His substitute, Li Ssŭ-yeh, is a quiet little man."[28] Poor Li: Chiang was a hard, almost impossible, act to follow.

"My preparations for the desert campaign are getting on well. Hassan Akhun turned up with the eagerly expected 12 Keriya camels, splendid animals fit for hard work. To my great satisfaction there also came Kasim Akhun, my old Taklamakan guide of 1900 & 1908. With him the proposed crossing from Maralbashi to Mazar-tagh [on the Khotan River] will, I hope, be practicable. . . . My main intention is to reach Khotan through the desert."[29] This was the "pleasant task" he had formulated in 1908 when, at the considerable ruins near the oasis of Maralbashi, his topographical instinct had connected these two lost, isolated, rocky points. Stein's confidence in his ability to cross and crisscross the Taklamakan, he would learn, was misplaced.

With his devoted veteran assistants, his twelve splendid camels, and his need to pose ever-harder tests to prove his mastery of the Taklamakan, Stein began his winter's program. From Maralbashi he summarized the first leg of the journey in a letter to Publius, 19 October 1913: "I have made my way here along a route which has seen no traffic probably for a thousand years—along the foot of an absolutely barren range, the southernmost of the T'ien-shan, which has never been surveyed. . . . For three days there was no drinkable water—a good chance for testing the efficiency of our water-tanks, etc. The whole was excellent preparation for the big Taklamakan crossing before us and I am grateful for the lessons it taught me, that is, when four of our splendid Keriya camels while supposed to be grazing on such scrub as there was, bolted. Luckily, the remaining animals just sufficed to move our camp and for Kasim Akhun of Tawakkel to set out after the fugitives. After a day and a half's search over waterless ground he tracked them and brought them in safely & relieved my care."

Tanks, mussucks, compressed lucerne for the camels—Stein had prepared carefully for the trial ahead. Three marches took them across the Yarkand River through a grazing jungle where water was taken on for the long journey and six additional camels were recruited to lighten the heavy loads of water. Stein could be proud of the long camel train and the hopes it carried with it. Prepared for great difficulties, he faced the true sea of sand. "On the very first day we encountered closely packed sand dunes recalling a choppy sea. By the second, all trace of vegetation dead or living was left behind, and a succession of trying Dawans with not a patch of level sand between them faced us." Progress became painfully slow; by evening the supporting camels showed signs of exhaustion, and as they struggled across *dawan* after *dawan,* on the third day the Maralbashi camels broke down. The line by which they steered was a diagonal across the high ridges with sand-choked valleys between: they barely made nine miles a day. Stein calculated "that the average total of ascents on one day's march amounted to close on 4000 ft.—no small task for camels."[1] From the top of a high ridge, Stein looked over the endless waste of sand piled into mountainous waves; he realized that the heavily loaded camels could not get through without serious losses. "I was forced to renounce, however reluctantly, this 'short cut' to Khotan."[2]

It was hard to admit defeat. 25 November 1913, to Publius: "I hastened to gain the Yarkand R. by a route further east and felt doubly glad for the timely return when we were caught by a wholly unexpected Buran and the temperature dropped to 24 degrees. But by then we were moving through the sombre riverine jungle and were thus to some extent sheltered besides having unlimited fuel. What this unusually early winter would have meant to us among those awful sand Dawans, I do not much care to imagine. Chance favoured us in catching guides in succession, hunters or else shepherds at the rare points where we touched the actual river course; . . . fortune also favoured us by providing a short-cut to the Khotan R. through a change in the terminal delta which took place some three years ago. I was heartily glad when the long, weary seven marches to Mazar-tagh were finished. . . .

"A Beg sent ahead from Khotan with ponies had faithfully awaited us for a fortnight. My renewed visit to the ruined fort keeping watch on the lonely Mazar-tagh hill was rewarded by an interesting discovery: the remains of the Buddhist shrine I had vainly looked for in 1908 as well as some refuse heaps previously overlooked. These yielded additional spoil in Tibetan records. Four forced marches brought me to Khotan. [In the twenty days] since Nov. 1, we had covered 450 miles, not bad considering how much was over drift sand. . . . You can imagine my joy when on approaching Khotan I was met by a big cavalcade of Indo-Afghan traders with brave old Badruddin Khan at their head and heard that they had just welcomed Macartney outside the West gate. Quite soon I was shaking the hand of my friend and helper to whom I had said goodbye, for years it seemed, only seven weeks ago. . . . Here my band has been joined by Ibrahim Beg, the steadiest of all my old retainers, and Khabil, a very clever young Sarikoli, who sold his cattle at Tashkurgan after first following me to Kashgar in order to be free for my service. He promises to be a most useful follower and talks besides his own Sarikoli,

Turki & Persian, also a little Chinese and is rapidly picking up Urdu from my Indians." Best of all, Stein could camp with dear old Akhun Beg and enjoy his new airy pavilion.

To Publius, Camp Niya, 8 December 1913: "I have moved east as rapidly as my tasks would permit. I should have found it hard to pass by my old Niya site; so I received here news of some more ruins discovered among its sand cones with relief. Whatever the result of our quest might be, it gives me the chance again to spend a few days among the fascinations of my own little Pompeii. . . . The Keriya Amban gave me quite an enthusiastic welcome. It was curious to find that he knew *Ancient Khotan* from Kashgar, just as it was a surprise to see *Desert Cathay* among the books of an enterprising Armenian trader at Khotan."

Endere, 23 December 1913: "My visit to where the Niya R. ends proved interesting & fruitful. In the course of four days' hard work, I completed the clearing of certain ruins & secured a final lot of those quaint Kharoshthi documents on wood. Want of adequate labour in 1901 had obliged me to leave untouched the large ruins we had christened the 'Yamen.' I was able to make good the omission of clearing a few rooms deep in the sand. We finished by the light of a bonfire in the evening & as we set out for the four miles' tramp across dune to camp, it was an impressive leave-taking from this most fascinating of sites. Perhaps the most striking discovery was a large fruit garden and vineyard showing all the details of disposition in wonderful clearness. When shall I again be able to pick up curling vines which yielded grapes about 250 A.D.?" The magic quiet that Niya had cast on Stein's antiquarian soul was in sharp contrast to the pull of Miran, whose "fine Hellenistic frescoes I had to leave behind in 1907."[3] To get to Miran he had to start from Charklik.

He wondered how things were at Charklik, a curiosity sharpened to caution. "I left Charchan for Charklik on New Year's Eve in peace and comfort," he wrote Publius on 9 January, but "just as I started news reached me of a band of Chinese 'revolutionaries' having captured the ·magistrate of Charklik and set up a district chief of their own. . . . they have always been careful to show attention to Europeans so I had no special reason to feel uneasy. But I confess, the prospect of having to exchange polite visits with such people was not altogether inviting. It was a real relief when, on nearing the Charklik border, I met an old acquaintance & heard the news of the dramatic turn events had taken. Until then we had marched by the deserted old road & not met a soul for six days. . . .

"It is true, the meek Turki people who had left their chief to the tender mercies of the Karadaz [professional gamblers], were quite resigned to their new masters. But the late Amban, aware of his danger, had implored military help from Karashahr weeks before. Eleven days ago a small detachment had arrived near Charklik, too late to rescue the Amban who had been put to death a day or two earlier, but in time to restore order. The Begs & people left the usurpers in complete ignorance of the Tungan [Muslim Chinese] soldiers' approach—it was a quarrel between Chinese, not their own—and punishment thus came suddenly. The Chinese officer had brought only 20 men, but he set upon the revolutionaries while they were sound asleep. The Tungans did their work with natural zest & practically the whole band was killed. . . . The old order reigns again. . . . Things have been pretty bad at Keriya & Khotan and the larger body of troops, 300 strong, moving westwards, means a great deal in a region practically all desert—it makes it difficult for the local Begs to furnish all I need in the way of supplies & transport for the next 2–3 months." Charklik was quiet, but difficulties were in the making for Stein.

"I doubt whether any archaeological work has ever been tried on such forbidding ground," he wrote Publius early in March, and he was not boasting. "I have tracked the ancient Han route step by step, as it were, around those awful salt-encrusted shores of the dried up Lop-nor, ground compared with which the desiccated dead region of Loulan seemed almost enticing and homelike."[4] Two months had passed since he had left Charklik for Miran, two months of prodigious effort, two months of anxiety, but also two months of successful topographical and archaeological work. The weeks had been fruitful but also very trying.

A letter from Macartney had warned that the Urumchi authorities were objecting to Stein's surveying strategic sites. "Mac. had at once applied to the Peking Minister [British] by telegraph for a recommendation to the Central Chinese Govt. of my work being purely scientific in character."[5] This "difficulty" was disquieting. Fortunately, the work Stein had planned would take him far from official reach and give time for the matter to be put right at Peking. Stein was not the man to be scared away; but he was apprehensive; his anxiety found an outlet in his concern for Lal Singh who, it had been planned months before, was to rendezvous with him at Loulan. In this troubled state Stein had started for Miran and the sanctuary the desert offered.

He was determined to retrieve "those fine frescoes poor Ram Singh could not secure. Removing them was a very difficult job as the wall behind had to be quasi-tunnelled and the brickwork sawn away before the brittle plaster surface could be taken away. Thanks to previous experience all eleven panels were safely recovered, 'backed,' & packed. Six boxes of formidable size and weight. The problem of raising adequate food supplies for my large party was resolved by persistence & the good-natured attachments of the friends I made seven years ago during my stay at Abdal. Miran has once again become an inhabited place since almost all the Lipniks of Abdal, not so long ago nomadic fishermen, have settled down near the fields they have learned to cultivate. For a fortnight my men had warm if humble quarters; I myself had quite a spacious sort of barn for myself and my many boxes."

Deeper into the desert, even beyond such a Lopnik settlement, Stein's search took him. He had twenty-six camels—"what trouble the raising of this big convoy has cost only the gods know"—to carry supplies for "close on to 40 people to sites some 100 miles from the nearest water, to assure an ice supply for a minimum of four weeks and to extract the maximum of work from man and beast for every one of those precious days."[6] In this statement to Andrews, Stein gives a dimension to his talent, a mixture of caution and drive that distinguished him and made his desert forays both possible and productive.

His letter to Andrews from Altmish-Bulak tells of his advance to the lost Loulan site, north of the Lop-nor wilderness. "I visited two old forts which Tokhta Akhun, my old Lop follower, had discovered since 1907 in the desert west of our former route. The position of these ruins, a sort of western bridgehead for the Tun-huang Limes, proved of the same period as Loulan, i.e., 3rd to 4th cent. A.D., and subsequent search revealed at some distance a settlement which yielded records in Chinese, Kharoshthi, Early Sogdian and Brahmi as well as interesting remains of implements, fabrics, etc. The position of these ruins & of the ancient river courses traced near them is very important. . . .

"It is truly a land of the dead. You will appreciate my satisfaction at being able to follow tracks, however hard erosion may have made them, to places which no human foot has disturbed for 1500 years or more. The terribly eroded ground, so difficult for the camels, yielded an abundant harvest of small finds worked in stone & bronze. . . . But the exciting time followed when . . . on the top of a high erosion mesa, or terrace, and thus practically safe from erosion by the driving sands, we discovered an ancient burial ground with pits full of the most unexpected remains. Household implements of all kinds, worn out articles of dress & the like seem to have been deposited here along with human remains collected in confusion from earlier graves. The harvest in beautiful silks, brocades, embroidered pieces, carpet, rags & such would have delighted your eyes. I think they belong to the time of the Later Han dynasty [A.D. 23–220] and clearly agree with the two fragments I had found by the Tun-huang Limes. What a wealth of beautiful designs & colours is now revealed! I hope it means the opening of a new chapter in the history of textile art. There were complete bronze mirrors, too, with decorative borders & a host of other things.

"My satisfaction was equally great when, that night, I found myself in a ruined fort of remarkable preservation; its construction—fascine walls, etc.—proved to have been built exactly at the same period as the border wall west of Tun-huang (c. 100 B.C.). So the westernmost *point d'appui* of that ancient route pushed from China for the conquest of Turkestan has been found. It now remains to follow its line eastward through the unexplored desert to the present Tun-huang route.

"The exhaustion of men and beasts obliged me to turn northward to these salt springs of Altmish-bulak where the camels can get a drink after nearly three weeks & plenty of reeds & thorns to keep their teeth going. I was relieved by meeting brave Lal Singh who for 400 miles had made his way along the dried-up ancient river-bed & traced more cemeteries on his way. He brought the fine camels of Abdul Rahim, a hunter from Singer, who knows this dead hill region of Kuruk-tagh better than any man living. His presence will be a great help further on. . . . This tiny oasis of reeds and tamarisk scrub amidst utterly barren wastes of sand, clay & gravel is a perfect little paradise for camels. Though only about a mile in circumference there is good ice at the springs & fuel in plenty. How I enjoyed my first real wash after three weeks. . . . I am allowing man & beast a four days' rest here & then we shall start by different routes into the unexplored desert north of the extreme parts of the salt-covered Lop-nor depression."

A fortnight later, 7 March 1914, he told Publius the next stage of his exploration. "I arrived here [Kum-kuduk, Tun-huang Route] after a very successful crossing of the unknown desert north of the ancient Lop-nor. . . . Yesterday we struck the lonely desert track which connects Lop with Tun-huang; it seemed to me truly like a return from the land of the dead. Any hour may bring my heavy baggage with all its bustle & care to this solitary rendezvous place. With that splendid hunter Abdur Rahim, who has known the wastes of the Kuruk-tagh since childhood, I moved north to the foot of those truly 'Dry Mountains.' His five big camels were a great strengthening of my column, not so much in numbers as in stamina.

You may judge of their fibre from the fact that the baby camel to which one of them gave light at the foot of the main Loulan Stupa, has got through all these waterless & forbidding wastes almost on its own legs. . . .

"It was such a relief to feel that I was beyond the reach of outside interference and threatened obstruction; whatever might come thereafter, I knew that human agency could not stop the exploration of the unknown shores of ancient Lop-nor and my attempt to trace the route by which Chinese policy & trade had forced its way to the Western Regions beginning in 120 B.C. But the task of tracing that earliest route in this wilderness more barren than, perhaps, any on this globe, was difficult beyond any I had so far attempted. The heavy loads of ice & fuel taken for safety's sake, meant I could scarcely expect the camels to keep fit for more than 10–12 days. The approximate, calculated distance was likely to take up this time, even if no natural difficulties were to present themselves. And there remained the problem how to hit the line of the ancient route. Much, if not most, had to be left to good fortune. . . .

"After crossing a terrible maze of steep "yardangs," [wind eroded terraces] we gained the terminal point we had discovered before of the old Chinese route. I succeeded in ascertaining the initial bearing of the route by remains of a 'Limes' type watch-tower and an outlying indigenous camping-place on a towering terrace at the very edge of ancient vegetation. By the compass I followed up this bearing. . . . Luck favoured us in what seems a fairy-tale: successive finds of early Chinese coins, arms, ornaments & the like guided us with something like uncanny clearness. If I were superstitious, I might have thought myself led by the spirits of those brave, patient Chinese who had faced this awful route for four centuries with its hardships & perils.

"You can picture my feelings when I found the old track clearly marked for over 30 yards by some 200 copper coins which must have got loose from their string on the load of a camel or cart and been left strewn on the ground during a night march. They were all fresh from the mint. A similar find of bronze arrow-heads a little further on, all unused as issued from an arsenal, proved that they had all been dropped by a convoy coming from China. Other exciting experiences of this sort, separated by days of long dreary marches, gave assurance that we were somehow tracking the right line. Such incidents made me feel as though I were living through in reality some of the experiences which I remember having read as a little boy in one of Jules Verne's fascinating stories—The "Journey underneath the Earth," or whatever it was called.

"After 8 days, one of which was spent on the terribly crumpled-up salt-cakes of the ancient sea bottom, worse to walk on than glacier ice, we reached the last offshoot of the desert range which flanks the easternmost extension of the old lake. Here, sure of a guiding point to the Tun-huang track, we found the old route still clearly traceable on the ground not subject to erosion. On the ninth day we reached the first scanty vegetation by the lake shore. Yesterday, the tenth day, a long march brought us safely across a wide belt of salt encrustation with patches of actual salt bog, to the caravan track and the well, known as Kum-kudak. Here, to my relief, I found Lal Singh already arrived. I had sent him from Altmish-bulak by another route along the desert hills northward and his survey has very usefully complemented my own work. Scarcely more than 24 hours after my arrival, the baggage turned up. So our concentration over big distances in the face of various doubts & risks has proved effective.

"To the hard-tried camels I must give a day or two of rest—and we humans, too, feel it welcome. My feet have stood the last five weeks' tramp, close on 400 miles, excellently. . . . With the baggage train came a ponyload of Daks [brought by the faithful Turdi]. Macartney announces receipt of a wire from the Peking Legation to the effect that the Chinese Foreign Office has telegraphed instructions to the Provisional Govt. to treat me favourably and not to hamper my archaeological surveys. [Macartney also told Stein that by the end of January, Lecoq was at Chini-bagh on his way home. He had indeed planned to descend on Miran, but illness and fear of Chinese obstruction had stopped him. Lecoq, but two years older than Stein, "felt he was getting too old for work in Turkestan."]

"It is a great relief to know that this intervention was so prompt & effective—and that my work has not suffered a day's loss in the meantime." Relieved, successful—"I have managed to accomplish a long-cherished task"—refreshed, Stein pushed eastward "exploring unsurveyed ground along the foot of the Kuruk-tagh and around that strange terminal basin of the Su-lo-ho. . . . Exact leveling carried on from the eastern edge of the salt-encrusted Lop-nor bed over some sixty miles [five weeks of painstaking routine surveying] has proved beyond a doubt that the Su-lo-ho waters form part of the Tarim Basin. At Toghrak-bulak I entered once again that ancient *Limes* in the desert which I feel tempted to claim, as it were, my own command. During three delightful days I visited the ruined towers & stations en route which I had explored seven years ago."[7] Like homing pigeons the party reached Tun-huang on 24 March. "Our brave camels after all their desert trials had trouble extricating themselves from the bogs of the marshy delta of the Tun-huang

River. When at last the edges of cultivation were reached, I breathed with real relief. Eight weeks of struggle with nature lay behind me and all its objectives gained."

To an armchair explorer, Stein's piece of exploration could be matched by other heroic odysseys; to an archaeologist it might well seem perverse—so terrible were desert wastes traversed, so tense the gamble of the heavily laden camels' pace against the countdown of days. When Stein wrote in his 7 March letter, that the journey has been "in some ways more fascinating than any I have ever made," he was, as often to Publius, talking aloud to himself, clarifying his image of himself as an "archaeological explorer," and the emphasis is on "archaeological," not on "explorer." Sven Hedin was an explorer who discovered many sites on his expeditions in Central Asia; Lecoq, the archaeologist, worked within the oases and declined the gambit of the desert. Stein, though he gloried in his triumph over hostile terrain, went primarily to read aright the details in a significant chapter of the grand story of mankind's past. Could an area so altered over time still bear witness to what it had once meant in Eurasian overland contacts? His crossing of that moonscape brought him a deep understanding of the colossal organization required and the dreadful cost in human life when, via the Loulan route, the Han Empire, in 102 B.C., sent an army of 60,000 men to establish its hegemony over the Tarim's northern oases as far west as Fergana—from which a bare 10,000 men returned. It was his ability to set down question and answer that marks Stein as a forerunner of today's anthropological archaeologist.

Struggle and anxiety were over but pressing chores remained. The page proofs on *Serindia* had to be corrected and sent off—he apologized for not being able to work on them during his desert travel.

He visited the "Caves of the Thousand Buddhas" and was welcomed by "Wang Tao-shih as jovial & benign as ever. He had suffered in no way from the indulgence he showed in a certain transaction [Stein was still guarded] & only regrets now that fear prevented him from letting me have the whole hoard in 1907. After Pelliot's visit a year later the remaining collection was sent for from Peking; Wang or his shrine never seeing a penny of the compensation which was grabbed in the official channel."[8] There, more than ever, Stein missed the gaiety, the goodwill, the bubbling energy of Chiang; his successor, Li, was dour, graceless, disinterested—by contrast Wang seemed friendly and communicative. "He had been acute enough to keep back abundant souvenirs of the great hoard when its transfer to Peking was ordered. And from this cache I was able to acquire four more cases full of MSS. Of course, it needed a good deal of negotiation but in the end I succeeded even though Chiang's help was sadly missed."[9]

On his way to An-hsi, "the cross-roads of Asia," Stein methodically tracked and examined the Han limes. At An-hsi, he again stored his heavy load of finds under the reliable Ibrahim Beg's care and then continued tracing the *limes* "along the whole line north of the Su-lo-ho . . . where previously I had ascertained its existence by probing at a few points and by a more or less lucky conjecture. The finds of ancient Chinese records were more numerous than I had reason to hope for in the refuse of ruined stations."[10] By the beginning of May he was once again at the "Spring of Wine" oasis at Su-chou, "delighted to let the eyes rest again on its green trees and fields. . . . It was a great relief when a hearty welcome from its Taotai assured me about the prospects of further work and that the funds transmitted through the Legation are available at the Yamen—this saves me from a good deal of care. No objection has been raised to my moving north along the Etsin-gol (formed by the joined Su-chou and Kan-chou rivers), and so I hope to try this new field on Mongolian soil. This is the old home of the Indo-Scythians and Huns—of course, these in their nomadic existence are not likely to have left behind much of ruined structures. But it is an experiment, this extension north of the Kansu oases, and if it succeeds may mean an archaeological summer campaign instead of my usual mountain travel."[11]

Work along the Etsin-gol was an approach to the "ruined town of Karakhoto where Col. Kozloff, a very discerning Russian Traveller, came 6 years ago upon the plentiful ruins of the Hsi-hsia, or Tangut Kingdom. But, of course, much is left to chance." Above the lakes where the Etsin-gol terminates, he reached the Tangut Mongol grazing grounds. "These Southern Mongols' way of life helps one realize better than any book could the conditions prevailing when the Indo-Scythians and Huns respectively held this ground. I arranged with their chief, who paid me a long visit, for the labour & transport we may need during our stay. I could free our brave, hard-tried camels of their loads so that they may graze at ease & enjoy their much-needed holiday until late in August."[12]

The ruins of Khara-Khoto: Marco Polo's city of Etzina. Its walls "still rise in fair preservation amidst the solitude of a gravel desert, girdled by living tamarisk cones & two dried-up branches of the river. Each face of the square-built town wall measures about a quarter of a mile. An impressive sight, abandoned in the 14th century or so. Clearing refuse heaps with Mongol labourers unaccustomed to spade work, was not an easy task; but somehow it was done & yielded plenty of records in the old Tangut tongue, Chinese, Tibetan, Uighur. A well-preserved Persian leaf shows how far east the influence of Iran had spread also in the Muhammadan period. Col. Kosloff who first visited the site six years ago without recognizing its identity, had hit upon a small dome-shaped Stupa filled with fine stucco images & a big deposit of Tangut MSS & prints. Of the structure, said to have been almost intact, nothing was left but its base, and in the debris which covered this we found everywhere remains of smashed statutes, frescoes. . . . I recovered a box full of MSS and painted texts, mostly in the almost unknown Hsi-hsia script of the Tanguts. . . .

"As interesting to me were the abundant observations gathered about the progress which desiccation has made here since the 14th century. It was as if I had been enabled to see what the Loulan site was like about the 9th cent. A.D. . . . A thousand years hence work on the Etsingol will probably be almost as difficult as it was in the awful but fascinating desert of Lop. We could get water from about 10 miles and our tanks and Mussocks were constantly on the move, for our Mongols had to be kept ever soaked with tea, little wonder with temperatures up to 96° in the shade. . . .

"Our camels will get their summer holiday and I am to leave all spare baggage & men with them. They are to rejoin me by August 20th at Mao-mei, higher up the river, when we shall have returned there from our surveys in the Nan-shan. From Mao-mei we shall journey across the Pei-shan to Barkul by a hitherto unexplored route. This will mean a short-cut for us back to Turkestan."[18]

These were his plans—but again an accident changed them. From Kan-chou he started for the cool mountains through the beautiful "green gorges of the Nan-shan to the head of the easternmost affluent of the Kan-chou River. . . . At O-po, a small post about 11,500 ft. above the sea, the 16th of July was a day of rain and much mist. So I walked most of the way until late in the afternoon when a swollen stream forced me to mount my Badakhshi horse. It had been a perfectly quiet horse to ride till we got near the mountains where the presence of herds of brood mares had caused him some excitement. Once, before I had realized his new temperament, he had managed to get me out of the saddle. This time no mares were in sight and the heavy rain might well have dampened his ardour. Yet scarcely had I mounted when he began to prance about & rise on his hind legs. This time I stuck to my seat—only too well. While thus pirouetting about, he suddenly overbalanced himself & fell backwards on me.

"It was a nasty fall but luckily the ground was sodden & soft and I did not lose consciousness for a moment. Only the strain of the horse's weight on my left thigh was severe, and on rising, the poor brute managed to give me a kick, luckily on the fleshy part of the back of the left hip. This was a piece of luck in bad luck. Also fortunately, Lal Singh and Afrazgul were with me and put me on my legs. I felt, of course, badly bruised & quite unable to ride; but I soon ascertained that there was neither fracture nor dislocation. To get myself carried to the camp was a very painful business. It took hours to fetch men & my campchair & as I had to give up my attempt at hobbling along after half a mile, those hours in the rain passed wearily enough. There were Tangut herdsmen of Tibetan speech & race within easy reach; but nothing would induce these queer stolid people to approach or give shelter. At last help arrived with the badly needed furs, and after 10 P.M. I was glad to get into my little tent.

"Of the fortnight that followed, there is little to relate. The bruised parts were very painful and there was nothing to do but get them fomented as my handbook advised and to contain myself in patience and stay quiet on my back. . . . I managed to write a full account to the Surgeon of the Lanchou China Inland Mission; his reply assures me of complete recovery of the limb. Even before his reply came I had begun walking, first on crutches, then with sticks. After five days of this I could cover 2–3 miles a day & was starting slowly on my way down."[14] The letter to Publius, not written until a month after the accident, carried reassuring particulars. "Massage has proved most useful and luckily one of my Yarkandis who served as Macartney's Hospital Assistant for a long time, is quite an expert at this. I must mention also that Afrazgul & my young Sarikoli, Kabil, made most devoted nurses & looked after me splendidly." Stein had learned from his first accident to respect the body's timetable for healing.

Lal Singh was otherwise engaged. Stein had sent him to explore the whole drainage area of the Kan-chou River eastward to the Pacific watershed to link this up with his former survey. On his safe return, the party started for Kan-chou. As Stein was hobbling along he met "a young American called Gilbert, who returning from a trip to Urumchi, had heard of my mishap and promptly started to see if he could help. He brought no medical knowledge, but he was an entertaining person to meet and full of Chinese experience picked up during protracted wanderings. It was a pleasure to have his company for two days or so and to offer him hospitality in Indian fashion which after months of travel from one Chinese inn to another was obviously appreciated. He brought the first news of the dire catastrophe which has befallen Europe."

World War I had begun. A bare fortnight after it started, the news reached Stein in the distant Nan-shan Mountains: Stein must have felt how incredibly fast the world had shrunk even as he felt how far he was from his dear English friends and from Hetty in Vienna. Pages and pages are filled with his analyses of news recently received but already passed into history. Despite his origin, he shared the emotions and hopes of sensible, right-thinking sahibs: confidence in the British navy and belief in the eventual victory of the Empire. Emotionally, he was mobilized and a participant—the crisis permitted him to voice his worries about Hetty and her family and to seek to establish communication with them through some neutral country—and to take it as his duty to carry out his schedule of planned work.

The last day of August he arrived at Mao-mei, the oasis he had first visited when tracking the Han limes along the Etsin-gol. It had served as a bastion guarding Kansu against Mongol raids from the north. To Publius: "My camels are in very good form and will, I trust, face the long journey through the Pei-shan [Dry Mountains] bravely. About 25 marches are counted to Barkul. . . . I hope our two columns (Lal Singh is to follow another route) will get through all right to reunite at Barkul." The trip across the Pei-shan, a wilderness of rock and gravel, posed a problem: riding a horse was too hard on his deeply bruised muscles; reclining on the back of a camel was easier on the leg but generally more uncomfortable—he had to hold on constantly to prevent a spill. A pony litter was improvised out of two rough roof posts and rope netting, and in this makeshift hammock Stein traveled safely.

The four-hundred-mile area of the Pei-shan that he had to cover was "one huge waste of detritus and gravel from which there emerged at intervals low, picturesque ranges of completely decayed hills. The Pei-shan had been old & decayed long before the Himalayas had begun to rise, when even the very 'mature' T'ien-shan was still in its babyhood. Nowhere have I seen a landscape bearing the mark of extreme geological age. . . . We started on the 3rd [September] Lal Singh to the north, I with the bulk of the party along an unsurveyed desert stretch of the Limes. After five marches we were to meet at a well which, owing to the vicinity of occasionally-worked coal pits, was a relatively fixed point for our two Chinese guides. Even before we reached it, we had ample opportunity to realize our guides' wholly inadequate local sense: whenever the track got indistinct, or another route joined, our guides got bewildered and took wrong turns. Our knowledge of the proper direction and the fortunate encounter with a small Mongol camp—the only people we met—helped to bring us safely to the well of Ming-shui. . . . The ground covered, I must confess, did not make orientation easy. . . .

"From Ming-shui onwards we could sight, far away, the snowy peaks above Hami, a most reassuring beacon. Soon our hapless guide lost the route and his attempt to regain it, persisted in with the frantic obstinacy of a child, cost us a day of useless wandering—particularly trying on our hard-worked animals—and at the end of it he ran away from our caravan. Twenty-four hours without food or water brought him to his senses; it was a relief when he turned up again just as we were starting off on a line dictated by topography. . . . Yesterday morning [26 September] things looked a little gloomy after a camp where the only water available was that always carried for safety. I sent a man up a craggy spur to look out. Relief came with dramatic completeness: to the north a vast gravel basin spread out and far away, on the opposite side, fully 16 miles off, there just showed the dark line of trees. Emerging from our gorge, a wonderful sweeping panorama lay before me.

"To the West, the Karlik-tagh covered with fresh snow; still further, the snowy range above Barkul, our next goal; north and northeast, a seemingly endless succession of barren, low chains in red and brown with vast desert valleys between—my first sight of the Altai system and the Dzungarian plains. Never had our camels moved so fast & so straight as towards that little dark streak in which I located Bai [the northwesternmost hamlet of the Hami oasis]. Our poor bewildered guide persisted that it was all a mirage of the Gobi. Well, it proved Bai all right. We had a most hospitable reception from the folk of this little oasis, some 30–40 houses, and what pleased me especially was a little orchard near the Beg's house in which to pitch my tent.[15]

The district magistrate of Barkul, a little, wholly Chinese town, was a lively, scholarly younger edition of Pan Darin. "He has given me most useful information about sites at Guchen, my next goal, & is quite enthusiastic about my Tun-huang finds. If I could have company like his for a few months I might really learn a little Chinese! . . . All Ambans seem to have been instructed to let me have as much money as I may need. Luckily I still have ample stores of silver, now double in value as against the hopelessly depreciated provincial paper money." On 18 October he reached Guchen, a big trade center at the north foot of the T'ien-shan, whose grazing lands contrasted sharply with the barren region along the south foot of the mountains. The town was filled with Muslim Kazaks, refugees, driven south by the Mongols' risings. "They are fine burly figures, showing their wealth in good clothes & huge fur coats. Were it not for the scattered Chinese garrisons, it would not take them long to raid along the trade route to the south as their predecessors had done. For the present the newcomers behave themselves. Guchen, as in ancient times, is closely connected with Turfan on the other side of the mountains. After visiting an old site near here I shall move across the Karlik-tagh range to Turfan."[16]

Pei-t'ing had been visited by Hsüan-tsang. The site, mentioned in the Chinese Annals, was better known by its Turki name, Bash-balik (Five Towns). Devastated by the Tungan rebellion, it was too far gone to furnish Stein with anything more than its exact location. Within its shattered walls it was honeycombed with pits, showing that farmers of nearby villages had been "long accustomed to extract soil for manuring,"[17] a practice familiar to Stein from the Punjab, where old villages and towns were similarly mined. His archaeological conscience satisfied, Stein turned south to cross the mountains to Turfan.

"After the long rest afforded by my pony litter, my leg stood this exertion quite well. In five days I made my way across the T'ien-shan. The contrast between the north and south sides of the range was most striking—north, picturesque gorges with fir-clad slopes recalled Kashmir; south, only barren rock & detritus with practically no moisture. On the 12,000 ft. pass the snow lay hard frozen; in the Turfan depression it is still sunny autumn with the maize & cotton crops standing and full foliage on the trees." Crossing the mountains, going from winter to autumn, from moisture to aridity, Stein's party encountered "four respectable Turfan outlaws armed with Mausers of the latest pattern which the Chinese troops are selling quite openly. These worthies had repulsed the troops who attempted their capture and were now comfortably established in the mountains, receiving all they needed in food, etc., from wayfarers on the much-frequented route. . . .

"I have found comfortable quarters and have been able to plan my excavations for the autumn. My heavy baggage from Su-chou and An-hsi has just safely joined me [29 October] and in a day or two I expect the two Surveyors whom I dispatched by different routes. I propose to work near Kara-Khoja, the old Uighur capital. Conditions are quite tame, with labour & supplies in abundance. The ruins are extensive & by no means yet exhausted."[18] Winter was setting in, the season for archaeological activity. "The ruins close by have suffered terribly by locals digging for 'khats' & earth to be used as manure since Lecoq & Co. exploited the most promising positions. . . . Fortunately new ground has been broken since untouched ruins above ground have given out: there are many old tombs on the absolutely dry Sai, and several of these which the local people have dug into during the last few years have apparently yielded stucco sculptures & MSS of Uighur times. The lead thus given is useful as proving that religious objections could not be justly raised. I propose to leave the Kara-Khoja cemeteries for later. Up to the present [12 November], the prospecting in the old towns has yielded some good frescoes, MSS fragments of various sorts including Manichaean, and a very interesting hoard of metal objects, including bronze mirrors, various large vessels, etc."[19]

The Turfan depression had long interested Stein: geographically and historically it commanded his consideration. Below sea level (the lowest elevation in China) and bone-dry, Turfan was, nevertheless an especially fertile oasis, enjoying a nine-month, two-crop growing season. Once surface water from the mountains to the north had been available to irrigate its fields, but since the eighteenth century its agriculture had depended on *karezes,* underground canals skillfully tunneled out to tap the subterranean water. Some of the Turfan settlements, hugging the mountains, straddle the border between the nomadic pastoralists of Dzungaria and the settled agriculturalists of the depression. These two distinct worlds, the nomadic and the settled, connected by easy mountain passes, have had, since antiquity, economic, ethnic, and political ties—they are two distinct ways of life joined together in a symbiotic relationship.

The towns of Turfan were particularly rich archaeologically. They had been part of the Uighur empire, founded in A.D. 847, which, until 1031, when their eastern holdings were seized by the Hsi-hsia (Tanguts), included Kan-chou and Tun-huang, as well as Turfan. They retained their western domain. They rightly judged the Mongol power, and the Uighurs, realistic and astute, tendered submission: they not only did not lose their empire, but their literate and capable aristocrats, themselves once pastoralists (they preferred horsemeat to the common fare of mutton and fowl), manned the Mongol bureaucracy. By acknowledging Mongol overlordship, they gained access to the vast Mongol empire. Their oases preserved both Manichaean and Buddhist cults as well as literary and artistic elements from India, Persia, and China which, for over a thousand years, found their way into the Tarim Basin. The Uighurs' late conversion to Islam maintained this rich pre-Muslim amalgam to within four or five centuries of the present. This was the unique cultural banquet that had drawn the German archaeologists to Turfan.

Even if Turfan had become a ruined site, Stein would have visited it— its special appeal was the incident of Hsüan-tsang's stay there on his way to India. The king of Turfan, enchanted by the Chinese pilgrim's learning, refused to let him leave, reluctantly relenting when Hsüan-tsang threatened a hunger strike. Thus Stein's patron saint had peaceably conquered the royal will. The capitulation was complete: the king gave the monk letters of introduction to the rulers of the oases along the way, thereby providing the assistance that made the pilgrimage successful.

Of the Turfan sites he surveyed, Stein chose to begin his work at Murtuk, where he commenced to remove what remained of magnificent frescoes decorating a series of cave temples. "Grünwedel had removed only a portion. I heard some time ago from Foucher that many of the panels while stored at Berlin had been destroyed by rats or mice which attacked the backing of *thick glue* used to strengthen the plaster as well as the straw in the latter."[20] Stein was aghast when he saw what remained of those splendid temple paintings. "Lecoq's & Grünwedel's assistant, an old sailor employed as a 'Techniker' in their Museum, had ruthlessly hacked the frescoes to the right and left of the pieces they had chosen. One must assume that neither of these two learned men supervised these operations." Against the double threat of vandalism and carelessness, Stein decided to save what was left. Then confident that his sapper and draughtsmen were thoroughly expert and that the job would take some weeks, Stein used the time to make a quick trip to Urumchi.

"I wanted to consult the doctor of the Russian Consulate about my leg & to see dear old Pan-Darin again, who is now Provincial Treasurer. . . . It was delightful to meet Pan again & our talks were long. The Chinese dignitaries from the Governor of the Province downwards exhausted themselves in attentions, dinners, etc. No thought of last year's attempt at obstruction! My best Christmas cheer was the doctor's repeated assurance that I could reckon on my leg getting quite as strong as before. Urumchi had plenty of snow & what with the cold & the strange Kazak & Mongol figures about I felt half transported to the Siberian borderlands. Altogether an interesting experience."[21]

When Stein returned to Murtuk he was pleased with the "successful results. . . . Sixty cases full of fine fresco panels had been safely removed & packed, & the work laid out and started since my return [7 January

1915] will probably add another 30–40."[22] He cautioned Andrews against repeating what he had said about the Germans: "I have no desire to cry over spilt milk in public or add fuel to the recriminations which go on among those Berlin Museum champions. It is a comfort to know that I had spared enough time to save so much from inevitable destruction and return to India from where the main elements of this art came."

Stein next turned his attention to Astana. "With the rescue of the Murtuk frescoes well advanced, I began exploring the old cemeteries on the Sai [gravel plateau] above Astana, a village adjoining Kara-khoja. There had been a certain amount of exploitation here too; yet the number of rock-cut tombs is so great that we still had a large field for systematic search. The tombs are of all dates, mostly of the 7th cent. A.D., at the beginning of the T'ang domination of Turkestan, and this increases greatly the interest of observations and finds. These have been abundant owing to the absolute dryness of the place & the custom, Chinese in origin, of depositing cherished objects & miniature reproductions of all things the dead might need more or less in the Egyptian fashion. . . . Among the queer jumble of fabrics for wrapping up the bodies, silk prevailed; and from such we collected specimens of brocades & embroidered work. Many are clearly of Sasanian type [Persian]. . . . Printed silks too, with portraits of the dead after the fashion of certain mummy types, & a mass of stucco figures representing mourners, divine guardians, caparisoned horses, etc.; clever naturalistic work. Byzantine gold coins were used to appease the Charon of the local underworld & their discovery is important proof of the trend of the trade at that period."[23]

The tombs were not simple pits cut into the rocklike gravel; they were abodes for the dead, as traditional in plan and as carefully made as a house. The tombs "were approached by a trench, about three or four feet wide at the bottom, leading down . . . to a depth of from twelve to sixteen feet. At its end the trench gave access to a narrow rock-cut entrance, about three feet wide and only three to four feet high; from this the rock chamber was gained by dropping down to the floor, a foot or two lower. The entrance had originally been walled up with rough brick work, through which the first plunderers had broken a hole sufficiently large for a man to crawl through. The tomb chambers were either square or oblong, the largest measuring eleven feet square. . . . The height varied from five to six feet."[24]

"It was not exactly pleasant work crawling down through passages half filled with drift sand and supervising the search. But we had an excellent local factotum—and the bodies were quite mummified and discreet. I satisfied my conscience by having all approaches to the tombs duly filled in again." The scraps of silk, mainly used as face covers, were numerous and of great variety, chronologically intermediate between those recovered at Tun-huang and the far more ancient ones unearthed in the grave pits at Loulan. Chinese silks, the industrial product most prized by the West, used Persian and Byzantine designs for that market. Perhaps the most touching of the wide range of grave goods recovered were the remains of fine pastries that reflected the personal preferences of the deceased.

By the beginning of February 1915, Stein sent a convoy of "45 heavily laden camels under Ibrahim Beg's care for Kashgar, a two months' journey. There are 141 cases of antiques, besides those already sent to Chinibagh last winter. . . . I am particularly glad to get the convoy off now because an enquiry made from Urumchi a week ago . . . though polite in form, suggests no helpful intent in that quarter. Well, I could assure the local Amban in good faith that I had practically finished with his district & was soon to move westward."[25] It almost sounds as if Stein was content to turn to surveying projects.

His over-all plan had several parts, all coordinated to complete the surveying of the Turfan depression, east of the Tarim and north of Loulan. Afrazgul was to "track the ancient route discovered last year east of Loulan to the exact point where it crosses the bed of hard salt marking the prehistoric sea bottom, and then follow the latter's shoreline towards Abdal. Then crossing the desert west of Loulan he is to rejoin me early next month [March] on the Konche-darya. I longed to do it myself . . . but my 'archaeological conscience' demanded inspection of a site west of Turfan. [Yar-khoto, its name described its situation: a plateau islanded above *yars*.] I was then to move by a new route through the western Kuruk-tagh . . . and, after picking up Afrazgul, to move to Korla along a portion of the ancient Chinese route skirting the course of the Konche-darya. There still remains a month before the Buran season opens." Lal Singh was also to make a desert tour; the second surveyor was to stay behind to complete the large scale map of the Turfan district and then rendezvous at Korla, bringing "with him the hapless Li Ssü-yeh and other impedimenta. Just now, "he wrote Publius on 15 February, "it is opportune to disappear from the Chinese ken."

Stein himself headed south for Singer, the only spot in the Kuruk-tagh that was permanently inhabited. There he was to pick up his guide and visit two anciently occupied sites. The Kuruk-tagh rose in range after range: the first, a wave of fine detritus and sand; the second, higher, less sand, and the beginning of rock; then down a twisted gorge still without vegetation to a salt spring marked by the skeletons of sheep left to die. During the seven days' journey they found drinkable water at only one place, but snow, in sheltered shaded places, provided water for the camels. From the mile-high, third wave of coarse sand and gravel, Stein looked down on a broad basin of reddish clay and sand; at the heart of the Kuruk-tagh, an area four hundred miles from east to west and two hundred across, lay the tiny oasis of Singer: a single homestead whose irrigated fields produced the grain sold to traders following the direct route between Turfan and the Lop tract. The founding of Singer is a frontier saga. A hunter of wild camels had first settled there some seventy years before. He died young, but his son, also a greater hunter and equally enterprising, made Singer a permanent settlement; his four sons knew that region as their home. The three oldest hunted camels and took on special jobs—the ones who had guided Sven Hedin to Loulan accompanied Afrazgul and Lal Singh on their difficult journeys.

With the youngest, Stein followed the "western portion of the earliest Chinese route to the Tarim Basin. . . . I also explored two ancient burial sites which proved to belong to the autochthonous population of hunters & fishermen in Loulan about the first centuries of our era. . . . I also filled in the last blank in our survey of the ancient dried-up river [Konchedarya] which once brought life & traffic to this region."[26] Stein told Publius where he had gone and what he sought. "Afrazgul's route would have to cross mine and we did all we could to make him aware of our presence by cairns put up along the gravel-shore, by big bonfires lit every night, etc. When I returned to this little salt spring, a haunt of wild camels & did not find Afrazgul though nearly a week overdue, I felt rather uneasy. You can imagine my relief when about midday [15 March] I heard the faint sound of camel bells. There was Hassan Akhun, my camel factotum since 1900, proudly leading the seven big camels specially chosen for this expedition. They looked rather gaunt, my fine camels, but fit all the same, and soon Afrazgul turned up too in full health & spirits. . . .

"Afrazgul's natural gifts for this kind of work will make him a valuable recruit for the Survey of India. I shall do my utmost to secure for him a start such as he richly deserves." From Camp Gerilghan, Konchedarya, 25 March, Stein continued his long letter to Publius. "At Yingpen, where the bed of the 'Dry River' contains a small, spring-fed marsh, I passed three days exploring a small fortified post . . . badly dug about by treasure seekers from Korla; but the refuse heaps were untouched.

Bits of Kharoshthi records on wood & other evidence conclusively proved that the station belonged to the period when the earliest Chinese route crossed the Lop desert from Tun-huang. A series of graves were tenanted by Chinese officials or soldiers who had died at this post."

Stein turned to the northwest. "The desert route still occasionally used now I knew to be marked by a succession of ruined watch towers. I was scarcely prepared for such striking evidence of their belonging to the period when the earliest route from Tun-huang to the Tarim Basin was in use. Their construction was practically the same as 'my' Limes far away to the east of Lopnor; the same skillful Chinese engineers must have been at work here in raising these massive high towers which have stood the vicissitudes of some twenty centuries. Only at one point could we get water. We managed by limiting water rations and once by sending the camels & ponies off by night for a good drink some 10–12 miles away to the west. Refuse layers had survived at all towers & the fragments of Chinese records, the utensils, etc., have borne out the dating.

"I have felt on my own ground, as it were, here & am sorry that a long march tomorrow is to take us past the last towers of this line to the newly-founded agricultural colony on the Konche-darya."

As March ended, the parties were reunited. Stein headed westward for Kashgar and his strenuous, adventurous, roundabout way back to India.

There was nothing haphazard about Stein's route home: a look at the map makes it clear that it traced the outlines of Afghanistan. Like a devout Buddhist who circumambulates the cella, his thoughts and prayers fixed on the sealed sanctuary, so by pony, yak, camel, goatskin raft, and on foot Stein circled Afghanistan, heartland of ancient Bactria. Again and again he looked across the border and longed to read in its ruins its ancient past. But access was barred to him. He was a patient man, or perhaps, he was a man forced to be patient who must believe that all things come to him who waits. At Kashgar his return journey began; it came to an end when he reached Kashmir.

Before Kashgar, his stay around Kucha was extended by information about a considerable number of sites not mentioned when he paid a rapid visit in 1908. "I have been able to collect a mass of exact data on the present water supply of the oasis; our surveys will remove any doubts still remaining about the reality of the diminution of the irrigation resources within historical times. . . . The heat has increased rapidly and mosquitoes have a high time. Fortunately water of sorts is to be found near almost all of the ruins though irrigation has become impossible long ago. . . . As Kucha is watered by two snow-fed rivers rather wayward in their supply, the necessary measurements and enquiries cost us a good deal of trouble. . . . It was fascinating work tracking down those old canals which no water could reach now under any circumstances. . . . I have visited interesting cave shrines at Kizil, a link of importance between the 'Thousand Buddhas' of Tun-huang and the caves of Bamian, north of Kabul. When may the time come for seeing that Bactrian prototype?"[27]

Kashgar was a good five hundred miles off, and Colonel Percy Sykes, who was replacing Macartney away on home leave, had already set the date for his holiday; still Stein followed the ancient northern route he had not traveled before. His goal was a brief visit to the cave shrines locally known as Ming-oi (near the village of Jigdalik). The few pieces of paper and scraps of palm-leaf and birchbark manuscripts laboriously retrieved satisfied him that it was highly probable that here was the source of Hoernle's Sanskrit and Kuchean documents: they tended to corroborate the account that had been given to Macartney when the documents were offered to him for sale. The cave had special significance for Stein: he felt that the finds from there had started him on his Central Asian expeditions. Continuing toward Chini-Bagh, Stein found that the route of T'ang times led through Maralbashi and along the now-dry Kashgar River.

At Kashgar (reached on 31 May), his homeward journey took the shape he had hoped for. He received his Russian visa and special permission to cross the Western Pamirs and make his way through the big valleys of Wakhan, Shighnan, Roshan, Darwaz, and Karategin. A jubilant letter from Marshall told him that the government of India had decided that his collections of 1906–8, as well as his new ones, were to be installed in the museum to be built at Delhi. His best news was a letter from Hetty received "after nine months' worrying uncertainty; sent via New York and Shanghai. My joy was tempered by the news that she had been very seriously ill at Christmas with Influenza. I am glad to say that my nephew has been spared service so far. But my 'Dr.' [a Ph.D. in philology] niece had worked in a hospital since the autumn which shows her fitness in a new line."[28]

It took a month to do the necessary repacking. His finds, twice as many as from his Second Expedition, filled 182 new cases, and most were tinned to safeguard against moisture. He also included eight mule trunks containing developed negatives, books, equipment, and personal items for the caravan. It was to start from Yarkand on 1 August under the care of Lal Singh, to be delivered to the Baron in Kashmir.

With light heart and light step Stein started for the cool mountains to spend a week clearing his "desk" of reports and official letters. "I pitched camp at the head of a green side valley; snowy crests look down upon me; a young fir forest, a huge collection of Christmas trees, closes in around my tent. Below, a wide grassy terrace is occupied by 'my' Kirghiz."[29] Afrazgul was to accompany Stein. "I said goodbye to my trusted old companion, Lal Singh—and to my brave camels who had enjoyed a well-earned month's holiday with excellent grazing at the foot of the range. They had been brought up here where I could see them on my way, and it was a comfort to see how sleek & well-rested they looked after the hard work they had gone through. I hope they will requite our care by fetching a good price when Lal Singh sells them at Yarkand."[30]

As frisky as a pack horse freed of his load, Stein started across the Ulugh-art Pass, 16,200-feet high, to reach the Sykes's camp. "They gave me a most cheering welcome and our common day of halt passed only too quickly. Both Col. Sykes and his sister had enjoyed their Pamir trip hugely. He had met Col. Jagello, the Russian Commandant of the Pamirs, & prepared him for my visit. So things could not be better. . . . I parted from my kind hosts on the morning of the 22nd [July] refreshed by the contact with England on the edge of the Pamirs."

Though the war was raging and Stein's letters reflect the convulsions of that vast and terrible bloodletting, it is a reflection distantly seen, always expressed, but pale by comparison with the remote, serene world through which he traveled. From the first luck was with him. The customs officer recommended by the Russian consul general made his entry a mere formality; quite by accident he halted for the night where Colonel Jagello also stopped. "He was rushing down to the railway & Tashkent on a few weeks' leave. I could thus assure myself that the necessary instructions to the Kirghiz, etc. had been issued which are to help me across the Western Pamirs & down Wakhan." Stein followed the big Alai Valley, the route indicated in Ptolemy's *Geography* based on a trader's account of the early silk trade with China, the route leading "from Baktra to the land of Seres [China]. Local information has helped confirm this important identification first made by Yule: Here at Daraut-Kurghan was where the 'Stone Tower' reported by the agents of Mäes, the Macedonian trader, is likely to have stood. It is good that I have to practise what little Russian I picked up by my Kashgar lessons. Of course, Turki & Persian amply suffice to help me with the local people."

The journey so auspiciously begun took Stein through a region still rarely visited by people outside the Soviet Union. He described it to Publius. Camp Bashgumbaz, Alichur, Pamirs, 23 August 1915: "A day's halt at this Kirghiz encampment provides me with the comfort of a warm, roomy felt-tent gay with rugs & carpets. . . . I have had a most successful journey along the whole western rim of the Pamirs, passing in succession the high ranges between which all the great affluents of the Oxus make their way westwards. In the course of seven days . . . we crossed three passes up to 16,000 ft. Then I descended to the little settlement of Iranian hillfolk (which you may find marked as Tash-kurghan on the Murghab, one of the main feeders of the Oxus), the easternmost inhabited point of the difficult alpine gorges which make up Roshan. I could study the ethnography of its people least disturbed by foreign elements, familiarizing myself with these homely peaceful Tajiks & their humble way of life.

"Progress beyond was rendered unexpectedly slow by a great physical change which no map as yet indicates. A big earthquake four years ago had brought down whole mountain-spurs & thrown huge barricades of rocks & detritus across the main valleys. Already in the Tanimaz valley we had to scramble by the roughest of tracks for hours over hillslides; the course of the river had been completely blocked by an enormous barrier. A big lake has formed extending 20 miles and still constantly spreading & rising. And even below this barrier the track along the steep mountain sides had been carried away by rock-shoots in most places. Though the direct distance from Tash-kurghan was scarcely more than 15 miles, it took us two and a half days to get through to the western end of the new lake.

"The view of the deep green waters winding between barren precipitous mountains was, indeed, glorious. But its enjoyment was somewhat impaired by the serious preoccupation as to how we should ever get the baggage around the cliffs which fringe the lovely large lake to reach the fiord-like side valley of Yerkh. For about a mile they seemed to bar all access. At the big barrier of rock debris some 1500 ft. above the present level of the lake, I met a small body of Russian topographers sent to make an exact survey of the lake which, when or if, it breaks into the almost dry valley below will threaten most of the Roshan villages with destruction. The Russians had been obliged to leave practically all their baggage behind at Yerkh & get around the worst bit of the lake 'shore' (wall, would be more appropriate) by using a tiny raft of goat-skins. Since I could not leave my baggage behind, the meek & obsequious Tajiks set to work to make the goat track along the rock projection & across sliding shale couloirs safe. It took them nearly a whole day, for the rock was unsound in places & there Rafiks had to be built in Hunza-style. By evening baggage & all were safely across & I could camp with a feeling of relief & modest triumph on the fine grassy plateau of Yerkh.

"My journey beyond, along the shores of the Yeshil-kol lake which a similar cataclysm appears to have created at some unknown epoch in the distant past, and over the wide steppe-like expanse of the Alichur Pamir was pleasant & easy. I shall always remember with delight the glorious solitude & purity of the scene: no troublesome intrusion of man & all his petty contrivances. Yet I was on an ancient route that had seen more than one historical Chinese exploit. Along it Kao hsien-chih had marched his troops before they turned into Gilgit & crossed the Darkot Pass in 749 A.D. Near the eastern end of the lake rises the ruin of a small shrine wherein the Chinese set up a stone stele recording their victorious pursuit here of the last Muhammadan rulers of Kashgar just a thousand years later [when the Khojas of Kashgar with their followers fled towards Badakhshan in 1759]. . . .

"I propose to move south to Lake Victoria & the Great Pamir where I hope to follow Wood [Captain J., who discovered Lake Victoria in 1838] & old Hsüan-tsang's footsteps down the Oxus through Wahkan where there are old fort-sites, etc. to examine."

In quick succession Stein sent two short notes to Publius; they carry the full flavor of his happiness. The first is dated 27 August 1915: "Camp, Victoria Lake, Great Pamir. I am using the last chance to send a direct Dak to India by my Yarkandi, Muhammadju, who is to return via Sarikol. It is delightful to have reached at last the 'great Dragon Lake' of my Chinese patron saint and to tread the track followed by Marco Polo

& Wood. Gloriously clear weather continues & one feels grateful for that at such altitudes." The second was written four days later: "Camp, Zang, Russian Wakhan, Sept. 1, 1915. From Victoria Lake three long marches, first along wide Pamir-like Basins & then in a picturesque valley both fringed by the big glacier-crowned Tange, brought me to Lashar-Kisht, the first Wakhan village & the site of a small Russian border post. The temptation of sending a fresh Dak across the Hindukush border, within full sight of here, to Chitral is too great to be resisted. A trustworthy Wakhi is to take it across the narrow strip of Afghan ground . . . and if the headman of the first Chitral hamlet does his duty & sends it to the Asst. Political Agent, Chitral, it ought to reach the Malakand within 7–8 days. How near the Frontier and my old haunts they seem! . . .

"Wakhan has proved a charming surprise, so green and balmy after those bleak wind-swept Pamirs. At Langar-Kisht I half felt in some high side valley of Kashmir, and today's march down the united course of the Oxus brought me past hamlets & meadows worthy of Kashmir—or Devonshire. The survey of two ruined hill forts of 'Kafir' [unbeliever: Zoroastrian or Buddhist], origin & curiously like so many 'Buddhist' fort villages I had examined in Swat & along the Yusufzai border, has kept me busy all day. Now I must hurry to send off my messenger who is to cross the river and the occupied strip of Afghan territory before the moon rises! A queer regime of *zulm* [tyranny] there makes this precaution advisable. The Wakhi people are most pleasant, so simple & courtly at the same time. When is the dream of my visit to Afghanistan to come true as has this of following the track of Hsüan-tsang, Marco & Wood?"

His passage through Wakhan was relaxed. To Publius, 15 September 1915: "I have had most enjoyable times passing pretty tree-ensconced hamlets scattered up & down amidst the big alluvial fans which descend to the Oxus from the high snowy ranges. Ruins of ancient strongholds at a number of places were examined. I could take my time & get thoroughly familiar with the valley as it stretches down to the Big Oxus bend at Ishkashim. Long days of surveying & clambering over precipitous rock slopes have convinced me that the ruins go back to pre-Muhammadan times, the largest probably dating from the Indo-Scythian period [first century B.C.]. Wakhan has always been a great thoroughfare between Bactria & the Chinese dominions, and with conditions being the same as in the days of Hsüan-tsang & Marco Polo, it was easy for me to make sure of all antiquarian details.

"On reaching the region of Ishkashim, which from early times has been a separate little hill-chieftainship, a buffer between Wakhan &

Badakhshan, I used my chance to record the local language. Its existence was recorded long ago but no specimens were obtainable. It proved an interesting Pamir dialect, full of archaic forms which carried me back to the Avesta studies of my youth. It is spoken only by some 150 families." Stein was always eager to collect vocabularies and texts for his distinguished friend Grierson, proud to contribute a few languages to the three hundred and eighty four contained in Grierson's survey. At Ishkashim, Stein was a welcome guest of Captain Tumanovich and his wife. "Their little house was cosy & bright and a fine specimen of the Russian Baby helped to make me feel as if I were in Europe again—or somewhere on the Frontier. . . .

"At Kharuk, the headquarters of Col. Jagello, Russian Commandant of the Upper Oxus & Pamirs, where the valleys of Shughnan debouch towards the Oxus, I was given the kindest welcome. To my surprise, it proved that Col. Jagello had published a kind of abstract of *Sand-buried Ruins* in a Russian geographical magazine. His help has been as effective as it could possibly have been & with the instructions he sent on to the Bukhara authorities in Darwaz, I hope my onward journey through the mountains northward will be equally easy." He marched "to the head of the Shakh-darra, one of the main valleys of Shughnan, gathering much ethnographic information & making myself familiar with what, from ancient times, has been an important route descending from the Pamirs towards Bactria.

"Then I crossed the Dozakh-darra, the 'Pass of Hell,' to the other main Shughnan valley, and found the crossing better than its name. . . . The descent along the excellent new road constructed by Russian engineers made a pleasant interval of ease. Then followed a bit of real mountaineering when we crossed the bold ice-covered range towards the Bartang River over a large & much crevassed glacier where step-cutting was needed. But the Shughnun men are all expert climbers & as we had plenty of them, the baggage & ourselves got across quite safely. The descent, that day and most of the next, led over moraines or steep slopes of rock debris piled up in wild confusion. What satisfaction it was to be able to face them with my leg for which a year ago even a long sitting was painful!

"The transport arrangements worked perfectly & fresh coolies awaited me at the snout of the glacier towards the Bartang Valley. For two days we had to clamber along a succession of precipitous rock faces, quite worthy of Hunza, over narrow ledges or else frail 'Rafiks.' It was quite a relief when at intervals stretches of water clear of cataracts & boulders allowed us to glide past these obstacles on a small raft of inflated goat-skins steered by excellent swimmers. The spurs on either side seemed to rise wall-like up to 8–10,000 ft. above the river, and the narrow gorge left between them appeared always to close up for good. It was like a glide down to a Hades of alpine grandeur & peace. In the midst of this forbidding rock wilderness we passed a few villages at the mouth of the side gorges. With their luxuriant groves of walnut & fruit trees they looked like a fit abode for the happy dead set behind impassable mountains.

"Yesterday [25 September] we emerged in the main Oxus Valley where the Bartang joins the Ab-i-Panja in the fertile little stretch of Roshan. I found a charming camping place in the old orchard adjoining the ruined fort of the former Mirs of Roshan." Stein spent a quiet day gathering folklore. At one time Chinese control had extended to Shughnan; all his informants confirmed this and rendered a short Persian inscription as fixing a boundary "by order of the Khaqan-i-Chin." He also learned that the population had been drastically diminished "owing to the Mirs' practice of selling women and children as slaves to increase their revenues."[1] He noticed how fair the Wakhis were: "Nowhere have I seen such an abundance of perfectly Grecian features coupled with light eyes & hair. If Roxane was really a princess of poor little Roshan, as philological conjecture has it, there would have been no reason to wonder at Alexander's choice."

"Soon after leaving Roshan, the clear weather we had enjoyed through Aug. & Sept., threatened to break up. Fortunately each time the clouds broke on the mountain ranges, I was moving down in the valleys where rain could not stop my progress; and the days when passes had to be crossed, a glorious bright sun shone out again. From the time I entered Darwaz by the big fertile valley of Wanj, I moved through Bukharan territory."[2] The *mihman* of the Fountain of Wealth, as the amir of Bukhara is styled, had issued directions as instructed, and everywhere the local dignitaries of all sorts and ranks presented themselves ceremoniously. "It seemed like the gorgeous display of garments in old Venetian paintings brought to life again. Silk robes in rainbow colours, in big bold stripes of bright yet harmonious tints, in brilliant check patterns of quite ancient design were to be seen daily on this old world hierarchy. So things may have looked in the India of the Moghul times. That the administrative methods were likely to be of a correspondingly old-fashioned type could not detract from my enjoyment. . . . During the last 10 days I had to cover over 360 miles and owing to difficult tracks, usually had to continue my rides late into the evening or night."

His clothes were worn and soiled, his baggage battered, but his spirit high as he came to Samarkand where he wrily dated his letter of 30 October "Camp, Grand Hotel." Nearby was the ancient capital of Sogdiana, Afrasiab, a city buried like Yotkan under piles of debris mounds. "I left Samarkand on the evening of the 25th [October] and had 2½ most pleasant days at Bukhara, the other famous center of ancient Sogdiana, seeing its old world splendour. I had a carriage of the Amir's & a local official to take me around. Here [Ashkhabad] I arrived for the passport formalities, etc. to be attended to before I can move to Persian ground. Carriages are secured and if I can be off tomorrow & all goes well on route, I may hope to reach Meshed by Nov. 3rd—the baggage is to catch up some days later."

His mountain-climbing holiday along the Afghan-Russian border was over. He headed for Persia.

The British consulate in Meshed was, in a manner of speaking, a frontline post: German agents were trying to slip through Persia to gain a foothold on the Persian Gulf. "There has been trouble in places about Central Persia, but now necessary arrangements have been made both from the Russian & Indian sides and the route I propose to follow is as safe as things can be under present conditions. I have lingered here for a week to get my expedition accounts made up. . . . & now arrears of 18 long months are off my hands, It relieves my official conscience."[3]

On 11 November he started for the Sistan. "I cannot hope to cover the 500 miles before Nov. 31st. I struck south through the mountains which tribes of Hazaras, honest Mongol folk, and Baluchi inhabit. They are keeping quiet now. The escorts which the chiefs provided were a treat for the eyes: picturesque-looking fellows provided with an abundance of modern weapons such as any Pathan might envy. Crossing the low-lying gravel wastes, my eyes were refreshed by the sight of a glorious ruin at Khirgird—a big Madrassah built by Shah Rukh [1377–1447, son of Timur, famous as a patron of the arts] and decorated with exquisite coloured tile work. Even at Bukhara I had seen nothing so perfect in design. Never has Muhammadan art so impressed me as here set against absolutely bare desert background. Further on we crossed a big stretch of gravel & sandy desert, the usual line of inroads of robber bands from across the Afghan border, without any exciting incident."[4] He reached the consulate for Sistan and Kain on schedule, 1 December.

His journey there induced a subtle mixture of nostalgia and frustration: he could look across to Afghan territory "where the rivers flow which I wrote about 30 years ago in my first published paper, 'Afghan-

istan in Avestic Georgraphy.'⁵ I still feel as keenly about this region as
when I was young." To reassure Publius of his safety on the Persian-
Afghan border, where German activity had been concentrated, he mentioned that "Sistan now boasts a British garrison consisting of half an
Indian cavalry regiment as well as Levies which make this an important
buttress. . . . My descent to the Sistan basin has been quite enjoyable.
It is, indeed, as I expected, a faithful *pendant* [counterpart] to the Lop
region. Though my archaeological instinct prevails, I am not altogether
sorry that dessication has not yet advanced so far here."

Almost at once he had an exciting find. His letters to Publius of 26
November to 2 December 1915 tell the story: "I left the Sistan 'capital,'
a place recalling what Loulan or the Miran site might have looked like
when traffic by the ancient Chinese route passed through it, on Dec. 6th.
Next day I reached the ruins on the rocky island of Koh-i-Khwaja which
rises as the conspicuous landmark of Sistan among the wide lagoon &
reed marshes of this *pendant* to Lopnor. The ruins were supposed to be
Muhammadan, but an instinctive reliance on the continuity of local worship, drew me to it as my first goal. The ruins were extensive & at first
sight a little puzzling. But a kindly chance helped me. The very first day
a little bit of frescoed surface was noticed behind a later vaulted passage.
Next day a fresco scene of unmistakeably Buddhist character came to
light & complete clearing has allowed me to recover panels of about 12
feet by 6. My joy at this rapid discovery is great. It is the *first* tangible
proof of the long-conjectured extension of Buddhist worship & art influence into true Iran and also the first find of pictorial art remains from
this ground dating back to Sasanian time.

"The fresco figures are particularly interesting as showing how late
classical art adapted to Indian subjects was modified in Iran whence
Turkestan Buddhist art has borrowed so much. You know how often I
had wished to recover such a link from the soil of Bactra. Well, Fate has
allowed me to find it. With this clue to hand, the true character of the
picturesque complex of ruined domes, vaulted halls, terraces, etc., built
on a steep spur of the rock island revealed itself rapidly. It became easy
to recognize the very numerous shrines built on ground-plans familiar
to me from Turkestan sites & the Thousand Buddhas, all in sun-dried
brick but remarkably massive. Then details of Sasanian architecture, so
closely akin to Byzantine, struck the eye when it knew what to look for,
until at last I could trace even remains of big stucco relievos on the walls
of the big temple quad. in spite of all the damage these had suffered
from the occasionally heavy downpours of at least 1200 years. In style

these almost-effaced relievos agree with the early well-known Sasanian sculptures of Naksh-i-Rustam in the Persis.

"Further search has shown what seems to be even older art-remains. In a long & rather ruinous corridor we found hidden behind a later brick vaulting, obviously necessitated by structural risks probably some time before the arrival of Islam, a painted dado of pure Hellenistic type. I wonder whether the decoration does not go back to Parthian times [ca. 200 B.C. to A.D. 224]. We are removing all fresco panels so far cleared. . . . It will be hard work to survey all the sites known within the short time available. There is little doubt that the war has helped me in Sistan, for so far no obstructions to my work by the Persian administration has been attempted."

He wrote Barnett that the frescoes had "suffered much from both rain and white ants. Later I was able to make reconnaissance of remains reported in the desert south. There I found abundant relics of stone & bronze age settlements at numerous wind-eroded sites and to my delighted surprise came upon a chain of ancient watch-stations undoubtedly marking a border & recalling very curiously the Han *Limes* west of Tunhuang by a variety of features. The exact age of this border line has yet to be determined. But various signs point to early origins, perhaps in Parthian times. It is a strange line, geographically anyhow, between the Han 'Great Wall' and the Roman Limes. It pleased me quite as much as that first discovery of Buddhist remains on Persian soil."[6]

His long journey was coming to an end on the same note on which it had begun: a telegram advised him that "under present conditions the Govt. of India did not think it advisable to approach the Amir about my proposed visit to ruins of the Afghan side. So my conscience was satisfied and I was free to return to [the Sistan consulate] to clear off numerous tasks. Sistan is now watching with interest the result of the attempt made by German parties to break through from Kirman to the Afghan border. One German officer has surrendered after getting separated from his party in the Lut desert & going without water for three days. The troops have moved up to intercept them—the southern route towards India is not likely to tempt them just *now*. I am leaving here [Camp Shahristan, Sistan] for the desert south where I hope to devote about 8–9 days to the further exploration of the Limes. Then I propose to push on to Nushki [the railhead]. There are 23 marches but I hope to cover them with little baggage on riding camels in about a fortnight, the rest of the baggage following."

"Camp, Shah-sandan, Sistan. Nushki route. . . . I managed with 20 labourers taken out in the desert for 'well-sinking' (to spare Persian susceptibilities) to clear portions of two debris-filled watch-stations on the ancient *Limes*, & to trace its line eastwards quite clearly into Afghan territory. There were finds in the soldiers' quarters recalling those on the Tun-huang border and at the Niya site. But with the annual rainfall of two inches which Sistan still receives at present, finds of written records could scarcely be expected. A complete excavation of all these border posts would take months. Yet the evidence already secured suffices for fixing their approx. date in the first few centuries A.D. . . . On Feb. 3rd I said goodbye to this little desert between the Mamun and the salt-lake of Zirreh & am ever pushing on as rapidly as my camels will allow me towards Nushki."[7] From there he could go by train to Quetta.

"In the utterly barren foothills that we skirted, in the long stretches of gravel 'Sai' or sand-covered ground that we crossed, in the desolate small posts we encountered—in all this there was much to suggest scenes that must have met the eyes of those ancient wayfarers while making their way through the Lop Desert past the foot of the 'Dry Mountains.' Even the great 'Salt Marsh' was recalled by glimpses of the salt-fringed sheet of the Gaud-i-Zirrah as I saw it glittering far away in the distance. At the same time the sight of the many hundreds of dead camels which the convoys of military stores moving from the Nushki railhead to Sistan for months past had left behind on their trail, poignantly brought home to me once more the vast amount of suffering which that far more difficult route from Tun-huang to Loulan must have witnessed during the centuries of its use by the caravans and military expeditions of Han times."[8]

Approaching closer and closer to British India and free of the exactions of running an expedition, Stein had time to consider a wholly unexpected problem. At the Sistan consulate he had heard the almost unbelievable news that Prince Louis of Battenberg, Queen Victoria's son-in-law, first sea lord and the royal navy's outstanding figure, had been forced, because of his German origin, to retire. If war hysteria could strike down a prince, what about Stein himself: he too was by birth an enemy alien? Suddenly, he felt stripped of his skin; he steeled himself against rebuffs, suspicion, hostility when he wrote to Publius 12–15 December: "I sometimes wonder what change in personal relations I must be prepared for on my hoped-for return to England in the spring. It is impossible for me to advertise my feelings about the country to sundry & all, & I do not expect to escape unpleasantness or even trying experiences." But Stein's sensitivity was needless: unlike the prince, he did not have to worry that anyone coveted his crossings of the Taklamakan. Reassurance was immediate. At Quetta he "had a hearty welcome from old friends."[9]

His homecoming gave his friends the opportunity to show their affection. "I was met at 6 A.M. (!) at the Delhi Station by Maclagen which is enough to show what the oldest of friends does for me. Things official have been made delightfully easy. The Viceroy gave me a good chance of talking over past experiences at lunch & showed much interest in the specimens of 'finds' I had brought and laid out in the Viceregal Lodge. There has been quite a gathering of friends at the extemporized little exhibition including 2 Miran frescoes. . . . New Delhi is a fascinating site to visit and its makers, the architects Lutyens & Baker [then laying out the capital and designing its noble official buildings] gave me a most warm reception." On 16 March, after a thirteen-hour tonga ride from Rawalpindi, Stein reached Srinagar and the "hospitable care of the Baron & Baroness [who had moved there after he had started on his expedition]. They both look very well and seem rejuvenated, as if the long years in grimy London which separates their Lahore and Kashmir lives had been obliterated."[10]

Stein's stay under the Baron's friendly roof was, in his lexicon, "a real holiday." Idle, he was not. While he and the Baron went over the collection that had arrived in Srinagar at about the time Stein himself had begun his Persian circuit, he was having typists transcribe his notebooks—that would insure that the originals would not be exposed to submarine risks. This was his sole concession to friends who urged him to postpone sailing for England, a caution neither his pride nor his habit could heed. He stated his position baldly to Publius: "I feel I may as well share the risk with others in their thousands, putting my reliance on the watchfulness of the Navy."[11] He could return with flags flying. The Allens, untouched by any prejudice, had invited him to stay at their home. "Your letters have brightened my return to India. How delighted, then, I was to read of the welcome you were both preparing for me in what has always seemed a true Home. God bless you for all your goodness."[12]

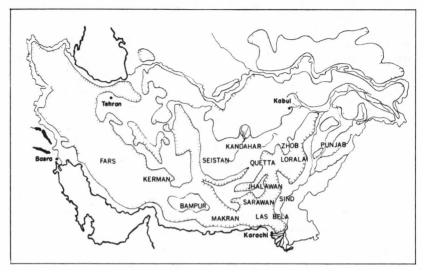

Regions of the Indo-Iranian Borderlands and parts of the Iranian plateau. From Walter A. Fairservis, Jr., *The Roots of Ancient India,* 2d ed. (Chicago: University of Chicago Press, 1975), map 10.

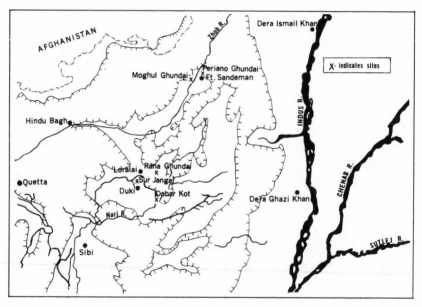

Location of certain sites in the districts of Zhob and Loralai, Baluchistan. From Fairservis, *ibid.,* map 13.

5

1916–26

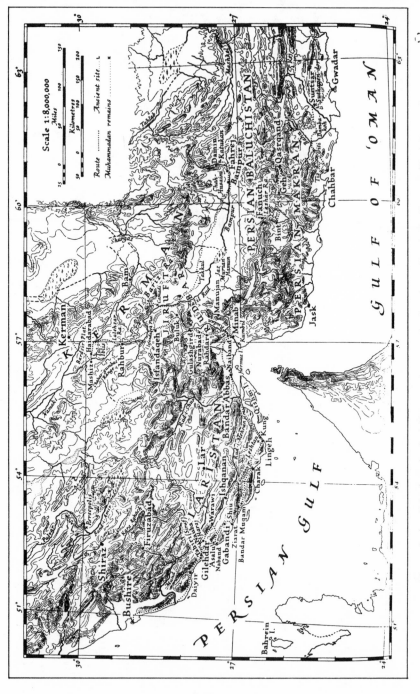

From Stein, "Archaeological Reconnaissances in Southern Persia," *Geographical Journal* 83, no. 2 (1934).

22

For the peripetetic Stein the decade after the Third Expedition might seem to have been a period of well-earned rest. Yet in those years he was enormously productive, preoccupied with writing and seeing through the press the Detailed Reports of both his Second and Third Expeditions: the multivolumed, monumentally splendid *Serindia* and *Innermost Asia*. As if this was not enough, he also conceived and carried through the beautiful work *The Thousand Buddhas: Ancient Buddhist Paintings from the Caves of the Thousand Buddhas on the Westernmost Borders of China. Recovered and described by Aurel Stein with an Introductory Essay of Laurence Binyon* (1921); his *Memoir on Maps of Chinese Turkestan and Kansu from the surveys made during Sir Aurel Stein's Explorations, 1900–1, 1906–8, 1913–15. With Appendices by Major K. Mason and Dr. J. de Graaff Hunter* (1923); and, for good measure, a handful of articles. Also in the public record were the additional honors awarded him.

Yet on the personal level there were events that occupied his heart and mind and cut into his working time. The "team" he counted on to order the vast, complex collections he had brought back began to fall apart: the "dear Baron," so dependable and so perceptive in handling artifacts from the past, appears to have had the maneuverability of a steamroller in his own affairs, and the Recording Angel, when finally she came to Srinagar after the war, did not transplant well. These episodes, protracted and trying, Stein could take philosophically, but not so the lasting sadness brought by the deaths and the hurt by betrayal (as he saw it) that reduced his circle of collaborators.

These assorted happenings—their beginnings, middle, and end—are recalled in his letters. They convey his changed response to situations outside his control. A tone of restraint replaces the frustration and impatience he had earlier expressed; now he accepted the mindless "constrictions" of babudom as well as the foibles and frailties of humans. Gradually, the tone carries over into his very real concern for the well-being of Ernst's family (the niece and nephew now grown and trained for distinguished work), struggling in an inflation-ridden Vienna, and for his own postponed (but for how long?) retirement from government service. From the summit of fame and esteem he had attained, he looked around to see where and how he might find congenial work. This was a nagging necessity. The bitterness he had once felt was concentrated on his not being able to visit Balkh: Afghanistan, long a buffer zone between Russia and British India, became a nation in 1921 and was forbidden to him.

Nineteen sixteen: his home leave began auspiciously. Enroute to London, he renewed old ties at Paris and sought to enlist new collaborators for the latest collections. As soon as he reached Oxford for his long stay with the Allens, he wrote Marshall: "Help is assured for the new materials when the time comes to work on them."[1] Grateful for Marshall's support, which had enabled him to undertake his Third Expedition before the Detailed Report of the Second was finished, Stein was at pains to reassure him. "I am gradually gathering up the threads for the completion of *Serindia,* my main task." But first he had to prepare the lecture he was to give to the Royal Geographical Society early in June.

At the British Museum he was gratified that his collection (from the Second Expedition (the material from the Third remained at Srinagar) was housed in new, attractive quarters. There, with selected items advantageously displayed, Stein received Austen Chamberlain, secretary of state for India. "The S. of S. expressed his full readiness to give moral support towards the grant for a special volume of plates. . . . It was a most pleasant two hours—and I had a chance [something Stein never missed] for urging all our hopes from B[alkh] to the Recording Angel's translation to Kashmir."[2] Chamberlain's moral support resulted in *The Thousand Buddhas,* its cost underwritten by the India Office, and accounted for Miss Lorimer's salary while she was employed at Srinagar. On Balkh alone Stein drew a blank; but then it was wartime and the Foreign Office had other, more pressing matters.

"When the world is so much out of joint & so much is at stake of the future, this return from my peaceful deserts to the turmoil of 'civilization' would have been a trying experience had your home not been open to welcome me,"[3] Stein wrote Publius when, after a four months' stay, he left his friends' home. Wartime England was a troubling place, and Stein needed a purdah so he could concentrate on writing of distant lands and past times. First he went to a country rectory as a paying guest. The landscape was attractive, but the padre was inept, his spouse incompetent, and the cooking indigestible. His second attempt was lucky. A select pension in a south Devon manorhouse owned by an architect who supplemented his professional income by accommodating "a few men who need quiet for work or rest. The place is run in a very efficient way, quite remarkable for a bachelor artist, & so far meets all my needs. (I meet the company only at dinner). . . . From my study I command a view of many miles towards Exeter & Torquay."[4] Soon he was calling it a "bright peaceful place" whose fellow boarders were congenial enough for him to join at lunch—"it is a short meal & costs me no sacrifice of time."[5]

Still he was restless: the war touched everyone and everything, and his situation as a naturalized subject made him feel vulnerable. "I have come to the conclusion that it will be best for me to return to Kashmir by the middle of May. . . . England cannot offer rest or real peace for work in such tragic times as we are living through, & there is no advantage to pay for this by reduced pay—and increased expenses."[6] *Serindia*, he could report, was moving steadily forward. "I am quite back again in 1907 experiences. Those of 1913–15 help much to correct & explain. My great aim is to get over Tun-huang & the 'Thousand Buddhas' before leaving. The rest is plain sailing; I am now at Miran. I get up always at 5 A.M. & work steadily from 6 A.M. to 1:30 P.M. Then generally follows a peaceful short walk on the moors, correspondence, & more *Serindia* work til dinner."[7]

From the Baron's letters Stein caught the first whiff of dissatisfaction. Stein was at pains to mollify his friend's attitude, advising him out of his own experience. "It never does to fight Departments, principles, methods & the like. It will be a great load off your mind when this housebuilding [the Technical School and annex and residence] is completed."[8] Or, "as to the petty antagonism between one's own subordinates, this is quite the *regular* experience and can be turned to good use. Nobody in an Indian establishment minds being watched & suspected as long as his superior encourages good work & is just in its appreciation. I can & need not write more to you about these common-place maxims for India."

Stein knew that in giving advice to a friend he must walk delicately. Vainly, he tried to curtail the Baron's grandiose building notions. "I shall prefer to use the large room in my 'private suite' which you mention for a study, not a bedroom. . . . The 'dressing-room' will do quite well as a bedroom. . . . You must not be unhappy if the work is not up to your standards. The great point is to achieve adequate shelter & space. After all, there is no need to build for generations ahead. Who can foresee the changes in Indian conditions by the middle of this century?"

The new year (1917) came, and, while writing, Stein lived "through happy times of exploration again & tried to forget the present."[9] But the present crowded in on his thoughts. "My anxiety about my sister-in-law continues & weighs on my soul at all times. Nothing has reached me, and I find it hard to believe that all replies to my letters since October have been held up. Not a single one of her letters before had failed to reach me." His plaints sound in his weekly letters: "The total want of news from her makes me feel very anxious about her health; for nothing else can, I fear, explain it."[10] His fears ceased when the interrupted exchange of letters (routed through a friend in neutral Holland) resumed. Hetty, with all her aches and pains, worries and lamentations, was holding up bravely.

For himself, Stein was worried about a joint on the foot whose toes had been amputated. "If you have a chance ask Professor Osler [Sir William, then Regius Professor of Medicine at Oxford] about a foot specialist in town," he wrote Publius, "there is a tendency for the toe joint to swell on longer walks & I feel I ought to ask advice."

Just then, real tragedy struck. "Poor [Raphael] Petrucci died on the 17th [February] from Diphtheria after having got safely, it seemed, over his very serious abdominal operation. It is a cruel blow of fate for all of us who are interested in Buddhist art & Central-Asian & Far Eastern studies. No need to tell you," Stein wrote the Baron, 28 February 1917, "what it means for me in particular. The typed copy of his Appendix was safely transmitted to him . . . & cheered him up a week or two before the operation. But, alas, the revision he was anxious to effect for the press, will never be possible as he intended it. Chavannes, fortunately, who was his Master in Chinese & collaborator too, has taken charge of the MS. for us. . . . For anyone else it will be a serious task."

Petrucci had been studying the paintings for five years; his work was divided between Paris and London. "Two chapters on the donors & on the mandalas [a visual symbol of the universe with its guiding deities] are ready for the Appendix. Chavannes will publish separately, under P.'s name and his own, a full account of the inscriptions on certain of the larger paintings which need full command of Chinese texts for their interpretation. The rest of P.'s programme will be taken up by [Arthur]

Waley, Binyon's Chinese-knowing assistant, and by Binyon himself."[11]

Even though the immediate problems caused by Petrucci's death had been resolved, the situation it revealed disturbed Stein. On the eve of his start for India, he wrote Barnett a long, thoughtful letter. "It seems very sad that after these masses of MS. [from the Second Expedition] have been safely deposited for fully 8 years [at the British Museum], I should still be unable to state any exact data about those which have an antiquarian or historical interest in my *Serindia* chapters dealing with the 'Thousand Buddhas.' Apart from [Lionel] Giles' short paper on the *Tun-huang lu* and a few tantalizing short extracts given by Pelliot from a similar local text, nothing has been published. A sad illustration of the want of those Sinologists trained for, & interested in, critical philological work on the Chinese side. . . . Kindly forgive my troubling you with all this, but you can realize best how much I feel the gap in my materials."[12]

He stopped at Paris on his way back to India. He wrote Publius on 4 October 1917: "Yesterday I spent almost wholly with Chavannes at his charming villa at Fontenay-aux-roses, settling problems of the *Limes,* arranging for Petrucci's Appendix, etc. The day before we were all gathered for tea in his garden. Chavannes looked overworked. . . . I told him of my wish to dedicate *Serindia* to him, the best of collaborators, & he seemed pleased."

During that stay in Paris the *affaire Foucher* had its beginning. Foucher confided to Stein his "hope of going out to India for congenial work in connection with the Archaeological Survey."[13] His proposal enlisted Stein's energetic support. Since the Frenchman had visited India with his aunt, Mme Michel, in 1895, they had remained friends; his subsequent publications had shown his "unequalled iconographic knowledge of Gandharan art," and his company had always delighted Stein. "I need not tell you how much both research and the Archaeol. Dept. would benefit by this plan." Once in India, Stein lost no time in visiting Marshall at Taxila to transmit Foucher's desire. Foucher is "anxious for a change on account of his health & also because he rightly feels how much conditions in [wartime] Paris hamper the work for which he alone is qualified. There are tasks which he had been invited to undertake years ago on behalf of the Archaeol. Survey. . . . Marshall, too, to whom F[oucher] on my advice had also written, welcomed the idea very warmly. . . . I eagerly wish matters could be pushed on sufficiently for him to come out in the early spring as he hoped."[14] Stein had no reason to be suspicious of Foucher's impetuosity.

This was good news; Stein then pictured for Publius the state of the house the Baron had been building since his arrival in Srinigar in 1913. The house "to my relief has now risen to its roof which is gradually being covered with self-manufactured tiles, the first ever made in Kashmir. [Quite appropriately, it would be called "Red Roofs."] Floors, ceilings, etc., will give work for another 7–8 months. The annexe may be completed by the spring when we hope for the R.A. to come and put the new collection systematically in hand. I fear the Baron by acting as his own contractor has burdened himself with a big load—and personal expense, too. We must only wish that they may enjoy the comforts of what has been so hard to achieve." As for himself, Stein, "after 4 days of hard driving managed to get the old shelter [at Dal Lake]," where he had lived in 1905–6, "into tolerable order for winter occupation with windows & doors that close and electric installation for heating."

Almost as though it was a present for his fifty-fifth birthday, a letter from Foucher told Stein that the Académie des Sciences had awarded him its 3,000-francs prize for geography. To Publius he confided that "the position the Académie occupies in all branches of learning gives [their prize] value. But what, I confess, I appreciated particularly on this occasion is that the award should have been made at a time when my name & origin might have been thought to stand in the way of such recognition."[15] When Stein received the sum, he sent half to Chavannes, asking him to give it to French war charities.

Early in 1918, Stein received two letters whose news he passed on to Publius: one from Chiang announcing his intention of visiting Kashmir—"It is delightful to think of seeing dear old Chiang arrive here next summer";[16] the other from Foucher, telling that he had accepted "an honorarium of Rupees 10,000 to write a detailed account of the Ajanta frescoes and to complete his and Marshall's monograph at Peshawar & Taxila. . . . The work would probably need an 18 months' stay as a minimum or perhaps 2 years. But Foucher would probably be glad for this as he seemed very eager to get away from Paris & all the turmoil of the 'Western world.' "[17]

Before the year ended Stein could triumphantly announce: "Foucher arrived quite safely at Colombo [Ceylon] at the close of Oct., learned the great news of Nov. 11 [the Armistice] at Lahore, his old haunt of 1896, and has since most happily been at work among Marshall's excavations at Taxila. It cost me an effort not to go down & meet him in our beloved Gandhara, but I hope to join him & Marshall at Sanchi (which like most of the great sites of India proper I have never been able to visit!) and eventually to spend some days with them also at Ajanta."[18]

Stein allowed nothing to distract him from his strict regime. Whether correcting the sheets of the new atlas—"it is no small pleasure to do one's marches thus again & to feel that the results are collected & preserved for what we human ants may call a long future"[19]—or whether, getting on with *Serindia* page by page, he relived the years spent in Central Asia. His work was demanding; it was absorbing; it filled his days with contentment. He contrasted his own simple way of life with that of the Baron and his wife. They were, he assured Publius, fairly comfortable and then, in letter after letter, he described how they actually lived despite their grandiose plans. "Their future bedroom is serving as a general living-room, one dressing-room for sleeping & another for cooking. . . . They have spent their 2 months' winter vacations at work ever since 1914, and most of the summer holidays, too." And finally—"The dear Baron, I fear has all the virtues of a 'willing horse.' Unfortunately I can never disassociate from the latter the Russian proverb . . . 'Some people are born with saddles on their back, others with spurs on their feet.' "[20]

Spring came and passed and summer was ending. "The housebuilding which for the dear Baron was the 'dream of his life' (to quote the Baroness) continues to be a nightmare. They are still limited to their bedroom with the kitchen in the dressing-room. But a year hence there is hope, perhaps, of the palace [Stein's nickname] being finished."[21]

"I wrote the last line of the text early on Sept. 2 [1918]." Stein had completed *Serindia* while still on the Marg. "The same evening I lit my big bonfire to announce the event to the Barons some 20 miles off & 6000 feet below, and—to the mountains. No, there was the faithful Nilakanth, Pandit Govind Kaul's son, too, to watch for it. He wrote me that as a boy he had seen the bonfire I lit some 22 years ago when my Rajatarangini translation was completed." And then, "I indulged in the longest walk & climb I ever allowed myself on the Marg, starting at 4 P.M. and getting back by 8, as usual."[22] Stein had finished a long, hard job and needed a long, hard stretch.

Two deaths in 1918 touched Stein deeply, the loss of men for whom he had the greatest admiration and respect. Born of shared work, their scholarly acquaintanceship had deepened into friendship. Acquaintances, friends, family: a qualitative difference marked off Stein's carefully graded categories. Family was the temenos enclosing the triptych of mother, "our dear angel," dearest Papa, and Ernst, and of living friends only Publius and the Saint could fill that sacred space. Chavannes and Hoernle were friends, and their deaths, though expected for months, were hard for

Stein to accept. When he speaks of them his pensive tone has the quality of an eloge—not a heartbreak.

In April, a letter told Stein of Chavannes's death. "I had my misgivings since last August and, alas, they proved only too true. But I am very grateful to Fate for having granted me those two sunny days in his peaceful villa while he was still as keen & brave as ever, ready to help with his inexhaustible critical knowledge & industry. It was such a satisfaction that I could ask him then to accept the dedication of *Serindia* which seemed to please him.[23] There, in commemoration, it is printed: "To the Memory of ÉDOUARD CHAVANNES/ who from his unrivalled knowledge of China's Past/ Never failed to Guide and Encourage my labours/ This record is dedicated in Admiration and Affection/ For the Man, the Scholar, and the Friend."

And Hoernle, with whom he had had contacts since the end of the last century? Stein read of his death in a Reuters' dispatch a few months after Chavannes's. He wrote Barnett: "He had always been a most devoted & painstaking collaborator and I owed him much gratitude for his kind help before I had first started for Turkestan. How great his services to Indology were in different and extensive fields, you know best."[24] Certainly Stein must have recalled that terrible afternoon when he visited Hoernle in Oxford to tell him that he had been fruitlessly spending his time and scholarship on forgeries.

How different the warm words, the personal sorrow expressed four years later at news from Kashgar of another death: "I feel the loss of dear Chiang more deeply than I can, or need say here. You"—he could speak frankly to Publius—"know how grateful I was for the invaluable help he had given me, for the boon of a friend so true & devoted. . . . Where can we hope for that 'other birth' in which we might work once more together, as we often talked of? I have written to the Vice Consul & offered a contribution towards the cost of sending dear Chiang's mortal remains back to Hunan, as was one of his last wishes."[25] Learned in the Mandarin tradition, with his lively conversation and his proven devotion to Stein's scientific interests, a scholar and a gentleman but not a Victorian scholar and gentleman, Chiang had found his way into Stein's "family"—he remained the "brother" with whom he had shared a great adventure.

The tenderness Stein could express surfaces at the news of the death of Dash the Great, the terrier, whom he could no more have kept out of his heart than out of his bed during the bitter winter nights in the Taklamakan. He had loved him foolishly, recklessly, when Dash was a frisky puppy, and he never wanted to do otherwise. Dog enough never to stop chasing rabbits, marmots, and automobiles, Dash was smart enough to be terrified by *rafiks;* small, compact, tireless—he left his pawprints along the Han limes. When man and dog reached England after the Second Expedition, their long separation, forced by quarantine, suggested that Dash be adopted by the Allens. Theirs, Dash knew, was a foster home to be accepted until his master's periodic returns. "How happy he still seemed last year on our rambles across the moors when he managed to catch his last rabbit." Stein sorrowfully told Madam when she wrote that Dash had been killed by an automobile: "That death for our old traveller, ignoble as its form must appear for one who roamed over such great spaces of desert and mountains, was at least painless. . . . His old master, too, would like to disappear in that way; but if the choice rested with him would prefer a sudden end by avalanche, Pathan knife, or the like to annihilating 'civilization.' "[26]

On the night of 20 February 1919, Habibullah Khan, amir of Afghanistan, was assassinated; his third son, Amanullah Khan, in the same breath proclaimed himself king and his country independent, demanding that British India recognize both. Half-hearted hostilities, the Third Anglo-Afghan War, lasting from May to August, terminated with a peace signed at Rawalpindi. Henceforth Afghanistan controlled her own external affairs; to this end the amir established diplomatic relations with outside countries. For Stein, these events brought life to his longtime dream and hope of reaching Balkh. True, the altered conditions did not declare themselves quickly—there was time enough for Foucher to insinuate himself in Stein's dream and hope. That was still in the future; in 1919 Stein's concern was directed to other matters.

"The main thing is to feel sure that you are both happy & enjoying the great change which has come over the Western world,"[27] was his New Year's greeting to the Allens. He himself felt fairly satisfied: the war was finally over; *Serindia* was completed; the R.A. was actually on her way to India; his retirement had been postponed for a year—his services on special duty outside the regular department's cadre had been sanctioned; and he was finally going to see some of the famous glories of India. A small disappointment was the last-moment change in the program. "Foucher, instead of going to Ajanta from Sanchi on the 25th [January] as intended will not be able to reach there until the middle of February. Either I must give up seeing Ajanta with him, a great loss, or else put in three extra weeks over visits, a sacrifice of time, such as I am

not accustomed to nor perhaps could justify officially!" Stein's disappointment was real: "The lesson obviously is that planning a satisfactory programme well ahead is far easier in the desert than where the human factor in the shape of valued friends comes into play."

Stein and Foucher met at Sanchi, the stupa planted by King Asoka
and subsequently enlarged and beautified. Swearing Stein to secrecy,
Foucher divulged his reason for not keeping the Ajanta visit. Stein
waited a month before writing the news to Publius on 26 February 1919
with a cautionary entre nous. Foucher had gone to Colombo for a secret
rendezvous with Mlle Bazin, his pupil. "He asks me to keep [his mar-
riage] quiet for another month so it will not pass to France where Foucher
is evidently anxious to explain things first himself." Two months later,
19 April, he is still whispering: "Foucher must have broken the great
news to Mme. Michel [the aunt] long before, but has not yet released
me from discretion—probably an oversight."

In the cave temples of Ajanta, whose walls glow with the exuberant
yet delicate painting made during the early centuries of the Christian era,
Stein found a "grand counterpart of the Thousand Buddhas. My visit
was meant only to secure some idea of the arrangements needed for the
planned reproduction"[28] of a volume to be similar to that of the *Thousand Buddhas*.

Back in Srinagar, Stein found the Andrewses tense and tired from
their unremitting building. "Unfortunately this renders the dear Baron
rather sensitive to the charges which the R. A.'s advent necessarily implies in the work at the Collection. I hope & wish she will recognize the
need of showing consideration for the Baron's experience & devoted zeal
while at the same time helping *me* by her ways of putting things into
methodical order." How awkward the situation, how futile were Stein's
efforts to lessen the tension. "I shall remain at Srinagar until the end of
April & this time [five to six weeks] will, I hope, suffice to smooth the
way to proper collaboration. The Barons need a good rest—& sometimes
I fear the risk of his repeating his Lahore experience & giving up what
after all might be quite a pleasant existence. This quite *entre nous*."

The R.A.'s skittish behavior became clearer when she told Stein that
she had received "an official enquiry from Dr. Thomas [the librarian of
the India Office] asking whether she would wish to become an applicant
for a Sub-librarian's post to be newly created there. The way in which
the letter was worded seemed to me to amount practically to an offer."[29]
Stein could not understand what had changed the R.A., what had transformed a devoted, capable assistant into a problem. First he was inclined

to believe Publius's suggestion that she was showing an accumulation of war strain; then he subscribed to the Baron's reaction that "the offer of the post was pre-arranged, probably before she started for Kashmir."[30] Whatever the explanation, the R.A. made it quickly apparent "that she did not care either for India or Kashmir and still less for her work here."[31] Stein's peacemaking effort collapsed; without further ado he tried to find a likely successor, someone with the R.A.'s "college training and museum experience."[32]

In all this, Stein's main concern was not with the R.A.—She was replaceable—but with the Baron and Baroness. Slowly, sadly he recognized that, though they were his friends and commanded his deepest affection, he could not change them. The Palace was an incubus they themselves had created. "The housebuilding is now nearing its end and doubts remain whether it will ever be possible for the Baroness to run it with servants. I hope they will secure the leave they plan for next spring & summer. But I fear also that it may end in a rash abandonment on their return. . . . It is a case of premature discouragement as it was at Lahore— following the sanguine dreams artists are tempted to indulge in."[33] Writing thus to Publius did not mean that Stein ceased to value the Baron's artistic qualities or that his strong ties of friendship were loosened; it meant merely that he was aware that such traits "deep-rooted in one's individuality cannot be eliminated late in life and must be accepted as determinant factors. . . . I close this after the State Dinner to their Bombay Excellencies, quite a pleasant function, which gave pleasure to the Baroness after all her domestic drudgery, & allowed them to see many friends."

However much Stein was concerned with the situation at Srinagar, from the moment the news from Afghanistan indicated changes, his hope that "Balkh may become accessible,"[34] had new tenure. "I am doing all I can to be prepared," he had written Publius on 16 May. "Will the new effort succeed better? That the seclusion of the 'Heaven-created kingdom' will become a thing of the past is scarcely to be doubted." Preparing meant enlisting the support of D. B. Spooner, the officiating director-general of the Archaeological Survey; it also meant learning Pashtu (the East Iranian, official language of Afghanistan). "I had tackled Pashtu seriously in May for the sake of a future field and since a fortnight have had the advantage of a very intelligent temporary [Pashtu-speaking] Pathan orderly to practise conversation. So a bad gap in my linguistic equipment is being filled in now. By coincidence this young Karakhel had spent a year in Darel with Raja Pushtun Wali & could tell me much of interest of that adventurous tyrant. What benefits his regime brought all vanished when he was murdered. I am sending photographs taken in 1913 to his refugee family."[35]

As the year was ending Stein could look forward to a reunion with Hetty: the government telegraphed him that his leave was granted. He had a pleasant and useful visit with the Marshalls at Taxila where he was told that "the Fouchers are likely to be enabled to spend another year in India. . . . I start for Europe on Dec. 27th."[36]

"The reunion with my sister-in-law was happily achieved yesterday on the Brenner Pass. She had to travel about 10 miles alone, & then in glorious midday sunshine between heights covered with snow, I could take charge of her, having travelled up the day before. Of how I spent the icy night in the ancient Brenner coaching inn, now begrimed with all signs of its use as a barrack, shall be told some day. . . . We propose to travel in a leisurely fashion around N. W. Italy before moving by the end of March to Lausanne where my sister-in-law would feel comfortable & await my return in September."[37] Their long separation was over; their lengthy reunion was seasoned with serious sightseeing, earnest conversation, and a variety of shared experiences—"I have been reading T. Hardy's *Dynasts* to my sister-in-law recovering from an attack of influenza." Slowly Stein was getting a full picture of the situation Ernst's family faced: the post-war inflation had wiped out Hetty's considerable fortune. "There is much I have to think about & arrange for my sister-in-law & her children's affairs," he wrote to Publius before starting for Paris and London.[38]

Paris did not seem the same with Foucher in Peshawar and Sylvain Lévi away teaching at Strasbourg. At London he consulted with Barnett about future collaborators: "For the Chinese documents from the Han *Limes,* etc., I am trying to get [Henri] Maspero, who, I was glad to learn at Paris, has at last been chosen as Chavannes' successor. For the Hsi-hsia (Tangut) stuff of which there is a good deal, I had a most welcome letter from B. Laufer [director of the Field Museum, Chicago] willing to undertake the cataloguing not merely of the Hsi-hsia materials but also of the Tibetan text remains."[39]

A note to Publius from Switzerland, where he rejoined Hetty, has Stein's unique quality—a kind of fingerprint, 7 August 1920: "Thanks to an old Swiss friend of my brother who had retired to this charming spot above Lake Thun, I had two exhilarating 'Marg' climbs. . . . In addition I managed to gather a good deal of useful information about grazing rules, etc. from the local Gujars [herdsmen]. That I could stand 12 hours or so of walking & climbing without serious inconvenience to my foot was, of course, reassuring."

By the end of the year he was back in India.

Nineteen twenty-one has the calm, the sweetness and light, the petty pleasures savored before the storm. Stein had made a brief but refreshing pilgrimage to Ujjain, one of the seven sacred Hindu cities, the famous old capital of Central India. "I have been 'gathering merits' by visits to a succession of Hindu Tirthas [shrines]. By securing a man of learning, a Sanskrit teacher of the local College, & 'purifying' my tongue in the sacred language, I was able to see & learn a good deal in these few fleeting hours. Ujjain is a delightfully old-fashioned place & does not know the globetrotter. So it was easy to acquire a number of interesting acquaintances among Pandits & priests, besides impressions of that India which has changed so little at the core in spite of all the foreign varnish."[40]

He finished *The Thousand Buddhas,* "including preface, dedication, etc., & feel so relieved that the bitter cold & rain does not affect my comfort."[41] Summer had not come to the Marg by June, yet it was where he chose to be. "The Andrews [whose] fine villa is now completely finished & offers comfortable peaceful quarters close to my new materials [Third Expedition]. All the same I was glad to return to camp life— on my beautiful alp."[42]

Alp and dog, the two objects of his unconcealed love, meet in an incident reported to Publius on 30 September 1921: "Dear Dash IV (who had a strong & warm coat made to measure by Nilakanth's tailor!) caused us all great anxiety by being missed when he ought to have come for our evening climb. The search made in the twilight was in vain; I had to go to rest with a heavy heart for Dash on his plucky hunts among rocks & ravines might well have been caught by a panther & ended his career. No sign of him during the night & my Pathan, who is devoted to the little fellow, came much depressed to waken me in the morning. I had just given the order for further search, hopeless as it seemed, when Bahadhur jumped up as if possessed—with Dash fondly pressed in his arms! He had found him ensconced in his basket. He seemed a little dazed & when I got him into my camp bed, two cuts on his forehead told the story: he must have lost his foothold on one of the rocks near the tents & after a bad fall lost consciousness. So he could never hear our shouts. Probably the cold of the night brought him to his senses &

then he clambered back, taking discreet care not to wake me by jump-
ing on the bed. He soon got back his old spirits but heeds my calls more
than before. He is, of course, being spoilt more than ever."

Stein enjoyed the autumn. Lord Reading, the new viceroy, came to
the Kashmir capital; they had a long talk about Afghanistan, the viceroy
promising to use all the power of the government to aid Stein's plans to
visit Balkh.

And then, as Stein had feared, the Baron decided to quit. It was a
foretaste of the bitter news that came with the new year.

It began hopefully enough. On 10 January he told Publius: "the Foreign Dept. is resolved to get the proposal [for Stein to visit Afghani-
stan] put formally forward as soon as the new Minister (Humphreys)
[Colonel Deane's son-in-law] reaches Kabul next month. Of course, they
cannot guarantee success, but evidently they are keen on securing it for
a quasi-diplomatic reason.

"The Afghan Mission while on the Continent offered a heavy welcome to *foreign* scientific expeditions. The French F. O., where Philippe
Berthelot, Foucher's old friend is a sort of permanent Under-Secretary
of State, promptly seized it in order to let Foucher proceed to Kabul.
Whether F. himself was keen on this seems doubtful. . . . But there appear to be reasons (connected with the attitude of the 'tante terrible')
which make it convenient for him to keep away from France longer.
Anyhow he was reported by the British Legation at Tehran in Nov. to
be planning an early start. Here in India nobody has had a line from
him since he left for Persia the end of April, and the same mystery is
experienced by his friends at Paris. That Foucher is specially going to
Kabul in order to anticipate my Balkh plans, I do not believe." Time
and additional evidence would shake Stein's faith in his friend.

Stein had set his heart on entering Afghanistan by way of Badakhshan
via Chitral, and though, in the following months, Humphreys did all he
had promised and more, "Mahmud Tarzo Beg could not agree to my
visiting Afghan Turkestan on account of the 'disturbed state of those
regions.' "[43] The alternative? "To go up to Kabul by motor car is not
quite what I should like. But as Marshall . . . rightly thinks, it would be
a great advantage to get into personal touch with the Afghans and, if
possible, overcome their suspicions."

"Yours and Madam's dear letters arrived to cheer me just when I
needed it," Stein wrote on 21 May 1922. "A letter received yesterday
from Humphreys will show you how things stand at Kabul. It is easy
to read between the lines that Foucher's reception at Kabul when con-

trasted with the reply received in April on my own plan reflects a 'diplomatic situation.' Suspicion and antagonism to British aims must remain the leading factor there." Having accepted the obvious political reality, Stein grabbed at the handiest person on whom to lay the blame. "I think the driving power lies in the ambition of his [Foucher's] wife and in certain unfortunate reasons which make it necessary for him to keep out of France as long as possible. It has now become clear why in the autumn of 1919 he went out of his way to get my plan of exploration on Afghan soil championed by Senart at the Oriental Societies' Joint meeting (my name being discreetly kept in the background), and also why on my return in Dec. 1920, he avoided meeting me at Muttra & Agra in a fashion which seemed inexplicable to others also. It seems a poor return for what I did to facilitate his employment in the Archaeol. Survey."

Unavailingly, Stein lifted his eyes to the mountains that watched over Mohand Marg. "Bitter as is the fresh failure of old hopes, it has not found me unprepared." There was no need to disguise his true feelings from Publius. Nor could he shut his mind to what Mme Michel wrote. He had seen Foucher's move to India as "a desire to keep away from France as long as possible in view of certain allegations held over his head by 'la tante terrible.' Whether true or not they are only fit to figure in a French novel of the baser sort. . . . To count on Foucher's possible scruples about 'playing Jacob,' as you [Publius] put it, would only make deception worse, for amiable as his ways always are, his friends had never reason to credit him with any great strength of character. All the same it is a sad experience. And having cleared my mind of oppressive thought I fear only to make you share certain bitter reflections—let me turn to more pleasant topics."[44]

By October the last hope was buried when Humphreys wrote Stein that the amir had given a total and definite refusal for him to visit Balkh and that he could not look "for a more gracious disposition for next year. He also reported that Foucher expects to visit all sites of importance south of the Hindukush *plus* Bamian this autumn & winter and plans to go to Balkh in the spring."[45] From Foucher himself Stein finally received a "long communication full of prevarication & cant. The convention he negotiated (as he puts it "under pressure" from the Afghans!) would, if recognized make all archaeological work in Afghanistan dependent on French consent. F's letter is a clever composition, full of shuffling & perversion, intended partly to throw a cloak over the manifest want of gratitude towards those who enabled him to work in India for 2½ years in the pay of Government. The cleverness of this 'document humain' bears out only too well the truth of a remark in Madame Michel's sad revelations: 'Foucher ment comme les autres respirent.' Personally, the experience is a very bitter one to me. But perhaps I ought to feel grateful that similar ones have been spared me before."[46]

Even as Stein was doing everything to keep his hopes from drowning, the man who, twenty years later, would be his *deus ex machina,* came to see him. "A most pleasant & interesting visitor straight from Kabul: Mr. Van Heinert Engert, of the U.S. Diplomatic Service & lately Chargé d'Affaires at Tehran, had gone to Kabul via Simla to return the visit which an Afghan Mission had paid to Washington last year. Now on his way back from Kabul, he came for a short visit to Kashmir & soon made the climb up to my Marg. I found him a delightful guest & man of a most scholarly mind. Originally trained for historical research at the Univ. of California (his home State) & at Harvard, he had for the last ten years seen history being made at Constantinople, Syria, Asia Minor & elsewhere in the Near East. You can imagine how much we had to talk about in the 2½ days which he spent here. As Mr. Engert explained, there is no special objection to my person, just a genuine enmity of everything British & an honest barbarian satisfaction at keeping Govt. in the position of a humble suppliant. In the end, he thinks, the request will be graciously granted.

"May his prediction come true!"[47] Stein also found Engert to be a practical person. He wrote the Baron: "Engert strongly recommends our having a couple of fire extinguishers at the Collection. What do you think of the suggestion and what would they cost? Of course, as a Californian with San Francisco experience he realizes the risk."[48]

"Last night's bonfire was blown about a good deal—like a distracted world."[49] September was ending; Stein's mood was autumnal: in only a matter of days he would have to descend from his beloved Marg. The Baron and his wife would soon be leaving Srinagar. "As regards your 'burning of your boats,' well, that merely meant your acting upon your decision of November last. You know how eagerly I wish that you may find it justified, though I shall miss you greatly in Kashmir."[50] Dal View, the cottage where he had spent many winters, was no longer available, and he was glad to seek shelter in one of Raja Hari Singh's villas, Almond Cottage. "The letting of a house is now a personal favour, all former tenants having been turned out two years ago."

That November he turned sixty. "The inspiriting prospect of another journey is not before me," he wrote Publius. "I am trying to think of other chances to work in the field while strength lasts. How I wished there were other ancient sites in a real dead desert left to explore. . . . I do not wish to end my days tamely in the study or poking into areas which others have searched over & over again. Well, anyhow till the end of 1924, Serindia sequens [his still untitled Detailed Report of the Third Expedition] will keep my hands occupied."[51]

He was very comfortable in Almond Cottage—"more comfortable, perhaps, than I ever was during a Srinagar winter. I have room for unpacking my buried book boxes, and the rough cataloguing of their contents is proceeding faster than I expected. It is a pleasant occupation after dinner to have many happy memories revived by tangible contact with relics of past studies."[52] He was sending all the books he was not likely to need to the Hungarian Academy "in whose fine Library I first started my Sanskrit & other philological studies as a school boy. The Librarians have ever since been kind friends to me."[53]

As 1922 ended, Stein exerted his muscles of self-control to climb out of the despair and bitterness that had possessed him. Death, even the death of his friends Chavannes and Hoernle, however much they saddened him and made him face the hard task of finding replacements, was easier than the defection of a friend. Had Foucher died he would have mourned him; but he was alive and in Afghanistan. "Nothing is to be gained by clinging to illusions, however cherished their object," he could write Publius half a year later, 21 May 1923: Morbidity had been transmuted to an elegaic acceptance. "My hope of reaching Bactria made me take to Oriental studies, brought me to England & India, gave me my dearest friends & chances for fruitful work, and for all this I must be grateful to Fate. The main question now remains in which direction can remaining years of fitness for congenial work be most profitably used."

"It is pleasant to begin the New Year's writing before sunrise." His first thought, his first writing was a letter to Publius on 1 January announcing his "election to the Académie des Inscriptions as 'Correspondent de l'Institut.' Ever since v. Roth, a truly great master, received it in 1884 and since I read in Miss Yule's biography how Sir. H. Yule greeted it on his deathbed, I have looked upon this distinction as the one which could, since centuries, claim undiminished international value. It has never been made an object of exchange in scholarly courtesies & that helps to remove an embarrassing thought which in view of Foucher's Kabul activities you will guess. But what is on my mind is that but for the fortunate chance of exploration, my work as a scholar would not justify this honour."[54]

New honors, steady concentrated work on "Serindia sequens," a quiet summer on the Marg, and the passage of days, weeks, and months brought a much-needed peace to Stein. "The last week, passed in quiet Report work, brought me news of another attention, la Grande Médaille d'Or of the Société de Géographie, Paris, 'pour l'ensemble de votre oeuvre,' as it says in the Secretary's letter following the lead of the Swedish one."[55]

By fall the Report, still untitled, had advanced to where Stein could lay out an elaborate itinerary: starting the end of February (1924), it would take him through Egypt and Syria on his way to Europe. "I am bringing all your letters since 1903 or thereabouts home for safe deposit which, I hope, will be in accord with Madam's wishes."[56] Stein looked forward to a real busman's holiday: sightseeing, visits to his many friends, work, and a happy reunion with Hetty. He also brought Dash the Less— how could he leave him behind? "That the voyage was made so enjoyable for me is largely due to Dash's revivifying company. . . . He is treated as only privileged dogs can be, freely walking over most parts of the deck & paying me visits at daybreak in my cabin. He quickly endeared himself to quite a number of fellow-passengers & introduced his master to pleasant acquaintances which he, with his unfortunate reserve would surely have otherwise missed."[57]

Stein's special brand of good luck began when he disembarked at Port Said—Dash the Less remained on board bound for England and quarantine—where he met Alpine Club friends about to start for Petra. Stein's excursion to 'explore' Petra with them got underway a few days later when at Luxor he received a telegram "from Philby telling me that he expected me at Amman by April 1st for Petra."[1] His sightseeing at Cairo, Luxor, and Jerusalem was of the briefest as he hurried to keep his appointment at Amman, "the little town which boasts of being the capital of Trans-Jordan."[2]

Philby was not at Amman but had arranged Stein's onward journey. A "motor trolley down that much fought over bit of the Hedjez Railway which stretches across the desert plateaus. . . . A lonely journey of some 150 miles through impressive solitude. . . . What a joy to feel once more in a desert. . . . The next morning Mr. Philby's car carried me across a roadless gravel 'desert' to the crest of the hill overlooking the great site of Petra. A three hours' pony ride brought me to that extraordinary cañon of fantastically eroded red sandstone cliffs. There Philby first greeted me."

What a meeting that is to look back on. Harry St. John Bridger Philby (1885–1960), impetuous, daring, unruly, then serving as British chief representative to the ruler of Trans-Jordania, the brilliant Arabist who would embrace Islam and become advisor to the successful Ibn Saud—Philby greeting Marc Aurel Stein, the Hungarian-born baptized Jew, knighted by the British. Stein was famous for his explorations of the Taklamakan; Philby was already attracted to desert exploration—he was the second Englishman to cross the "Empty Quarter," the Rub al-Khali, the dread, desolate waterless area occupying more than a fourth of Saudi Arabia. What a moment in their careers: Stein, his desert expeditions behind him, barred by political upheavals from Afghanistan and Chinese Turkestan; Philby's exploration just ahead made possible by the political upheaval in Saudi Arabia. And what a setting for the meeting of these two; the rose-red city of Petra, a long-abandoned, seldom-visited site still quick with "what the civilization of the Hellenized Orient meant in [Roman] Provincia Arabia."

Stein made the daylong return trip across 170 miles of stony wastes in the company of Philby. "A long talk on the way back to Amman has resulted in Mr. Philby asking to come with me from Aleppo or Alexandretta onwards. He is resigning his post (& probably also his I. Civil Service), partly owing to differences over Palestinian policy, & hopes after sending his wife & children home by sea, to rejoin me before I leave Syria. It will be very pleasant to have the company of this accomplished traveller & explorer of Arabia Deserta."[3]

Stein's sightseeing took him to Haifa, "at the foot of Mount Carmel & within sight of the purple sea," to picturesque Acre, "a sleeply little port with memories of Crusaders' exploits," to Damascus, "the Syrian counterpart of Samarkand," to Baalbek, "Roman grandeur in the planning and richness of Hellenistic art in the details . . . the most impressive monument of a civilization which has fertilized East & West," and, with hurried side trips to interesting sites, to Beirut, "this great successor of Tyre & Sidon." There the permission for a little tour of "4 or 5 days between Aleppo & Antioch was readily granted and arrangements for needful local help settled." At Tripolis, "quite charmingly medieval in its narrow bazars & vaulted alleys,"[4] he took the train to Aleppo, a "big town, still a great centre of commerce. . . . since medieval times the true terminus of the ancient Central-Asian trade route from the side of China."

So far Stein had been on well-traveled roads; at Aleppo his own jaunt "among the little known hills towards Antioch" began. An "18 miles' ride across the terribly fissured rocky hills all of volcanic material, allowed me to see the same day the wonderful remains of Kala'at, that strange agglomeration of churches & monasteries which in the 5th–6th cent. had sprung up at the spot which St. Simeon, the Stylite, had sanctified by his ascetic miracles. [At Turmanin] I had discharged the gendarme and the weak-kneed Armenian 'guide' whom the Bureau des renseignements at Aleppo had sent with me for a start, Both were ignorant of the ground & not particularly anxious to take further responsibilities in a region still visited by little bands who pass for patriot-rebels or bandits according to the light in which their activities are viewed.

"The readiness with which local escorts of armed villagers were forthcoming assured me all that was needed for guidance & hospitable reception, & I am delighted to feel that thus travelling alone, with the lightest of baggage & only my exceedingly poor Arabic to help, I managed to carry through my program completely. In the course of three days' hard riding & marching over that strangely broken ground of hills, I visited over a dozen ruined sites, still showing in abundance massive remains of late Roman & Byzantine times. . . . The amount of labour spent in erecting these churches, mansions, etc., of huge blocks of stone must have been enormous. How the people who lived then in this almost wholly barren region were maintained is rather a problem. . . .

"Those days spent with villagers little touched by Western influences taught me more of the life & ways of the people, . . . & the fact of my being alone, without any official attendant or even servant, in no way hampered the execution of my rather extensive programme. All the same I was glad when a march protracted long into darkness had brought me down safely over tracks which often reduced themselves to staircases of boulders, to the little town of Harim, famous for its big springs & the strong castle which guards access to them. Next morning I started off by the delightfully level 'road' which leads to Antioch, . . . a charmingly, sleepy town of true Turkish aspect, stretching between the orchard-lined river & the picturesque rocky hillcrest to which the ancient walls ascend." And so he reached "fever ridden Alexandretta" to meet, as per their arrangement, Philby's train bringing him from Beirut. "It was a delightful reunion."

"We put up for two days at Konia at a true Turkish hostelry in the centre of town, explored the numerous old Mosques with their fine sculptured facades & lovely coloured fayences within; thanks to Philby's remarkable conversational powers & sociable manners we were able to pick up quite a number of Turkish acquaintances. In the 'Museum' full of rescued fine sculptures I found some remarkable pieces. . . . Altogether it was a most profitable halt. There as well as all along we experienced nothing but kindness & friendly attention. . . . What pleased us was evidence on all sides of the order & fair economic situation produced by the present Turkish regime. Even the ravages caused by the ill-fated Greek advance had been satisfactorily repaired. In not a single instance did we meet with a request for 'bakhshish' [alms]." At Konya the two men parted: Philby went to Ankara to assay the political and religious climate a month after Ataturk had abolished the caliphate; Stein continued on to Istanbul for a "necessarily hurried sightseeing." Budapest, Vienna, and Paris had to be visited—his usual roundabout road to Oxford. He arrived there 23 May.

A year before, 21 May 1923, he had written to Publius and Madam telling them what Afghanistan had meant in the shaping of his career. Now he wanted to talk over the problems he must face. One effect of his failed hope of working in Bactria was that he had "little chance of being allowed to remain in the service . . . I am far beyond the age limit." At the end of the summer he had dusted off two cherished projects and proposed them to Marshall. One was to make an "archaeological survey of a tract on the northwestern border of the Hazara District"[5] where, from information he had obtained, Alexander's Aornos might be identified. The other, inspired by the naming of a committee to allocate sums from the Boxer Indemnity Fund, was for financial support for a "fresh journey of archaeological exploration" in Central Asia.

In his letter to Marshall, Stein mentioned that he had found a title for *Serindia* sequens: *Innermost Asia.* That, his immediate task, was approaching completion.

"I have great news to announce," Stein wrote the Baron in Delhi toward the end of October. "Publius was elected President of Corpus Christi, his old College, & you can imagine what this means for him & all who love him. It is the best reward which a just destiny could give him for all his lifelong devotion to scholarship & University education. . . . How great a share Madam had in this most auspicious choice you can realize. She will be the ideal hostess in the President's house, too. And this brings me to a practical & urgent request. They will have to furnish that large place, & it has occurred to me that some or most of the carpets you brought away from Srinagar & wish to dispose of would admirably suit their needs. So it might be a convenient & safe opportunity for any transaction the Baroness & yourself may be prepared for."[6]

At almost the same time Sir John Marshall had great news to announce to the world. "At a single bound we have taken back our knowledge of Indian civilization some 3000 years and have established the fact that in the third millennium before Christ, or even before that, the people of the Punjab and Sind were living in well-built cities and were in possession of a relatively mature culture with a high standard of art and craftsmanship and a developed system of pictographic writing."[7] Mohenjodaro and Harappa, cities of the Indus River civilization, became known to the West.

The discovery shocked scholars out of the idea that the Vedic Age was the earliest Indian horizon; perhaps it was this very idea that had kept people from seeing what was writ large and plain. Harappa was "before 1856, one of the landmarks on the middle Ravi plain of the Punjab. Made up of numerous mounds extending in a circuit of some three and a half miles, the site rose in places up to sixty feet. General Alexander Cunningham . . . even describes buildings found there without realizing their true antiquity. From 1856 to 1919 a systematic destruction of Harappa was carried on thanks to the inevitable stupidity of so-called 'practical' men. . . . The Lahore-Multan link of the Indian Railroad required a firm footing to pass over the muddy and oft-flooded plains of that area, and the presence of that great mass of ancient fired brick was too good to be true. It was a natural quarry and remained so until an enlightened government finally put an all too tardy stop to that usage."[8]

Discovery began in 1921 when three Harappa mounds were being excavated by Daya Ram Sahni, one of the Indian archaeologists Marshall had recruited and trained. Among the other small items recovered were two pictograph seals—seals such as General Cunningham had found fifty-two years before, though, unprepared, he had failed to hear their message. The same year R. D. Banerji was working at Mohenjo-daro, "the mound of the dead," on the Indus itself, the other large urban cluster of settlements that include some 150 village sites. Cutting through a mound, "he reached levels yielding seals exactly like those from Harappa and at once recognized their prehistoric character."[9] When, later, Marshall examined the seals from the two sites, he unhesitatingly announced that they belonged "to the same stage of culture and approximately to the same age and that they were distinct from anything previously known to us in India."

Impressive, hauntingly mysterious, the Indus Valley civilization contributed to the prehistoric discoveries of the 1920s: when Anderssen found the first evidence of China's Neolithic culture, the elegantly painted pottery of the Yang-shao sites; when Woolley uncovered the grim but spectacular royal tombs at Ur; when the unviolated storehouse of splendid grave goods in Tutankhamen's tomb was opened. Of a different kind was the publication of Childe's *Dawn of European Civilization,* which surveyed and gave significance to a mass of little-known archaeological findings. Europe's patrimony went far beyond Greece and Egypt: it was heir to the many inventions made in various parts of the Eurasian landmass.

Stein had followed the Indian influence far to the east; now he would search beyond India's western borderlands for "possible links in the region which separates the Indus Valley and Mesopotamia."[10] Thus, once again, Stein's luck came to the rescue. He was relieved of having to decide where and how he was to spend the coming years in congenial work: Marshall made it very clear that every capable hand was needed in the immense task ahead. The discovery of the Indus River civilization had made the headlines—it meant, as Marshall well knew, bread and butter for his Archaeological Survey.

24

"The remaining years of fitness"—Stein's oft-repeated phrase is mocked by what he managed to do in between the writing of his massive Detailed Report. To get the measure of the man's energy, a calendar of where he went and what he did in 1924–25, as told primarily in letters and cards to Hetty, is instructive. Nineteen twenty-four: 22 February, sailed from Bombay and rested aboard the ship; 12 March, he was at Mena House, at the Pyramids; 8 April, at Haifa; 15 April, at Aleppo; 23 April, at Istanbul (and there were the fast, strenuous side trips along the way); 4 May, on the railroad at the Serbian border enroute to Budapest; 11 May, a triumphant welcome there, though he had hoped to be "incognito"; 15–20 May, at Vienna with Hetty; 21 May, at Paris; and on 25 May, after "all this travel and bustle I was happy to find in my loved familiar house [the Allens'] a week-end of quiet!" The Allens, Arnolds, and Andrewses were all together. During June and July he worked with Andrews at the British Museum—on 5 July, he went happily to a garden party at Buckingham Palace.

On 12 August he started on another round: 17 August, he spent "a heartwarming" day at Basel with the Allens and Arnold; 18 August, he met Hetty at Bregenz for a shared vacation in rural Austria; 2 September, he was at Munich discussing dessication with Professor Rudolf Geiger, the climatologist; 6 September, he was back in London "facing a heavy work load"; 8 September, a visit to Oxford; 20–30 September, ten days at "Highbroom" with the Eversheds, then a return to London to do some work with the Baron; 4 October, back in Oxford to be reunited with Dash the Less, finally released from quarantine; 12 October, to Surrey to weekend with the Griersons; 15 October–15 December, at Highbroom

correcting proofs of *Innermost Asia;* Christmas in Vienna with Hetty;
29 December, at Budapest for New Year's; 6 January 1925, at Berlin to
see Lecoq and the German Central Asian collection; 8 January, at Leiden
and the Hague with the Saint and Vogel (the friend in neutral Holland
through whom Stein had maintained contact with Hetty during the war);
9 January, returned to Highbroom to stay until 14 April; then stopped
off at the island of Jersey to visit the Macartneys; 19 April, at the Hotel
Portofino, above Genoa; 6 May, he visited Rome; 9 May, Paris; 11 May,
at Oxford, at the president's lodging, Corpus Christi College—"the quiet
of their large hushed rooms do great things for my work"—11 May
until 23 August, he remained at the Allens; 23 August–20 September,
in Vienna with Hetty; 22 September, back at Oxford with the Allens;
16 November, via Paris to Marseilles to catch the boat for Bombay.

Only briefly did the pace slacken when, his vacation over, he reached
Delhi (22 November) for a month's stay. Immediately his routine re-
asserted itself: "Every morning by 9 A.M. I sought the spacious freedom
of my bright office at the 'Museum of Central-Asian Antiquities; then
work on [*Innermost Asia*] proofs with plentiful correspondence until
5.30 P.M.; next a walk to a ruined structure in the nearby Campagna refreshed me; a young Pathan well versed in Arabic, secured for peripatetic
practise of Pashtu conversation, made these walks a distinctly useful preparation; by 7 P.M. I am back at the office doing more work for an hour
or so, & then a brisk walk of 15 minutes takes me home [to the Western
Hostel] for dinner with the Baron [who was in Delhi for the winter
months installing the frescoes]."[1]

After this ordered month, Stein began a program of "rapid travel &
visiting," spending the Christmas-week recess on calls to old friends in
Lahore, Rawalpindi, and Jhelum—"delightful & refreshing in spite of
its five nights on the train"[2]—and a three-day "truly delectable Central-
Indian tour" to the glories of Khajuraho. "Its two dozens of beautiful
Hindu temples and the profusion of elegant sculpture which covers them
all from base to the top of the spire, made one realize that standards of
art quite different from those we inherited from the classical world may
lead to creations of absorbing interest to the western eye. A most comfortable camp had been pitched for us at the site by the Maharaja of
Chattarpur, a quaint philosopher-prince. The evening in camp brought
fine fireworks round temples & tanks. . . . At Datia we viewed the magnificent high palace built in Jahangir's time [1569–1627] & then the
Maharaja, his decendant, entertained us to a grand dinner party on New
Year's day."

A letter from the chief commissioner of the Northwest Frontier Province notified Stein that he had "completed arrangements with the Miangul of Swat who is prepared to take me up from his side to the Pir Sar spur facing Thakot on the Indus and a possible site of Aornos. . . . I wrote at once that the latter half of February or else March would be the time best suited for that short tour, but that I should be happy to set out at any time if the uncertainty of 'tribal politics' were to make such delay inadvisable."

At the Trigonometrical Survey, where Stein went to secure two assistants for the Swat tour—one trained in plane-table survey work, the other a general handyman—the chief commissioner "fully approved my wish to make the most of the opportunity & thinks that my tour, purely antiquarian as it is, will prove a distinct gain too in a political sense. Darel furnishes a most useful precedent."[3] Having tailored the projected tour to suit his taste, Stein hurried to Srinagar to have his camp equipment readied; then he went to see Marshall about his assignment west of the Indus Valley.

"Of the last eight nights, not less than six had to be spent in trains. I reached Mohenjo-daro, Marshall's 'Indo-Sumerian' site after 36 hours' dusty travel down the Sutlej & Indus 'valleys' ('tame desert' plains would describe them better) and found the remains very instructive. With some 800 men at work more & more of those massive structures in hard brickwork are being unearthed. Though most of what relates to their character & uses remains still very puzzling, there is plenty in the way of small carvings, painted pottery, jewelry & the like to give some idea of a civilization going back at least 1–2 millennia further than the earliest datable relics of India known before. I could also satisfactorily discuss with Marshall arrangements for my next cold weather survey in Baluchistan & Makran. . . . It is delightful to be once again in these old haunts and to have the trans-border plan before me."[4]

By 2 March his camp kit for the Swat tour was ready. He mailed two volumes of typed copy of his 1913–16 Diary to Publius. "As you know, the Pamir & Sistan chapters still remain to be written, and it seems a useful precaution that a copy of the original should be at Oxford—available for use by another hand if the need were, against all expectation, to arise."[5] Did he remember having mentioned to Publius that rather than end as a traffic fatality he preferred sudden death from a Pathan knife? Certainly, he was not morbid; he was jubilant with every right to feel jubilant. The Swat tour would be an adventure tailored to fit him. It would take him into a region no modern European had seen; he would be the guest of the latest successful chieftain to win a kingdom in the Hindukush valleys; and his objective was to identify the site of a spectacular victory of Alexander the Great. And the bonuses! Buddhist shrines glowingly described by Hsüan-tsang to examine, panoramas to exult in, relics of languages to sample for Grierson, and always Aornos to offer the classicists.

At Peshawar, on his way to Malakand where the tour was to begin, he managed to "find three narrow Turkoman rugs which may prove acceptable to Madam for the passage" in the new presidential house. He received auspicious news. "Fate seems to show favour at the outset: Shah Alim, Pakhtunwali's nephew & my factotum in Darel, now serves the Miangul [of Swat] & has been attached to me for this tour. Great luck."

His passion for possessing the past and his pleasure in its recapture are in his Personal Narrative, 9 March–16 May 1926. (It has been severely pruned but its flowers left blooming.) Stein's P. N.s have a conversational quality. Compared with his sober, factual, scholarly books and articles and with his intimate letters to Publius, his P. N., also sent to Publius, is Stein talking out loud—his way of remembering the human element.

Its beginning might be called (paraphrasing Yule's *Cathay and the Way Thither*) "Aornos, and the Way Thither."

"On the morning of March 9th I was whirled off in Metcalfe's [the political agent] comfortable car for the Malakand. Those picturesque hills of classical form and bareness greeted me as they had done in April 1906, on the start for my second expedition. But there was noticeable change in the landscape. Below in the gorge, once so barren, there cascaded a new river: the Upper Swat Canal which, brought by a tunnel over two miles long through the range, now carried fertility to the eastern half of the Yusufzai plain.

"The Pax Britannica had left its fresh marks on the once blood-soaked heights of the Malakand, too. The Political Agent's new house close to the crest of the spur overlooking the pass seemed full of the comforts of an up-to-date English home and Mrs. M's care had made it delightfully bright with carpets and flowers—and two splendid specimens of the British Baby. . . . On the morning of the 10th the start was made and it was the Miangul's motor car which carried us down into the valley and then to Thana, the biggest village within that strip of Swat which is 'protected territory' and controlled by the Political Agent. . . . To find myself in the peace & freedom of my little tent, the true home I had not occupied since the Marg, was a great joy.

"I had decided to stop at Thana for two nights in order to visit the ruined sites, those remains of Buddhist times which abound in the whole of this part of Swat. Its fertility and large population explain the care which all old Chinese pilgrims had taken to visit Swat and the glowing accounts left us of what by its ancient name was known as *Udyana*, the 'garden.' The first day took me up to the foot of the Mora pass where diggings done for Deane in 1897 brought to light a mass of fine Buddhist relievos brought to safety in the Calcutta Museum. Those old Buddhist monks knew how to select sacred spots—a glorious view down the valley, picturesque rocky spurs clothed with clumps of firs & cedars and the rare boon of a spring close by—give charm to the spot for those, too, who do not seek the future bliss in Nirvana.

"On the 13th, a day of threatening clouds & occasional drizzle, I made my way into the Miangul's dominion. At Kotah, the first village, I found my tent pitched outside a small fort of true medieval look; it had been recently built, visible evidence of the ruler's right perception of his subjects' still somewhat unsettled allegiance. To my great satisfaction it was Shah Alam, Pakhtunwali's nephew & my old guide in Darel & Tangir, who accorded me welcome there by the Miangul's order. An exile since Pakhtunwali was murdered in 1917 does not diminish his status with the Swatis who call him Shahzada (prince). Plenty of good talk we had together on the long tramps which I devoted to ruined Stupas and other remains during three days.

"At the village of Barikot, beautifully situated at the debouchure of three important side valleys leading up to the slopes of Mt. Ilam, my camp was pitched near where the Miangul has started constructing a 'Tahsil,' fort, jail and school—all due prerequisites for consolidating his rule. Some day when his conquests will include the high alpine valleys of the Swat Kohistan he will be able to establish a hill station such as no other Indian prince could hope for. But I am quite content to see Swat in its happy transition from turbulent faction strife to the settled conditions under a capable 'tyrant.' Two days hence I hope to see the Miangul—or 'Badshah' (king) as he now claims to be titled.

"My arrival at the king's seat, Saidu, took place five days later than I had foreseen due solely to the great and varied interest of the 'sites' which I was able to explore. Coins of Indo-Bactrian Greek mints and of the great Indo-Scythian rulers were brought to me in numbers as soon as the first offerings of this sort had been rewarded. They had all been picked up on the steep hillsides which are dotted with massive remains

of stone-built houses & towers. They helped to provide much-needed
chronological indications for this type of ruins so abundant throughout Swat.

"Then we left the good mule road and turned up a verdant narrow
valley. All over the little terraced fields, bluebell-like flowers provided
what looked like a carpet for a large relievo carved on a detached rock
showing a seated Buddha. The pious zeal of the Pathan invaders—Swat
was occupied by the Yusufzai clans only in the 16th century—has done
what it could without too much trouble to deface the sacred image. Yet
the serenity of head & pose remained. And just to delight my archaeological eyes, too, there rose a big Stupa with carefully constructed masonry and in almost perfect preservation. It had not been dug into for
'treasure.' It still raises its fine hemispherical dome, with the stone facing
practically intact, to some 45 ft. above the circular drum. Together with
this and the triple base, the whole structure attains a height of fully 100
feet. I am scarcely mistaken in the belief that apart from small votive
Stupas it is probably the best preserved of all the ancient shrines which
Indian Buddhist worship has raised over supposed relics of its hallowed
founder. Only the huge circular stone umbrellas raised above the dome
had fallen & now lay in a heap on the base of the Stupa. The biggest of
them measures 14 feet in diameter: to raise it to that height would have
been a mechanical task worthy of some Egyptian builder.

"*Bir*[kot] ('Bir castle'), an ancient stronghold on a bold conical hill
rising within less than a mile from where my camp stood, is a very conspicuous landmark for many miles above and below the village. Two
very narrow rocky crests which might have given a chance for assailants,
were carefully guarded by towers & bastions. On one we discovered an
antique 'ammunition dump' in the shape of big rounded stones brought
up from the river rubble to serve as balls for slings or catapults of large
size. Abundant pottery debris points to great antiquity of occupation, &
coin finds confirm this.

"From Bir-kot I moved my camp on March 19th to Udegram village
higher up the main valley. On the way I saw the big Stupa of Shankardar,
heard of by Col. Deane, & seen by the force which moved in 1897 up for
a few days' punitive expedition. At another point where the rock passes
close below bold cliffs, clambering over rocky ledges where boots had
to be discarded, I was shown at a small natural cave a curious relievo: a
royal personage dressed in the typical Central-Asian costume of the great
Indo-Scythian Emperors, surrounded by divine attendants. The whole has
suffered badly by iconoclastic zeal; but it serves to support what sculp-

tures far away at Muttra & elsewhere suggest about a kind of Caesar's cult paid to the great Kanishka & his immediate successors. . . .

"I moved up to Saidu, the hereditary seat of the Badshah, now being rapidly developed into the capital of Upper Swat. Here his holy grandfather, the great Akhund of Swat, had lived as spiritual leader, and he, too, like so many holy men in the West had known how to choose the right spot for pious devotions while alive, and for local worship thereafter. Amidst a cluster of trees and pilgrims' resthouses there rises the domed structure which shelters the remains of the holy Akhund, or Teacher, under whom Swatis and Yusufzais had for years fiercely fought Sikh aggression, while Ranjit Singh, a rival, was growing old. Around on prominent hilltops there showed high towers meant to give a safe refuge in case of sudden invasion. Now white terraced mansions have risen since the Badshah in the last four or five years made himself the sole possessor of his grandfather's sacred inheritance.

"By the side of the Chief's residence I found a comfortable big tent pitched for my quarters. I am to be free to extend my tour wherever my interests call me within his present borders. Only I must allow him to continue treating me and my party as his personal guests. The Miangul is a person of remarkably active habits. The present month of Ramzan, with its strictly observed fast between the first sign of dawn and nightfall, leaves but little time for rest & sleep for the Faithful. Yet early as my starts [to visit ruins] were in the morning I always found my host about taking the fresh air on the terrace before his Darbar rooms or attending to business of sorts.

"Miangul is a strongly built person of middle height with a fine head of strikingly intelligent features. His age is believed to be about 45, but from the plentiful grey hairs in his flowing beard he might well be taken to be older. His figure is sparse like that of all Swatis. His quickness of eye and limbs impressed me as befitting the role he has to play on his newly-founded throne. Like many strong rulers in these parts he is accustomed to dispose of affairs by word of mouth and is apparently little hampered by the fact that his knowledge of Persian [he spoke practically nothing but Pashtu] would scarcely allow him to scrutinize such papers as have to be issued in his name.

"Yet the amount of business which he has to transact as supreme arbiter of disputes, as lord of the manor in widely scattered landed property and as chief organizer of his armed forces must be great, all tasks claiming a record of some sort besides that of a reliable memory. Of the amount of such work it was easy to judge merely from the crowds of

Jirgahs (tribal councils), local khans and other individual applicants whom I saw gathered early in the morning in front of the Miangul's terrace. It seemed a very convenient substitute for that *takht,* or raised seat of judgment, which Indian as well as Central-Asian tradition necessarily associates with the function of a ruler.

"When returning in the evening, it was quaint to meet the same miscellaneous host of Khans, with their followers in carefully separated groups, and humbler folk walking slowly on the wide road which the Miangul has made between Saidu and Mingaora, a flourishing big place by the river some two miles off. They had taken relaxation from long waits and pleadings by watching the football game played by the schoolboys in a field laid out halfway, and were now coming back in the hope that the Miangul's fleshpots would soon enable them to break the day's fast. It is a rule that all those who seek justice from him must be entertained as guests, if need be for three days. A wise practice it seems, both for softening the feelings of the disappointed litigants and for expediting judgment. Anyhow, the fiscal expense implied by this method of legal hospitality was obviously in the Miangul's mind when on my request to be allowed to pay for all supplies furnished to my party he smilingly referred to its smallness as compared with the hundred or two of the 'guests' his kitchen had to entertain daily. I was told by others of the reasonable arrangement which divides the 'guests' into three distinct classes: one, partaking of the ruler's own table, another having meals with his Wazir, and a third, made up of the common herd, being served straight from the big kitchen. I have little doubt that this third is quite as exacting in quantity & not less critical as to the cooking.

"That all those parties taking their cases before the Badshah walked about fully armed need scarcely be mentioned. Even the humblest Pathan cultivator aims here at the possession of a rifle, revolver or pistol. Obviously it is a good investment to acquire an up-to-date weapon and the need of understanding their sights has, I imagine, something to do with the knowledge of European numeral figures of which I found curious epigraphical evidence in sgraffiti on boulders, etc. How they might puzzle an archaeologist in the distant future!

"Since leaving Saidu my escort has been increased to a detachment of the Miangul's own guards, about 25 men, under the charge of the 'Sipahsalar,' or commander of his forces, a pleasant enough man and brother of the Miangul's capable Wazir. He had had our camp pitched in the centre of the large village, on the terraced court of its principal landowner. At first I did not much relish the cramped space, the noise made by a large company of cheery men-at-arms, etc. But when the continuous downpour soaked everything outside my little tent, I felt glad enough for the convenient darkroom which my excellent handyman had managed to improvise in the Khan's 'best room' and armoury combined. Advancing 'civilization' had brought there even something like a table and two iron chairs, all most useful for work.

"On the 31st of March I moved my camp to Charbagh, a large village that might well figure as a third claimant to the dignity of having been the old capital of Swat. Here, too, ruined Stupas and debris-covered sites of ancient occupation could be traced in the valley which heads towards Ghorband and the Indus. As I looked towards the easy Ghorband pass and the wide open valley leading down to Charbagh, I thought of the relief & delight with which the old Chinese pilgrims travelling from Darel through the difficult gorges of the Indus must have greeted this view marking the end of their troubles. Pious Fa-hsien, who made his way down the Indus to Swat in A.D. 400, has left us a graphic description of this terrifying 'route of the hanging chains' with all its dangers & fatigues. No European has ever passed that forbidding route through the Indus Kohistan, but I could hear details about its 'Rafiks' and risks direct from Shah Alam. Splendid cragsman as he is and as nimble with hands & feet as any of his people, he talked with undisguised aversion of his seven days' exhausting journey down that succession of precipices above the swirling big river confined to a narrow channel. I wish I could do it yet with him. But there is no chance of the Pax Britannica opening a way in our time through those lawless Kohistani communities whose small settlements lie wholly outside the sphere of political or military interests.

"Shah Alam, together with Raja Pakhtnuwali's two sons, is now eating the bread of an exile from the Miangul's bounty. He does it with an innate dignity & with firm hopes of a turn in Fortune's favour. Since the downfall of his uncle & master, he has made an attempt to regain Tangir & Darel, in the name of his cousins, but it failed. He had found refuge first in the mountain tracts of Kandia. So I guess the support he now receives in Swat may be due to reasons not altogether altruistic. His present host is not without some cause credited with the ambition of extending his sway northward, and the claims of an exiled pretender might prove yet convenient for use. History in Kashmir and elsewhere records many examples of this; so there would be nothing new in the story.

"Yesterday [2 April] I began the journey up the river which is to carry me to the present border of the Badshah's sway in the Swat Kohistan. Beyond it lies Kalam and other alpine valleys, at present still independent, which may become a bone of contention between him and those ruling Dir & Chitral. Of course, I shall take care not to pass beyond the territory actually held by my host. The people of the Swat Kohistan are of the same Dard race which down to the Pathan invasion inhabited the whole of Swat; and specimens of the two Dardic tongues still spoken up there, together with anthropological data, will be useful.

"For the first two days of the move up the Swat valley the journey brought no marked change in the scenery. Though the hills on either side rose in height and approached the river much closer, enough room was left on its banks for wide terraced lands fit for rice cultivation. The abundance of water coupled with adequate warmth of the summer allows two crops to be raised in the year. The first is always wheat, oats, rapeseed, or a sort of lucerne, not unlike the familiar *bidar* of Turkestan. . . . The ruins of Stupas and monasteries, modest in size, were very welcome. They definitely proved that the prevailing belief of Buddhist remains not extending beyond Churbagh near the river's big bend was wrong.

"At Paitai where the camp was pitched on the more open east side, I had much antiquarian satisfaction. Fully twenty-eight years have passed since I received from Col. Deane the impression secured through a native agent of his (then not yet turned into a forger of such things!) from an inscribed stone. It showed two big footprints marked with Buddha's emblem, the wheel of universal sovereignty, and below them a line of bold Kharoshthi characters. I sent the impression on at once to Professor Bühler. In my accompanying note I expressed the belief that the inscribed stone, said to be situated at Tirat, a village of Upper Swat, was probably the same which Fa-hsien and Hsüan-tsang describe as showing the miraculous footprints of Buddha. Professor Bühler in the notice he promptly published of the inscription, a brief paper printed a couple of months before his tragic death on the Lake of Constance, proved this belief to have been right.

"Curiously enough local worship had received fresh nourishment by a recent event. Between the Stupa mound & the road I found a tomb marked by votive offerings of varied sorts. It holds the remains of an unfortunate Afridi who had come as a trader of rifles some 2 or 3 years ago and been murdered not far off by some men from the Kohistan. His tomb was appropriately placed here & is now receiving due worship as the resting place of a Shahid, or martyr. If life is of little account as yet in this region, there is anyhow some compensation to be found in the ease with which a violent death makes martyrs out of sinners who suffer it.

"Crossing once again to the left bank [we] reached [by dusk] the large village of Churrai, the gate to Swat Kohistan. Churrai, with its closely packed dwellings and lanes scarcely wide enough for two ponies to pass, proved quite a busy place of local trade. From these come all the closely woven & gaily but tastefully coloured woollen blankets which India knows as 'Swaiti rugs.' The chance to collect my first specimens of Torwali, a Dard language, detained me for a day. Its busy work was pleasantly varied by a walk through those narrow lanes where plenty of doors and shop recesses showed excellent carving in wood.

"From Churrai to Baranial, the chief place of the Torwali speaking hillmen, the valley rapidly contracted. Two days of torrential rain that kept us at Baranial were used to record Torwali stories. Torwali, like the other dialects of Dard stock spoken by independent tribes on the Indus gorges is bound to be swept away in the course of time by the ever-advancing tide of Pathan speech. The total population of Torwal can scarcely exceed 6000 souls; this comprises also quite a number of immigrant families speaking Chitrali or Pashtu. And in addition to all this mixture, there are plenty of my familiar Gujars grazing their cattle high up in the side valleys & cultivating on tenure such little plots as Torwalis can spare them.

"When on the day of my arrival I visited the village, I was greatly struck by the amount of fine woodcarving old & new to be found in this quaint rabbit warren of houses. The lanes between them are incredibly crooked & narrow. But a curious method of 'town planning' secures unexpected open spaces by the flat roofs of certain portions of these houses, most with two stories, being combined into little piazzas. These roofs are made accessible to fellow-citizens or strangers by primitive stairs or ladders. The houses are built mainly of timber with wattled walls which recalled those I had excavated at Niya & other Taklamakan sites. My general impression was that the methods of housebuilding in this mountain tract cannot have changed greatly since the times when Lower Swat with its fertile lands enjoyed its flourishing civilization of Buddhist times or when, even earlier, Bactrian-Greek chiefs with Hellenistic traditions in speech and material culture held sway.

"This earlier phase was vividly called up before my eyes by the remarkable variety of decorative motifs of purely Graeco-Buddhist art which the plentiful woodcarving in the doors, pillared verandahs, etc., of Baranial houses displays. Many of the most frequent designs—floral scrolls, bands of ornamental diapers, etc.—which I know so well from the Gandharan relievos, have survived practically intact in this far off mountain corner. Nowhere in Chitral, Darel or elsewhere is evidence of it so striking. I did

my best to secure specimens in old carved brackets & panels thrown away
as useless lumber, by photographs and by a local craftsman's drawings.

"At Baranial the hardy Swat mules had to be replaced by human transport—a road made practicable for ponies & mules would not have been
safe for laden transport after the damage which the preceding heavy rain,
with a few avalanches in places had caused. The two marches to Peshmal,
the Badshah's last village on the Swat river led up and down along steep
spurs and across picturesque valleys all flanked by mountains still snow-
covered. No archaeological remains were to be looked for, but it was delightful to feel that I was moving between mountains on which no European's eyes had rested before. To the east the high watershed dividing
the Swat river drainage from the Indus gleamed like walls of crystal—as
the old Chinese pilgrim Sung Yün wrote.

"Deodars and then firs & pines of magnificent stature became more &
more frequent. There was unfortunately much evidence also of the destructive way in which the Kara-Khel sept of Nowshera, those saintly
timber-contractors, who for years had free hand in their dealings with the
Kohistanis of Swat and elsewhere too, had worked here. Some of the
more accessible spurs had been badly denuded, while in places huge trees
had been felled & left unsawn owing to some squabble with the Torwali
woodcutters. This haphazard exploitation has been stopped since the ruler
of Swat established his hold on the Kohistan. He realizes the value of its
timber resources & is setting about to utilize them in a more intelligent
fashion.

"On the march to Peshmal, a rude Gujar hamlet, we had to keep low
down by the river. No more villages were to be seen here only scattered
homesteads clinging to the green slopes. It had been pleasant to note as
we moved higher & higher in the valley, how the number of fruit trees
increased. We had been travelling with spring in our wake and at Churrai
the apricots and apple trees were already in full bloom. But [at Pesh-
mal]—from six to seven thousand feet—the big walnut trees still stood
bare, only little flowers in sheltered spots announced the coming of spring.
Next morning with a temperature not very much above freezing point,
an early start with the Sipah-salar & his host enabled us to push to the
ridge which marks the boundary of Torwal. I had been promised a full
view of Kalam, the big village situated where the two main branches of
the Swat river meet. I was not disappointed. Nestling below an amphitheatre of mountains all still carrying snow lies a fertile little plain which
seems to mark the basin of an ancient lake scooped out probably by glacier
action. To eyes grown accustomed for days to deeply eroded gorges, this

plain looked quite imposing. Above it, only a little over two miles from our ridge extended the long lines of houses, built in tiers, which form the chief place of Kalam. Once again it had fallen to my lot to approach that huge barrier of the Hindukush, that true divide between India and Central Asia, by a new route.

"Reluctantly I hurried back to Chodram to reach it in time for due celebration of the 'Akhtar,' or Id, which marks the close of Ramzan. To-day's [15 April] halt in honour of the great feast has allowed me to give a good treat to my Swat friends and hardy escort. It was a pleasure to see how quickly the big heaps of pilaw disappeared among three scores of mouths which had for a month not touched food or drink in daylight. Two days hence I hope to turn off from Khwaja-Khel to the mountain tracts above the Indus, in some ways still less known than the high valley just explored.

"On April 16th the return journey down Torwal was resumed. A short halt at Churrai allowed me to acquire two fair specimens of those gay-coloured Swati blankets. Unfortunately aniline dyes have begun to affect this home industry with the usual sad results. I soon had occasion to feel grateful for the warmth these heavy woollen rugs gave me.

"I was now anxious to get to work east of the Swat-Indus watershed. On the second day, after a sunny and warm afternoon I was little prepared for the heavy downpour which followed a thunderstorm late in the evening and accompanied by violent wind continued all night. My fear of the tent collapsing through the wooden pegs giving way in the sodden soil of the terrace on which our camp was pitched, was not realized. The day which followed was wretchedly wet. Once again I admired the cheerful goodwill of our escort. Few of them have more than a single suit of thin cotton garments, with perhaps a large sheet of the same material to sleep under. That all of them sport solid leather bandoliers or more elaborate Sam Brown belts neatly packed with cartridges does not give protection from cold & damp.

"Our goal was the main village of the side valley of Lilaunai. A steep descent to the confined bottom of the valley confirmed the first sombre impression. Swat and adjacent tracts are singularly chary about arbours or orchards such as the living might enjoy. The Pathan system of *vesh,* or periodical redistribution of lands among the families belonging to the several clans, may account for this. But in this respect the dead fare better than the living: around and over their graves thickets of Ilex, wild olives and other trees are allowed to grow up. Time turns good and wicked alike into 'Shahids' or holy martyrs, and the sanctity thus bestowed upon burial places protects the trees from the villagers' reckless axes. What was to be seen at Lilaunai after we had reached it did not help to relieve that gloomy aspect. The village, [said] Sipah-salar who knows all his master's conquests like his pocket, had suffered from internal feuds. Ruined homesteads, roofless houses, cut fruit trees, met the eye in more than one place before we reached the Badshah's newly-built fort which now assured peace—of sorts. The Papini Saiyids, who hold Lilaunai, claim saintly descent; but this does not prevent them from cherishing their enmities dearly & from being given to violence otherwise.

"I was welcomed by the report that the pass to the head of Kana had been made practicable for loadcarrying men. The baggage had to be divided into small loads so it was nearly 10 A.M. by the time a start could be made under a brilliantly clear sky. The snow-covered pass came into full view; at last towards 2 P.M. we reached the crest through a narrow track made by men sent ahead—they had beaten a path through 4–5 feet of snow.

"It was strange to be met towards the close of the climb by sounds of weird music produced by drum, fife and bagpipe! The little band of the Khan of Bilkanai had brought up the three performers with a large posse of carriers to relieve the Lilaunai men struggling up in the snow. The effect was distinctly encouraging, and long before I had finished with photographic work and my modest 'tiffin,' all the loads had changed hands and were rapidly being passed down the steep slopes towards Kana. So eager were the Lailaunai men to get clear of the bitterly cold wind blowing over the crest, the headmen would scarcely wait for the special reward I wished to distribute among the lot.

"For over two miles down progress meant a succession of jumps from one deep hole of hard trodden snow into the next—or immersion in soft snow. Then came a mile of wading in slush and at last dry ground was reached and the scramble down over trackless stony slopes, tiresome as it was, seemed a welcome change. We gained the bridle-path winding along the cliffs which line the bottom of a narrow deep-cut valley. The big spurs on either side of the valley bore clusters of storied dwellings. Little crowds of villagers who had descended from their heights were watching our progress—our visit was a great & unwonted event for this far-off valley. It was with a feeling of relief that I turned at last into the main valley of Kana.

"It proved unexpectedly wide and open. Even more unexpected was the grand fashion in which I found our camp laid out at Bilkanai. Amir Khan, who holds upper Kana in a kind of feudal tenure, to show due attention to his master, the Badshah's chief commander and his host, had enclosed the terraced field where our tents were pitched with lines of young pine trees carrying strings of gay paper flags in true Indian fashion. Neatly laid out paths between the tents made the whole scene contrast strikingly with the grim old fortress-residence of the Khan under the protection of which the camp was pitched.

"On April 24th after recording specimens of the Dardic dialect spoken by Duber men whom Shah Alam had collected, I started down the valley. The quest of Aornos was now drawing me eagerly to the south. But for the attraction of this goal, I should have gladly spared a little more time for Kana, so bright was the scenery this fertile open valley presented, so cheering the welcome its Khans and people accorded us.

"Kana, occupied by the Jinki-Khel clan, does not know that mischievous system of *vesh,* or periodical redistribution of all lands, which elsewhere prevents anything useful connected with permanent occupation being undertaken such as the planting of fruit trees, arbours, or systematic extension of cultivated ground. But Kana had not escaped that bane of local feuds and vendettas to which the *vesh* practice contributed so largely. The numerous ruined or deserted dwellings at Bilkanai supplied visible proof of this.

"The two Khans of Kana had attached themselves to the cause of the Miangul long before he gained overlordship. When in 1922 the Pathan clans east of Swat formed a confederacy to resist the Miangul's 'forward policy' the Khans had kept aloof though they could not prevent their tribesmen from sharing in the fray and defeat and they themselves had been besieged and had suffered losses. But with the Miangul supreme, their reward came: the Khans had been allowed to keep their forts and to exercise their old quasi-baronial powers.

"Naturally enough both were eager to demonstrate their loyalty by a special display of hospitality. So all through my move down Kana we were treated to gay gatherings of their armed retainers with fluttering banners and much martial music performed on drum and bagpipe. It is difficult to express clearly and briefly what it means for one with historical instincts to be in close, constant contact with men whose ways of thought and doings reflect conditions which in the West passed away centuries ago. To be able to observe the ease with which all things really needful could be arranged for the favoured few under a quasi-medieval regime, was exemplified in my own case. The advantages thus enjoyed were great indeed and might well explain and excuse a certain reluctance on my part to return to 'civilization' and all its exactions.

"By the afternoon of April 25th we had reached the lower end of the valley of Kana. We had to leave it by two slightly trying river crossings— the river of Kana and that of Ghorbund into which it empties, both far too deep and rapid to be forded without serious risk. They are bridged both above and below the confluence where rocky defiles confine them; in both places the bridge consisted of a single big rafter, sagging a good deal and swaying. I confess, it was with relief that I found myself safely on that side of the Ghorband river which is reckoned the tribal area of the Azi-Khel centred at Chakesar. A long ride up the valley brought us by nightfall to the village of Upal: from this place, some 5000 feet above sea level, the search for the likely location of Aornos was to be started. This could only be done on foot. So the mules which had brought our baggage & our riding ponies were discharged and dispatched across the Upal pass to await us at Chakesar. I was pleasantly surprised at the ease with which the hundred-odd load-carrying men required were collected by the morning of April 26th. Occasional labour of this kind forms part of the conditions of tenure upon which the Gujar and other 'Faqirs' are allowed to cultivate the land owned by their Azi-Khel masters.

"It was a day full of eager expectation for me, but not free from anxious uncertainties either. For two preceding nights I had carefully read over again the descriptions which Arrian and [Rufus] Curtius have left us of Alexander's great exploit at Aornos. Detailed enough as regards incidents of the operations by which the formidable rock fastness was captured, the topographical indications seem singularly wanting in precision. How I wished that the great Alexander had brought in his train some man of letters accustomed, like a Chinese Annalist, to keep his eye open for topographical facts and to record them and essential military events with impartial clearness! The best I could hope for was to find a ridge or plateau which by its size, position and configuration would account for the character of the operations recorded to have led to the conquest of 'the Rock.'

"All the more grateful I felt when what proved a long day of physical toil soon brought an encouraging omen. Some ruins of heathen times, had been reported by the grey beards of Upal on a little spur visible from the village & about 1000 feet about it. So I soon left the stony ravine in which our long baggage column was moving and climbed up the spur of Chat.

"I had my reward for this detour; for the little plateau on the top showed not merely well cultivated fields of a dozen Gujar households but also the remains of a small circumvallation whose characteristic masonry of the 'Gandhara' type could be recognized at once as dating from the

Buddhist period. Search in the fields close by soon brought to light fragments of decorated pottery of the kind I had learned to associate with ancient structures in these parts. It was a pleasant surprise when Mahmud, an intelligent young potter from Upal acting as our guide, picked up before my eyes a well-preserved bronze bracelet, of unmistakable antique shape and showing a snake head. It came from below the ruined walls. The rain of the preceding days had loosened the ground there and prepared the discovery.

"The reward I paid on the spot as always in such cases made Mahmud beam with joy. He was not aware that that ounce of silver was meant to be a return also for an important bit of information he had given me in his talk while climbing. I had been cautiously testing his knowledge about localities further east and in particular about the high ridge of Pir-sar [Holy man height, or Guru peak] having been proposed as most likely to answer the description given of the place to be looked for. Of course, Mahmud knew nothing of the great Sultan Sikandar [Alexander the Great]; nor did my repeated careful enquiries among local Pathans, Gujars, Mullahs reveal the slightest indication that folklore or quasi-learned tradition in this region connected the exploits of the great hero Sikandar of Muhammadan legend with Swat and the adjacent hill tracts. But in the course of my talks with humble but sharp-witted Mahmud, I heard for the first time the name of Mount *Ūna* mentioned. It was believed by all people, so he said, to be the highest peak on the range which stretches from the pass of Upal to the Indus, and just below it lay Pir-sar, a big alp cherished by the local Gujars as the best of their 'summer settlements' both for grazing and cultivation.

"It did not take long for my philological subconsciousness to realize that Ūna (pronounced with that peculiar cerebral n which represents [in Sanskrit] a nasal affected by a preceding or following r sound) would be the direct phonetic derivative to be expected according to strict linguistic evidence from the Dardic or Sanskrit name which Greek tongues had endeavoured to reproduce by *Aornos*. But such philological indication could count only if actual topographical facts about the Pir-sar should prove to agree with the details concerning the siege of Aornos which have been handed down to us from reliable sources.

"My eager wish to reach Pir-sar, if possible, the same day made me think little of the steep climb of some 2000 feet which carried us from Chat to the crest of the range. The onward march proved far longer and trying than I could have foreseen. The track, if such it could be called, led along the precipitous and rocky south face of the range and kept close to the crest with all its ups and downs. It was 6 P.M. when we passed a fine spring gushing forth and after half-a-mile's scrambling further we reached the open crest of a rocky spur which descends straight from Mount Ūna. From here the long and remarkably flat ridge of Pir-sar was first sighted. It seemed near enough as I looked across the deep valley which separated us from it. In the end three more hours were needed to reach the goal.

"First we had to get past the stern massif of Mount Ūna. Climbing along its crest we reached at last a small shoulder, some 200 feet below the actual summit, where it was possible to take to the well-wooded side of the mountain. Here the descent began along cliffs and steep gullies where the snow still lay thick. It brought us to the small, tree-girt alp of Burimar which had seemed to link up with the wooded height. But now as we passed down the grassy slope of Burimar, the fact was revealed to my eyes that a deep and precipitous gorge, previously masked by close forest, still separated us from that height. It was a surprise, far from pleasant at first sight, for tired men now that dusk was rapidly coming on.

"As under Mahmud's guidance we were stumbling down among rocks & pines into that narrow gorge until its bottom was reached nearly 1000 feet below the alp, a thought soon began to cheer me. Was this not the deep gap on Aornos which at first baffled the Macedonian attack? In the growing darkness there was no time to examine the ground to consider the expedient by which Alexander had managed to bring his slingers and archers after days of toil sufficiently near to the opposite side to reach the defenders with their missiles. When at last we arrived at the bottom and I found that the little pass there formed by it was less than 40 yards long and only some 8 yards across, I began to grasp what the 'mound' had meant which that great commander had caused to be built with cut trees and stakes across the gap in order to render the assault possible.

"We were now indeed on the slope of Pir-sar itself—and did not have to fear the stones and other missiles which an enemy holding the heights would find it easy to hurl down. Yet the ascent was exacting enough for its fatigue to remain well-fixed in my memory. Fortunately towards the end the moon rose sufficiently high to make footholds less difficult to find. But I was heartily glad when at last by 9 P.M. we had reached open ground at the end of the ridge. The fact of our having to tramp a mile more over practically level ground served to impress me still more with the remarkable natural features to which this ridge owes its present attractions for the Gujars and its fame as an ancient mountain refuge. In the centre of the long flat plateau I found our big camp pitched and escort and load-carrying men gathered around the big bonfires, glad for the protection they gave from the bitterly cold wind blowing. It was long after midnight that I could seek warmth myself amidst my rugs. But the growing conviction that Aornos was found at last, kept my spirits buoyant in spite of benumbed hands and weary feet.

"Violent gusts of wind shaking my little tent left but a poor chance of sleep before I rose next morning at daybreak. The view spread out before me was extraordinarily wide & grand. With the icy blasts blowing down the Indus from the snow-covered ranges of the Kohistan comparatively so near, it was difficult to enjoy it fully until midday had brought calm & warmth. It was the same throughout the three days which we spent on that exposed height. But all the same it was a most fascinating time of work, favoured by continuous clear weather. The task implied a detailed survey of Pir-sar and the whole ground which surrounds it, supplemented by whatever information could be gathered about practicable routes, cultivation, water-supply and other local conditions. That some of the details recorded, especially in the rhetoric description of Curtius bear distinct marks of exaggeration can no longer be doubted. But considering the forbidding difficulties of the ground and the great distance which separates this mountain tract from the Peshawar valley and from the resources for an army to be gathered along that highway to the Indus, one might well feel surprise if such exaggeration had been kept out altogether from the story as handed down to us in accounts composed centuries later.

"The longer I studied the ground with all its formidable obstacles to the movements of large numbers of men, the more I felt amazed at the immense energy & ambition which made Alexander try his strength at the conquest of so inaccessible a stronghold, and quite as much too at the devotion and power of endurance shown by his Macedonians. But then the whole of Alexander's triumphant achievements from the Mediterranean far into Central Asia and India is full of incidents which testify to such combined boldness and skill as would be sought in the divine hero of a legend rather than in a mortal leader of men.

"At no point of the site did I feel more the awe-inspiring effect of the great achievement than when I stood on the fir-clad hill which rises some 250 feet above the end of the plateau and guards the approach from the gorge. On its very top I found, deeply buried in the deposit of long ages and overgrown in places by big trees, the remains of ancient walls unmistakably marking an oblong fortification. It was here, I believe, that Alexander was 'the first to reach the top, the Macedonians ascending after him pulling one another up,' as Arrian tells us. Some three hundred feet below I found a small flat shoulder, exactly corresponding to the 'small hill which was on a level with the rock.' To it a few Macedonians forced their way on the fourth day and fell. But it was their audacity which caused the Indian defenders to abandon further resistance. As I visited the scene all the points of the record became clear.

"During the three days of my stay there was plenty of life on the Pir-sar plateau, Khans from the villages below towards the Indus coming to visit us, strings of Gujars coming up with supplies, etc. The Sipah-salar and all his host bore up bravely with exposure & cold. But there was no attempt to disguise their relief when the completion of my tasks made it possible to fix the start for the morrow.

"It was hard to take leave of Aornos on the morning of April 30th. A long march brought us down to Chakesar by evening. Here I felt brought back to the present and a somewhat turbulent recent past. Regard for the exposure & fatigue which those happy days on Aornos and the marches to & from it implied made a two days' halt advisable. It felt warm at a little below 4000 feet and flies as well as mosquitoes were celebrating spring revels all around my small tent. For me it was welcome to record a hitherto unknown Kohistani tongue, spoken only by some 60–80 households of the small Dard community of Batēra. Even Polyglot Shah Alam who first told me about it, found that form difficult to interpret. Fortunately we found two intelligent visitors from Batēra speaking Dubērī as well as Pasthu in addition to their own tongue. So the taking down of linguistic specimens for Sir George Grierson's learned analysis did not prove too troublesome.

"Chakesar, close to the point where four much frequented routes from Kana, Ghorband, Puran and the Indus valley meet, has plenty of good cultivatable land held by the virile Azi-Khel clan. This accounts for the local importance which the place claims not merely as a petty trade centre but also as a seat of Muhammadan theological learning. The visit I paid to the four 'Schools' or Jumats, established in the shady courts of different Mosques, was quite interesting. Half a dozen 'big Mullahs,' headed until a year or two ago by a famous expounder of the Law, are attracting students from all parts of the Frontier region. Since Abdul Jalil's death the fame of the small medieval University has become somewhat dimmed. Yet over 120 'Talib-ilms' [students] were still to be counted, all of them as well as their teachers maintained in food & quarters by the Pathan landowners of the place, with some contribution from the tithe now levied by the Badshah.

"The amenities of Chakesar—good rations, helpful local patrons, and an equitable climate—are appreciated by wandering students. Quite a number imbibing instruction in Muhammadan law & religious tradition, in addition to Arabic grammar & belles lettres, were said to have come from different parts of Afghanistan. When I enquired for one with whom I might converse in Turki, a jolly-looking greybeard was produced with evident pride: he came from far-famed Bukhara. At first he was much embarrassed when I started talk with him, for he had left Central-Asian parts some 11 years ago. But gradually Turki came back to his tongue. For fully 27 years he had been a wandering student and thought himself still far from that standard of knowledge which would allow him to settle down as a guide to others in matters of holy law.

"Another aspect of the Talib-ilm's life was presented by the ruddy-faced Özbeg youth, Magdum, from Afghan Turkestan. After studying at Kabul & Jalalabad, for the past year he had found instruction and feeding to his taste at Chakesar. He was pleased to hear himself called up in his Turki mother tongue while sitting with other students. But for some little time speech neither in Turki nor in any other tongue came from his lips though they kept steadily moving. At last the laughter of the crowd watching brought the explanation: the young fellow's throat and mouth were choking with the rice he had pushed in from the big platter containing the morning meal that he shared with five others. Eating meant competing in quickness of absorption &, he had taken special care to store away what he considered his due share before abandoning the treat to them. When he had at last struggled through these additional mouthfuls, we had quite a cheery talk about his past studies & his hope of soon continuing them at Delhi.

"The day's march led us across the Kaghlun pass into the wide fertile valley of Puran. The gorge through which the ascent lay was clothed with big clematis in full bloom; rhododendron trees in full bloom furnished the escort with glowing crimson nosegays stuck into the muzzles of their rifles. The descent to Aloch, the chief place of Puran, was done in a drizzling cloud. The tents came in late and were pitched in pouring rain. Two pleasant marches took us through fertile Puran and the adjoining Mukhozai hill tract never before seen by European eyes. At Chauga village we halted for the night. Here the Badshah's invading force commanded by his Sipar-salar, my present protector, four years ago was hemmed in by the confederated tribes of Puran, Chakesar, Ghorband and Kana for seven long days of intermittent fighting. With justified pride the alert commander-in-chief pointed out to me the steep rocky spurs on either side of the valley from which he at last succeeded in driving the greatly superior force of the besiegers. Their hold had slackened, probably through their supplies from the homes failing, as is usual with Pathan tribal gatherings in such cases.

"On May 5th, over the 7000 foot Nawal-Ghakhai pass, we reached the northeastern border of the big territory of Buner. With pleasure I greeted again the passes & valleys so well remembered from more than 28 years ago when doing my 'archaeological tour with the Buner Field Force.' Far away to the east I could take a farewell view of the long stretched range with the rugged height of Mount Ūna, or Aornos, standing guard over its Indus end. My camp stood for two nights in the picturesque valley of Gokand and a daylong ride down the valley was rewarded by the discovery of interesting remains below the village of Bagra which in 1898 had been beyond my range. In the little side valley of Top-darra I discovered a fine Stupa remarkably well-preserved in spite of the luxuriant growth of vegetation which had managed to effect a lodgement in the solid masonry of the hemispherical dome and its triple base.

"I made my way across the Rajgalai pass, on 7th of May, to Pācha, a large village with Buner's most famous shrine, holding the remains of the saintly hermit known simply by the name of Pir Baba [Holy Father]. On the way I stumbled, as it were, upon the remains of another Stupa which owing to the decay of nearby Gujar hovels had remained apparently unknown to the local informants. [A fine drizzle made it] a somewhat exhausting experience of an aerial Turkish Bath. There was no ignoring the fact that over the big open valleys of Buner hot weather was beginning to set in. So I felt quite glad when by May 10th, I was able to start for the last task of my exploration programme: the visit to the top of Mount Ilam. Legends of ancient date, as the account of my Chinese patron saint proves, cluster round the peak—its conical head of precipitous crags rises to 9200 feet in noble isolation about the wooded spurs which radiate from it.

"The ascent was made in two days. The first brought us to the small village of Ilam-Kile, nestling in alpine seclusion. From there the final climb of some 3000 feet to the top of the peak was accomplished before midday of May 11th. The climb took us steeply over and between big masses of much-weathered rock, often carved into quite fantastic shapes by prolonged water action. In places deep gullies still filled with snow, were crossed. Then at last the tower-like mass of rock forming the main summit was gained: four isolated crags like pinnacles of a square church tower and, on the easternmost of these, a small platform artificially enlarged. Here local worship continues the Hindu belief that locates the throne of Ramachandra, an incarnation of Vishnu; an annual pilgrimage brings to it Hindu families in considerable numbers from the villages of lower Swat as well as from adjacent parts of the border.

"But Mount Ilam rarely fails to wear a cap of cloud after midday, and it was only the following morning that descending over precipitous crags for some 200 feet I found a small tree-girt hollow which holds a little spring and a string of round pools fed by it, worshipped by Hindu pilgrims as sacred features of the site. My eyes rested particularly on a stretch of big flat-topped rocks alongside. There could be no doubt that Hsüan-tsang had them in mind when he referred to the large rock tables shaped as if by the hand of man which were to be seen on 'Mount Hilo.' The morning of the 13th of May saw us start down to the valley which leads to Saidu which we reached the following day where a hearty welcome awaited me from its ruler.

"A suite of little rooms, rather stuffy & dark, quite close to the Badshah's airy reception hall, was set apart for me and there I passed the 15th occupied with farewell visits from those who had shared my peregrinations and with the distribution of suitable presents. The Chief Commissioner had been kind enough to provide them for me out of his own Toshakhana [official gift fund] in special recognition, as he put it, of the 'political' value of my tour.[6] Both in my own quarters and in the private apartment of the Badshah I had long talks with the ruler who like an old friend discoursed freely on the trials & struggles of the past and his plans for the future. On the morning of the 16th I said goodbye to my kind host & protector and in the course of less than four hours his motor car carried me with Sipah-salar & Shah Alam to the Malakand. Sooner than I wished the grim mediaeval-looking towers and *enceinte* of Malakand were passed and under the hospitable roof of Mr. Metcalfe, the Political Agent, perched high above the pass, my Swat expedition found its end."

At the year's end Stein answered a postscript Madam had added to Publius's letter; she was sounding him out about their desire to promote Stein for the Boden professorship. "Let me assure you at once that I deeply appreciate even the suggestion that I might possibly be considered worthy of such a great academic distinction as the tenure of the oldest Sanskrit chair in England. . . . Let me confess to you in all truth that I do not feel equal to such a charge and could not accept it with good conscience even if it were spontaneously offered. For fully 27 years I have drifted almost completely away from the Sanskrit studies of my youth. I need not blame myself for this. A kind Fate had given me chances of working in other fields, appealing more to my taste and also my abilities. I have used those chances as well as I could."[7] Stein knew that he had not kept up with the advances being made in Indology. "If I were now to attempt the preparation of lectures which could serve as guidance for research and advanced work, it would mean such efforts in absorbing the results achieved by others that I could not possibly produce anything really useful myself.

"I must regret my present limitations, but in my 'present birth' I have vowed myself to original work and . . . full freedom to devote my labours to such subjects as really interest me." Many expeditions and the hope of more to come, far from academic study and reference libraries, had taught him that his true love was "in the field where geographical & archaeological observation enables me to clear up the past within my modest limits." Modesty, expected of eminence, dictated his language but not his refusal.

Could Stein, who over the years increasingly recalled a person, place, book, and occurrence, fail to remember those early years when he longed for a professorship in Europe? Could he forget the years when he was harnessed to the irksome academic schedule of Oriental College and had to wangle time for his Kashmiri verification of the *Rajatarangini?* Or the nightmare months when he felt trapped in the Calcutta Madressa? The reasons he gave Madam and Publius were eminently reasonable, but his refusal had deeper roots: he had pledged himself never to become tied to the academic schedule. Time and his extraordinary fieldwork had vindicated his preference. In the quarter of a century that had passed, Sanskrit had become a tool he occasionally used, not an end in itself. He had moved outside the confines of classical philology into the wider world of Eurasian interaction.

6

The Borderlands and into Persia
1926–36

25

Age did not slow Stein down. Though most men in their mid-sixties turn to leisure and the quiet pursuit of hobbies, he maintained his strenuous pace—a tight schedule of expeditions and writing. To him, as to all scholars, the detailed reporting was an integral part of research. "I am hard at work daily from 7 A.M. to 6 P.M. pushing on as well as my poor pen, and my inveterate aim at critical soundness, permit the account of my Swat expedition."[1]

The tours made in this decade took him into British Baluchistan and Makran on a Fourth Expedition to Central Asia, and four times into Iran. Fieldwork sandwiched in between the vast, preparatory correspondence and the subsequent reporting was the tasty, nutritious filling to sedentary chores, the discoveries were the mind-filling condiments that gave his work wide appeal. Most of the tours were quite short; none was easy.

Stein's muscles remained as firm as his enthusiasm. His heart would know terrible desolation when, in quick succession, death took first the Saint, then Publius, then Hetty; and Stein voiced his need for a living object, close at hand, that he could love. "What I miss greatly is a Dash. Never since 30 years have I been without one on the Marg," he had written Publius in 1926, "and every little nook among rocks & thickets recalls Dash IV who was so happy here during three summers," Dash the Less, who had gallantly traveled on the boat to England after Stein disembarked at Egypt.

It will be remembered that in 1904–5 Stein had reluctantly acceded to Marshall's request that he investigate a mound "at a little place called Nal" in Baluchistan. The tour, he had felt, interrupted his effort to finish

the Detailed Report of his First Central Asian Expedition and his preparations for the Second. He had, as he then told Publius, "rushed at great
speed through a considerable part of the barren uplands known as the
Loralai District."[2] As a classicist and student of Vedic and Avestan writings, he had considered Baluchistan's "maze of rocky gorges and desolate
open valleys" devoid of what was germane to his studies.

The meaning of Mohenjo-daro and Harappa suddenly gave significance
to the Baluchistan mounds—mounds conspicuous in height and liberally
peppered with cultural debris. In them Stein would seek for evidence
which could bear on the genesis of the Indus Valley civilization. As the
sands of the Taklamakan had yielded proof of distant Eurasian contacts
and communication, so the Baluchistan mounds might indicate how, in
the third millennium B.C., the cities of the Indus Valley had come into
being through impulses started in Mesopotamia.

The Indus Valley civilization indicated a culture with a comparatively
high standard of domestic comfort: "houses built with burnt bricks and
provided with baths, hypocausts [heating vents under the floors], elaborate drains, etc., [and] the lay-out of paved streets."[3] Metal crafts, bronze,
copper, and gold jewelry, were fully developed; engraved seals indicating
the rise of a true glyptic art; figurines of stone, terracotta, and bronze—
some of a Mother Goddess and others of bulls sacred to the Hindu god
Shiva anticipated Hindu divinities; cultic objects like the linga and yoni—
all bore witness that distinctly Indian forms of worship antedated the
coming of the Aryans. Again and again Stein would be struck by the fact
of the "great antiquity of religious notions which are different from the
Vedic texts, and manifestly pre-Aryan, but still predominant in the popular cults." It was to document whatever links there might have been between the early culture of the ancient Near East and the Indus civilization
that Stein made his archaeological reconnaissances into the area connecting them.

Baluchistan was one such. Now the westernmost province of Pakistan,
it was part of imperial Britain's 1200-mile-long Northwest Frontier Province, extending from the Hindukush to the Arabian Sea. Stein had already become familiar with much of its northern sections; now he was to
examine the country south of the Khyber Pass. Topographically, it falls
into four sections: the northwest, including the rich Quetta valley, Zhob
and Loralai districts, has the river-systems of the Zhob, the Pishin, and
the Anambar; to the south, tucked away within surrounding hills, is the
flat Kacchi plain; central Baluchistan (formerly Kalat State) includes the
Sarawan and Jhalawan districts; and far to the south, fronting on the

Arabian Sea, is the Las Bela district abutting on the Makran. This coastal area, 350 miles long, cut by utterly sterile ranges and their almost equally arid valleys, was in ancient times Gedrosia, the poorest and least known of the Achaemenian Empire provinces. Gedrosia, the scene of Alexander's disastrous overland march from the Indus to Babylon, the forbidding coast of the Ikhthyphagoi, the Fish-eaters, along which his fleet had sailed, appealed strongly to Stein. His search for prehistoric remains would allow him to follow in the Macedonian's tracks.

The tour began, 7 January 1927, at Dera Ismail Khan, from where he was taken by automobile almost to the Afghan border. "I am still within the reach of the British guardians of the border. My first move brought me to Draban where I found numbers of sites with prehistoric pottery, etc., to explore on the surface. It was a lucky hit to pick up the thread last dropped in distant Sistan. Yesterday was spent profitably in similar reconnaissances around Tank & today I had a most enjoyable journey here to Miranshah with a visit to an interesting site of Buddhist times. Tomorrow I am to ride across the hills to Spinwan; thence my real progress into Baluchistan is to begin. I could not have wished for a more interesting time on a part of the border which in 1904–05 I had barely been able to skim here & there."[4] How different is Stein's tone from his earlier experience, a difference due to Stein's reading the message of Mohenjo-daro and Harappa.

Northwest Frontier: not just heroic words conjuring up a romanticized past but the new British presence in Baluchistan at the time of Stein's tour. The Wazirs, a settled tribe, occupied North Waziristan, while the Mahsuds, famous as border raiders, predominated in South Waziristan; for both the blood feud was an honored institution. The British "circular road," a modern limes which since 1923–24 "keeps a hold on the heart of Wazir & Mahsud country,"[5] brought Stein to Spinwan. Near the post he surveyed an ancient mound "built up with the debris of centuries, probably long before history dawned in this region with the advent of the Persian Great King's rule." He spent the night at the district commissioner's home, which offered such luxuries as electric light and a large bathroom with tiled walls—far different from the gardens where he had camped in Turkestan. "Then the car carried me down once more through the valleys drained by the 'Zam' of Tank, territory which the valiant but ever troublesome Mahsud had won only a few centuries before from Wazir clans. . . .

"The opening of this southern link of the Limes has cost a good deal of fighting and aeroplane bombing, but still more of gold which for the last few years has been steadily poured into this poor tribal land. The Mahsuds while being slowly weaned from the traditional raiding on weaker neighbors, have been kept at nominal peace by contracts on road-making & plentiful allowances for guarding the roads and posts they helped to build. But in their playful mood they managed to blow up one or two fine iron bridges which span the torrent beds, and armoured cars are still usefully employed to remind them that the road has come to stay.

"At Jandola, a dreary place and an old post of the Waziristan Militia abandoned when the Wazirs rose in arms against the Afghan invasion of 1919, I found a nice Powinda puppy of the powerful breed which follows those nomads on their annual migrations. If 'Spin Khan' [White Khan] grows to full size, Pashtunis will be sorry when found by him near my tents on the Marg." Making his way from post to post, Stein came to Sarwekai where he found his party and baggage safely assembled.

Accompanied by an army captain and protected by armed pickets guarding the heights, Stein went forward to meet a "large posse of Wazir headmen clad in a picturesque mixture of tribal rags & old uniforms. There my escort left me and I went on ahead. It was a thoroughly unconventional gathering which greeted me in genuine tribal fashion. The 'Political Tahsildar,' an imposing looking & very capable man, had no easy task in introducing to me, one by one, dozens or more of Maliks. Singly, they could all probably lay claim to raiding exploits, but collectively they assured perfect safety and proved very pleasant company."

In the care and company of the Waziris, Stein made long marches through lonely wastes of rock and rubble. During the day an icy southwest wind blew and the nights were bitterly cold. He saw where scattered summer villages had been burned by the Powindas in retaliation for losses suffered from Waziri raiders. Looking about, it must have seemed as though man rivaled nature in creating that area of desolation: "A trying march in the face of a pitiless wind brought us to the valley of the Gumal, the ancient Gomati of the Rig Veda, the last of those rivers mentioned in a famous Vedic hymn, which I had wished to see since my early student years. Our way led over ground very difficult for the camels where it passes through the boulder-filled gorge of the Kanser-warai. Yet it is on the regular route which the Powindas, numbering some 40,000 people, follow year by year when coming with their camels, cattle, etc., from the Afghan highlands down to the Indus plain. There is no water to be found anywhere before the Gumal is reached, and I thought of the careful planning it must need to pass such a host encumbered with women & children over this ground. It is a recognized hunting ground for Wazir raiders such as the cheery men under whose protection I was moving. . . .

"I sighted a party of horsemen coming to meet us from the Baluchistan side of the Gumal valley. They were the men of the Zhob Levy Corps sent to meet us. The meeting of my Wazir protectors and the Levies who guard the border against their visits was not altogether free from a certain reserve—to put it discreetly. That night special sentries had to look after the safety of my tents, etc., in case my tribal friends, now free from responsibility, were tempted to show their skill. On Jan. 25 after I had taken a hearty farewell from those hardy Wazirs and after a long march under the protection of the Zhob Levies, a car carried me over a winding road to Fort Sandeman."

A week later, 6 February 1927, Stein wrote Publius from Fort Sandeman. Only four miles away was "the big mound called *Periano-ghundai,* the Witches' Hill. Because of its plentiful remains of good prehistoric painted pottery on the surface I selected it for my first trial digging in Zhob. It is far too big to be cleared completely without hundreds of men and months and months of work. But the trenches opened from its slopes with the 50-odd men we could raise, brought to light quite enough to give some idea of that civilization of the late stone age during which the debris accumulated. We found fine decorated cups & jars used with the burial rites, pots in which burnt bodies were deposited, stone implements & remains of the mud-brick walls which sheltered the living. So a useful start has beeen made." By 13 February he felt satisfied that "the links with the corresponding remains on the lower Indus & far away Sistan & Persia were clearly established."

Two days later he was clearing burial remains from another mound, the Mughul ghundai, higher up the Zhob valley. 20 February 1927: "The remains are ancient, too, but belong to a period when iron was known. It all means gathering useful experience on ground which is new. This has always been a very barren region but evidently in early times had a population which, if scanty, was less primitive than the present semi-nomadic one of Pathans." A scant week was all Stein could spare for excavation. "The saving of time is important and in the Loralai Agency for which I start tomorrow a large number of ancient mounds is awaiting examination. It will not be easy to do justice to them all before the close of March when the end of the financial year stops (awkwardly enough) the available exploration grant. Thus rules the Babu wisdom, and against it all the intelligent goodwill of Govt. is of no avail."

The pressure of scanty time was unrelenting. He moved to the big valley of the Thal to excavate a huge mound he had examined in 1904. Then his probing of the Dabarkot mound—rising more than 100 feet— had showed that none of its widely scattered debris related to Buddhism. But now his focus had changed, and his trial excavations spoke eloquently to him. Enthusiastically, he wrote Marshall of the "layers upon layers of debris marking occupation from prehistoric times onward. The materials for the study of successive types of painted pottery were abundant. Structural remains all in stamped clay or mud bricks were difficult to interpret; but beside them we found a well-made drain of big bricks, recalling in a small way what I saw at Mohenjo-daro.

"The last week [10–16 March] was spent in clearing the remains of a small stupa which so far has escaped all notice. It is probably the first ever discovered in Baluchistan. In the center of the circular base we found the relic deposit resting undisturbed on the live rock: it has small gems set in gold setting & pearls. More useful was the discovery of potsherds inscribed in Kharoshthi and Brahmi, mostly of the Kushana type. They will help to fix the chronology of the painted pottery brought from a considerable number of mounds in this area. I am now engaged in rapid visits to as many of them as can be reached within the available time."[6] (Six months later he could inform Marshall that Sten Konow, the Norwegian scholar, had deciphered and translated the potsherds from the monastery attached to the stupa. "They record provision made for the supply of water . . . and the name proves the extension of Kushan domination to that part of Baluchistan. Both records belong to the 2nd century A.D.")[7]

The final two weeks of his tour were spent near the Quetta cantonment surveying ancient mounds. "Most of them show pottery remains extending over a great range of time. Experience makes rapid diagnosis now a comparatively easy matter."[8] Stein's sharp, trained eyes had quickly adjusted to the new inquiry.

Duly installed on Mohand Marg, Stein had the relaxation and companionship he had craved: "Did I tell you that Spin Khan, now so very big he can scarcely pass under the table, has a very lively little companion? An engaging airedale pup who does as a substitute for a 'true' Dash?"[9] He settled down to a daily menu of concentrated work: he had promised Marshall to have his Baluchistan report finished by August. "Pushed on under some strain, it is happily nearing completion," he wrote Publius on 24 July. "It has not been easy in a tent to lay out, sort, & arrange into containers hundreds & hundreds of small antiques. In the course of writing up results many observations of detail, puzzling on the spot have fallen into order & explained themselves. I think something like chronology of prehistoric & later ceramic ware has been evolved for this region." *An Archaeological Tour in Waziristan and Northern Baluchistan* was completed on time.

While writing up his material, Stein was also projecting his tour for the next cold season. Two reasons made Las Bela and Makran his choice. A survey might extend Hargreave's important discoveries at Nal and Mastung, in the Kalat State, further west; at the same time it would examine a hypothesis that contact between Mesopotamia and the Indus had followed the coast. Stein himself was partial to the Las Bela-Makran region because it had never been systematically surveyed for archaeological remains. When he outlined his proposal to Marshall, he projected a five months' tour that would cover an area of 200 miles from Kalat in the east to the Persian border and 250 miles from Kharan in the north to the sea coast. It would start in November.

Stein was expanding his Swat P.N. into a volume to be published as *On Alexander's Track to the Indus*,[10] when he had an unexpected and most welcome visitor. Musa Akhun, the brightest of his Turkestan helpers, had come enroute to Mecca. Musa Akhun's garb and news of Central Asia fanned Stein's continuing hopes for another Turkestan expedition. While collecting his gear in Srinagar, Stein found his equanimity shattered. "Spin Khan and Dash, alas, were bitten by a stray obviously mad," he wrote on 23 October, voicing his despair to Publius. "Vaccine has been wired for. But even if they are saved I shall have to travel without them. Where to leave them is still a problem. Does fate wish to deny me a travelling companion?"

A veteran of desert travel, Stein welcomed and utilized whatever technological advance promised to expedite and ease the work. Still committed to traditional transport for cross-country marches—camels and donkeys— he suggested to Marshall that using two lorries as a movable camp would be advantageous. They had, he would learn, certain disadvantages.

"It was good to be marching again, restful to advance through barren mountains where there was peace and freedom to let thoughts range far into past, present & future," Stein, under the shade of a stunted tamarisk, could write Publius early in December.[11] He had been held up at the fertile oasis of Mastung waiting for camels that had been ordered to meet him. "I was very fortunate in being able to get clear of troublesome proofs of I[nnermost] Asia Index (full of misprints, mistakes & omissions hard to catch & put right)."[12] Four long marches into sadly thirsty mountains and drainageless basins, a "tame" desert almost empty of human activity, brought him from Surab to Nauroz-kalat, the capital of Kharan. At that little oasis he was welcomed by the great-grandson of Azad Khan "whose memory is fresh as a famous freebooter. He ruled his deserts for some 70 years & raided far into Afghanistan, Persia and the Indus borderlands. The descendants of the slaves he brought back have been returned only last summer."[13]

Stein carried a tank of water in case the Dodge lorries were stopped by rough going and unavoidably delayed. Panjgur, the capital of the Makran, "a narrow stretch of date palm groves, was the headquarters of the Makran Levy Corps, composed of wild looking Brahuis and Baluch Makranis. . . . Mr. Medcalf of the Telegraphs & his plucky wife gave us a Christmas dinner quite orthodox in its creature comforts." The party passed lonely border posts and a series of mounds whose potsherds connected his Sistan finds with those of Zhob, visited the year before. "Even if labour were to be found in this drainageless basin visited only by nomads, I should have to forego excavation to husband my time for the more inhabited valleys near the Gedrosian coast. There heat sets in by the end of February and the distances to be covered are great."[14]

As if to punish Stein for calling it a "tame" desert, the Makran showed its temper. Approaching Turbat, the region's largest settlement located in the least barren part, "some hours of heavy rain which in this land of utterly bare mountains means a big flood, cut me off from my camp. Within a few hours the river became quite unfordable. Fortunately, the Adjutant looking after 'road' repairs was on my side of the torrent and I could seek shelter in his tent & pass quite a comfortable night on two chairs."[15] That was just the beginning. The next day (11 January) Stein had the baggage started ahead on the camels while he planned to save time for writing and follow later with the two lorries. "Well, we had scarcely started when it began to rain and the detritus surface soon became very soft. By the sixth mile one lorry stuck in what was a shallow dry bed. Attempts to get it out of the spongy rut failed & after an hour or two of vain efforts during which the other lorry brought to help also got embogged, we found ourselves in a steadily rising flood.

"Fortunately we got stuck near one of those little rock islands which jot these wide valley bottoms. To that we managed to rescue my yakdans & bedding in time. Then sitting on the serrated little ridge in safety, we watched the flood spreading over a quater of a mile in width. Of course we were drenched to the skin long before the flood receded & we could leave our rather uncomfortable perch. Then we settled down on the soaked ground as well as we could for the night. A deeper flood bed cut us off from Turbat. A good fire was kindled with petrol saved from the motors & the men got dry & I got a change from the bedding. The night passed without too great discomfort.

"As the morning wore on help came from Turbat (though no food), So with the little baggage carried on two camels (in truth more reliable means of transport) we tramped 10 miles off to where the baggage train had halted. The men had regaled themselves on a pot of dates & I, with a good tablet of chocolate, did not mind missing dinner & breakfast. Later 40-odd men pulled the lorries out of the flood-bed & back to Turbat."

After this exercise in patience and four long marches, Stein reached the mound known as Sutkagan-dor. He had only to look at it to recognize its importance. "Interesting finds in the way of stone implements, ceramic remains & the like indicated its early chalcolithic period. Complete excavation would take months & far more labour than could be collected over the whole valley. Comparing the fair stone masonry walls, up to 30 feet high which enclose the main site, with the wretched mat huts which the present scanty population of Makran lives in, one realizes the ups & downs of civilization. Yet people living in such primitive habitations & clothed in a way to match, get their matches from Japan or Europe and their headmen are supplied against raids from across the Persian border, with 0.303 bore rifles!"[16]

Gwadur, the little port on the Arabian Sea, belonged neither to India nor to Persia but to the sultan of Muscat (a close relative of the sultan of Zanzibar). "In keeping with this quasi-African association, it had heat and a profusion of flies." Yet for Stein it had the special aura of places associated with Alexander: Aornos had witnessed the Macedonian's triumph, Gwadur his agony when he led some 12,000–15,000 soldiers accompanied by women, children, and camp followers through its hot, lonely, barren land. Arrian had described its natives as a savage, hairy lot whose nails were like claws. Among quite different inhabitants Stein spent three days (25–28 January 1928). "My men were much pleased to see the great water & enjoyed their fish treats. . . .

"A bitter north wind which set in as we marched along its great bay and still blows has given us a taste of the delectable climate of this coast even in winter. On the Gwadur hills the heat had been suggestive of the tropics and ever since we left I have used all the warm things with me & not succeeded in keeping the cold blast off under a blazing sun. . . . The green ocean which kept in sight to the south gives colour to this dreary coast and history invests it with romance. It was a pleasure to rest my eyes on headlands which Alexander's fleet had watched as their ships hugged the coast on their long voyage up the Persian Gulf. Gwadur is likely to have been touched by Alexander himself as he came down at more than one point to revictual his fleet with what grain could be requisitioned in poor Makran. It was a fine exploratory performance, this march through Gedrosia, but a great strategic blunder, too, attended by severe human suffering & losses."

February was half over when Stein returned to Turbat to work on the Shahi-tump mound which earlier he had marked as an important site. 24 February 1928, to Publius: "On the very top we unearthed burials which can scarcely be later than the time of Mohenjo-daro. The designs of the painted funerary pottery vessels, of which we found dozens and dozens, show descent from those I discovered in Sistan and Zhob. The chronological evidence is important & probably throws back the latter finds beyond the 3rd millennium B.C. The complete bodies brought to light had no 'treasures' (where should they come from in this poor land?) but are valuable as illustrating the customs & civilization. [More would have been retrievable] if only Makran had *less* than 7 inches of rain per annum. Fortunately I got up to 200 men for work so the task was pushed on quickly,—there is still an extensive area to be surveyed. Last evening [23 February] all finds were safely packed & I could prepare for the move eastwards. Some 500 miles have still to be covered between this & Kalat."

Stein's precious mail followed him into the Baluchistan wastes: Publius's letter of 31 January found him at the Kulli mound on 1 March. A messenger sent from Turbat "had covered close on 100 miles in something like 26 hours, with a single change of pony. Not bad going."[17] The Kulli mound was a site that would give its name to a culture. Stein recognized its importance and worked strenuously in the time he had supervising the excavation during the day and spending long hours at night writing up notes and sorting finds. "Seven days hard digging at Kulli in spite of the heat, glare and daily dust storms, mere scratching as it had to be, brought to light interesting relics of about the 3rd millennium B.C. The stone-bult dwelling houses let one realize something of the conditions of daily life."[18] Better than anyone else, Stein knew the pitiful inadequacy of his trial excavation. "The site might keep one busy for months and I still have a big area to cover before regaining Quetta. Here at Kulli I discovered quite a little town site."

Fifty years later and after considerable work done on that ancient horizon, Fairservis dates Kulli to around 2500 B.C., lists its early phases as pre-Harappan, and "in its latest phases [is] contemporary with the mature Harappan civilization of the Indus River Valley. Here we have clear signs that . . . town life was achieved."[19]

For Marshall, Stein summarized his findings in the great drainageless Kolwa trough (where Kulli is) and farther east in the valleys of Jhao and Nandara. "Prehistoric mounds of all sizes are scattered over this wide area in such numbers that it cost a great deal of hard work to survey them all and to carry out trial excavations at a few particularly striking ones. The extensive use of stone materials, brought in some cases from hills miles away, has preserved structural remains. Two or three might be described as small towns. . . . The pottery throughout shows close resemblance to the funerary ware of Nal. . . . It has become rather trying here owing to heat, glare, insect pests and other delectable features of the Makran spring, and I am glad that the remaining portion of my programme is now taking me to the higher valleys of Jhalawan. I hope to complete the tour by the third week of April and shall then hasten back to Kashmir."[20]

At the Nandara site Stein was puzzled "by what seems to have been stone funerary piles built up in the 'best room' of the dwellings. It is always good to meet new features not recognized before."[21] Here, too, is his first mention of the *gabarbands*, massive stone dams intended to catch rainwater. Later archaeologists would call such water-control systems "in the southern reaches of Baluchistan . . . a triumph over the exigencies of the region. The use of stone both as boulders and in cut blocks gave these farmers a fine advantage in salvaging the soil and water resulting from the short but violent rainstorms usual to these regions. . . . The gabarband system had its beginning in Baluchistan at least as early as the period of the Nal settlements [2500–2000 B.C.]"[22]

On his way to Kashmir, Stein paid a thirty-hour visit to Marshall at Taxila. Among the topics they discussed were questions concerning publications and, most particularly, Stein's retirement. After eleven years of successive extensions, Stein was scheduled to retire at the end of November when he would have completed his sixty-fifth year.

During that summer he wrote up his tour: *An Archaeological Tour in Gedrosia*. Looking back on his fieldwork he told Barnett that it could be "only a pioneering survey. Complete excavations at particular mounds would take months or several seasons. They will have to be left to others in the future."[23]

In planning his retirement, Stein mulled over two tours. Both appealed to him. The one, to the Indus Kohistan, where the river makes a great bend and cuts through the mountains, would be under the guidance of Shah Alam. But the latter's stated aim of regaining control of Tangir invited tribal complications. It is sad, he wrote Publius on 29 July 1928, that "his desired company stands in the way of that cherished plan. The chance is not likely to come my way thereafter." Time would prove him wrong: he would make that tour in 1941–42.

He outlined the other tour for Publius. "On the evening of Nov. 30 (when my time of service will have ended) I will set out for Karachi, reaching Basra on the 6th."[24] Woolley, who was then engaged in his successful excavation of Ur, and who had worked on Stein's collection as a young man, assured him that the great sites of Mesopotamia were accessible from the nearby railway. "Then I shall go to Baghdad, a base for Samarra & Ctesiphon [the winter capital of the Parthians]; subsequently by car along the Euphrates to see the Old Roman Limes stations. Palmyra I wish to see before going into Syria and I am anxious to see more of the hills about Antioch, full of early Christian churches, etc. From Alexandretta I propose to sail for Athens and by April I hope to have a reunion with the Hierarch for a short tour along the Dalmatian coast."

Letters sent enroute fill out the schedule. He stayed with the Woolleys for four days. "All W. had to show me was full of interest from the big storied Ziggurat down to the tombs he has succeeded in clearing to a depth of over 30 feet. His work is as thorough, capable & guided by intuitive knowledge as one expects in his case and he has fully deserved his rich harvests. All his results have been achieved with a minimum of staff. From Ur I visited Eridu which tradition in Sumerian times believed to have stood before the flood. It still awaits proper excavation & as it is some 15 miles away from water, Mrs. Woolley suggested it might suit me!"[25] From ancient Lagash, "a terribly dug about site," a thirty-five-mile drive over an "utterly barren plain of alluvium to ancient Larsa gave one an idea what a once fertile deltaic region teeming with humanity, has come to since irrigation stopped. . . . A further drive in places between rudimentary dunes brought me to Warka, a very large site. The Germans have set to work with a light railway, movable huts, studio, etc & are fitted with adequate financial backing. But somehow Woolley's improvisation & grasp of opportunities appeal more to my taste."

After visiting the famous sites along the Euphrates, he examined those along the Tigris. "The ruins of Samarra are of enormous extent, stretching about 20 miles along the river. Those which yielded most [to the archaeologists] are sadly upheaved. The heavy winter rains rapidly destroy whatever is exposed after long burial, & trenches [World War I], British mostly, run through the ruined area in many places, have done the rest."[26]

At Mosul a new experience awaited Stein: "Four long excursions of over a thousand miles in the cockpit of a little Bristol Fighter (6th Squadron, R.A.F.) allowed me to visit such very interesting sites as Hattra, Samarra, Arbela [Erbil] and the Jabal Sinjar,"[27] an escarpment that for more than a century, was important on the Roman-Parthian border. Stein found this introduction to aerial reconnaissance expeditious, enjoyable, and useful: in time he would apply it.

"From the Sinjar hills I made my way along the line of the Roman border to the Khabur River where many big mounds mark an ancient line of communication from the north to the Euphrates. I soon reached Dura-Europos, that most fascinating site of a Macedonian [Seleucid] & later on Roman colony, abandoned just about the same time as our Niya site. There was a great deal to be seen in its Hellenistic wall paintings, reminiscent of Miran."

But visits, sightseeing, air reconnaissance were not enough. Stein had set his heart on following a still unsurveyed route from Dura-Europos across the desert directly to Palmyra. At its height, under the rule of its queen Zenobia (who died soon after A.D. 274), Palmyra had been a great and lucrative entrepôt between East and West, trading with India, Egypt, Rome, and Dura-Europos. "On the 3rd [February] I started with an imposing detachment of Méharistes, Beduins of a French Camel Corps guarding the peace of this 'tame' desert. The distance to be covered was about 140 miles & we did it on camels according to my programme in five days. I managed to carry a plane table survey along the route. I had splendid guides, living compasses, or else on those featureless plateaus one could not possibly find one's way to the rare waterholes. The winds were cold & cutting but I kept very fit and enjoyed it hugely. Of Palmyra's ruins, wonderful in extent & size, I shall have much to tell you."[28]

From Palmyra to its modern counterpart, Beirut, then along the coast to Haifa; he visited Megiddo "where a large Univ. of Chicago expedition has been at work for 10 years or longer on the mound occupied on its top by the Solomonic stables." And so he came to Antioch, where his enthusiasm for its many Byzantine churches brought him an offer from the French advisor to the Syrian High Commission: "He would be delighted," Stein wrote Publius, "if I cared to tackle such a task. But it would mean another Rockefeller to provide ways & means—and an obligation to tie oneself down for years & years. So it is better not to think of such a tempting offer seriously."[29]

At Ragusa (modern Dubrovnik) Stein had his long-planned reunion with Arnold. On being knighted, Arnold the Saint became the Hierarch; whether in thought or person he imparted his own joy and excitement in intellectual exploring. Of the four close friends only Arnold was portly: "I found the Hierarch beaming in radiant rotundity," Stein reported to Publius.[30] It was during that fortnight spent together that the Hierarch prepared Stein for his death which followed in a year.

In July, Stein came down to London from Oxford—in addition to his usual rounds to the British Museum and the India Office, he attended the annual garden party at Buckingham Palace. There he received a cable-letter of the utmost importance: "Langdon Warner has written me he can get funds and from people who will give me a free hand and not pester me with lecture tours and lionizing."[31] It had been sent by Charles Rockwell Lanman (1850–1941), a Harvard University professor, a Sanskritist who had also studied with von Roth at Tübingen and who thus was doubly bound in friendship to Stein. In a few days Stein received Warner's letter.

On his trip through the Middle East Stein had visited the digs supported by Yale, the University of Pennsylvania, and the University of Chicago. Now, retired yet still possessed of worthy projects and the strength to carry them out, Stein seemed to be returning to his earlier pattern of seeking a "patron." The direction his search took is clear: he, who invariably used classical terms, who found the desolate Makran coast touched by the glory of Alexander, invoked not Croesus but Rockefeller; thus he had discerned where funds might be found and had enlisted "dear old Lanman" to be his intermediary at Harvard University. Again his luck was with him. Stein had indeed found a generous patron.

On Ancient Central-Asian Tracks is Stein's account of his three expeditions. The book originated as the Lowell Lectures given at Harvard University in December 1929. Popular in intent, they have a unity made possible by the vantage point gained by the passage of years; they are fitted together as the several parts of a single sustained exercise. His Fourth Expedition, undertaken the year after the lectures, was a fiasco. He never wrote it up.

Its exclusion from his published reports invites the biographer to consider it. How does it fit in with the others? The First Expedition had proved Stein to be as brilliantly effective a fieldworker as the *Rajatarangini* showed him to be a scholarly Orientalist—a pioneer in a vast new field for research; the Second, extending his investigations to Tun-huang and the Aksai Chin, made him a hero and brought him knighthood; the Third, extending his field of operations still further—into Mongolia on the east and Iran on the west—gave substance to his role as "Archaeological Explorer." The Fourth has a different libretto; it adds nothing to the exploration of Asia and its complex pre-Muslim history, but is rather an excursion into Stein himself: it locates where he stood in his career and the inflexibility of his innermost attitudes and habits.

The beginning coincides with a critical point in Stein's life. In 1928, Marshall relinquished the office of director-general to give full time to his excavations at Taxila and Mohenjo-daro. Macartney, after twenty-eight years at Kashgar, retired and returned to England. Stein himself, having talked and hoped and saved and planned to enjoy "freedom" from official duties, achieved it. To Stein, addicted to routine, the retirement that brought changes in work habits and work friendships was disquieting.

There is a poignancy in his insistent desire to return to Central Asia, almost claiming it as his own. There he had realized schoolboy fantasies: like an Alexander the Great, he commanded the past of an area he had added to Indian studies. More than ever he yearned to recapture those wonderful years when he had conquered high, snowy peaks, crossed mountain passes and waterless deserts, and found treasures that dazzled the astonished eyes of Europe. A tent, he had long ago decided, was his preferred home. He might liken himself to the nomadic Kirghiz, but there is the impression that he shared much with Baden-Powell, who dreamed up the Boy Scouts.

Stein's success in persisting deprived him of those qualities of accommodation that might have restrained him from seeking a Fourth Expedition. He lacked what might be called "cultural sensitivity." He had smarted at the indignity visited on Battenberg but was privately amused by the makeshift headgear worn by the queueless Chinese to celebrate their revolution. He never appreciated that the queue was an infamous kind of brand imposed on the Chinese by their Manchu conquerors in 1644. Macartney was also amused; but then his father had come to Nanking to subdue the Taiping rebellion and had stayed on to serve the Manchus in China and London. Both men merely thought of it as a Chinese custom that added a quaint and suitably exotic look to the scenery.

Then there was the matter of language. Though Stein constantly deplored not knowing Chinese, his ignorance encysted his insensitivity: he had no firsthand access to the tradition, the high quality and riches of Chinese scholarship. Stein was mainly interested in China as the recipient of Buddhism, the Indian-born religion, in its assimilation of the Gandharan prototype and the cultural transformations as it moved across Asia. In sharp contrast was his attitude toward the Kashmiri pandits whose language he delighted to speak and with whose sacred literature he was familiar. Just as he misunderstood the fierce resurgence of national pride on the part of the queueless Chinese, so he misunderstood the Chinese resentment at having pieces of their past carried off to foreign captivity. A measure of his insensitivity was his belief in the preeminence of Western scholarship.

Considering the Fourth Expedition, it is impossible to resist using as a metaphor Stein's mishap when his horse reared and fell back on him. "I stuck to the seat—only too well." Earlier he had had warnings. His dogged persistence, honed sharp by his long apprenticeship in learning not to take No for an answer, had hardened into willfulness. He could not forget the endless, time-consuming correspondence needed to get funded;

it had been his greatest problem. He could forget the momentary "diffi- culties" he had bravely ignored, surmounted, or had had cancelled: when on the Third Expedition Urumchi ordered him to cease and desist, he had disappeared into the desert to allow Macartney time to have the prohibi- tion rescinded; when the kindly amban at Turfan had made discreet, per- tinent inquiries, Stein had complied—but in the way a man asked to lock the stable door obeys after first letting the horse out. A less willful man would have heeded such incidents as warnings. But Stein persisted. The horse would have to fall on him.

The story belongs only tangentially to Central Asian research, but it does belong to Stein's biography.

It began with a routine inquiry; its escalation promised much to both Harvard University and Stein. "I received by last mail [3 August 1928] an invitation from the Lowell Institute, Harvard University, to give a course of six lectures on my C-A explorations (the fee being $600). Of course, I cannot accept for the next session, but there possibly might be an occasion for me to cross the Atlantic in the autumn of 1929."[1] Thus to Publius; to the Baron, he added significantly: "I would be interested if suggestions made to me from the U.S. regarding help towards explora- tion on my old field can be realized."[2]

In October Stein confessed that "the hope of obtaining support in the U.S. for a C-A journey is very problematical. It would mean a long public- ity campaign for which I have neither the taste nor special qualifications. The Chinese obstructionist attitude is in any case a very disturbing factor. A German geographical party just returned . . . [which] has told me of very trying experiences with the Chinese. And I have now learned that Yang, the capable Governor who kept order in [Sinkiang], has been mur- dered & several of his chief assistants with him. I only wish & hope that my old friend and patron, Pan Darin's son, who had inherited his father's goodwill towards me & had risen rapidly to influence by honesty & sound judgement of things Western, may not have been among them. Unless there is a *definite* prospect of support in the U.S. I shall not be able to spend the time for lecturing at Boston & elsewhere. By next spring things may be clearer."[3]

The dialogue thus begun between Harvard and Stein moved toward an arrangement both desired: Stein could tell himself that what had hap- pened to the Germans would not happen to him—he had friends in Sinkiang, and as he had coped with "obstructionist" tactics before he would know how to deal with them again. He could not imagine it other- wise, and what he wanted more than anything else was an expedition in the old style and with the old objectives. Publius had sensed Stein's un-ease at his impending retirement, yet even to Publius he could not admit the uncertainty that confronted him. He had written on 9 October 1928: "The intelligent support of Government while it was Indian, though not 'Indianized' gave me those chances of congenial exploratory labour in a big & new field which I could not have hoped for either with my modest resources or by help from private sources. But conditions have necessarily changed." Hargreaves, the officiating director-general with whom Stein had worked and whom he respected, did not seem able to cut through the red tape as Marshall had done. "The sacrifices of time exacted by the Babudom which now holds a poor minor department tightly in its coils have become too serious for one of my years."

"One of my years!," a desperate lament from a man dreaming of a strenuous tour.

Harvard approached Stein toward the end of what, in the United States, might be called the "Museum Period." For almost half a century noble buildings, opulent indices of the economy and secular cathedrals of civic culture, were erected to receive paintings, statuary, tapestries, armor, arti-facts, and elegancies of the past, large and small, with emphasis on Euro-pean masterpieces as well as assorted objects brought back by expeditions unearthing numerous sites of antiquity. Oriented toward democratic ends, the museums were aesthetic expressions of an acquisitive society. Wealthy men and women, advised by experts, indulged in splendid hobbies that were both edifying and competitive. Collectors moved in rarified worlds of finance and society; they were engaged in filling the spacious halls that, once they existed, needed to be filled.

Boston already had examples of Indian and Japanese art (thanks to Ananda Coomeraswamy and John La Farge). The time was ripe for Chinese art. In 1923, two years after *Serindia* was published, Langdon Warner, curator of Chinese art at Harvard's Fogg Museum, traveled to that repository of Chinese painting, the Caves of the Thousand Buddhas at Tun-huang. Though Stein had removed much of the manuscript hoard, he had respected the frescoes, contenting himself with photographs; War-ner surreptitiously lifted pieces off (using glue and heat) and brought them back to the Fogg. The second Fogg Museum Expedition, 1925, the pre-lude to Harvard's interest in Stein, was an exercise in frustration and futil-ity. In *Buddhist Wall Paintings* (Harvard University Press, 1938), a slim volume of text and photographs, Warner tells what befell him and his associates at Wan-fo-hsia, the Valley of the Myriad Buddhas, which Stein had visited after leaving Tun-huang. According to Warner, they were

prevented from reaching Tun-huang by the first rumblings of the upheaval that pitted all China against the foreigner.

Warner tells what happened. "The local authorities permitted us to
spend seven days at the Temple of Mo Ku T'ai Tzu, two long miles from
the rock-cut chapels we had come to visit. . . . The situation was one of
extreme delicacy on account of the presence of a dozen villagers who had
left their ordinary employments, some fifteen miles off, to *watch our movements and to try by a thousand expedients to tempt us into a breach of the
peace which would warrant an attack or forcible expulsion from the region.* [my italics.] It took unwearying politeness in the face of nagging,
treachery, and even open hostility, to avoid physical violence."[4] His five
associates were "deeply chagrined at the frustration of all their ambitions
for spending eight months of intensive study on the Tun-huang frescoes."

Was it then or much later that Warner realized that he was the reason
for that hostile surveillance? "The local authorities cannot be entirely
blamed for their refusal to take our word that we would remove none of
their treasures from the spot, a promise that was kept in letter and in
spirit to the great detriment of knowledge."

A sad story and, in view of his past actions, a pitiful admission. One
further element needs to be added. When Warner's party arrived in
Peking (supplied with barrels of glue), he was advised "at the earnest
request of the National University at Peking" and by "foreigners best
qualified to know conditions [to] include in my group Dr. Ch'en, . . .
who was to devote himself to our interests on the road and aid us in deciphering inscriptions when we reached our goal. We found ourselves,
therefore, seriously crippled when, on the second day at the chapels he
insisted on leaving posthaste for Peking to attend the sickbed of his
mother. . . . I reluctantly gave him my fastest mule-cart, my best driver,
his salary, and an extra sum of money large enough to enable him to
hasten his way to the coast. That was the last I saw of him. . . . I did discover by chance on my way back that his haste had not been so compelling
as to prevent his tarrying a fortnight at Lanchou. . . . The loss of our only
Sinologue at this moment was indeed serious." There is no need for a
Sherlock Holmes to see that it was Ch'en who alerted the local people to
watch the Warner party; he himself had been coached on how to behave
when the Chinese government, unable to deny powerful representations
from the United States, handled the matter with politic effectiveness.

Harvard's unabated desire to acquire unique examples of Chinese art
made them turn to Stein. During the spring and summer of 1929 the plan
took shape: Stein was to give ten lectures (between the fifth and twenty-

seventh of December) and, in addition to a handsome fee, was to receive financing to lead an expedition for Harvard to Central Asia.

Between planning and execution, came the first sign of trouble. In October the New York stock market crashed.

For a time it seemed hardly more than a sudden, brief indisposition. "Professor P. W. Sachs, one of the Directors of the Fogg Art Museum of Harvard, has taken charge of the task of raising funds for my proposed enterprise and being a rich man himself, apart from a distinguished art critic & collector, commands needful connexions, etc. He is managing our project with experienced care & discretions."[5] His lectures successfully given, Stein reported as he was about to leave Boston: "Sachs was able to inform me that the whole of the money for my future explorations, whether in China, Persia or elsewhere in Asia has been secured: £20,000. One half is to be contributed by a big endowment [Harvard-Yenching] left to Harvard recently for researches bearing on China. . . . All my conditions are agreed to."[6]

At the end of April 1930, Stein arrived at Nanking, the capital of China's new regime. (In the meantime he had returned to England, soon recrossed the Atlantic, went by train to Vancouver to catch a boat for Japan, enjoyed an intense, brief sightseeing tour there, and then spent five days in Shanghai, "that huge suburb of London, Paris and New York combined.")[7] His strategy called for his securing a special passport which would authorize him "as far as the Central Government was concerned, to carry out archaeological exploration accompanied by topographical surveys within Chinese Central Asia including Inner Mongolia." He presented himself to the Chinese officials in the company of a member of the British legation and a representative of the American ministry. Using all the prestige available, he hoped to counter "nationalist agitation that had fastened upon foreign scientific enterprise on Chinese soil as mere exploitation of the nation's inheritance and contrary to its interests."

His P.N. gives the arguments as they developed. "In addition to the grievance of not having a full share in such [recovered] material, there was also felt a supposed slight in Western-educated Chinese savants not being associated with famous discoveries. All this had resulted in a self-constituted 'National Council of Cultural societies' at Peking, endeavouring to lay down by decree that no foreign scientific expeditions were to be permitted except with a Chinese co-leader and a staff of Chinese savants: that all objects collected were to remain in China, etc." Stein's own position was equally adamantine. "I had made it clear from the start with all those who would help at Peking, Simla, Kashgar, the Foreign Office, Harvard, etc., that it would not be possible to take up work if such conditions were imposed. To drag a party of Chinese savants about with me in a waterless desert and inhospitable mountains, to have to settle my plans with a Chinese co-leader necessarily ignorant of local climatic conditions, could only imply a waste of time, energy and money."

Politely but firmly Stein rejected the Chinese requests, enumerating the factors that enboldened him. "My intended start from, and return to, the far-off Indian side; the practical isolation of Sinkiang from true China; my being still favourably remembered by old Mandarin friends there and the prevalence in its administration of notions averse to much of Young China. What really mattered most was a formal approval by the Central Chinese Government which would save the face of the provincial authorities in the event of nationalist objections being raised by my work among unemployed local office-seekers & the like."

Stein could congratulate himself when on 7 May, against his "real expectation," his passport was handed to him; it vindicated both his scientific position and his familiarity with Chinese official etiquette. "I have made quite a number of interesting acquaintances among Young China and begin to feel a little more hopeful about its future."[8]

Returning to India by boat, he disembarked at Calcutta. He visited his old residence, "the Principal's slum-girt palace opposite the Madressah. The gatekeeper to whom I talked did not quite believe my statement that I had once lived there. For me that glimpse of the 'City of Dreadful night' sufficed; I left with feelings of gratitude for having been allowed to escape from it 30 years ago."[9] Calcutta was the past from which he had fled; Central Asia the past which he hoped to recapture.

Back on Mohand Marg, Stein was finishing his preparations for the expedition when on 9 June he received the news that his dear friend, the Saint, the Hierarch, had died.

Was it grief that made him deaf to the warnings sent from Simla? The "Central Research Institute," he was told, had learned of Stein's passport, and its members were mounting their opposition. It is almost as though Stein needed strenuous activity to handle his grief; only the vast solitude of the Taklamakan could match his feelings of desolation and loss. "Anyway," he wrote Barnett, explaining why he went ahead with his expedition, "the attempt to carry out my plan must be made & I propose to start early next month."[10]

The expedition left Srinagar on 11 August 1930. Stein chose the Gilgit Road, the fastest way, to Kashgar. At Gurez, his surveyor's assistant was stricken with pneumonia and had to be left behind. On 25 August he was at Gilgit. There he was informed that the Chinese—the chairman at Urumchi under orders from Nanking—had slammed their back door shut, the very door which he had planned on entering unobtrusively. It was disconcerting: Stein felt that if he could but get to Kashgar he could straighten matters out with Pan-Darin's son. But first he had to get there. He used the tactic that had served him before. "This breach of the agreement reached in May is being protested at Simla and there is hope that the effort will succeed in removing the obstacle," but he admitted to Publius that "the delay is awkward. Indefinite waiting is not advisable as the passes towards Kashmir would close by the end of August."[11]

Simla's protests were effective; Stein reached Kashgar on 8 October. He was welcomed by Captain Sherriff, the new consul-general, by "dear Old Pan-Darin's son who had been indefatigable in his efforts to smooth my way ahead and played the part of a devoted councillor & interpreter,"[12] and by Hassan Akhun, his trusted camel factotum. There was also a new voice, that of the Urumchi chairman's English-speaking special representative; aloof, authoritative, it was the voice of New China.

Stein was not the easy target Warner had been; the Chinese officials had to be less brusque, less direct, since Captain Sherriff had the wherewithal to negotiate. "Telegrams sent to the great man at Urumchi indicated my programme for the winter's travel & work & recommended its approval. It took a week before the 'Chairman's' reply was received. To our disappointment it indicated dilatory intentions; for it suggested paying a visit to Urumchi, a matter of some 5–6 weeks of travel along a familiar high road."[13] Stein, keeping Publius fully informed, told him that Captain Sherriff had telegraphed a reply indicating "how unfavourably such dilatory action must affect the India Govt.'s action with regard to the Chairman's requests for the supply of arms from India. Only a small installment had so far reached the province, and the further supply would obviously have to depend on the attitude now shown by the Chairman." The chairman wired his consent to Stein's taking the long, roundabout way to Urumchi, visiting the Khotan-Keriya regions, and then crossing the Taklamakan to Kara-Shahr.

Five and a half months later, after a two-thousand-mile march around the Tarim Basin, Stein finally accepted the intent of the chairman's action. Instead of "wasting time for a visit to Urumchi"[14]—as he told Barnett— he would return to Kashgar. He added the harassments he had endured: his passport had been recalled, and he had been prohibited from doing any digging. Stein's spirit, though sorely dented, was not broken. "My visits, seriously hampered as they were, did not remain wholly without profit."

Niya was where he gained the "profit." He had spent a week there, "revisiting the ancient dwellings familiar from former visits and satisfying myself by long tramps across the solid ten miles over which the remains extend, that none of any importance had escaped us. Thirty years had scarcely seen any change in the condition of the ruins. Though the presence of the rather truculent Chinese emissary of the Keriya Amban frightened the dozen labourers I could secure, through the devoted exertions of my Pathans & some old followers, I was able to search at places I had kept *in petto* before. Thus was secured a useful addition to our collection of documents of the 3rd century A.D. . . . I said farewell for the last time to that favourite ancient site where I could live more than anywhere else in touch with a dead past."[15] Was it there that Stein finally laid the Hierarch to rest? At Niya it would seem Stein accepted the inevitable and decided to turn back for India.

He deposited the antiquities that had been quietly unearthed at the consulate, "pending a decision by the Chinese authorities as to their disposal."[16] He brought back photographs of all the manuscripts and wooden tablets made for him by Captain Sherriff.

Retracing the route that had first brought him to Kashgar thirty-one years before, he took the road for India. "It has been a very trying experience for me, but looking back I do not feel I ought to blame myself for having taken the risk."[17] Brave, if futile, words from a man who had had the wrappings of his self-invented myth—that Chinese Turkestan was his: its past, its peoples, its most secret places—stripped from him. Like Warner, the tyro, he had been expelled.

As if to ease his hurt, Stein was in Gilgit soon after some villagers, taking some soil for manuring from the mound of a small stupa "had come upon a mass of remarkably well-preserved Sanskrit MSS. on birchbark. A rapid examination of the MSS. from the unearthed wooden case lodged in the Tahsildar's [collector's] office has shown that many of the large 'Pothis' may date back to the middle of the 1st millennium A.D. and that the deposit cannot be dated later than its closing centuries. The manner of the deposit, the writing, etc., correspond to those of Buddhist Turkestan. No single 'find' of such magnitude has ever been made there, and in India I do not know of any find of this kind." He could almost console himself with the thought that the whole venture had not been in vain.

"If the discovery had been made when I passed through Gilgit in 1900,

I might well have undertaken the complete clearing of this Stupa and of three smaller ones still intact which closely adjoin it. But under the changed conditions"—Stein was learning the lesson just taught—"the matter must be left to the Kashmir State authorities." Happily he could add: "*Entre nous,* I was able to acquire specimens which had passed into villagers' hands & thus to save them from dispersal."

Stein had kept Harvard fully informed of the events. "I have reason to feel gratified by the unfailing support & confidence which Professor Sachs extends to me. It is no small thing to be assured at the same time of ample financial resources & of a completely free hand how to use them for cherished aims. Every letter I have had from Sachs has emphasized this ideally helpful attitude, and you will readily understand how much it has helped me to decide upon turning to alternative plans."[18] Sachs' admonition had been: "look forward not backward."[19]

Summer on his beloved Mohand Marg, with its view of the mountains, mighty and immutable, restored his well-being. "I am pleasantly occupied with revising & completing my Lowell Lectures for a small book. It is a congenial task to revive those phrases of my three successful expeditions."[20] Writing *On Ancient Central-Asian Tracks* eased the failure that had marked his Fourth Expedition.

27

In 1916, when Stein had sought early Buddhist traces in the Sistan, the war had stilled the long rivalry between Great Britain and Czarist Russia that had reduced Iran to an unhappy, unstable buffer zone. This situation changed when Reza Pahlevi successfully established his rule (1925), though it took years for him to control regions whose mountainous territory offered sanctuary to local dissatisfactions. Permission to travel in Iran was difficult to obtain. Thus, when in 1931 Stein thought it worthwhile to extend his Baluchistan reconnaissances westward into Iran, he could only make tentative plans. "The possibility must be reckoned with that the new regime in Persia may look upon foreign scientific enterprise with the same jealousy & suspicion as happily 'modernized' China."[1] Understandably, Stein was not courting another fiasco.

Stein hedged his bets. He prepared for Persia and/or for the Northwest Punjab, the region between the Jhelum and the Indus. The latter, essentially an antiquarian inquiry, would not involve costly excavating. Tehran's permission did come at the end of October; it stipulated that Stein be accompanied by Persian officials. "They are to keep note of my 'finds' with a view to final disposal under the new Persian law allotting a half share to Persia."[2] Experience suggested that he not start until all arrangements were completed: "I do not wish to face prolonged waiting on the Makran border until officials arrive in the traditional leisurely fashion." While waiting, Stein completed *"In Memoriam,"* the Hierarch's obituary for the British Academy that he had promised to write.

Stein's Punjab tour—it lasted only a month—was to ascertain where Alexander the Great had crossed the Jhelum (Hydaspes)—"stole his passage across the river in flood"[3]—to confront Porus, the powerful Indian ruler of the kingdom east of the river. His baggage, however, was packed for tours planned for the next four months. First he surveyed ground "over which Alexander's advance *cannot* have lain, as learnedly assumed by Vincent Smith on the strength of Abbot's poetic topography. I believe I have new topographical data which help to fix the position of this crossing."[4] His search brought him to the Salt Range he had visited in 1889 on his first archaeological dig. At Murti, "I discovered Hsüan-tsang's site of the 'White-robed Heretics' & located the ancient capital of Simhapura he had visited near the present pilgrimage place of Ketas."

Stein's observation and consideration of the topography led him to doubt the opinion of a "succession of writers, some amateur students on the march & others, men of learning working in their study,"[5] which placed the meeting of the two armies in a confined marshy area. Stein placed it near Jalalaphur. His conclusions confirmed "Arrian's detailed account of the place. . . . Definite evidence has been supplied by an interesting discovery—the ruins of a remarkable stronghold on the ancient route which it defends. It lies about 150 stadia, about 10 miles, from Jalalaphur, the distance Arrian indicates between Alexander's camp on the river in the face of Porus and the promontory where he crossed. Forgive these antiquarian details. But they have been so much on my mind for days that I could not refrain from putting down the result here."

The antiquarian tour stopped on Christmas day. Word came that his escort would meet him at Dashtiar, the Persian port to the west of Gwadur—his farthest point in 1928—by 9 January, 1932.

Being led by a native escort—the very idea had irked Stein when the Chinese stipulated it as a condition for his Fourth Expedition—provided him, as he quickly told Publius, 21 January 1932, "with useful & pleasant company. A couple of men come from the Kashgai hills & can talk homely Turki besides their Persian. So I begin to feel as if Kashmir & the Punjab had only been a short break in Central Asian wanderings. The 'tame desert' surroundings help to strengthen the impression." Good relations continued. Stein's only concern was that the size of the escort was a burden to the region's scanty population and their meager resources. "We are on good terms with the Sirdar [chief] of the Baluch tribes from Bampur to the coast, whom our Commandant is taking along. His presence & feudal authority are distinctly useful."[6] Before the tour was over, the "genial Commandant has become very keen on things archaeological & searches sites with zeal."[7] The commandant was also attentive to the preferences of his distinguished guest: Stein found his camp had been prepared in a "fine suburban garden of the true Persian type"[8] in a small oasis encircled by stony wastes.

The tour (January–April 1932), sufficiently arduous to continue the illusion of Central Asia, was to cover some 250 miles. Beginning at the flat deltaic tract at Dashtiar, he went along the coast to Damba-koh, a site already known for its thousands of burial cairns, which proved to be of the comparatively modern Parthian times. "Since it is important to dive further back in time, I intend to move northwards into the valleys & plateaus about Bampur where links may be hoped for with the prehistoric remains of Makran explored in 1928."[9] The problem of water was eased by a rain storm: "pools left behind in the otherwise dry riverbeds have saved recourse to distilling water from the rare & usually rather brackish wells." At Giti they came across the ruin of an ancient fort and picked up pieces of fine glazed ceramic ware "mostly in brilliant green hues" dated to late Sasanian times. "Clearing a rock-hewn trench we recovered enough of some large bowls, etc., to permit complete reconstruction."

Where to go in that difficult empty terrain? So far, the Persian Makran was proving as free of ancient settlements as the Baluchistan Makran, and "the thought of what the climate may become" encouraged Stein to turn inland and continue the search on cooler ground among the hill ranges. His eventual goal was Kerman. "But it must all depend on what our surveys may show in the way of sites claiming excavation. There is a great attraction in being on ground which in an archaeological sense is still unexplored." At Geh, a pleasant oasis with fine date groves, they waited for fresh camels; then they made their way northward through the mountains to Bampur, the headquarters of Persian Baluchistan.

The route led from oasis to oasis—from Geh to Bint and thence "up the narrow stony defiles of a usually dry riverbed, long and trying in places for the camels,"[10] to Fanuch on the edge of the Bampur basin. That 120-mile trough had been a line of communication between the Persis (Stein always called ancient Fars, the area west and north of Laristan, by its Greek name) and the lower Indus. "Finds of chalcolithic painted pottery were made. They had been deeply buried in the alluvium and [were] disclosed by accident."[11] From quick probes made in the dune-covered area to the north and from specimens brought to him from abandoned settlements elsewhere, Stein decided that the Bampur trough offered the best opportunities for archaeological work. The trough, as drainageless as if it were in Chinese Turkestan, did not pose serious difficulties about water; but, as the fierce heat settled, by early April fieldwork would become difficult.

Stein first followed Bampur stream to its headwaters; it fulfilled his expectations. "We were able to trace & partially explore quite a series of early sites that yielded a considerable quantity of interesting finds including well-preserved painted vessels with a great variety of designs. The links with my Sistan (1916) and Baluchistan-Makran (1927–8) are striking."[12] Then, retracing his steps, he moved down the far-stretched Bampur basin across some 160 miles from the "dying Bampur River to where the Halil Rud, descending from the mountains of Kerman, approaches its termination in the Jaz Murian marshes."[13] He found no prehistoric sites on the lower reaches of the rivers. Once at the Halil Rud, Stein felt he had reached the limits of the area he had set out to explore. The prospect of the onrushing heat and sandstorms gave him a valid reason "to resist the temptation of settling down to prolonged excavations. It is good that our further journey towards Bam and Kerman will take us steadily to higher &, I hope, cooler tracts."[14]

A succession of large sites was located as they made their way up the Iranian plateau. Rapid excavations at some of them indicated settlements occupied from prehistoric to early historic times. He described the sequel to Publius, 19 April 1932: "Then followed three refreshingly cool marches to the Bam-Kerman highroad. It meant crossing several passes between 6000 & 8000 feet, and what a delight it was to see fresh verdure at the bottom of valleys and to pass snow-fed clear streams. At Bam are the familiar fruit trees already weighted heavily with their coming produce. Pomegranate bushes of great height still bear their large brilliant flowers. Coming from the sunburnt wastes of the land one can appreciate the Persians' love of their orchards." The aridity and heat already burning the Baluchistan area can almost be felt when Stein adds, "the coolness gives some chance for badly blistered lips to recover slowly."

From Kerman, Stein started for Europe. At the British consulate he left his camp gear, two Kashmiri attendants, and the ever-devoted Mohammad Ayub Khan. It is well to be reminded of what travel involved before air flights. The fastest way from Kerman to Bushire was a six-day, roundabout lorry trip via Yazd, Isfahan, and Shiraz. The lorries, weighed down with six bulky cases of antiques, moved slowly on indifferent roads and suffered frequent punctures. "At Persepolis we were the guests for 64 hours of that most learned archaeologist, Prof. Herzfeld; and then an 18 hours' [boat] crossing to Basra."[15] There Stein caught the "Taurus Express" for the long, dusty train ride (before even the thought of air conditioning) to Istanbul (Stein always called it Constantinople). "There

I spent most of my time in the magnificent museums & enjoyed more than
I can put into words, looking over the wonderful collection of Persian and
Perso-Chinese paintings guarded at the Old Seraglio, the discovery of
which filled our Hierarch's last happy week with such joy. He had planned
the publication of a great selection from that pictorial treasure trove. I am
to take my sister-in-law to her son in Berlin and a couple of days must be
spent there, too, for the sake of scholarly interests. I have planned to
spend June & July in England, then take my sister-in-law in August to
some peaceful spa in the Tyrol & return to Oxford & London for the first
three weeks in September."[16] Nothing arose to make Stein alter his
schedule.

The funds Harvard provided paid for Stein's archaeological tours, and,
as before he had reported to Marshall, now he sent "progress reports" to
Sachs. The latter, he told Publius, "recommended foregoing or postpon-
ing the call for the third (and last) installment of my Boston supporters'
grant." Stein was not surprised: the news he had from Europe told him
how the depression was being felt. More than ever he was grateful that
his careful planning and thrift had left him with sufficient funds for one
or more explorations in southern Iran.

The worsening financial conditions, his retirement, and his peripatetic
life, which took him far from the money market, made Stein turn to a
suitable advisor for help. The obvious, the preferred person was Madam's
brother, Louis Allen, a man with connections in banking circles. From
references made in letters as to their meetings, it is clear that the role of
Louis as a friend and trusted confidant grew with the passing years. Perhaps it was Publius's illness—the beginning of the long downhill fight—
that brought the two men together. Louis Allen (Publius's cousin as well
as brother-in-law) had a special role in Stein's intimate circle.

Stein returned to Iran by another route. From Baghdad, where he left
the train, he crossed the Iraq border to Kermanshah. There, and at
"Hamadan, the Ekbatana of the Medes, I found the kindest welcome
under the roofs of the Managers of the Imperial Bank of Persia. I owe it
to information sent ahead by the Bank's City Manager to whom your
brother Louis had recommended me. So you see how blessed I am in his
friendship however far away."[17]

At Tehran, "the quaintly modernized capital" Stein had long and serious conversations with Godard, the French director of antiquities. "He
stands ready to help with official arrangements. He and his artist wife had
done good work years ago at Bamian and Kabul with Foucher!" By the

end of October Stein was back at Kerman to find "the Surveyor & my Kashmiris all well & to receive a very boisterous welcome from Dash VI."

Stein's P.N. of his second tour in Persia has a distinctive quality. There is a modicum of archaeological minutiae; it is more of a travel account calculated to enliven a sick room: Publius was very ill. While Stein waited for letters with news of his friend's condition, his letters convey his concern, hope, and the kind of news—the closest thing to chatter Stein was capable of—that might give momentary distraction to his ailing friend.

From Kerman he headed back for the Makran coast to continue from where he had left off. "My route from pleasant Rabur, ensconced in its homelike orchards, led to the high plateau of Isfandeqeh; then we descended by an unsurveyed route through utterly barren hills to the wide valley of Buluk where I discovered two early prehistoric mounds marking settlements on ground now utterly devoid of running water. Also burial cairns in great numbers exactly like those traced in far-off Baluchistan. Some modest excavations sufficed to establish the approximate period."[18] The word "modest" was a euphemism; the real situation he reported to the Baron. "The mounds would have invited trial excavations if the Persian officer in charge of our escort, some twenty rather 'tame' conscripts, had not objected to any but the merest scraping. In the same way, surveys were to be prohibited."[19]

These crippling strictures were eased after the commandant made a "hurried night's ride back to Kerman and received a more reasonable interpretation of his orders." Stein feared it might be an omen of more obstacles to come.

Following the Rudan River to the Gulf, Stein found another prehistoric mound, "actually being brought under the plough. I arrived just in time to establish the extension of that early civilization (about 3rd millennium B.C.) close to the Gulf Coast. The oasis of Minab, poor as its date groves and its ill-fed people look, was as refreshing to the eyes as it had been to those of Nearchos and the men of Alexander's fleet. The coast, a desolate belt facing the Straits of Hormuz, had extensive debris areas that all dated from Muhammadan times. "Still it was a satisfaction to feel that I once more was on ground associated with Alexander's story."[20] At nearby Bandar Abbas an accumulation of mail, larger than usual, awaited him. "A goodly number of friends kindly remembered my 'coming of age' [Stein's seventieth birthday] as dear old Lanman put it."

There is something touching in Stein's transforming his P.N. to a travelogue. "The port of Bandar is not a place of any resources. Even water is wanting & has to be brought in kerosene tins by women & donkeys from wells at nearby Naiband. I got quarters in the half-ruined Mess House, all that remains of a proposed small British cantonment (1916–20). It was started during the war at no small expense on this bare sandy coast, 2 miles east of Bandar Abbas. It is in ruins but has drinkable water & enjoys cool breezes. The port itself could scarcely have looked less inviting when it was created 300 years ago; when the heat sets in by March it becomes half-deserted. Today [13 December] I paid a visit to the famous island of Hormuz, some 10 miles out, whence the Portuguese dominated the trade up to 1622. It is a terribly barren rocky island, but the castle built, I believe, in Albuquerque's time, is a very imposing structure even in its ruins. There are great cisterns, essential at a place where the few brackish wells run dry for a great part of the year."[21]

Continuing westward along that grim coast where it had not rained for two years, the party welcomed brief rain storms that brought a blessed relief from the heat. But the coolness "has not driven off the mosquitoes and with them malaria. I myself have kept perfectly fit with plentiful doses of quinine and the use of a mosquito net, as has Muhammad Ayub Khan, my ever cheery & active surveyor."[22] Or he relates incidents "to give some relief to this poor P.N." In one, delayed by wily camelmen decamping, they had to cross tidal creeks flooded by the recent rain. "Having secured a guide, we started in late afternoon when the tide was running out. But it took nearly six hours to cross that continuous expanse of mud, quite liquid in places and very sticky in others. The camels slipped & floundered all the way; for the little donkeys we were riding it was hard work. The young moon gave some light & assisted guidance. So our caravan never got bogged down. . . .

"Another recollection is of the 25 miles' march along the wildly eroded & utterly desolate coast. Most of the way lay up & down masses of fallen rock under precipitous sandstone cliffs washed by the sea at their feet. Until recently this wholly uninhabited coastal strip was a favourite ground for petty robbery & now sees a good deal of smuggling carried on by the same gentry from behind the coastal range. So it was scarcely surprising that our escort with the baggage came upon a party of three hiding in a small ravine. The two who had guns promptly bolted; the third, unarmed, was caught while trying to get away with a donkey. He, of course, duly pretended to have been the victim of the others. . . .

"The presence of our escort officer is useful not so much for protection as for securing local assistance." With mishaps and adventures the party made its way to Tahiri, now a humble village port but once the site of Siraf, which, by the "testimony of Arab geographers, during the first centuries of Islam, was the chief emporium for the trade with India & China & a place of great wealth. By the 10th century its decay had already set in and a big earthquake then completed its ruin. I am encamped on a small terrace cut from the hillside below the Shaikh's castle. . . .

"The very narrow strip of ground rising from Tahiri Bay to the sandstone ridge some 150 feet above must have once been crowded with the high houses of merchants, etc. In the small valleys behind there spreads a great necropolis of rock-cut graves & grottos: hundreds & hundreds of tombs cut out of the steep hill often arranged in regular tiers and approached by a broad flight of stairs; many deep wells & vaulted cisterns scattered about help to give some idea of the amount of human labour which must have been spent in surroundings as uninviting as this arid Gulf region. Islam prohibits depositing with the dead anything used by the living and so no antiquarian harvest could be expected. The grottos holding the bodies of unbelievers were found open and completely cleared of whatever they may have once contained. Coffins, especially, would have been duly appreciated for their wood—fuel on this coast is as scarce as water. . . . The kilns where the uncoloured pottery was produced was located by me. May you kindly excuse my wearying you with such scrappy archaeological details. What I long for is a Dak which may bring good news."

The journey into the interior was by a steep track that rose over a "sharply tilted wall of chalk rock leading up some 2800 feet to the foot of the pass. The poor little donkeys had difficulty & the going was bad enough for us men too. Where the track winds up almost vertical rock-walls at the stretch appropriately called the 'Six Screws,' it looked dangerous for the laden animals. But somehow, the little donkeys accustomed to this toil managed to get up without loss. Remains of a walled-up bridle-path, of caravanserais and of cisterns met with again & again were evidence of the traffic that passed when goods from Shiraz and other great industrial places of Old Persia passed down here to Tahiri-Siraf." Up and down the serpentines; up and up and up to the watershed and beyond to an open plain where a "small rivulet passes some date groves before it dies away in the thirsty gravel glacis."

The Gilehdar tract: Stein found relics of a "very early, probably, neolithic culture. The painted ceramic ware looked older even than those found by me at Makran sites, while the stone implements are more plentiful." Again considerations of time and of the areas still to be investigated made longer excavations impossible. His goal, the town of Lar, an important center between Shiraz and Bandar Abbas, led him to the large village of Warawi, the seat of Sohrab Khan, headman of that small tract. Warawi was on the threshold of Laristan, a region "known for its predatory propensities." Stein had a foretaste of these when one of his "protectors" tried to lift his toilet kit holding his sponges and brushes. "Sohrab Khan, an elderly man of tall stature and active ways, gave us a friendly welcome; apart from an ample flow of rhetoric he uses his physical strength to keep his obstreperous crowd of tribesmen in order. On hearing of the attempted theft, he went for the evil-looking offender and belaboured him with his august hands in a very effective way. So all seemed well for the time."

At Warawi Stein received disquieting news: the chief of the settlements through which their way led sent word that he could not provide protection and transport; nor could Sohrab Khan give help beyond his own limited area. Unwilling to run the risk of being looted, Stein took the advice to heart and returned to the coast by the direct route from Warawi. "The track which we followed on February 2 and 3 proved more difficult than that up from Tahiri." To add to his discomforture, Stein realized he was two steps back from where he had begun the march into the interior. It was then in the hope of saving time and energy that he decided to go by boat to Daiyir, the little fishing port north of Tahiri.

"Well, use of the country craft provided an experience of Persian Gulf navigation. The boat was a rather old & battered affair but of the type which carries what trade there is from the Persian Gulf as far as Karachi and the Oman coast of Arabia; 45 feet long and 8 feet across, it sufficed for taking our party and impedimenta for what was expected to be a single day's journey." A south wind promised a quick passage. Collecting the crew, taking on water, and loading the baggage took four hours. Soon after the boat started, the wind suddenly veered to the east. By nightfall, they had sailed some thirty miles and passed well beyond Tahiri and Stein lay down to sleep expecting to find the boat at Daiyir by daybreak.

"The god of the local winds had decided otherwise. About 2 A.M. I was awakened by the violent movements of our craft. The northwind, or Shamal, which blows with great force—sometimes for days, sometimes for weeks—had sprung up. It needed the skill and energy of our crew of coast Arabs to shape a course towards the coast and prevent us from being blown down too far. The single big sail was reefed with difficulty amid continuous raucous shouting in gutteral Arabic; water coming in through leaks in the bottom even before the start, now gathered rapidly—two men bailed constantly. Uncomfortable hours passed. Daybreak showed we were off the village we had passed a month ago on the way to Tahiri. It was a great relief when the anchor could be cast about a half mile from shore. The Shamal increased to a gale. I thought that the rope of palm fibre which kept the boat to anchor might snap and let us drift before the wind. There was no way of knowing how long the adverse wind would last. I decided to seek *terra firma*. It took another three hours to get shallow boats from the fisherfolk and bring us all with our baggage to safety. . . .

"The voyage, short as it was, made me realize what voyages in such boats up & down the Gulf and on to India must have meant in old times. Ships of the same small size and archaic build, through all Islamic times travelled as far as China. I now can better understand the hardships endured by the men in Nearchos' fleet. How glad Nearchos must have been when he succeeded in bringing his fleet safely to Hormuz and could get away from its dirt and overcrowding to seek Alexander's camp, be it only in ragged garments and uncouth appearance." Stein's emotional identification tinged all his Alexander quests.

"How are we to reach Bushire without too serious loss of time? Neither camels nor donkeys are left in this narrow strip of coast after two years of virtual famine; and Daiyir, where we might find them is still 30 miles off. Westernization has not yet thought to provide this far-stretched coastline with wire & telephone." Mercifully, help was at hand: a trusted messenger bringing Stein a huge dak from Bushire carried back his need for aid. By 19 February he arrived at Bushire and waited there at the British consulate, anxious to learn where he might proceed now that Laristan was barred as an unsafe area.

At Bushire, albeit in comfort, Stein waited, eager too for a reply to his telegrams to Tehran. "You will realize," he wrote the Baron, 17 March 1933, "that I am getting rather restive over this enforced inactivity, however pleasant the hospitality." A month passed. He went up to Persepolis on Herzfeld's urgent invitation and luckily met M. Godard there. From him Stein learned that Tehran wanted to stop his work and not merely postpone it. Godard "has offered fullest support for a resumption of my investigations in the autumn and has strong hopes that if the application is made after a couple of months it will succeed as a matter *ab novo*. Meanwhile a cable has informed me that the Wali of Swat is ready to welcome me back to his dominion. In 1926 I had been obliged to leave a considerable portion of it unvisited and April-May is still a pleasant season for filling in the gaps. Then I hope for the peace of the Marg to prepare preliminary reports & get ready for resumed work in the south of Persia by October & November."[23]

"May God grant to dear Publius full recovery," so he ended his letter of 1 April. The mail in April and May from Oxford had an ominous ring. In June Publius died. "I will lift up mine eyes unto the hills whence cometh my help," Stein wrote Madam from Mohand Marg. He was more than ever grateful for its peace and beauty.

Just a year before, Publius was convalescing from an operation and seemed well on the road to health; the good reports encouraged Stein to leave England for Iran. At that parting, the friends made a pact: each side would write truthfully on matters of health and well-being. From the letters, first from Publius and increasingly from Madam, Stein could follow his friend's decline. In May, even before the news of Publius's death, Stein acted on the pact and described an accident that cut short his Swat tour. "The horse which had carried me for ten days quite safely over bad ground, stumbled suddenly on one of the Wali's good motor roads and coming down on both knees caused me to fall off. It meant a crack near the end of the collarbone which, though really not painful, needed a little attention & rest to heal quickly."[24] This pact makes it possible to follow Stein into surgery as into Swat.

Madam, who had shared Publius's Erasmus work and was a friend of Stein's in her own right, continued both the Erasmus publications and the lifeline of weekly letters. "When the mournful news reached me, I thought at once of many things which your goodness allowed me to encumber you with all these years," Stein wrote her.[25] "I could wish for nothing better than that you permit them to remain under your care provided your new home would allow of some minor space where to put books & boxes—and provided you would kindly allow me the privilege of bearing a small share of the rent in compensation for the space taken up. I do not know how to estimate this share, but I feel strongly that it should not be less than the Annuity (£80) which my will contains for you. It was meant as a posthumous small contribution towards the expenses of the Erasmian peregrinations. The will dates back to 1910 and since then my savings have doubled, while, alas, the Hierarchic couple for whom a similar provision was made, has passed away. So it would only be right to let you have their share. . . . Please let your brother Louis, who knows all about my financial affairs, advise you if need be."

Stein went to Iran for the third tour (November 1933) to continue his search for prehistoric sites. For the first time in his long career of archaeological investigations, he was on his own financially. "In view of the chaos affecting financial conditions all over the world," he told the Baron, 28 December 1933, "it behoves me to be very careful about the use of the savings of a lifetime." This did not involve a deep change: he had practiced thrift all his life. But he was properly appreciative when "the British School of Archaeology in Iraq [the Gertrude Bell Foundation]"[26] contributed £500 toward his tour into Fars. He was uneasy lest the "tribal unrest" that had deflected him from Luristan, the region due north of Bandar Abbas, might also hamper his exploring Fars.

Fars, "the heart of Persia,"[27] was more or less unexplored archaeologically. Historically, it wears crowns of glory. Fars was the cradle of the Achaemenian dynasty (539–330 B.C.) whose kings ruled the greatest empire of the ancient world; its capital, Persepolis, though broken, still spoke of imperial might and splendor. Darius the Great had built it; it had been sacked and burnt by Alexander. From Fars, too, came the Sasanian dynasty (A.D. 224–651), equally mighty and notable for a courtly elegance and technical excellence. Geographically, its oases—Firuzabad, Qir-Karzin, Jahrun, Khafr, Fasa, Servistan, Shahristan, Runiz and Istahbanat, Shashdeh, Darab, Barnavat (their names sound with an epic cadence) are the lifeblood of Fars. Shaped like an open Navajo bracelet with semi-precious stones, each in its own rock setting, the oases enclose the huge elevated Niriz basin filled with the flawed glittering jewels of salt lakes and marshes.

Stein's party was composed of an inspector of antiquities, a military officer in charge of a dozen hardy soldiers, the devoted Muhammad Ayub Khan, two Kashimiri servants, Ali Bat, a capable cook, and Ali Malik, a Mohand Marg *dakshi* acting as a personal attendant. Starting from Shiraz, 21 November 1933, a 1300-mile tour began; going counterclockwise, it visited each oasis. Stein liked his Iranian companions. The inspector was not just pleasant but was keenly interested in "old things"; the officer was "a quiet congenial person ready for hard travel if need be. He joined us at Shiraz at very short notice, the elegant young officer originally detailed to accompany us having suddenly declared himself ill after learning that the ground to be covered was far from motor roads."[28] For transport he had eighteen Shirazi mules. "I never experienced the least trouble on their account, however bad the stony tracks were or however difficult to secure adequate fodder. Nor was a single mule ever found to suffer from serious sores."[29] Praise for the mules was also praise for the muleteers.

On the way to Firuzabad they met large parties of Qashgai moving to their winter grazing grounds. The P.N. tells the tale. "Travelling households with their flocks of sheep, donkeys and ponies all driven along by hardy women and children. Of the men only a few were to be seen; but babies in plenty carried on the backs of their mothers or else packed in bags on donkeys. They were not crying like the young lambs of which six or eight tied close together in sacks, formed the load for a donkey. A bitterly cold wind swept the pass."[30]

Arab geographers had written of the Sasanian ruins in the wide and graciously fertile valley of Firuzabad. In its center the city of Gur had covered an area a mile and a half in diameter; its earthquake-shattered palace was said to have been built by the founder of the Sasanian dynasty—"three large domed halls which form the central part of the palace built with a solidity which rivals the way the Roman master builders handled their concrete material"—and the so-called Castle of the Princess atop a precipitous rocky spur had a royal residence when danger threatened; its noble central hall had a dome that measured close on 50 feet.

Of prehistoric ruins there was no knowledge. It was, therefore, especially satisfying when Stein thought a hillock surrounded by cultivated fields was man-made. A trial trench disclosed painted pottery and worked flint blades on top; underneath an abundance of pottery and stone implements from the fourth millennium B.C. "Neatly cut stone seals with geometrical patterns recalled Makran and Baluchistan sites of the same time." A systematic search would have required months, a length of time he did not have. He took the road for Jahrun, the next oasis. It was reached after a tiresome march—hard ascents and troublesome descents over a terribly stony range, then across valleys empty of permanent habitations with Qashgai tombs here and there. "Like the Turks of Central Asian stock, Qashgai seek no dwellings while alive yet provide them for their dead." They came upon a silent village, its protecting wall and towers still intact, a sad monument "to the influenza epidemic which had swept away the major portion of the population."

Jahrun delighted him. "Nowhere before had I seen a compact town—it is credited with some 2–3000 houses—so completely hidden away among its groves of date palms and fruit trees. Seen from afar the eye sees nothing but the tops of palm trees with their gracefully waving big leaves." The oasis lacked any prehistoric sites. It "depends for water wholly on well and qanats, i.e. subterranean canals, and in my tours in Baluchistan and Makran I had found that under such conditions prolonged prehistoric occupation leaving its traces in the shape of mounds could scarcely be expected."

After a detour into the Khafr oasis there followed days of rough marching across a high range to the broad basin of Fasa. Close to its main town was an old site, the seventy-five-foot high "Mound of Zohak." "The villagers coming for work were in excess of the 60–70 we could supply with implements and watch. It was gratifying to find that all the men were steady, hard workers, quick in learning what was aimed at and reasonably careful about "finds." In the course of two days' hard work, trial trenches were carried from a height of 25 feet on the slope to an average depth of 11–12 feet, ending at the mound's foot well below the level of the surrounding fields." The results were puzzling. Stein finally surmised that the small prehistoric potsherds picked up from the surface and not met with again must have been dumped on the top of the mound with earth dug up elsewhere—but from where he could not determine.

Fasa had the ruins of an impressive Sasanian stronghold, the Qaleh-i-Gabri, the Castle of the Fireworshippers, whose "defences and ruined dwellings extend for three miles along the crest of very steep rocky ridges," a high massive barrage—"even the tradition of this imposing engineering work is lost"—and an extensive burial site. "Bone fragments from bodies which, in agreement with the ancient Iranian practice, had been exposed to beasts and birds of prey, [were] under stones heaped up in pyramids about 15 feet high whose recesses held interesting archaeological finds. We gathered at different cairns beads, iron arrows, and javelin heads, a dagger, a knife and, most important of all, a well-preserved Sasanian coin." At last, in the southernmost part of the oasis, they found half a dozen prehistoric mounds that gave evidence of having had prolonged occupation.

On 5 February, Stein reached the oasis of Shash-deh where he met the Ainalu, a Turkish tribe who had settled in the eastern portion. "My desire to use my Eastern Turki secured me a particularly hearty welcome. I managed to collect specimens of their language both in brief stories and a few songs. In spite of vast distances and dialectic differentiation in the course of many centuries, speakers from certain parts of northern China are still intelligible to dwellers by the Aegean and Adriatic." The Ainalus, though prodded by the government to settle down in mud-brick houses, preferred their traditional black tents. Near their encampment was the mountain fastness where Rustam's son, as told in Firdusi's great epic, suffered captivity. Inevitably, it invited Stein's inspection. A ride up a rock-lined valley, then a "steep ascent of 2000 feet to a narrow crest overlooking a maze of deeply eroded ravines, and finally a distinctly difficult climb by a narrow fissure in the bare rock to a plateau which rises about

250 feet above the steep slopes of scree. Helped by Muhammad Ayub Khan and a few agile Ainalus, I managed to climb up safely but in some places wished I still had my toes."

There was also the "town of Ij," said to have been destroyed by the Mongols. Stein decided to visit it while waiting for permission to enter the oasis of Darab, adjacent to still-forbidden Laristan. The site of Ij was awe-inspiring: it seemed impregnable—a shelf hung halfway between a deep ravine and lofty cliff—and invulnerable to a siege. Water was stored in forty huge, deep tanks cut into the rock augmented by eight "hanging reservoirs" clinging to the sheer rock like gigantic swallows' nests (forty-four yards by eighteen and twenty feet deep). Stein was deeply impressed by the hydraulic skill used to assure the water supply and by the engineering skill visible in a carefully constructed aqueduct that carried the water from a little brook along a succession of cliffs, through tunnels, to replenish the water in the hanging reservoirs. Reluctantly, he left the fascinating town of Ij, and on 18 February he started for Ishtabanat, crossing the watershed toward the drainageless Niriz basin.

"Ishtabanat, pleasant little town ensconced among orchards where almond trees were already showing their pink blossoms, could not detain us long. Two long easy marches in open valleys brought us to the half ruined village of Runiz-bala." His discovery of painted pottery of the chalcolithic (Bronze) period was dwarfed by that of an unmistakable relic of Zoroastrian worship, a four-foot-square monolith standing ten feet above the debris-covered ground of a small rock plateau. On its top a circular hollow a foot deep "proves this great block of stone to have served as a fire altar, such as the coins of all Sansanian monarchs represent as an emblem of their adhesion to the Law of Zoroaster. This great stone is, as far as I know, the first altar discovered intact up to the present. Its position close to fine springs filling a large pond, has its significance: such founts of life-giving water must have been an object of local veneration long before Zoroaster's creed arose." Close to it was the inevitable shrine of a Muslim saint, "another instance of that continuity of local worship which I had often traced from Kashmir and the N.W. Frontier far into Central Asia."

Permission to visit Darab was granted, and Stein quickly turned eastward to utilize what remained of the cold season for work on that lower ground. His search for evidence of prehistoric occupation was successful: a small, low mound near the main town was excavated with a large number of willing laborers. There he found relics similar to those found all along his route. Near the ruins of a Sasanian palace a more extensive mound yielded a harvest of painted ware "with a great variety of bold patterns." Then, without a break in his P.N., Stein adds "from an accompanying abundance of worked stones there came to light here at the deepest levels such a quantity of coarse ceramic ware, exclusively handmade and primitive, as to make an early date very probable for the layer of painted pottery immediately overlying it."

Stein had reached the neolithic level, the period antedating what, using the current British terminology, he referred to as chalcolithic. The term "neolithic" was indeed as old as Stein (having been coined in 1865), but it was not appreciated as a critical stage in social change until V. Gordon Childe (1892–1957), the single most influential mind working in prehistory, synthesized its revolutionary role. At that time, Stein was in Iran and not alert to the significance of Childe's findings. Because Stein had been trained as an antiquarian and had but recently been alerted by Mohenjo-daro and Harappa to the impressive prehistoric presence in India, he did not sense the importance of the neolithic evidence he stumbled upon. It would seem that for him finding coarse, handmade, primitive ceramic ware and an abundance of worked stones was primarily a way of dating what lay above that level. Research into the neolithic— and earlier stages—would soon begin to activate archaeologists. Stein, however, lacked the concept that would have transformed a prosaic, tedious excavation into an exciting quest into man's even more distant past. The relevance of Stein's work in the emerging research into prehistory was that he gave proof of the interconnections between the civilizations of the Mesopotamian and Indus River valleys.

Stein's P.N. of his Fars tour makes his preference clear: it is more of an antiquarian's account than an archaeological record. "While these trial diggings [at a third mound] were carefully watched by my ever-alert surveying assistant, I was able to devote a good portion of my time to the study of the palace ruins"; he also snatched a few days for a more extensive survey of the town of Ij. When this was done to his satisfaction he took a new route to Darab.

From the Kham Pass he could make out the broad Darab valley and the waves of hills diminishing in height as they flowed toward Laristan. "I could not help feeling glad that I was not called upon to face the heat and the brackish water of the poor barren villages in crossing the wastes of Laristan." He made his first camp at the village of Madehvan. Potsherds brought from two nearby mounds had whetted his appetite to explore the area. At one of the mounds, "two days of steady digging with plenty of willing and efficient labour has allowed us to carry trial trenches through a thick chalcolithic layer with abundant painted pottery and worked stones down to deposits of an earlier prehistoric period which knew only coarse handmade ceramic ware. Curiously, the two strata were separated by another which yielded only flints used as knives or saws as well as some bone instruments."

At Ij, the head of the local subsection of the Ainalu tribe, "a man of intelligence as well as tribal authority," directed Stein to a mound called Tul-i-rigi: "we found layers of painted pottery overlaid in places by fractional burials." Two days of heavy downpours brought a flood, but the local people, gratified for this abundant "water of mercy," attributed it to the presence of Stein's party. Another mound, the largest one near the village, had been "left undisturbed for the last five thousand years or thereabouts. When another case had been filled with specimens of that varied ceramic ware which our trenches had yielded in abundance and the level was reached where only coarse handmade potsherds were to be found beside the usual worked flints, it was time to move eastward."

Stein was eager to survey a series of old remains mentioned by Arab geographers and abandoned in the late Middle Ages. "An imposing circumvallation that encloses a steep rock hill some 250 feet high marks the site of old Darab-town, appropriately known as Darab-gird, the Circle of Darab. Its huge circular fosse, close on a mile in diameter in front of a 40-foot high earth wall implies an immense labour—digging and raising such masses of earth—with none of the mechanical appliances now available." He could still trace canals laid in regular lines that had carried water to tanks in various quarters; and an arched aqueduct across the fosse that had brought the water of a canal from the distant Rudbal stream.

From Darab-gird he went to study and photograph a twenty-foot relief carved into a cliff rising sheer above a spring-fed pool. It showed Shapur (who died about A.D. 272), in full armour and mounted on a charger receiving the surrender of the Roman Emperor Valerian who begs for mercy. Below the horse of the king of kings lies a slain foe, while behind him are ranged Persian nobles and soldiers. "In this sculpture the national pride roused by victory over mighty Rome cannot fail to convey the full significance of that historical event even to the modern beholder." Stein then examined a rock-cut shrine, variously called the Mosque of Stone, or the Princess's Palace, or simply the Caravanserai, in whose shape he saw a correspondence to that of a cruciform church: "We know that the Nestorian and other Christian Churches were widely established throughout Persia from Sasanian times down to the Middle Ages."

He was equally responsive to the visible prosperity of the village of Deh-Khair, "accounted for by the fact that the village owns its own land and what matters even more, its own water, instead of being, as most villages in Fars, occupied by rackrented tenants of absentee landlords." His next camp was at Khusu, a village on the old caravan route into Lar, from where samples of painted pottery had been brought to him. Four days of digging at the mound brought a "harvest of interesting ceramic material. At one point our trial trench was carried down to a considerable depth until the layer was struck where the wheel-made chalcolithic ware gave way to a mere coarse hand-made pottery, probably neolithic."

Leaving Darab (6 April), they crossed the watershed separating the Rudbar, which flows into the Persian Gulf, from the streams emptying into the drainageless Niriz Basin. The ascent to the 7300-foot pass led through "almost continuous plantations of fig and almond trees laid out with much labour on the slopes of bare limestone detritus. Those trees can manage to grow at higher elevations without irrigation." As they left the region of fast-encroaching heat and crossed the wide, waterless Darab plateau, they were pelted by a violent rainstorm with gale-force winds that inflicted an icy discomfort. An easy march then brought them to Niriz, headquarters of a district much of which "is taken up by large lake-like depressions, dry for the greater part of the year and covered with a crust of pure salt."

At a site near the town of Niriz, he picked up "that coarsely painted ceramic ware dating from a late prehistoric period with which I had become familiar in British Baluchistan and which had so far eluded me on Persian ground." As always, the present interested Stein as much as the past: among the local officials who called on him—in true Persian custom, the visits became little tea parties—was the chief of public instruction, an informed and very helpful man who had only two schools to supervise. "He had sad stories to tell of his past employment in the Lar tract where most of his time seemed to have been spent amidst lootings by rebel tribesmen and protracted sieges shared with the small garrison in Lar town. His younger brother had been killed on the road together with his wife. His two children, a pretty little girl and a lively small boy, were being brought up by their uncle. On visiting him in his house, I was struck by the contrast which the neatly kept interior, rooms opening on a miniature formal garden in the centre, presented to the filth of the narrow lanes through which the modest residence had to be approached."

By 13 April—"economy in the use of time still available had now become essential"—Stein started for the hill tracts of Banavat, the final part of his program. He left the salt-encrusted lake of Niriz, "a miniature edition of the great dead sea of Lop-nor," and marched northward, to the elevated tableland of central Iran. In what was the favorite summer grazing grounds for nomadic tribes, he could not expect to do archaeological work; but always appealing to Stein were the geographical and antiquarian investigations to be carried out in that little-visited region. The village of Mishkan was his goal.

Stein was delighted by the plantations of almond trees, the vineyards enclosed by walls, the use of every bit of ground clear of bare rock, and the closely packed dwellings of Mishkan. "In the space unfit for cultivation, the thrifty, hardworking folk have located their 350-odd households." Mishkan reminded Stein of villages in the Tarim Basin. It had been founded two generations before when the failure of an adequate water supply forced the people to abandon Dohu, some four miles away, and move nearer the two qanats that were still functioning. Mishkan, like Deh-Khair was happily without rackrenting landlords. "But sickness takes its toll even in such healthy surroundings. On a visit to the home of the local headman, it was sad to be told that the baby by the father's side was ailing and the only one of seven children still alive."

Approaching the long, narrow Banavat oasis, they saw, within the encircling mountains, an eight-mile stretch of fields, orchards, and vineyards. All had previously belonged to the family that for generations had supplied the governors-general of Fars. The present head was now a détenu in Tehran, and the villagers rejoiced to be "spared the periodical visits of the great man and his locust swarm of followers." Near Mung, the easternmost of the Banavat villages, Stein investigated a site in a desolate valley surrounded by deeply eroded barren hills where the owner of the village had noticed a place strewn with worked flints. The small terrace was bare; nothing else remained. But Stein's lively interest in these finds prompted two villagers to search closer to the stream where some pottery, exposed by water, had attracted their attention. They had "hit upon a large amount of vessels, more than a dozen intact, varied in shapes but all, with one or two exceptions, handmade ware of coarse, archaic type. I lost no time in visiting where the find had been made." It was a neolithic burial place. Careful excavation disclosed bones of a "partial burial," the remains of a body that according to custom had first been exposed to the beasts and birds of prey.

By 27 April, they started up the Banavat Valley for Dehbid, where a geologist of the Anglo-Persian Oil Company had noticed a mound that seemed worth examining. The mound was ancient. It presented difficulties: manuring had cut its sides into perpendicular walls. It was dangerous: a recent landslide had buried two donkeys and almost trapped the men loading them. Excavating cautiously, Stein recovered an abundance of painted ware and, associated with the pottery, bronze needles, celadon beads, button seals with geometric designs similar to those previously found in British Makran, and a totally new specimen—a "well-preserved skull of a young man or woman with a neatly drilled hole left by trepanning, a practise traceable from very early times."

On the last lap to Shiraz, Stein stumbled, as it were, on a beautifully decorated tomb; its "exquisite tile work adorning the walls of the tomb-chamber, the high porch and the niches and squinches of the outer hall" partook of the beauty of Timurid art. "The sight of what vandal hands had done to the mosaic of enamelled tiles was even more depressing than the decay wrought by neglect and time. One could wish for a Persian Lord Curzon who would get the place put on a 'List of Protected Monuments.' " At Pasargadae he pitched his tent opposite the tomb of Cyrus the Great (reigned 559–529 B.C.), "which has stood all the ravages of time better perhaps than any ancient monument east of Egypt." With a monument well known, often described, studied by scholars, the only task left to Stein was to make the first "detailed topographical plan of the whole site," a task that "fully justified my three days' halt."

For Stein, his visit there seemed a reward for all the potsherds from prehistoric sites he had excavated. "Alexander the Great had visited the tomb on his return from the Indian expedition and the disastrous march through Gedrosia. He found that the rich contents of the tomb which Cyrus' successors on the throne had duly guarded, had been plundered during his absence by those in charge, and had had the offenders duly punished. To me it was not a little satisfaction to let my eyes rest on the structure which the Macedonian conqueror had viewed with the respect of the truly great for the great." At the end of his tour of Fars, the cradle of Persia's ancient might, Stein rejoiced at again being on the tracks of his hero, Alexander.

More such moments were to come. Stein left the wide plain of Pasargadae by the ancient route leading to Persepolis, the road built to give safety and ease to the king of kings when he journeyed annually from his winter palace to his summer seat. A short mile from the tomb, the river enters a gorge "where the greatness of Achaemenian power has left a most impressive mark in the road that passes for some 308 yards along an almost vertical face of limestone by means of a gallery cut into the solid rock. The road uniformly 5½ feet wide and quite level, is protected on the side where it overhangs the sheer precipice above the river bed far below. The cliffs above the road are throughout hewn into vertical walls of rock, up to 70 feet or more in height. Chisel marks show the care and precision with which this stupendous work of rock-cutting was done. Where the gallery winds around a projecting portion of the precipice, rows of regularly placed footholds are still to be seen. I felt thrilled by the thought that I could touch the smooth, even rock floor which the feet of Alexander's horse must have trod as he moved down from Pasargadae to Persepolis."

"To me it will be a great comfort to be able to talk of Publius," Stein wrote Madam as he was about to start for Europe.[1] He also told her he had been named recipient of the Huxley Medal given by the Royal Institute of Anthropology and had agreed to give a lecture on the occasion of its presentation to be held 28 July 1934. He had already sent her all his "carpets, big, small, Turkish, Kirghiz, Persian—they number well over two dozen,"[2] to be used as she liked either at Barton House, the country cottage, or at 22 Manor Place, the new home Madam was building at Oxford. Stein spent as much time as he could at Barton House, "grateful for its peace and charming walks" and happy to enjoy the "delightful ease and regularity of Madam's protecting regime."[3]

He had made it a custom when in Europe to spend August with Hetty. When their joined vacation ended, Stein escorted her back to Vienna and returned to Barton House. Briefly: he was called back to Vienna—Hetty was very ill. When it seemed that she was out of danger, he went to Budapest for a scheduled lecture. He was there when on 11 October 1934, "almost 32 years to the hour after Ernst,"[4] in her eightieth year, Hetty died. He returned for his last visit with her; for forty-eight years she had been enshrined in his heart as "family." "Surrounded by a profusion of flowers, she was laid out in the dignified yet simple fashion which she would have approved. Her face looked so serene and contented as I had seen it only at rare times of relief from anxious cares. So it will live in my memory."[5]

An impressive service was conducted by the pastor of the Reformed church; her ashes were placed in Ernst's grave in the Protestant cemetery. Four years later, when Hitler absorbed Austria, almost one hundred years of assimilation would be nullified by the fact that Ernst's parents had been Jews.

Madam stepped gracefully into the voids left by Publius and Hetty. To the last she remained Stein's close, thoughtful, devoted friend—almost family. Habit and performance had won her that. As if to mark her new estate, in May 1935 she had addressed him as "My dear Aurel"; politely he followed her example and wrote "My dear Helen." Just once. Then he implored her to permit him "to cling to the respectful title which your incomparable goodness has endeared to me for so many years."[6] There was to be no change.

Equally firmly, Stein stayed close to Ernst's family. His niece Thesa, and her husband, Gustav Steiner, both practicing physicians, continued Hetty's role but in a different key. Gone were Hetty's anxious cares. Again and again Stein wrote of enjoying his "niece's spacious apartment & lively company." Where Hetty's claim was to their joined mourning for a precious figure forever gone, Thesa and Gustav Steiner, happy in their marriage and work, invited him into their active, interesting lives. Oxford and Vienna were the line of Stein's emotional axis; thus, Christmas was spent with Madam and New Year's Day with Thesa and Gustav.

A pleasant surprise ushered in 1935. "The Society of Antiquaries, that dignified, ancient body, wishes to present me on April 30th with its Gold Medal, a wholly unhoped-for honour."[7] The announcing letter was received in Italy where Stein was spending a few days with an old friend, de Filippi, a member of the Alpine Club, on his way to sightseeing around Naples and Sicily. Pompei lived up to his greatest expectations. "You will smile," he wrote Madam 28 January 1935, "at my saying that I felt my happiest recollections revived of those days which saw me unearthing those sand-buried dwellings of the Niya site or cleaning those Hellenistic frescoes in windswept Lop desert. It must suffice to assure you that it made me forget the fatigue of constant peripatetic sightseeing, broken only when I sat down in the peace of the Doric temple (6th cent. B.C.) remains, to eat my lunch. I had the wonderful site almost to myself. . . . The order maintained now among the ruins is as impressive as that which prevails in Naples, once so noted for its beggars & dirt. Be praise to Mussolini for it all!"

Pompei, Herculaneum, "the great treasure house, the Museo Nazionale" at Naples—all prelude to three weeks spent in Sicily. He wrote Madam, 20 February 1935, on the journey from Messina to Palermo, "we passed Milazzo, the ancient Mylae, and I thought of my uncle Sigismund who,

as one of Garibaldi's officers shed his blood in the battle which finally
freed Sicily in May 1800 from Bourbon troops." Palermo, Monreale,
Agrigento, Syracuse: "I little dreamt when I read, as a boy, Aeschylus'
Persians, I should behold the very scene where the drama was first played."
Back to Rome and then to Paris: "Long talks with Sylvain Lévi, that foun-
tainhead of true learning on all that concerns Greater India."[8]

By April he was back in London where, in addition to attending the
ceremony at the Society of Antiquities, Stein arranged for the India Office,
"ever a place where learning was patronized,"[9] to underwrite the volume
of Chinese documents that Henri Maspero, Chavannes's pupil and successor, was editing. May began the journey back to the Marg with the
customary visits to Vienna (where he laid a wreath on the graves of Ernst
and Hetty) and to Budapest. With little thought for rest he had maintained a tight, fast schedule, seeing new sights, visiting old friends, arranging for work in progress and future efforts. "I rested in bed until
7:30 A.M.," he wrote Madam from the Orient Express. "When did I last
see the sun so high on rising?" He had more much-needed rest on the
long train trip to the border of Iraq.

There the tempo changed. "I was to be carried by aeroplane from Mosul
straight to Baghdad," from where he wrote Madam, 16 May 1935, "because a rising south of Baghdad had damaged the railway line & stopped
communication to Basra, I was to be brought there after a two days' halt
by another machine of the R.A.F. Could anything more convenient have
been wished for?" Stein transformed the plane trip into more than a con-
venience. "I found a big plane waiting for me, one of the troop-carrying
kind capable of taking 20 fully equipped men. With only its small crew
there was abundant space to walk about in it. I was invited to take the
second pilot's seat and in this open, airy position could see everything
below. A brief landing at the great ruined site of Hatra, once a Parthian
stronghold, allowed me to get a glimpse once more of its huge circumvallation. Then flying at 5000 feet I could fill my eyes with all those instructive features of a semi-true desert. Over the great ruined city of
Samarra, which a Caliph's whim had created in the 9th century A.D. and
then again abandoned, we flew. One felt as much at ease as if rolling in a
first-class car over the smoothest motor road in England."

Stein, who had been satisfied to go at a camel's pace in the Taklamakan
and, just recently, at a Shirazi mule's pace in Fars, now began to focus
his attention on using the airplane for future tasks. His first flight made
him consider planes as an aide to certain kinds of archaeological reconnaissances.

By the beginning of June he was once again on the Marg, hard at work.

To reassure Madam, concerned at his being all alone, he enumerated his household there. "It consists of Alibat, the cook, & cheerful Alia; 3 Manygam Dak men [who twice weekly brought mail and supplies]; a trusted Gujar watchman; an old retainer who looks to the milk supply; a surveyor and Brahman clerk, each of the last two with his own cook; and 2 Forest Guards to prevent illicit grazing complete the camp this year & prove efficient. So you see the Marg is not too lonely a place for me."[10] There in July he completed the typing of *Archaeological Reconnaissances*, dedicated to Publius's memory, and a month later his account of "six months' travel & digging in the Persis."[11] By that time, too, all arrangements were completed for a start from Shiraz on 8 November for his fourth archaeological tour in Iran. It would last a full year.

Aboard the boat plying between Karachi and Bushire, Stein reviewed "the whole of that forbiddingly arid coast along which I had marched in less comfort in the cold weather of 1928 [Baluchistan Makran] and 1932–33 [Persian Makran]."[12] At the Bushire anchorage, five miles out to sea, he was greeted by the inspector of antiquities; his "future companion presenting himself with a letter from M. Godard. This assured me that all arrangements about escort, etc., had been made by the Tehran Ministries (Foreign, Public Instruction, War) and full instructions issued ahead to all the provinces I was to visit." The provinces, all in western Iran, included those lying between Shiraz and the last Kurdish frontier in Azerbaijan in the north.

Mirza Bahman Khan Karimi had been selected because he was a graduate in geography and history; he impressed Stein as a "very alert and obliging young man."[13] Karimi's account of the year spent on the tour is the only one left by a companion of Stein. Ch'iang Ssŭ-yeh, alas, left no record of their extraordinary travels and discoveries; nor did the previous Iranian inspectors. Karimi's "Condensed Reports,"[14] primarily devoted to the objectives and sites, sound a personal note; his jottings have a "Through-the-Looking-Glass" quality.

From the coast, close and humid under gathering clouds, the party started for Shiraz across the plain "and then up the 'ladder' (*climax,* as the Greek geographers called the succession of steep passes up to the plateau). It was hot and terribly dusty but the lorry and car managed to climb that forbiddingly bad road through the ranges of rotten limestone without accident or serious delay." At Shiraz everything was in readiness: "our 16 mules & 2 ponies are duly arranged for with the men who were with us before, and, after much haggling, at a rate reasonably lower." Stein was paying all expenses out of his own pocket. The escort had been mobilized; the weather was truly delightful; and Dash was ready for the adventure.[15]

Just before leaving, Stein had news of the death of Sylvain Lévy. "A letter from that dear old friend written a fortnight before his lamented death told of a happy tour in Italy. Bright light for me far away from a star has ceased to shine. My own letter from Karachi could no longer find him. It touched on scholarly interests which had been dear to him to the last. I feel the loss of this close friend of 48 years."[16]

Stein took the ancient road "connecting Fars with Khuzistan, the Elam of the Bible, or Susiana as known to the classical world from its famous capital, Susa."[17] A week later, 25 November 1935, he wrote the Baron from Ardakan, a small village in the Mamasani highlands, that had prospered as the fence for the raiding Mamasani tribes. Now the substantial homes of its headmen (held as détenus in Tehran) housed government officials. "The marches up to this 7000 foot plateau were quite interesting and enjoyable in spite of bitter winds and some rain. On the open plateau a curious small mound was sighted and proved to be chalcolithic up to its top. Two days hard digging yielded painted pottery. . . . Tomorrow we shall start down towards Behbehan by the route which is likely to have served the annual royal migrations of the Achaemenians from Susa to Persepolis. Karimi, our Inspector, is a very intelligent & willing assistant, ready to face some hardship. Still it is good to know for the sake of all the men that our further programme is to take us to lower elevations and warmer ground."

Jottings from Karimi's First Report: "It must be said that during this journey Sir Aurel Stein did not follow the roads but used impracticable footpaths to approach the mounds. He is always busy drawing the layouts and plans of all routes and mounds. This is a very tiring and extremely difficult trip. At Ardakan Sir Aurel Stein made some trial trenches at twelve different points. Sir Aurel Stein believes this site was a large ancient fortress. Body discovered with grave-goods and stone weapons and no trace of any metal. In closing this report I add that this journey on horseback in winter, sleeping at night in tents in the desert and spending each day in a place more or less safe becomes more and more difficult as it continues. It would require a young man of iron to endure all these hardships in a damp, cold climate."

From Behbehan, the next village, Stein wrote Madam on 20 December: "My stay here has been prolonged not for any need for rest but by several days of plentiful rain. . . . A small prehistoric site discovered close to Behbehan may detain us for a couple of days. Then I hope to start northward. My plan is to reach Malamir on the Karun by routes as far as possible unsurveyed. There Sir Henry Rawlinson nearly a hundred years ago observed numerous ruins."

Jottings from Karimi's Second Report: "We left Ardakan for Behbahan. Sir Aurel Stein wants to find the ancient route from Susa to Persepolis. There are two—the first, the more direct, leads past Ardekan and the second, is longer, more circuitous. Along it we found the remains of bridges and serais solidly built of stone and cement. . . . From Shiraz to Behbehan took one month. From here we shall follow the Royal Way and, if we do not tarry anywhere, this second trip should last only ten days. While waiting, Sir Aurel Stein is always busy drawing plans of ruins and mounds, of roads and secondary paths. He also takes down much information on the manner of living, the customs and habits of each province and region. . . . After the tents have been put up we make trips around to examine the countryside."

In one of the small mounds examined in the vicinity of Behbahan, Stein "found a quantity of ceramic ware . . . resembling those characteristics of pottery found at chalcolithic sites at Fars. It sufficed to prove [that] the same prehistoric civilization . . . once included ancient Elam."[18] Labor, he was pleasantly surprised to find "in this poverty-stricken region is available at low rates—lower even than it would be nowadays in India. The Bakhtiaris, reputed for their lawlessness and looting propensities until recent times, are a cheery lot, hard workers, too, in the digging." With the help of Behbahan's local administrator, Stein secured the escort and guides for the party's northward march through the unsurveyed Kohgalu hills. On Christmas Day the party started for Malamir.

Jottings from Karimi's Third Report: "During this trip [through the Kohgalu hills] we endured many difficulties in crossing inaccessible mountains. The roads in the mountainous area are so difficult that several times we almost fell into an abyss. As for myself I fell from my horse and hurt my right leg. In a ravine leading towards Malamir are stones on which there are Sasanian bas-reliefs illustrating the Sasanian kings with their captive, the Roman emperor Valerian. Sir Aurel Stein says they have been seen and described but before cameras existed to photograph them. . . . The writing on them, almost effaced, Sir Aurel Stein says is in Aramaic characters. . . . After marching several days we reached Bakhtiari territory and were received ceremoniously by an assemblage of chiefs. In that area one continuously comes face to face with Sasanian remains. . . .

"Six months ago"—here Karimi tells of an exciting find made at a little village near Malamir—"a villager of Shami was digging foundations for house walls and discovered a large statue dressed in a mixture of Iranian and Greek-style clothes. The statue carries a sword on his belt and wears a necklace; it is nine feet tall [Stein gives its height as 6 feet 4 inches] and two feet wide. Its torso was damaged during its discovery. It is made of bronze and, I think, the head has traces of gold. There were also two marble Greek heads, one may be that of Aphrodite, the other of a Parthian king or general. Both are superb. Next to the large statue was a small one; and two arms—all of the same metal. These should be taken to Tehran as soon as possible and placed in the museum."

Stein tells the story. Shami "lay at an elevation of about 3600 feet on one of the terraces which stretch down from a rugged high mountain and offers some ground for cultivation. A short while ago an order had gone forth to the scattered camps of semi-nomadic Bakhtiaris that all claiming arable land should settle down near it in permanent habitations. So one of the Shami people had started to build himself a dwelling with the rough stones lying plentifully about on that terrace. In digging down for a foundation wall he had struck the bronze statue at a depth of a couple of feet."[19] A week of steady work at the site disclosed the outlines of a quadrangular temple (76 by 40 feet) that had geen gutted by fire and sacked. In its center was an altar of burnt bricks, approached by a brick pavement. After its destruction the sanctuary, a Parthian shrine (247 B.C.– A.D. 227?), had been quarried for metal and other useful items. Yet Stein still recovered numerous objects of Hellenistic-Iranian worship. He marveled that "a locality so restricted in space and resources as this outlying valley of Shami, could afford a temple so amply provided." ("The large bronze statue of Shami," Frye suggests, may belong to a "cult of the heroicized dead king.")[20]

Jottings from Karimi's Fourth Report: "I submit the following to you. We spent 6 days seeing and studying the Elamite monuments at Malamir and to make excavations in two mounds nearby. Known to the Arab geographers as Izeh, it lies across the caravan routes from Ahwaz to Isfahan. The Englishman Leyard [discoverer of Nineveh] spent many years in these mountains and wrote a book on its fauna and the Malamir monuments. The excavations made brought to light vases, jars and other objects. Sir Aurel Stein says that the civilization of Malamir is prehistoric and dates to about 5000 B.C. while its pottery resembles that of Susa I. . . . We [the main party] took the Susa road and in two days reached the banks of the Karun. We had to cross the river, a dangerous and difficult undertaking. There was no bridge and the only way was on *kaleks,* inflated skins under a crude raft. After a thousand hardships and difficulties, accompanied by terrifying fears, we entrusted ourselves to the *kaleks.* Then having surveyed the ruins of two Sasanian bridges we had to recross the river. Sir Aurel Stein had all the finds packed in several boxes to be taken to Ahwaz. Before finishing this report I must inform you of the following: one cannot call this tour a promenade. It should be called a journey of difficulty, of pain, of bitterness, of danger and illness. One must accept the rigors of winter on snowy mountain passes and the hazards of kaleks on the rivers."

Karimi's terror at crossing the Karun on kaleks was no milksop reaction. The sight of the rampaging river gave Stein concern for his party; he was vastly relieved when the main group rejoined him on 17 February.

At the same time Stein made a quick trip to Ahwaz, the headquarters of Khuzistan province, to store the boxes and arrange for a military escort. He was about to enter Luristan. His two days spent at Maidan-i-Naftun, the great oil fields of the Anglo-Iranian Oil Company, allowed him to visit the site known as the "Mosque of Solomon." To the Muslims it was a platform for prayers, "but as jets of fire fed by gas escaping from the underlying oil-bearing strata were always known in the immediate neighborhood it is highly probable that the site served Zoroastrian worship in connection with pilgrimages to the sacred fire."[21] He dated it by the abundance of easily picked up Parthian and Sasanian coins: another instance of what might be called his "Law of the Continuing Sanctity of Sites."

With permission to enter Luristan and an escort provided for safety, the party headed northward. The Karkheh River, draining most of the province, was unfordable; no kaleks were to be had; the party made its way over difficult and previously unsurveyed ground that led through a maze of deeply eroded, winding gorges. "So we came to the long, winding valley of the river known from here onwards as the Saimareh [which] was to see us for close on two months." Then, when finally they were forced to cross the Saimereh, only one small kalek capable of carrying a single passenger and three mule trunks was available. It took seven crossings, each crossing costing an entire day.

Yet once this had been a main road with bridges built to facilitate traffic. Of the many ruined ancient bridges, one, the Pul-i-dukhtar, Bridge of the Princess, had a single arch almost seventy feet high. "I was able to visit the point known as Pul-i-tang, where the rock-lined bed is so narrow that a jump across it could be risked by the lithe-limbed Lurs from bravado, as Sir Henry Rawlinson witnessed. New settlements of makeshift houses were the result of the "measures adopted by the present strong regime to assure peace and security for traffic, all evidence of the forcible settlement of the nomadic Lurs." Stein, however, was sensible of the hardships such a change inflicted on a nomadic population accustomed to seek

warmth on the plains of Khuzistan during the winter months and summer pasture for the rest of the year. "The ups and downs of civilization
are aptly illustrated by the fact that Luristan nowadays does not know the
art of the potter. Its people are content to drink from skin bags, and the
same bags when inflated, [are] used for crossing the river. Only on occasion was it possible to collect a number of bags sufficient for more than a
single frail craft."[22]

Jottings from Karimi's Fifth Report: "What was added to the journey
from Salehabad were violent winds and torrential rains. After Susa we
returned to Salehabad and for three days we traveled under winds and
rain without a guide. We were lost in a place where there was neither
village nor people; on the evening of the third day we reached the path
leading to Jaidar from where we went to see the bridge, Pul-i-dukhtar
built by the Sasanians. Nearby we stayed two days excavating a Guèbres
[Zoroastrian] cemetery. There is a marked difference between the construction of these tombs, the position of the bodies and those of Shami.
The Lur graves have been rifled for objects [Luristan bronzes]. At least
95% of the ancient graves have been searched for treasures and then filled
in with earth. We saw many caravanserais; despite our looking we found
neither mosque nor baths. We excavated many mounds and graves [he
enumerates the places]. On our way we passed a superb woods burnt by
travellers. It would be useless to repeat all the sufferings and unheard of
hardships which we endured on this dangerous and almost unbearable
journey during which we were constantly exposed to the violent winds
and torrential rains of that mountain region."

Stein thought the same journey idyllic. "Spring had advanced with us
ever since we entered Luristan. . . . Two delightful marches along a range
clad with plentiful oak forest at elevations between 5000 and 6000 feet
brought us to the wide, trough-like valley of Koh-i-dasht. Owing to its
central position it had served as a favourite base for a succession of Lur
chiefs who, helped by tribal feuds and intrigues, used at times to domi-
nate Pish-i-koh, or the northern portion of Luristan."[23] The exactions
made by the tribal chiefs had depopulated an area with a fertile expanse
of arable land. The chief who had led an uprising but eight years before
was now a détenu at Tehran; his cousin and enemy, installed by the capital
as local administrator, guided Stein to a grave-strewn area. He recovered
a few iron bangles and bronze pins, but the graves like those previously
examined showed signs of having been thoroughly rifled for bronzes then
commanding high prices from dealers. Rifling was easy and foolproof:
the experienced local searchers had only to probe under the rocks and
shallow soil covering the burials to test for grave-goods. From large pots

left behind as valueless, Stein dated the burials to about the first millennium B.C.

Jottings from Karimi's Sixth Report. "From Koh-i-dasht to Alishtar we saw the largest, the most splendid of the bridges over the Kashkan: nine massive piers, from 60 to 80 feet high, support arches that clear the flood stage; its length was more than 200 yards. The lower half of the piers is solidly built of cut stone brought from a distance. A Kufic inscription fixes the date at 1000 A.D., yet it appears solid enough to be only 100 years old. Sometimes the people use dynamite to break up the piers to get the stones. It is imperative to stop this destruction and preserve this precious monument for posterity."

From Alishtar, Stein went along the foot of the high unsurveyed range that trends northwestward, separating Luristan from regions belonging to ancient Media. A number of old mounds were traced, and several graves yielded interesting relics: "engraved cylindrical seals used as part of a necklace and betokening Mesopotamian influence, and large bronze rings. These latter, far too heavy to wear, were probably meant to serve as currency."[24] Stein was disappointed that the burial sites even in the high valleys of Delfan had also been ransacked for their Luristan bronzes.

Jottings from Karimi's Seventh Report: "The tumuli in the Delfan district had been rifled as was evident from the pottery broken and scattered about. Muhammad Ayub Khan is always drawing plans. This Surveyor during the entire journey was on lofty heights, working and drawing his maps. . . . At Luristan there are numerous sites where if serious digs were made one could certainly discover precious objects and important monuments. We saw copper and bronze objects on the natives' arms which they had found either in the tombs or simply when working in the fields. In my opinion, Sir Aurel Stein does not want to excavate but rather searches to clarify the geography. The importance of the site's location, the height and depth of strata, the plans of excavated sites are used by him to indicate communications between civilizations."

After three months of hard travel in regions innocent of all amenities, Stein reached the small town of Harsin. "Set amid luxuriant orchards and groves of old plane trees, the life-giving water of its canals which irrigated these is derived from a single magnificent spring gushing forth under high limestone cliffs."[25] Kermanshah was reached on 3 June, and thence by car, two weeks later, the provincial capital, Senneh. From there, with a new gendarmerie escort, he continued northward to study the Caves of Karafto.

As he had been impelled to locate Aornos, so, too, the problem of Karafto appealed to him. W. W. Tarn, an eminent Hellenistic scholar and a fellow member of the British Academy, had directed Stein's attention to the caves. A Greek inscription mentioning Herakles, imperfectly copied more than a hundred years before, seemed to corroborate Tacitus's mention of a shrine in far-off Persia sacred to an oracle serving the legendary hero. The caves were on a high plateau in Kurdish territory and were reached "by climbing up the face of a precipitous rock wall with which a great ridge of limestone breaks off towards a lonely valley."[26] A high, natural cave had been augmented by "a series of large apartments and passages cut through the rock and disposed in three stories. [The inscription was] above the entrance to the highest flight of rooms, airy and well-lit, which would have served as a comfortable place of residence."

The name of Herakles, in a well-known cultic formula written in a script of the late fourth or early third century B.C. (much earlier than the campaign Tacitus mentioned), attested to the worship of the hero-god. (Herakles was considered by Frye as "a kind of power symbol.")[27] A creepy local legend told of the mounted god who hunted at night, being furnished at certain times by the priests with saddled horses and quivers of arrows ready for use; when, in the morning, the horses returned, the quivers empty, the god's victims were searched for by the priests. The tale had inspired numerous sgraffiti. (Stein made a squeege of the inscription and sent it to Tarn.) "A sgraffito of rider & horse over the inscription makes the identification certain. My topographical data explain how the place came to be known in connexion with one of Rome's Parthian campaigns."[28]

Jottings from Karimi's Ninth Report: "We went to visit the famous caves of Karafto. The temple of Herakles there is very had to reach. One gets there by a tortuous climb and stone steps; many rooms have been formed in that grotto, arranged with windows that open to the outside but with narrow, dark corridors. The name Herakles can still be seen cut into the face of the stone. The ancient caravan route—Sakiz, Zinjan— passes nearby."

From Karafto, Stein went westward along that caravan trail. At Sakiz, because the areas westward along the Iraqi border were still barred to him, he headed north; its high valleys and comparative coolness appealed to him for work during the heat of August. From a large mound near the hamlet of Dinkha, Stein recovered examples of finely burnished, thin, black pottery previously not found in Fars and Khuzistan. Near the defile where the Gadar River emerges into the Solduz plain, Stein explored a small mound that yielded an "abundance of worked stones, including fine obsidian blades . . . and a well-made bronze adze,"[29] in addition to pottery whose design and execution was linked to sites in Fars and further east. Riding over the wide Solduz Valley, still scarred by the destruction left by the Russian occupation, Stein was guided to a large mound near the village of Hasanlu.

The important mound near Hasanlu, into which Stein cut the first trial trench, was a rich climax to his fourth archaeological reconnaissance in Iran. In twelve days' work he found "the debris of dwellings . . . of a period when utensils of copper and bronze had completely superseded stone implements. . . . complete pottery vessels of varied shapes [were] found in small kilns. Among them a highly burnished black or red ware with a very fine smooth surface. . . . an assortment of hard pebbles which had been used as burnishers were found lying on a small palette-like trough ready for use by the potter's hand. . . . a large jar holding dozens of roughly cast bronze rods which may have served as ingots and the stone moulds for producing such ingots and broad chisel-like implements."[30] He also found graves with funerary furniture, food and drink for the dead, large jars with the elongated spouts similar to the bronze ones found in Luristan, and personal ornaments worn by women—necklaces of stone beads, of copper, shell, glass, and gold foil.

There is a tone of hail and farewell as Stein ends his account. "Before turning southward again, I looked across that inland salt sea [Lake Urumia]—its wide, utterly lifeless expanse—from the barren coast. It was curious to feel here at the northernmost limit of my journey that I had been brought within reach of Tabriz and the Caucasian regions. The cold nights which set in suddenly with the end of summer served as warning that it was time to regain Kermanshah and prepare for the proposed campaign in the autumn. . . . By September 27 we reached Saujbulagh and thence motor transport allowed us to regain Kermanshah after three days' travel."[31]

His stay at Kermanshah dragged out a full three weeks. He was delayed by the impossibility of securing mule transport—Muhammad Ayub Khan hired some at a nearby area at somewhat higher rates. Then he was immobilized by the expected visit of the shah—the roads were guarded and all traffic stopped; then days passed trying to round up a five-man escort; and finally he was stopped from his plan to return to the south along the Iraqi border. As though to assure himself that his tour was not over, he made a circuit westward into the valley of the Mahi-Dasht where, unexpectedly, he found numerous mounds. He almost sounds pleased that he cannot continue digging. "Many mounds are still occupied by villages, at others it was possible to establish very early origins by plentiful surface finds."[32] Or, commenting on the steepness of some mounds that made quick excavations impossible: "I did not feel altogether sorry for this as the high rates of labour due to the harvest season and the demand at the Shah's building operations at the newly established sugar factory would affect my budget. But our survey was profitable." He also admitted—"in compliance with your wish for truthful accounts"—that he was recovering from a cold. He told the truth but not the whole truth.

Barred from the Iraqi border, he went off on a sightseeing tour. "At Behistun I examined Darius' great inscription,"[33] and he continued on to Hamadan. There he reported his illness more fully: "a wretched attack of dyspepsia which growing insidiously since my long stay at Kermanshah manifested itself plainly. Trying to diet & sustaining myself on milk and similar slops, I felt weak when I returned to Kermanshah. The experienced American Mission doctor found nothing seriously wrong, gave me useful dietary advice & let me start northward for Hamadan. Here I am enjoying a good rest and am provided with all the things which can cure this wretched gastritis. My cook & Alia are looking devotedly after me. In about 5–6 days I hope to start for Ahwaz." He was still not telling Madam the whole truth.

He was, however, preparing her for a possible delay in reaching Oxford. "If advised at Vienna, I might take a proper cure at some spa to get the door closed to dyspeptic trouble." The doctor at Hamadan thought him fit to travel south. On 10 October he wrote Madam: "So we all started on the morning of the 10th in brilliantly clear wintry weather, packed, snugly in a bus-like motor, baggage, and all the 'Camp' together. The shaking of the motor provided needful exercise without any exertion on my part. Nor did I mind the delays caused by having to pass a constant stream of tribal migrations. It was a wonderful sight to meet the endless succession of bands moving down with their thousands of sheep, goats and cattle to their winter grazing grounds. The contrast presented by the women in their vivid local dress riding on ponies and donkeys to their menfolk wearing the imposed second-hand European garments, tattered & discoloured, was striking."

At Abadan, Stein found a good home for Dash. "Poor little fellow, he seemed to foresee the separation ever since we came down. But I trust he will get over the grief & enjoy life even in what must seem suburban surroundings to him—as well as his master."[34] A month earlier he had parted from Karimi. "My jovial fat Persian 'Inspector' beams with joy at the prospect of soon being relieved from further hardships of travel. I do not blame him for this. . . . Anyhow he will be admired in elegant Tehran circles to have stood camp life so long. I myself find it a little hard to take leave from it."[35]

7

The Promised Land at Last
1937–43

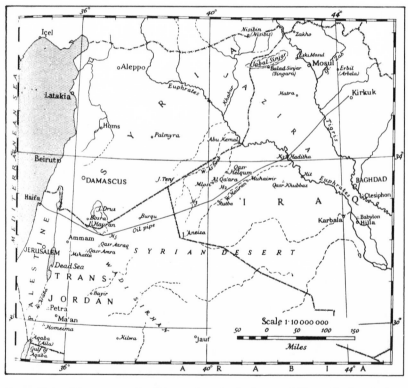

From Stein, "Surveys on the Roman frontier in Iraq and Trans-Jordan," *Geographical Journal* 95, no. 5 (1940).

Stein had a lifelong love affair with Clio. To the very end the muse was his mistress. Oldham, his friend from the Bihar tour (1899), hailed him in his memorial as "a great scholar and man of action," noting his excellence as "explorer, archaeologist and geographer."[1] To this unique combination he might, with equal validity, have added "historical topographer." Stein's passion for history expressed itself in varied ways; yet, strictly speaking, he was not a historian.

As a Sanskrit scholar, Stein proved his devotion to history: in his editing and translating of Kalhana's *Chronicle;* in his reconstruction of the ancient geography of Kashmir by on-the-spot investigations, he made his debut as a historical topographer. Similarly, his training in classical studies directed him not to textual criticism but to geographical problems—identifying places crucial to Alexander's eastern campaigns. Stein's command of Sanskrit, Greek, and Latin can be seen as tools serving geography.

In his Central Asian expeditions, he attained fame as explorer and archaeologist. Buddhism's expansion eastward he saw as a kind of tracer bullet that revealed the lost transcontinental movement of peoples and pilgrims, ideas and goods. The mass of documents recovered from the rubbish heaps of the limes, from Tun-huang, Niya, Miran, and other places, furnishes the raw stuff for ongoing historical studies: of the different schools of Buddhist thought, of Han and T'ang economics, of law and administration, of linguistics, popular literature, painting, and so forth. His plane table surveys of deserts, oases, and mountains (assisted by a series of capable Indian surveyors) provided the sand-buried sites and the limes with their geographical raison d'être.

Stein's attachment to history accounts for a certain lack of intensity in his Baluchistan-Iranian tours, which sought prehistoric evidence. The mounds, pre-Vedic, pre-Hellenistic, pre-Buddhist, had an intellectual but not an emotional attraction; again and again he chose to survey Achaemenean, Parthian, and Sasanian ruins leaving the excavation of the mounds to assistants; his enthusiasm mounted whenever there was a direct association with Alexander the Great. We have already noted the astute remark of Karimi, who accompanied him on the Fars tour, that Stein "does not want to excavate but rather searches to clarify the geography." Comparing these probings into prehistoric Harappan civilization with his Central Asian expeditions suggests that without the adrenalin of written evidence the muscles of his imagination slacken. Those of the imagination but not of the body. Nothing could halt his strenuous activity—not the loss of toes, attacks or dyspepsia, or old age.

"Afghanistan in Avestic Geography,"[2] his first paper, written at twenty-three, made the alliance between geography and history—a harmonious quartet of ancient history, philology, Afghanistan, and geography. Many of the problems he dealt with later emanate from this geographical interest: desiccation and its impact on human settlement (starkly exemplified in the Lop desert); the raising of Turkestan irrigated fields by alluvial deposits, an insight that enabled him to locate ancient Khotan under its centuries-thick accretion of soil; the continuing sanctity of holy places despite changes in religion and worship; the placement of limes—Chinese and Roman—to guard caravan routes.

His subsequent training in topography, the scientific description of a region's geography, ratified this alliance: a magnificent corpus of maps bears witness to this. His later archaeological explorations may seem more limited in scope, but they always include a trained surveyor as a necessary adjunct. However great was the military value of his government-supported, Middle East aerial reconnaissances, Stein's reasons for the work were purely antiquarian: to situate the Roman-Parthian border as mentioned by Latin historians. In his last years his geographical interest was given over to his concern with historical topography.

On his way to Europe early in 1937, Stein stopped in Florence to visit his Alpine Club friend, Filippo de Filippi. The Italian had been a member of the team that had tried to climb K_2 in 1909 and four years later had approached the Karakoram by walking to Leh "up the frozen Indus through deep narrow gorges, always watching the ice for breaks, always aware of the swift-flowing water under his feet."[3] Stein's special reason for the visit was confided to the Baron. 17 January 1937: "For more than two years symptoms of an enlarged prostate gland have asserted themselves in a way which makes an early operation as a thorough means of relief very desirable. It is a trouble very frequent in men of my age. After consulting with my dear friend, De Filippi, himself a medical man, I have come to the decision that it would be best to have the operation done at a well known Vienna institution as recommended by my niece's husband, an experienced doctor. My return to Europe, earlier than planned, allows a convenient measure of time during February-March before setting to work at Oxford & the Brit. Museum."

The surgeons had their own schedule, and not until 17 April, after the second of a two-part operation, could Stein reassure Madam: "I woke up & found it hard to believe that I had been relieved of that trouble borne for too long."[4] It was June before, well but still weak, he arrived at Oxford; at Madam's command and under her care he regained his strength at Barton House—"she wishes me to remain here a full month before going to London to be present at the allocation of the Iranian finds."[5] (Previously, the Tehran Ministry had agreed to his suggested division of finds.) By autumn he was visiting family and friends in Vienna and Budapest, after which he returned to Oxford, staying alone, while Madam and her ailing sister Olga took a long winter's cruise.

By mid-December Stein was traveling again. He first stopped at Brussels to visit his nephew Ernst, the target of Hetty's love and bereavement, who had proved his scholarly capabilities and been appointed full professor of Byzantine history and literature at Louvain's Catholic University; he had recovered from a "breakdown" and now "was happier and healthier."[6] That his nephew's salary was a pittance was unfortunate but not as unfortunate, in his uncle's eyes, as his conversion to Catholicism. (When in May 1940 the Germans conquered Belgium, Ernst, the grandson of Jews, fled for his life; later Stein would become deeply concerned with getting funds for him and his wife Jeanne.) From Brussels Stein went to Vienna to spend Christmas with Thesa and Gusti—he felt close to them after the solicitude and care they had shown during his long stay at the Rudolfhaus Clinic.

Then he started eastward to begin an utterly new kind of reconnaissance. The idea for it came from Stein's reading "La Trace de Rome dans le desert de Syrie," Père A. Poidebard's account of his aerial survey of Roman limes along Syria's eastern border. The article told how Poidebard had carried out his pioneer survey under the joint auspices of the Académie des Inscriptiones and the French High Command in Syria between the years 1925–32. Stein was enchanted: here was a new way to pursue historical topography; in an airplane Stein could accomplish in weeks what would otherwise take years. His long experience in such matters led him to begin promoting the idea well in advance. His opening salvo was an article on Poidebard's results for the *Geographical Journal* (January 1936); on his way back to Europe, though ill, he took time in Baghdad to enlist the support of the British consul and the director of antiquities for Iraq.

Then, at Mosul, he tested the idea by a trial flight over the desert east and south of the Sinjar hill chain. "It was possible to recognize at least two Roman roads & perhaps one or two mounds crowned by the remains of fortified posts."[7] He used his long weeks of convalescence to send memoranda to influential friends. By October (1937) the air ministry notified him that the air officer commanding in Iraq would cooperate. To secure a surveyor he was advised to contact the Iraq Development Co.; and by December it was arranged for him to have luncheon with Lord Cadman, chairman of two "mighty bodies, the Anglo-Iranian and British Oil Development Companies. I found him fully informed. So it was a real relief when that very affable, great man told me that I could rely on the requisite aid forthcoming."[8] A nod from the Zeus of the Iraq Petroleum Company provided what Stein required. "The concession of the I. P. Co. extends in parts across the Syrian border [so I can link] up my work with ancient roads traced by Poidebard on the Syrian side."[9]

In the flurry of new contacts, new procedures, techniques, and horizons, Stein celebrated his seventy-fifth birthday. In Hungary, articles noted the half-century of his departure for India, and a dinner was given in his honor. Most deeply felt was Madam's gift—Publius's watch.

At Beirut, early in January 1938, Stein and Poidebard met. The latter related "his experiences on the Syrian Limes, in the submarine exploration of the ancient harbour of Tyre and also of the Roman borders of Tunis, Algiers & Morocco! P. is a man of much humour & experience of the world. His advice for the tasks ahead is a great asset."[10] An even greater asset was having his path smoothed by Foreign Office directives; but to Stein it was evident that his luncheon with Lord Cadman "did more than 2½ years' correspondence."[11] His preparations completed, Stein retired for a month's study of Arabic to a quiet "Syrian Marg."

On 27 February he wrote Madam from Baghdad. "This morning I managed to assure myself of all needful arrangements by visits to the Iraq Co.'s local Manager, to the Director of Antiquities and to the Saint's old pupil, Mr. Edmonds, advisor in the Ministry of Interior. The transport was duly inspected and contact secured with Ilfifat Husain the Surveyor. If the cook secured at Beirut—none was reported to be available at Baghdad—arrives, I hope to be off to Mosul from where it will not take long to start work from a base camp at Sinjar." At Baghdad Stein happened to meet Skliros, the director general, who had come out on business. "The special attention I received was directly due to the interest this farsighted executive is showing in my scholarly enterprise."[11]

At Mosul the new venture began. Stein's party included a pilot, a photographic team, also of the R.A.F., and three noncommissioned men to handle the wireless. The last used their skills when they moved into camp (9 March) to rig "up an electric light for me—when did my little old tent think of this luxury?—and like them I can get fresh bread & vegetables flown out."[12] Stein was provided with a fur-lined flying-suit. "I do not mind the blast in the cockpit. It allows me to see a great deal of ground across which the Roman Parthian border must have laid & to get air photos of two ancient sites." He had made a successful 250-mile test flight, wisely choosing to aim his initial venture close to the Syrian border so as to extend eastward the Roman routes Poidebard had traced. Stein's decision to survey the region between the upper reaches of the Tigris and Euphrates was an attempt to translate the terrain into the language of the Peutinger *Table*. (This thirteenth-century copy of a road map of the Roman Empire had cartographic distortions—a compression of north-south distances and a gross extension of east-west ones.) On the *Table* Singara was shown as a "great stronghold on the Mesopotamian *Limes.*" Its identification gave Stein satisfaction and encouragement.

Motor journeys supplemented air reconnaissances. An eighty-mile ride to the foot of the Sinjar hills "enabled me to trace remains which together with the configuration of the ground offer indications as to the main line of the Roman Limes, likely to have followed to the Tigris. Outside the picturesque town of Balad ('town') of Sinjar we found a pleasant camping place on a little ridge jutting out into the wide alluvial fan formed by the small stream which issues from within the ancient circumvallation. The hills have been for centuries held by Kurdish-speaking Yazidis, those queer, supposed 'devil-worshippers,' who have long suffered from Muhammadan neighbors whether Turkish settlers [to the north] or the Bedouins of the Jazira [the plain to the south]. They are, in spite of their rather mysterious creed and strange uncouth looks, quite decent folk & good cultivators. Nowadays they are gradually reclaiming cultivable land long abandoned to the 'tame desert' which spreads down from the foot of their hills into the wide plain between the Tigris and Euphrates."[13]

Stein's careful examination of the walls enclosing the modern shrunken town and a large ruined area alongside, enabled him to determine the Roman origin of some sections; similarly he dated portions of a great canal cut through the live rock. An abundance of coins of the Roman emperors of the third and fourth centuries corroborated the fact that this was indeed the Roman Singara, for more than two centuries an advanced bulwark against Parthians and Sasanians. Far away to the south on the plain "are numerous mounds, large and small, dating from far more ancient times when all cultivable alluvial land must have been densely occupied. Now it is beyond the fringe of the Yazidi hamlets and is grazed over during spring and winter by the great Bedouin tribe of the Shammars. Their chief had arranged for whatever help we may need among them. His eldest son, who drives his motor car and speaks English quite well, still lives, I am glad to find, in tents like his father. I paid my return visit to his camp & found all his people still living in their black woollen tents, just as their ancestors did in Roman times."

By 4 April he could write to Madam that the venture was "bearing good fruit. I can report the discovery of 3 Roman castella which settles the previously unsuspected line of the Roman road connecting the well-cultivated centre of the Sinjar range with the great stronghold of Nisibis [Nusaybin] in the north. Elsewhere the links with Poidebard's world were duly secured." Six weeks later he had traced the southern extension of the limes "in the desert south of the Sinjar range. As far as Hatra lay a no-man's-land, a desert where water was difficult to find since prehistoric times & where practically no mounds exist. Yet an important passage lay through it: Ammian's account [Ammianus Marcellus, A.D. 330–395] of the Roman retreat in A.D. 364 shows under what difficulties it was followed. Hatra, which owed its existence solely to the caravan trade once passing across this great waste, was already then in ruins."[14] Using an automobile for closer inspection, Stein found a "series of Roman stations along the route connecting Nisibis with the vicinity of ancient Nineveh & further down even leading to Hatra."

This completed his first aerial-automobile survey inside Iraq. To the Baron, eager to have artifacts, Stein wrote on 12 April 1938: "The clearing of any of the Limes posts must be left to others with more time & less years. The posts are all of a size & likely to have been thoroughly looted by the Arabs of anything of material value at the time of abandonment. Also the climate would not have allowed anything of records to survive."

At Karachi, once again on Indian soil, he was happily reunited with

Dash, and on the morning of 1 June, the regular date, he started up Mohand Marg. "My legs after this long separation from real mountains felt the stiff climb a bit. . . . I allowed myself three hours to do the 3000 feet," he wrote Madam, adding "I mention this merely to assure you that I did not altogether pretend to ignore my years where physical exertion is involved."[15]

Stein lost no time in settling down to work on the report of his fourth Iranian journey; it would be called *On Ancient Persian Tracks*.

12 May 1938: Hitler annexed Austria. The Jews in Vienna knew their fate—they had been close enough to watch the dress rehearsal of Hitler's racial policy. To Stein the news recalled his painful emotions during the Battenburg affair. Only deeper. Then he had had nothing to fear; his loyalty to Britain was unequivocal. But in this? Nazi policy denied the baptism and Protestant upbringing of Ernst and himself and would have labeled them what their parents had been: Jews. Only Publius and Madam seem to have known this. It was she who almost immediately raised the delicate subject. Gratefully, he answered her letter. "With your intuitive sympathy which never fails, you have rightly gauged the care which Thesa's and her husband's position involved for me."[16]

Like others at a distance Stein could not grasp the terrible reality of Nazi theory and policy. He told himself: "Gusti, steady & brave in spirit, would probably have preferred to face the storm and might have weathered it in the end. But Thesa's imaginative temperament could not be equal to such a prolonged strain. It is not for me to judge of their decision towards emigration, only to lighten it as far as it is in my power. . . . Meanwhile I think it will be best to refrain from discussing it with those who can neither help nor fully understand the true conditions." How secretive the situation was, how, perhaps, Stein phrased it to himself, is clear in his 19 August explanation to the Baron. "You will be surprised to hear of the hoped-for early arrival in England of my niece & her husband. They are obliged to leave Vienna as he, being a non-Aryan, has lost his position as consultant for neurology on the panel of State employees & insured people. This means the loss of most of his income. My niece, though born and brought up as a Protestant, has been affected in the same way through her marriage with a non-Aryan."

Waiting for news, Stein turned to his work to get some hours of peace from his thoughts, while in England, Madam and her brother Louis, immediately, quietly, and effectively went about securing Thesa and Gusti's exodus from Vienna.

Leaving the Marg at the end of September, Stein received word that Thesa and Gusti had reached England and were being temporarily housed by Madam. "What gratitude I owe you for the hospitable welcome you gave them I cannot put into words. And what it meant for them to come into such a haven of peace & kindness after the months of mental suffering and anxiety they had gone through! It was indeed a catastrophe which forced them to face a new life. I feel it may have been for their good. You know"—and his values of respectability surface unexpectedly—"I have always felt that the mental atmosphere of Vienna was far from healthy, that there was a good deal of rottenness under that attractive surface. Anyhow I wish that [the forced emigration] may prove justified in the case of Th. and G."[17]

Madam must have passed on some of what she heard from Thesa and Gusti. "Your letter [he wrote Madam] gives me the first inkling of those incredible persecutions. Brief reports in such newspapers as reached me seemed exaggerated."[18] Belief or disbelief had not entered into his response. Madam, better informed, had rescued Thesa and Gusti. Her prompt action made it possible for Stein to care for Ernst's child. "As I wrote you before, it means merely repaying of what I owed to my brother in my youth."

In May 1940, the Nazi's would swallow Belgium, and Madam, Louis, and Stein would be concerned in saving Ernst's son's life.

By the end of October (1938) Stein had returned to Baghdad to continue his search. He found all arrangements in order and the cooperation with the R.A.F., the Iraq government, and the Iraq Petroleum Company assured. East of the Tigris, in the north, almost at the Turkish border was the village of Zakho. "It is pleasant to be in Iraq Kurdistan again and among the Kurds, akin to the folk I got to know on the Persian side of the Zagros range."[19] The village itself, on the Khabur River with snow-covered mountains in the distance, contrasted most pleasantly with the flatness of the desert. From there he discovered two Roman castella beyond the area permanently held as limes, probably dating from Trajan's transitory conquest. The emperor (who ruled A.D. 98–117) captured Ctesiphon, the Parthian capital, and from it carried off a princess as well as the king's golden throne. Trajan reached the Persian Gulf but, with his death on his way back at Hatra, Rome abandoned his eastern gains.

To examine Eski-Mosul, Old Mosul, the eastern terminus of the main limes, Stein crossed the Tigris, and, satisfied, he recrossed the river. He headed south, glimpsing the vast, shapeless mounds marking Nineveh, and so, passing Erbil, reached Kirkuk, headquarters of the Iraq Petroleum Company's vast oil fields. To Stein, the great flat plain where Alexander finally defeated Darius in the battle called after Arbela [Erbil] was thrilling and informative. "I was glad to be able to view that historic site where access was gained for Hellenic culture into Iran & the N.W. of India as well as Turkestan. From a high mound rising near the centre of the plain, probably prehistoric, the last Achaemenean King of Kings may well have watched the varying fortunes of that fateful day."[20] He paused, almost in reverence, before "the ancient town of Arbela still occupying a great circular mound."

A very different experience awaited him the next day (28 November) when, after a long drive over soggy ground, he reached the river. "The ferry refused to work in the darkness, so the night had to be spent in the shelter of a neighboring Kurdish hamlet. On the way to it we were held up by a party of armed villagers who had laid an ambush for an expected party of robbers and whom we at first, *vice versa,* took not without some reason for bandits. Anyhow, it all passed off quite pleasantly, though in that unwonted company the night did not provide much rest. The next morning, I managed to find the ruins of a Roman watch tower station near an old Turkish fortress for which I was searching."

Heavy rains made search from a plane useless and by automobile hazardous. As soon as it was feasible, Stein began his exploration westward toward the Euphrates. "From Hit, a place known already to Herodotus, I travelled along the ancient caravan route towards Palmyra and to my delighted surprise came upon a well-preserved castellum with a massive barrage, a fine specimen of Roman engineering. Nothing had led me to expect evidence of Rome's protection extending through the desert so far south. At another old caravan stage I found a large reservoir also dating from Roman times. Nowhere else is there water for some 120 miles, so such provisions were needed for water at all seasons."[21] From that taste of the desert to the west of the river he returned to the Euphrates.

Along a stretch where the river has cut its way through a gap in the barren plateau of limestone, he made his camp at the village of Ana. The settlement, a succession of picturesque groves of date palms and fruit trees, lines the river bank for some six miles. "An island in the middle of this long stretch, filled with closely packed houses, is undoubtedly that of Anatho which a Greek record of the Parthian Stages mentions in the time of Augustus [63 B.C.–A.D. 14] and which three centuries later surrendered to the Emperor Julian [A.D. 331/32–363] on his fateful Mesopotamian expedition. Constant occupation has left little of antiquity. All the more pleased I was when, on an excursion down the river, I surveyed remains on the island of Telbis, conclusively identified with the island of Thilabus mentioned in the classical records. Owing to the difficulty of readily finding a boat to visit the little deserted island, the massive remains of an ancient fort had apparently escaped archaeological notice. Julian's river fleet had found the small fort too strong for attack & had passed by."

K₃, one of the pumping stations on the Iraq Petroleum Company's 600-mile pipeline from the Kirkuk oil fields to the port of Haifa, offered a convenient base for flights tracking the old caravan route toward Palmyra. Nineteen thirty-nine began auspiciously with a number of flights that allowed Stein to identify places mentioned in classical records. "A letter of a Roman general recently excavated [at Dura-Europos] has been particularly useful. It refers to Roman stations where a Parthian envoy to the Emperor Severus [ruled A.D. 193–211] was to be hospitably received. Two on the Iraq side I can now fix by archaeological indications."²²

Dura-Europos was a kind of echo of the rich mixture of cultures Stein had found in Sinkiang. Its discovery, then recent, rich in inscriptions, reliefs, paintings, and architecture, was as astounding as it was unexpected. Located in the Syrian desert, it was a frontier city that served as one of the ports of entry for the continuous interpenetration of ideas and religions, styles and cultures between East and West. The wall paintings of its synagogue and Christian church show Biblical themes, Hellenistic styles, and Persian idioms. Beginning modestly as Dura, a Babylonian town, it added Europos when the Seleucids made it a military colony (about 300 B.C.) and honored it by naming it after the Macedonian town where Seleucus was born. Two hundred years later, it was known as a prosperous Parthian caravan town and remained so for the next three hundred and fifty years, until it was conquered by the Romans (A.D. 165). As a Roman military frontier colony, it was maintained until conquered and destroyed by the Sasànians about A.D. 275. From then until the 1920s, Dura-Europos lay under its protecting sand, lost, forgotten.

On 14 January Stein could write Madam: "I have completed my task along the Euphrates by air and on the ground," and ten days later, 25 January: "The ancient trade route from Palmyra towards Ctesiphon, of which I have reached the northern end as far as it lies within Iraq, continues to offer plenty of interesting archaeological observation. Again and again the Roman road could be seen on our flights and photographed. On the ground it would be invisible." In Iraq conditions had been ideal: the two pilots assigned to work with Stein had been helpful and interested;

the aerial photographers were equally interested; and the small Vincent plane served his purpose. "Of moderate speed and skillfully piloted, it gave me while standing in the observer's cockpit just that look forward which search for archaeological objects called for."[23]

Stein appreciated how ideal conditions had been when, in mid-February, he began work in the Trans-Jordan portion of the desert. There he had not only to reckon with inclement weather, he had also limitations placed on his flights: "[military] operations in Palestine necessarily restricted the employment of aircraft and pilots for archaeological work."[24] In addition, the available planes were too fast for careful observations.

H_4, his next base, had been placed at the eastern edge of a desolate lava belt, the southern limit of Poidebard's Syrian limes which had protected an ancient caravan route into Central Arabia. The Romans had built a massive barrage to provide water for a military tower and three forts. So far so good; he continued his reconnaissance and moved his base to H_5, the westernmost of the pumping stations, established in the very center of that lava belt. Just to its south was the ruined fortress of Qasr Azraq, most stratigically located. Freshwater springs and wells provided life-giving water; its runoff into salt-encrusted marshes supplied salt that caravans carried northward to Damascus and southward as far as Jauf, an important trading center in Saudi Arabia.

Stein understood that Qasr Azraq, with its unfailing springs, was the nexus of the Roman posts leading into Arabia. In the hope that the British authorities would help obtain permission for his following the line into the Wadi Sirhan across Ibn Saud's border, he made a quick trip to Amman. While awaiting the answer, he completed the survey around Qasr Azraq. "I made my tour from north to south through the outlying portion of the Trans-Jordan desert. Not very far from the head of the Wadi Sirhan, but still within the Trans-Jordan frontier, I found a ruined castellum occupying the top of an isolated volcanic hill, which because of its position was remarkably well preserved; parts of it would only need a new roof to be livable. A short Latin inscription recorded its construction. [Subsequent reading gave the date of A.D. 201.] Then beyond this point the utter aridity for close on 200 miles explained the absence of Roman posts. It has now become even clearer that the outermost line of the Arabian *Limes* must be looked for along the Wadi Sirhan."[25]

Straight as an arrow, the Wadi Sirhan pointed toward Arabia. As Rome's struggle with Parthia had been to wipe out the Parthians as middlemen in her trade with India and China, so the extension of her political power into the Wadi was to safeguard from tribal blackmail and blockade the goods she craved from Arabia. Rome could not pierce the barriers of waterless desert nor impose her will on the powerful Red Sea Coast tribes who controlled the flow of such precious commodities as ivory from Africa and gold, frankincense, and the hardened droplets of myrrh from Southern Arabia. In hoping to trace these caravan routes Stein belonged to the scholarly effort engaged in piecing together the economic life of antiquity. April came and almost passed; still he waited to be told whether Saudi Arabia would grant him the permission he sought.

At Aqaba, where he went to find out if word had come, he received a wireless message telling him not to expect further flights. "The wording seemed to indicate as the cause more trouble absorbing the R.A.F.'s attention." With the permission still not forthcoming, he turned to his last task in that region. "It was my aim to make a proper survey of that portion of the road first opened by the Emperor Trajan from Aqaba towards Petra and Syria which had not yet been determined in detail. So I moved down first to the extreme southeast of Trans-Jordan by the old pilgrims' road towards Mecca, along the now derelict Hejjez railway. It had seen most of the fighting in 1917–18 and blown-up viaducts, wrecked stations, burnt rolling stock, etc. graphically tell its story.

"While my camp stood for two days at Aqaba by the shore of the Red Sea, I was able to visit the site of the ancient port of Alia which had been its Roman name; also to examine what is left of the Roman post by the road which led up the sandy bottom of the great rift valley towards the Dead Sea. It had been observed before from the air but never visited. It felt unpleasantly hot down there and when my keen surveyor had measured the ten miles' base for the plane table survey to be carried over the ground traversed by Trajan's *Via nova,* I felt glad to leave Aqaba and its poor palm-groves behind. The survey of that ancient road right up to Petra, the southernmost part of Rome's *Limes* system in Asia, has proved a congenial task. To follow it was fascinating, if somewhat strenuous; much of the tracking had to be done on foot or by camel. At a number of points we came upon the Roman milestones and the meter of the Chevrolet car showed that the distances had been correctly measured by those who made the road. Incidentally, I was able to determine the position of stations which the Peutinger *Table* shows along the road and to prove the accuracy of the distances it records.

"Needless to say that busy as I am kept by tasks, my thoughts are constantly turning to the course of grave events in distressed Europe. I feel glad that I have been able to extend my Limes exploration to the geographical limit of my programme."[26] Stein was well aware of how ominous events were. He had, a few weeks before, filled out a card "specially issued to retired members of the Indian Service who may be able to return to India in case of war. I thought it but right to point out that my familiarity and topographical experience with much of the N.W. Frontier & beyond might be utilized with advantage. I trust the entry about my age might be benevolently overlooked."[27]

Aware of the likelihood of the tragic clash, Stein brought his program to a close the beginning of May. "I propose to travel to Beirut and spend some days there for needful consultation with Père Poidebard. I shall probably take passage to Marseilles, hoping that the Mediterranean will still remain clear of submarine & bombing attacks."[28]

By the end of May he was at Oxford; on 3 September England and Germany were at war; the blitz began and darkness descended on England.

30

"Olga looks to me distinctly improved and is now fully reassured about the requisite suppression of light after dusk. The shutters have been treated very efficiently with felt strips, the method used when I had to pass Kashmir winter in a study provided with three ill-fitting doors and five windows."[1] Much of the time Stein stayed at Barton so that Olga, Madam's sister, an invalid, would not be alone. A few days later, when he and Madam's brother Louis dined together in London, they "escorted each other safely home. Our night sense is greatly improved & traffic is reduced."[2] When each night was a milestone in a succession of will-breaking, nighttime bombings, Stein adapted his own formula for Madam: "Let Erasmus divert your thoughts at times from all your cares."[3]

11 November: Stein, on his way back to Asia, wrote from Paris. "It was hard to take leave of you, dear Madam, and harder still to think of how the time before we can happily meet again may affect you. Let me once more give my earnest prayer that you may spare your health and your over-abundant energy for the sake of Erasmus and a little also for the sake of those to whom your keeping fit is a *conditio sine qua non* for their mental comfort & peace. Please do not let your eager, brave spirit overtire your body." The full measure of Stein's voiced dependence on Madam—without peer now that Vienna was gone—gets added substance: "I am sorry to find," he adds immediately, "that I left behind the brown leather writing-case which Publius gave me while I was at Merton. I feel quite ashamed to have overlooked this faithful & useful travel companion. Forgive the trouble I give you by asking you kindly to post it by parcel post." Strange: never before in all his comings and goings had he forgotten anything.

For both the parting was painful, as though both feared what indeed it was—final. Stein, disciplined by work and habit, could not change his schedule. Nor would he: his prolonged visit would add to Madam's load. Olga's heart condition was worsening; Maud, Publius's sister, already showing the first signs of cerebral sclerosis, could not be left alone; and however Thesa and Gusti tried not to remain dependent, Madam was concerned for their well-being. Perhaps contemplating such encumbrances on Madam gave Stein the strength to leave. Also his recent operation had taught him the time, care, and energy demanded by invalids; it steeled him in his resolve to keep going as long as he possibly could, to hope for a landslide or a Pathan's knife, to die, as the saying is, "with his boots on."

His train journey east has a Proustian quality, of times remembered. "I passed through Italy, along the foothills past Brescia, Verona, Vicenza! Happy memories were revived of 1909, 1920 and 1929. My thought turned to that old dream, never realized, of an Italian journey with you and Publius, an Erasmian pilgrimage. At the station of Venice there had ended ten years ago that delightful Adriatic tour which together with Olga and Maud I had enjoyed in the blissful company of our beloved Saint. Then in the dim light of Trieste Station there came back the thought of the last night spent by me on European soil before my venturesome start, just 52 years ago, for India."[4]

Was it the remembrance of that first step toward the unknown that made his loneliness palpable? "It seems hard to believe," he wrote Madam from Basra on 1 December, "that three weeks have passed since you & Louis bid me godspeed at Victoria. That day so full of your kindness had prepared me for a long separation. Yet I find it hard to carry on without any fresh news from you since that kind greeting from Paddington cheered me at Paris." Long days on the train, impossible to discipline by his work schedule, left him prey to an overwhelming sadness.

On his way to Basra Stein had stopped at Mosul "to search for a post on the Roman route from Sinjar south to Hatra which I had missed locating last year. On the third day of my stay the search proved successful. There is new cultivation since order & peace have come to Iraq and near a large spring which had attracted one of these recent settlements, our Arab guide duly conducted us to a small site where amidst low debris mounds the remains of a modest Roman castellum could be readily recognized. It lies close to where I had suspected its location, but where an air reconnaissance made a year ago had somehow failed to reveal it. So Iltifat Husain, the faithful Surveyor, and myself could turn back well satisfied with this last search from my old base at Mosul."[5]

In the mail awaiting him at Karachi was a gracious invitation from the viceroy to be his guest before proceeding to Kashmir and a letter from the surveyor general offering to provide Stein's proposed Indus Kohistan tour with "a veritable small survey party for the last bit of unmapped ground within the Himalayas of India."[6] Stein interpreted the largesse in personnel as the Survey of India's concern for the defenses of that area.

Once in India his journey into the past continued. At Lahore he put up at "Nedon's great hostelry—how much changed since it first received me in January 1888. The Mall with its endless row of shops attests the industrial & economic advance. How strange it would look if one were to ride now along it on horseback! Mayo Lodge, now a native girl's-school, looked sadly changed. A high wooden wall around the uprooted front yard hides the house so effectively that it took a little time to find it! When I strolled in the evening into the Lawrence Gardens, I found Montgomery Hall no longer full of gay people. Evidently the reduction of the European element causes the place to look 'triste' and empty. Two badminton courts have replaced the dance floor. . . .

"Raja Sir Daya Kishan Kaul, my old pupil [from Oriental College], was extremely attentive & arranged for transport from Rawalpindi & accommodation in his summer house, Nagin Bagh [Srinagar]. It offers a sunny retreat and two small stoves in the rooms I occupy make it quite habitable, a change from Nedon's where in spite of a big fire it felt like an ice cave. My two servants promptly joined me even though my old Pandit's son was after his wont belated in arrangements. Dash gave me a rapturous welcome."[7]

As if to negate his happy homecoming, Stein makes a most unusual demand on Madam. "When on the voyage out I wished to vest myself in my 'tropical' grey suit, I found the trousers badly eaten in places by moths. I had worn the suit once or twice in July & then stupidly left it hanging without any protection in the wardrobe for the rest of the summer. Possibly the 'Valet' service might be able to repair the damage in some way. Forgive my encumbering you from such a distance with such a commission." Back and back his memories must have carried him until, once again, he felt like the homesick schoolboy in distant Dresden.

Stein, who had felt encouraged by his proposed Indus Kohistan tour, was taken aback when the wali of Swat telegraphed him that, owing to climatic conditions, the tour should be postponed. "Full confidence can be placed in my shrewd old patron's judgement"; he interpreted the message to mean, "when the tribal thermometer again stands at fair his previous offer of access will hold good."[8] When, soon after, he was shown photographs of the aerial survey of the Indus gorges taken as part of trans-frontier areas, he was further reassured: owing to their immense depth no reconnaissance from a plane would suffice; the task would have to await his tour.

As 1939 was ending Stein became involved in the "Woolley Report," an evaluation of the Archaeological Survey made for the administration by Sir Leonard Woolley. It was highly critical and took no cognizance of what had happened since Marshall's retirement. His successor, Rai Baha-dur Daya Sahni, the first Indian to hold the post (1931–35), had the misfortune to deal with an administration notable for its unimaginative and indifferent attitude toward archaeological work; what is more, the worldwide depression justified the massive cuts it had made in the depart-ment's budget. Such a combination of severe blows "had a stunning effect on the Department, and it almost came to the verge of a breakdown."[9] In so oppressive a climate, K. N. Dikshit, his successor (1937–44), needed an equally powerful voice to speak out in the department's be-half. Stein told Madam: "I had been asked to answer [the Woolley Re-port] by the Honourable Member (an Indian) for Education, etc., and could not well refuse this rather embarrassing request in return for what was promised towards getting at last that portfolio of paintings (after 15 years' delay) published."

Whether Stein would have accepted without so enticing a quid pro quo, it is impossible to guess. A delicate matter countering Woolley who had recently been his host. He then went on to outline the facts to the Baron, who had worked long and hard over the portfolio: "The Director Gen-eral and the rest are naturally nervous about the effect of Woolley's out-spoken report, especially upon the Govt. of India as far as the grant of money is concerned. W's report is a remarkable piece of work in many ways, considering the short time [three months] during which he had to make observations all over this vast continent and his naturally inade-quate knowledge of Indian conditions. Many of his remarks are just, some of his criticisms less so. The main fact he could scarcely be expected to dwell on in print, viz. that while Marshall worked under all the difficul-ties inherent in bureaucratic mentality, administrative machine and want of an example of state-aided research in archaeology at home, etc., he did achieve great things; but he could not create foundations sound enough to assure safe and satisfactory progress under much weakened & less well-prepared successors. Complete 'Indianization' is now practically achieved. History will judge whether it will be of benefit to India."[10]

That February, out of the blue, Stein was invited to accompany two nurses being flown to Gilgit. On 15 February he told Madam of the flight and his immense delight at the great adventure: "two most impressive & instructive flights, each of some 300 miles [were] accomplished within a little under three hours in a very spacious & comfortable aircraft. The route there took me across the broad valley of Buner, seen in 1898, and we struck the great river well below where the bold spur bearing the long, flat plateau of Alexander's Aornos juts out into the valley and makes the river bend round it in a loop at its foot. It was quite exciting to recognize the site from a height of about 12,000 feet. Then as the flight led up the river's course, hemmed in from both sides by high mountain walls, the scenery was inexpressively grand. The plane carefully navigated by two pilots had to keep close to the many bends of the tossing green river many thousands of feet below. From Aornos to the great bend of the Indus is a succession of gorges never passed by any Europeans. Without counting the many sharp twists, it cannot be less than 100 miles. It is here that the route of the 'hanging chains' described by the Chinese Buddhist pilgrims had led down by the river's right bank.

"Most stretches of the bank are flanked by extremely precipitous slopes, but here and there I noted small patches of easier ground, probably in part cultivated, at the mouth of side valleys. Communication between some of these though difficult, is practicable. On both sides of the Indus Valley great mountain ranges rising, as the map shows, up to 18–19,000 feet were in full view. [Mont Blanc is not quite 16,000 feet.] It is between the big spurs that descend from them that the river has cut its tortuous path. Near where the Indus makes its great bend to the South, the largest of these side valleys, Kandia, the principal Kohistan settlement, was recognized by me. Then after the plane had turned definitely eastwards, as the light resting on the wing showed, we passed high above the mouth of the Tangir and Darel valleys, well remembered by me from 1913. Beyond, the huge mass of Nanga Parbat [Naked Mountain], glacier-clad for many miles on its flanks, came into view and absorbed all attention. I had seen this grand ridge with its ice pyramid crowning one end from different sides before, but always at a distance, in the case of Mohand Marg, very great. Now it looks overwhelming when passed so near its northern side for what seemed quite a long time.

"Then where the route from Kashmir, remembered from 1900, strikes the Indus, there came a definite turn to the north which meant approach to the wide side valley of Gilgit. With much snow-free ground seen below, the scenery might have looked much tamer had there not shown to the east the great massif of Haramuth, a worthy rival of Nanga Parbat in height and in armour. I had never seen it before when passing up or down in the valley below. It was towards 1 P.M. that we descended to the large patch of the sandy plateau where the Hunza river meets that of the Gilgit, affording a safe landing place elsewhere hard to find in these mountains.

"Quite a gathering welcomed our arrival: the whole British station comprising half a dozen officers and two of the Mem-Sahibs, glad to meet the arrival from that outer world which even during the summer and autumn could otherwise be reached only by a fortnight's travel to Kashmir."

In the long, spacious, sheltered Gilgit Valley, once a center of Buddhism, the climate is excellent, never too hot or too cold. Apples, walnuts, and mulberries flourish, and especially the versatile apricot. Delicious when ripe in summer, still delicious when dried for winter consumption, the apricot's almond-like pit gives pleasure as a nut and nutrition as a source of oil or when ground as flour for bread. "Between the fruit orchards climbing high above the valley walls, are sloping fields of wheat, barley, buckwheat and millet; lower down maize, potatoes, turnips, grapes, cherries and melons grow."[11]

In his 15 February letter to Madam Stein related the rest of the story of the excursion. "The aircraft had to start back early and the four miles or so to the landing ground had to be covered on horseback with a long suspension bridge across the river to be passed dismounted. The sky was clouded as we started, but though a good height had to be kept until clear of the mountains, we were under the cloud ceiling. The route was somewhat changed and instead of seeing Aornos again it took me up the Ghorband valley and then down that of the Swat. When passing high above Saidu, the Wali's residence, I dropped my respectful Salaams to my old patron." He called the flight "an experience which cannot be surpassed"; it is even now considered one of the two of "the most dangerous as well as spectacular civil air routes in the world."[12]

The day after this flight, thanks to the admiration and friendship of Sir George Cunningham, governor of the Northwest Frontier Province, Stein was driven from Peshawar on a day motor-car excursion to the Khyber. Again his letter of 15 February: "I am to start with the Political Agent, not by the tame great road followed also by the railway to the Afghan border, but along the old caravan track through the Mullagori country." For him that region had special meaning. "I had left the Khyber unvisited all these years because I had wished to wait until my way would lie there into Afghanistan—and that day so long hoped for has not come. Yet now, when I have seen so much of the Frontier from the Pamirs down to the Arabian Sea I felt the time had come to fill in this gap also without fear of seeming to 'globe trot.' "[13]

In April Stein received Madam's sorrowful letter telling him of Olga's death.

June found him back on the Marg—"I experienced no trouble over the 5000 foot climb. To keep fit is essential for my work & more important than life's length."[14] There, without distraction, he could turn to his tried antidote for grief: work. He finished reading proofs of the Introduction and Index to *Old Routes of Western Iran* (most of the copies were destroyed when Macmillan's stock was burned during the blitz). Then he began the full statement of his work on the limes—it ran to about 120,000 words—expanding a preliminary report already sent to the *Geographical Journal:* "Surveys on the Roman Frontier in Iraq and Trans-Jordan." "I have just finished an interesting section. Some topographical observations—I should not call it a flair—had enabled me on the ground to determine the much-questioned location of Alexander's passage of the Tigris and of the (so-called) battle of Arbela. No texts were then at hand to check my identifications. Now close examination of Arrian's & Curtius's records has to my special satisfaction entirely confirmed them."[15]

Work was the vitamin essential for his health. Finished with all the travel notes and comment, he reached the bottom of the trunk wherein his ideas were stored. 24 August 1940, to Madam: "I sat down to a long-postponed obligation, a kind of In Memoriam of Captain Anthony Troyer. Did I talk to you at one time of his very interesting life? He served as a young Austrian officer in the wars of the French Revolution, attracted the notice of Lord William Bentinck [1779–1839], before Genoa, was brought out by him to Madras and remained there until 1816 in charge of the institution for teaching surveying to Officers of the East India Co. By a lucky chance in 1902 I secured exact data about T.'s Austrian career at the Vienna War Office; so chance helped me again about him in India. An old acquaintance from the [Trigonometrical] Survey of India, now retired, had got interested in T. as the founder of the modern methods in India. His carefully collected records from Madras most happily complement my own. So I have been able to write the life story of Troyer with the knowledge that he really did useful work while in India besides preparing a very poor translation of the Kashmir Chronicle [*Rajatarangini*] to which I devoted so many happy years of my youth."

Summer was ending, his writing tasks completed, and still he had no clear program for his cold weather tour. When the Wali of Swat advised that tribal conditions still did not permit the Kohistan tour, Stein's alternative project began to take shape: the Rajputana desert. But that was still months off. Just then, with time on his hands, learning that a new edition of his *Rajatarangini* was to be reissued, he decided to illustrate it with photographs of the ancient sites he had traced almost fifty years before. "I am now provided with a useful photographic assistant at the expense of the [Kashmir] State and can thus get my negatives developed over night, so as to make sure of the result. The [Kashmir] Prime Minister has agreed to meet out-of-pocket expenses estimated at Rupees 250. A pleasant holiday accorded by Fate to complete the work of my youth. 'My' village provides me with hardy ponies for transport and a few old retainers; so all moves will be done with ease and free from the troubles which used to attend the tracking with coolies in the old days."[16]

"I thoroughly enjoy my tour," he could tell Madam by the beginning of November, "strenuous as it is. The photographs are proving a complete success. A hardworking Punjabi Brahmin assistant develops my negatives night by night—a help such as I never enjoyed before. It pleases me greatly that my observations made on my hurried vacation tours of 1889–95 prove so exact that they can be reprinted after all these years without any change. All those who helped me with local information have now passed away and this makes me remember how far back that happy time of work now lies. It was an amusing experience to find that I have become, as it were, a historical record."[17]

There was still time, after the Kashmir trip, for writing tasks. "My paper on Captain Troyer is going through the press at Calcutta and I hope to prepare another on the caravan city of Hatra before I start southward."[18] When not absorbed in writing, concern for his niece and nephew filled his thoughts. Lest there be serious delays in communication while he was working in the Rajputana Desert, he gave Louis Allen full power to deal with his money so that he would be able to assure Ernst's children of their living allowances, in his son's case made difficult because of restrictions on sending money abroad. "How petty such cares must seem by the side of those which concern the future of the Empire & the cause of civilization."[19]

"After all the efforts made by me for the last 16 years, the Govt. of India has at last found it possible to arrange for the publication of the long planned portfolio of reproductions of the Buddhist wall paintings recovered by me on my Turkestan expedition. The Baron has spent so much time on their safekeeping, etc., that this publication will give him great satisfaction."[20] There is a noticeable upbeat to his spirit: he had good news about his forthcoming winter tour in the Thar Desert of Rajputana: the native states of Bikaner and Bahawalpur indicated they would welcome him.

Early in December he left for the plains and his winter's work, going via Jammu not only because it was the most direct route but also to settle pending matters. In addition it was a return to the past. 18 December 1940, to Madam: "My old clerk, Ram Chand Bali, now well up the State's official ladder, arranged for me to get through all my tasks. I visited again (after 50 years) the Raghunath Temple Library. Its 6000 old Sanskrit MSS. had been catalogued by me with the help of Pandit Govind Kaul & another excellent scholar friend in what seems now like a previous birth. It had been a dreary task but it saved the collection from being lost. I had a very attentive reception, had to talk Sanskrit again for an hour or so, thus purifying my tongue by use of the sacred language after all my peregrinations in the barbarian north & west. It was a quaint experience to find myself in the end garlanded in the traditional Hindu fashion for the first time in my life."

In his usual fashion he took the occasion to advance his projects. Jammu's prime minister assured him that he would urge the state to help support the new edition of the *Rajatarangini* if printing estimates from Oxford University Press were acceptable. Stein foresaw that preparing the two volumes with added appendices would give him congenial work for the summer on the Marg. His next stop, Lahore, gave him the chance to gain the support of the resident for the Punjab states in securing permission to extend his search into Bahawalpur from Bikaner. Then, as usual, he mixed business with friendly visits. "I called on dear old Mrs. Carmichael. I owe Dash VI to her selection &, as you know, much friendly help from her son at the Air Ministry"—young Carmichael had guided Stein's steps in approaching the Royal Air Force on his proposed aerial reconnaissances.

Jammu, Lahore, then to Bikaner City, which he reached 17 December after a ride in a slow-moving, narrow-gauge railway. The city "founded in the 15th century and representing a delightful bit of true old India—its many fine houses in the traditional style showing richly carved façades of red sandstone—is quite a show place. There is desert all around, devoid of water for the greatest part of the year, except what can be got from deep wells. But it is a 'tame desert' as far as I could judge from the

long railway journey, the dunes supporting low scrub and after the summer rains affording some grazing. I am told movements even by car will
not be difficult. What remains to be seen is how much chance the fairly
arid climate had allowed remains of prehistoric times to survive." Stein
was the guest of the maharaja, "a remarkable figure of the old chivalrous
type. I am assured by him of whatever help I may need in this territory."
Unsolicited help also came from the director general of archaeology: a
grant of 2,000 rupees which permitted him to hire his old surveyor, Muhammad Ayub Khan, now pensioned.

Stein, expert in reading desert topography learned in desert areas far
from Kashmir, now turned his attention to studying the riverine history
of India's Thar Desert almost on its doorstep. It was a kind of Sherlock
Holmes investigation around the fate of the Sarsuti River, the easternmost branch of the Ghaggar, a river that starts as a torrent in the outermost Himalayan ranges. Still a torrent, it enters the plains and becomes
first a modest rivulet, until, approaching the Bikaner border, it dwindles
to a sometimes-stream varying with the annual monsoon rainfall, feeding
only a limited belt of irrigation. Sarsuti, the Hindu derivative of the Sanskrit *Sarasvati,* had been a sacred river; according to Vedic pious fiction
it had once carried abundant life-giving water down to the ocean, a tradition supported by its wide, dry bed within Bikaner.

The automobile provided by the maharaja saved both time and effort
in surveying numerous old sites along the Ghaggar's dry bed. "I secured
a couple of days for trial excavations on a big mound [Munda] that gave
definite evidence of the early date of the curious painted pottery characteristic of many of these sites. I found it associated with small but interesting terracotta remains of a ruined shrine unmistakably showing the influence of Graeco-Buddhist sculpture so far away from the N.W. Frontier.
I must still survey mounds down that strange dried up river as far as it
lies in Bikaner territory as well as an old temple reported in the belt of
sand dunes to the southeast. There, too, this 'tame' desert has villages
maintained by water in tanks which rainfall fills in good seasons. When it
fails, as happens often enough, the people, hardworking & frugal, move
away with their cattle, camels & sheep to neighboring districts chiefly in
the easternmost Punjab."[21]

He tracked the remaining portion of the dry Ghagger in Bikaner by a
"convenient combination of rail & long rides on the back of camel &
horse. Thus a 40 mile camel ride to the Bikaner border and back, negotiated in one day, was interesting enough but I confess to have felt glad
when it was finished without my feeling too tired."[22] He noticed as he

made his way down the Ghaggar that the mounds diminished in number.

Satisfied that his survey in Bikaner was finished, he followed the desert's perimeter going via Jhodpur to the fascinating desert city of Jaiselmer. "It has no running water anywhere and only a very scanty and erratic allowance of the 'water of mercy.' There is a wealth of splendid ornamental stone carving in the old palaces and the mansions of trading families who once flourished when Jaiselmer saw much caravan traffic."[23] A five-day camel journey then brought him to the southern corner of Bahawalpur, a Muslim state between the Punjab and Sind. There where the territories of Bikaner and Bahawalpur meet, the dry river continued under a different name: the Ghaggar became known as the Hakra.

"I have just been able to connect my work on the Bahawalpur side with my survey on the Bikaner territory; the chance for a trial excavation is offered by a fairly large mound. Prehistoric pieces could be picked up on the surface."[24] Stein recognized plain ware and a few fragments simply designed with black bands or coarsely stamped with patterns as similar to those inscribed on Mohengo-daro and Harappa seals. Whereas along the Ghaggar he had found only a single, small, worked flint, stone implements were abundant along the Hakra. This marked difference suggested to Stein that "prehistoric occupation along the Hakra had stopped at its lower reaches after the branch of the nearby Sutlej had ceased to join it; for a time the floods of the Ghaggar may still have sufficed for cultivation . . . later settled agriculture became restricted to the Ghaggar higher up in Bikaner territory."[25]

His tour was finished by the first week in March, and he returned to Bahawalpur City. He had examined about eighteen mounds in an area of some 100 square miles, most of it overrun by dunes separated by the flat ground carrying the branches of the river; the configuration suggested an ancient delta. He described it in his letter to Madam of 5 March 1941: "At Derewar, where a cluster of old mounds had brought me, I found a big site [1300 by 900 yards at its base and more than 50 feet high] of what must have been the chief place on the central portion of the Hakra. There we luckily discovered what had escaped us before—the prehistoric burial place of the site. A short excavation sufficed to bring to light plenty of funerary furniture to settle the question of the period of occupation and with it also the time up to which the dead river carried its water towards the Indus. I am able to leave with the main object of this tour gained before the increasing heat makes work here too trying."

His tour was concluded. A surprise awaited him as he headed northward. At the Dera Ismail Khan station, friends from his 1933 Malakand visit, now assigned to the Frontier Province, greeted him. "Two long days were devoted to ancient sites I had surveyed in 1927; the Koh-i-Suleiman offered special interest now that I could compare its ceramic ware with material collected along the Hakra. Then I was brought [17 March] by car to Bannu, the Northern Waziristan cantonment. It was a convenient base for a renewed examination of the huge mound at Akra, marking the capital of this fertile tract in Greek & Indo-Scythian times. The visit offered a curious illustration of changed times: while electric light, hot & cold water in the bathroom have brought the old Commissioner's Bungalow up-to-date, an escort of some 20 military police had to be taken along when we went to Akra just nine miles away."[26]

Then a 120-mile automobile ride to Peshawar brought him to the home of Sir George Cunningham. His intercession with the military allowed Stein to move freely about the Khyber, the "Gates of India." "I traced the old caravan route which served trade and probably invasion long before the 'Gate' with its fine Afridis, blackmailing 'gatekeepers' of stories & films, came into prominence. It was a fascinating five days of congenial work collecting the evidence furnished by topography and surviving tradition among the tribes distinct from Afridis which show how the old road went towards Charsada, the ancient capital of Greek and Buddhist times. All useful material for a little study of historical topography."[27] Once Stein had wished to be titled "archaeological explorer"; now he alluded to himself as a historical topographer.

The Indus Kohistan tour, planned for October-November 1941, coincided with Stein's seventy-ninth birthday. In good time Madam broached the idea that he celebrate it with Thesa, Gusti, and herself. Perhaps she was activated by Stein's sadness at the recent loss of four old friends, a loss he could feel in the dwindling volume of his correspondence, and his mentioning that the Hackins, returning to France from Afghanistan, had drowned when their ship was torpedoed. He declined Madam's invitation—"however much I long to be in your dear presence"—choosing to realize the long-deferred, long-hoped for tour of the Indus gorges. "I should not mind the chance of being torpedoed on the way but there are tasks apart from the Indus Kohistan which oblige me to remain in India for the present."[28] What were these tasks? Did invoking them mask a deep reluctance, a dread that he might take the place of Maud who had recently died after "months of utter helplessness, a sad destiny for one who had been only too active before"?[29] What he really asked from Madam was her weekly letters and his freedom: Stein needed distance to be his own man.

That summer, on his beloved Marg, he wrote additional notes for the reprinting of the *Rajatarangini,* an occupation that happily transported him back to the years 1895–98. September he used for a little tour to photograph the great mountain mass of Haramuth, the 17,000–foot peak wreathed in cloud and legend. "The ruined shrines in the valley which I cleared just 50 years ago of some of its luxuriant jungle offered a delightful rest. It is a place where in old Hindu days, wise men used to retire to the forest to end their days there in pious contemplation. Though I shall before long enter my 80th year, I do not feel quite that sort of attraction."[30] Knowing his reticences as well as his confidences, Madam must surely have known what he was telling her.

31

Stein celebrated his seventy-ninth birthday on the Indus Kohistan tour. His letters, substituting for a P.N., described it. Jottings from them sketch in the adventure.

At the Malakand, 26 October: "The Wali of Swat greeted me as an old friend. In one of his cars I travelled to Saidu while he, remarkably active as he still is in spite of his white beard & one eye lost by a cataract, left for a tour of inspection over distant parts of his 'Kingdom.' All arrangements are made under the Wali's personal orders. So I can start promptly for high ground."

Camp Karang, 15 November: "I left Saidu to reach the mouth of the Kana valley I had first passed down in 1926 on my way to Aornos. It was very pleasing to find how well my visit was remembered by the old khans of that fertile valley. Excellent transport arrangements made it easy to reach in three days that Duber pass which then had marked the limit of the Wali's kingdom. The route led over wholly unknown high ground with delightful alpine scenery, just close enough to the tree line to afford ample fuel to give warmth at camps for the large numbers of load-carrying men & escort.

"A mule path, constructed for my special benefit, allowed most of the baggage to be carried on hired mules as far as the main village of Duber in spite of intervening steep ridges. In many places the track was far too uncomfortable for riding. A day's halt afforded welcome rest for tired legs. It also allowed me to start anthropometric work on specimens of Kohistani men; like other Dard-speaking tribes in the Hindukush region, the people are a sturdy, goodlooking race, often with fair eyes & hair.

"It was well to have had that restful halt. For the next day the ascent began to the Bisau pass, along steep, boulder-strewn slopes. The second day meant a continuous scramble over the rubble & boulders of ancient moraines left behind by glaciers which had once descended on all sides into the head of the valley and united in a wide ice-worn basin. Dusk fell; in spite of a comparatively gentle slope, progress in the light of two hurricane lanterns was necessarily slow. A narrow ridge between two former glaciers was the only spot where level ground for our tents could be looked for. It took fully four hours for me to reach it. After the constant scramble over rubble & boulders it seemed almost a relief to find at last some cliffs to climb up to. But our fleet-footed Kohistanis had reached the ridge some hours before, and by 10 P.M. I could tumble into my tent.

"The climb up to the Bisau pass continued an interminable succession of rubble slopes & moraines. It was a slow business. By 3:30 P.M. I gained the pass at close to 15,000 feet. Long before, we could sight the men from Kandia who had come to relieve our Duberis. An icy wind was blowing across the crest & I was glad when payment had been made to the discharged men & needful height observations & plane table work done.

"The northern side of the pass presented itself as a continuous snowy slope with black blocks of rock cropping out. The snow had become soft; though the men who had come up had marked something like a track, it was impossible to avoid slipping off it again & again. It was a comfort when, by dusk, a bit of fairly level ground was gained. Yet the most trying part of the descent was still to follow: for close on four hours it led down by the side of a snow-filled gully, invisible below the ground. Care had to be taken; slowly, but safely it was at last accomplished. The plucky Hakim led the way, lighting up sound steps with his electric torch while two hardy commanders of the Kohistan Levies took care not to let me slip. Once or twice there were stumbles among our little party, but by luck I did not share them, just as I was also spared the sight of the deep ravine by whose side the track lay.

"By 10 P.M. at last we were clear of the snow & could warm ourselves by a fire. But the end of the march which the sight of the fires had promised was not there. Owing to the absence of all shelter or local fuel, the baggage had been sent on to the first hut; after a much belated tiffin, I decided to move on even though some bedding had been sent up for me in case of my being too much behind. The track down led over an old moraine & was terribly stony, but by 1 A.M., in the light of torches, the camp was gained. With big fires all around it was quite a lively scene. The night at some 10,500 feet was bitterly cold. Timber taken from the roof of a hut supplemented what fuel had been brought up from the forest below.

"The warm sunshine which greeted me when I got outside my little tent rather late did not last long; the sun soon crept behind a high peak; and cold enough it felt in the shade, and dreary. The Hakim & all the men had followed the sunshine to a little plateau below and were greatly enjoying the lively airs which the little service band struck up to enliven an impromptu sword dance. The band had accompanied us all the way from Kana; it consisted of a clever Indian Gipsy playing alternately a bagpipe and fife, a drummer and a bell-swinger. The noise made is too great if one happens to be near, but from a distance my unmusical ear has begun to appreciate some of the lively airs.

"Late as our start was, it sufficed for us to reach, by nightfall the nearest little hamlet of Kacher. A day's halt was very welcome after that long, trying crossing of the Bisau. Two easy marches have carried us to this, a largish village on the main stream of the Kandia. Faithful Alia has not been well since those nights on the Bisau and since Kacher has been carried on a sort of primitive litter. Wrapped in warm quilts he ought to have been as comfortable as I was when taken across the Karakoram to Ladak; but being bandaged to the litter like one of the bambini of the Florence majolica plaques, he did not like it."

Still at Camp Karang, 1 December: "Here the fort fortunately provided warm quarters for Alia who was rather seriously ill with some internal complaint and fever. He has by now recovered fair strength. I myself, as the sequel of a bad cold found it wise to keep warm in my cosy little tent and thus to ward off an incipient bronchial catarrh.

"The Kandia river has cut its way through some forbidding defiles and I heartily welcomed the improvement which the Hakim had effected on the trying trail through them. Of course, even now the ladder-like paths up cliffs and the galleries built out in place along the precipitous rock slopes would be quite impracticable for laden animals. The Kohistanis, fortunately, are very quick footed carriers and cheery people." Thus and there Stein celebrated his seventy-ninth birthday.

Samp Seo, 15 December: "Since my last letter I have moved down the Kandia river to its junction with the Indus. The marches led along mountain sides where I had much reason to thank the Wali for the great improvement made in the footpath. Even now there are long stretches with galleries built out from sheer rock-faces a thousand feet or more above the Indus. There are constant ups & downs which, though safe, are a bit tiring. What these paths were like 'before they were made' could be judged from those to be seen along the opposite bank of the Indus still in tribal territory.

"The most exacting portion of the descent through the Indus gorges lies behind me. It was accomplished in four marches without undue fatigue, wholly as the result of the remarkable work of the Hakim in making a safe footpath where before only sure-footed hillmen could move with any assurance. This means constant climbs up to heights where the cliffs to be negotiated are less frequent & continuous, with corresponding climbs down to near the river. Between ascents and descents, 7–8 miles per day was quite enough for feet no longer young. It makes me quite proud that the longest of these marches, 14 miles or so, was managed by me in 10 hours. To shorten the daily progress would not have been possible, so rare are the spots where level space could be found for pitching our few little tents."

Christmas Day, Camp Patan: "My stay here coincided with the great Muhammadan *Id* [holiday] and now we are weather-bound. Also as Patan is the headquarters of the ever-obliging Hakim, and after two months' absence for my sake, it was right to let him have some time with his family and tend to local affairs. The long halt is not altogether unwelcome."

7 January 1942: "One day's march had brought us here from Patan and once again we are weather-bound. Today is the third since heavy clouds descended on the mountains and put them under snow although their height has grown less. The rain has scarcely ceased since our tents were pitched here on the 4th, and across the Indus big torrents are racing down the steep, bare slopes of the hillsides. The little patch of black ground around the tents is becoming a morass but inside my tent I am comfortable enough. The snow has caused the telephone lines to break down between the Wali's forts. How far away I have been from the troubles of this distracted globe you may judge that I learned only today of Japan's entry into the war."

Camp Ranitia, 17 January: "Warm thanks for your sympathy in my loss of dear Dash. [He had been killed by a leopard, waiting in ambush, and his end had caused Stein great grief.] I miss him much, but I have not tried to replace him. I feel at my age I may as well tread my path alone while in Asia.

"After leaving drenched Jijal I descended the Indus & for the next three days the clouds descended bringing fresh rain to the camp and snow within a thousand feet or so. But the tents are pitched on the roofs of houses & we have been spared having a little morass all around us."

Saidu, Swat, 27 January: "From Sholkara Fort which we reached on the 19th, I had one more interesting march down the Indus to Besham. There the direct route to Swat proper joins the Indus Valley. Before reaching Besham I had the satisfaction of sighting once more in the distance the heights of Aornos, now under heavy snow. Then before turning up the Ghorband for Swat, the route which the Chinese pilgrims had followed in A.D. 400, I found striking confirmation of the accuracy of their record. They mention to have crossed the Indus by a rope bridge just before leaving the river for fertile Swat, their blessed goal. Now below Besham, the exact position of that bridge could be unmistakably identified.

"From Besham it was possible in two marches to secure mules for the baggage & to cover part of the marches on ponies. Then we had once more a day of rain & snow; but for the crossing of the Karorai pass we fortunately had a bright day. Though it is only 8000 feet high, the snow lay all along the route so thick that without gangs of men who cleared a narrow footing ahead of us, the crossing would have been very difficult. Even so it was a rather trying experience—only by stepping exactly into the footprints trodden before us was it possible to avoid landing in deep, soft snow. After 10 hours of such tramping, I was glad to get into my small tent pitched on the snow-free roof of a hut at the first hamlet. A fierce gale blowing all night failed to bring the tent down. Next morning a short march brought us to the village of Shalpin whence by motor & lorry we came in comfort here to the Wali's hospitable guest house."

However he understated his extraordinary effort to Madam, a few months later he characterized his three months' tour as "distinctly trying work. It all had to be done on foot and the exceptional weather of this winter with its plentiful snow and rain added to the exactions. But I am glad though now in my eightieth year I was able to face them."[1] He may be forgiven his gentle boasting.

The Marg that summer knew Stein but briefly, just time enough for him to write the Introduction for the portfolio of wall paintings: "the last publication as far as I am concerned with my Central Asian expeditions."[2] He had been asked by Sir George Cunningham to lead a survey party into the Indus Kohistan, east of the river. A few months before, an invincible Japan had conquered an impregnable Singapore from its unprotected land side, and the governor did not want India's back door left unknown, unprotected. The plan called for Stein to be present at Naran, a small village in the Kagan valley, where the tribal headmen of the Jalkot tract gathered, as was customary, to "secure their supply of salt and other necessities while the mountain passes were open."[3] If the tribal assembly agreed to look after Stein, he was to leave with them about 20 July.

Stein kept the date. But the headmen did not. "The delay was reportedly due to the disturbing influence of a wandering faqir from the Kashmir side who, turned out for mischievous agitation, had moved into the Jalkot area. The holy man commands no following but causes dissent among the several tribal sections towards Government. Previously they had all offered to welcome any European visitor to their area. A capable Punjabi Surveyor has joined me, replacing Muhammad Ayub Khan, my old travel companion, whom a fatal illness of his son prevents from working." A month passed; the headmen did not arrive, and the faqir was still at large.

Stein met the stalemate by proceeding with an alternate plan: he would go into the Thor tract, a still unsurveyed, outlying portion of Chilas. After three easy marches, the going into Thor changed dramatically. "It meant crossing a succession of passes between 14,000 and 15,000 feet in height and quite impossible for riding or laden animals; where even my heavily nailed Alpine boots do not always give a sure footing. But assistance is always provided for by one or another of the Chilas Levies who accompany the party. As regard for my age justifies slow climbing, I can take my time over slippery rock debris. Chilas and Thor throughout are cut up by closely packed spurs with deep narrow valleys between."[4]

Stein was almost at the principal Chilas village when he received word that the Jalkoti headmen had actually met and agreed to expel the faqir. With that, the chief objective secured, the political officer thought it advisable not to press the tribe to accept responsibility for the Stein survey. In brief: the Jalkot tour was postponed for next year.

What had drawn Stein to that village in Chilas? It was, he told Madam, "a report of numerous rock engravings along the banks of the Indus of which I had not heard before; [it] was a welcome compensation for having to forego entry into tribal territory and also for the physical discomfort of visiting the hot Indus valley at this season. Apart from its summer heat, Chilas is notorious for the plague of its fierce local flies which love to suck the blood of newcomers. On descending from Thor I, too, did not escape their attacks. The journey up the absolutely bare bank of the Indus to Chilas village, some 24 miles, is dreaded owing to the heat."

Not heat, not flies, not the daily thousand-foot climb could deflect Stein from his chosen task. "There was satisfaction in my being able to determine by paleographic evidence that all these inscriptions, never studied or published before, are older than the 6th–7th century A.D. There is good reason to hope that both together will throw interesting light on the spread of Buddhism into the Hindukush as vouched for by that old Chinese pilgrim, Fa-hsien."[5]

Stein left for Kashmir following the route along the Kishanganga River. "On crossing a small stream whose bridge had gone, the mule carrying my two suitcases slipped and the drying of their contents, bedding, etc., cost a day's halt. More serious was the accident which caused the death of a mule. On a narrow path along a precipitous rock hillside, it lost its foothold owing to a projecting big boulder and, rolling down the steep slope, was killed. By a lucky chance its load, the suitcases with records, negatives, etc., things I could not have replaced, narrowly escaped falling into the rocky bed of the river below. Towards the end of 12 marches I was glad to find myself back in Kashmir. After all the barrenness passed, the 'kingdom' looked more verdant and fertile than ever. How grateful I must feel to the kindly Fate which allowed me to do so much of my work in Kashmir for the last 55 years."

That November Stein had his eightieth birthday. To please Madam he sent on to her a letter from Lord Linthithgow, the viceroy, who congratulated him "not only on attaining that great age but in attaining it with your faculties and your physical strength in every way unimpaired, and with the prospect of being able to add still more to the immense contribution you have made to clearing up the dark places of the past and the problems of Central Asia, and of India and its borders."[6] Another enclosure was from his old friend Dunsterville, equally energetic and adventurous, who wrote "we all think it time for you to stop exploring and come home again." Stein agreed with his friend; to him home was where his work took him.

Most happily, his work at that very time was offering him two cold-weather tours. It was flattering to be remembered for "so unimportant an anniversary"; it was gratifying to receive a present of 2,000 rupees from the nawab of Bahawalpur for expenses to trace the lowest portion of the "lost" Sarasvati river to where once it had flowed into the Indus. "It may take about a month and the Prime Minister [of Bahawalpur] has spontaneously offered to provide motor transport essential for the task. The cost & difficulty about petrol [rationed] might otherwise make me hesitate. The other tour, taken into view long ago, lies in the State of Las Bela along the Arabian Sea. There, too, a lorry would be indispensable. Fortunately my friend from the Persian Gulf is now Revenue Commissioner in Baluchistan and is likely to facilitate arrangements. I propose to start down to Bahawalpur about the beginning of December & hope to be free for the move to Las Bela early in January."[7]

The nawab provided a car built to negotiate sandy terrain, the low dunes along old riverbeds as well as those on the nearby once-cultivated area; he also chose the driver, an expert for the task to be done. "It enabled us to cover much ground (the distances are great) in fair comfort and thanks to the remarkable topographical sense of my driver, without loss of time. It reminds me again and again of those happy times in the Taklamakan and Lop desert."[8] Stein's reconnaissance carried him to the border of Sind, a fairly new province, and then back to Bahawalpur City along a portion of the lost river. He also made trial excavations at a prehistoric site he had noted before but could not investigate at that time.

Krishna Deva, a young scholar assigned by the director-general of the Archaeological Survey to accompany Stein, wrote in his report of 24 January 1943: "During the last week of our exploratory tour in Bahawalpur State, trial excavations at 3 sites [were made] all situated along the dried-up bed of the Hakra. Two, viz. Kudwala and Ahmadwala were found to be of Harappan civilization, characterized by the finds of black-on-red painted wares familiar from Mohengo-daro and Harappa; flint knives; large numbers of semi-burnt cakes of triangular, circular and oval shapes; perforated pottery; plain pottery with types similar to Mohengo-daro and Har.; pottery bangles and toy-cart frames; beads and bangles of shell and faience; etc. Two interesting finds from the former site are 1) a terracotta human figurine resembling in type figurines from South Baluchistan sites, and 2) a small pottery (?) designed for suspending things like toothpicks, ear-cleaners and tweezers, etc. . . . The 3rd site which was subjected to similar sounding yielded besides other objects, black-on-red ware together with polychrome ware of the Amri type bearing geometric designs painted in sepia on cream or greenish pale ground. As we came upon numerous mud spots [?] here we could not go much deeper and the actual relation of the two types of painted wares in deeper levels was left unexplored. Mud walls were also met with in the trial trenches sunk in the first two sites, some of which showed also walls of burnt bricks of the curiously modern size so well known from Mohengo-daro and Har."[9]

The second "minor cold-weather tour" began the end of January 1943 when Stein's party reached Bela, the modest capital of Las Bela State. He hoped before the heat made work impossible to pick up where he had left off in 1928–29. Thanking the director-general of the Archaeological Survey for his "contribution of Rupees 1500," Stein reported that "the first few days' survey sufficed to allow us to trace a number of ancient sites in the vicinity [of the capital]. Reserving trial excavations for later, I set out to the southwest mainly with a view to tracing the exact route followed by Alexander on his march into Gedrosia, thus linking up our tour with my explorations of 1928–29 in Makran. We first went to Ormara on the Arabian Sea coast, then up the Hingol valley to Jhau in Jhalan and thence back here to Bela. In addition, in Jhau some more prehistoric sites were traced. We now [4 March] hope to effect trial excavations at two mounds within reach of Bela. Altogether these four odd weeks, made somewhat trying by increasing heat and the mode of travel, all on camels, have proved very instructive in different ways."[10]

Jottings from Mr. Krishna Deva: "Ormara was really disappointing, since we had in vain expected it to be coeval with the prehistoric sites of Gwadur and Jiwarri, situated on the sea-coast further westward. The main interest of these hard journeys, however, was determination of the route which Alexander took for passage of his troops from the Las Bela plains across the western bordering hills into the Kej Valley of Makran. The topographical observations go against the accepted Malan route which is absolutely impassable for the wheeled vehicles which Alexander's army had. Sir Aurel thinks the pass which we traversed from Jhau to Bela to be the more likely route as this is fairly broad besides being along the shortest route from Bela to Makran.

"At Jhau, the group of closely situated mounds previously explored by Sir Aurel was revisited and subjected to a more thorough examination. Siah damb, the most promising, yielded an abundance of the typical Nal ware which had remained unearthed previously. Before returning to Bela we viewed the very numerous groups of crude caves excavated in the hard conglomerate cliffs along a spring-fed streamlet at Gondrani and a prehistoric mound with the Makran type of pottery on the bank of the Porali higher up. . . . Our last week at Las Bela we are concentrating on trial excavations at Karia Pir and Niai Buthi, sites respectively 2 and 5 miles from Bela. The latter appears to be a purely prehistoric settlement of great interest."[11] (In 1959–60, Fairservis followed in Stein's tracks and extended his trial trenches. Niai Buthi yielded "the familiar Nal ceramics at its earlier levels, in the top-most, signs of contact with the Harappan civilization. Between and amid are the pottery types which were first identified by Stein at the site of Kulli."[12]

These last tours restate Stein's old, abiding interests. Intellectually, he could appreciate the importance of the prehistoric sites, but his heart and imagination and energy were still caught up in his earlier loves: the Buddhist pictures and inscriptions laboriously scratched into the rock walls of the Indus gorges, the river lost since Vedic times, and the clarification of Alexander's tracks into and through Baluchistan. All declared his passion for historical topography.

Did Madam remember his previous mentions of Van Engert? "He is a very capable, very scholarly man of the best Harvard type who came to pay me a several days' visit on the Marg in 1928 and is fully acquainted with my old plans upon Bactria. In 1930 he gave me effective help from Central America by recommending me to the State Department at Washington for my fourth expedition to Turkestan-China. It was a very pleasant surprise to find my old aim so helpfully remembered by him after all these years."[13]

It was April 1943, and Stein, at Peshawar on his way back from Las Bela, was electrified to receive "a telegraphic invitation to Kabul, wholly unexpected, from my Harvard friend, Cornelius Van Engert, since some months American Minister there, accompanied by a statement that the Afghan Foreign Minister would welcome me to Afghanistan. After all previous failures to obtain access to the goal cherished since my early youth, this seemed distinctly encouraging." Unwilling to have his hopes raised in vain, Stein consulted his friend Cunningham, the governor of the Northwest Frontier Province, who immediately sent telegrams to the government of India and to the British minister at Kabul. "The former quickly brought a request about my intended work. I repeated at some length proposals I had repeatedly submitted since 1902 under Lord Curzon's aegis, and I made it clear that what I looked for was antiquarian exploration, not merely a rapid visit to Kabul as if I were a 'globe trotter,' which would give only a tantalizing distant glimpse of the 'Promised Land.' " Hindsight permits the question: the promise given to Moses?

Consultations, high-level communications, and more consultations at Peshawar: Stein did not reach Srinagar until mid-April. There for a week, as usual, he stayed with his old friends, the Neves. Dr. Ernest Neve, slightly Stein's senior, was a retired medical missionary who had been in charge of the hospital run by the Christian Missionary Society. Some years before, he had been relieved by the new head, Dr. Macpherson. The Neve home had a special place in Stein's life: he was always welcome there; from it he left and to it he returned from his tours. Both doctors kept a friendly medical eye on Stein's health; both knew well his history of dyspeptic attacks, and when he returned from his Las Bela tour they had tests made. "Neve read laboratory report on my digestive organs; gastritis recognized as the only cause." Stein wrote in his diary for 17 April. Reassured that it was the same complaint he had lived with for years, Stein stoically suffered but did not engage in hypochondriasis. The report gave Stein a clean bill of health and permitted him in letter after letter to assure Madam that in truth he was "very fit"; his diary tells another story.

His diary covers the months from 15 April to 17 October 1943. It lists when and how acutely dyspepsia—whatever the term meant in his case—plagued him. Sometimes it gave him sleepless nights, sometimes borborygmus and distension annoyed him or, more distressingly, was accompanied by severe belching and attacks of diarrhea. Worry, he noted, increased dyspepsia: worry over replacing Ali Bat, his cook for thirteen years, who died (in vain he had been given anti-syphilis treatment [1 June]); worry over getting funds to Ernst and Jeanne in their obvious distress; fretting at the wages sought by Ali's replacement—"tastes of inflation" (10 May); annoyance at the delays and the price asked to supply a new flap for his tent. A long, dolorous litany of dyspepsia.

The diary also gives trivial aspects not seen before. There is Stein the master of so many arts and skills showing ineptitude in a simple situation. "Lamp smoking up chimney after giving inadequate light" (9 June), with its sequel the next day. "Tried to fit Aladdin lamp with fresh wick, task not completed by the time darkness set in. Felt slightly feverish & went to bed after having sat without light." The diary gives an almost daily mention of the weather during his summer's stay on the Marg: spells of drenching rain and icy mists when, in his heatless tent his feet kept warm in a fur bag, he wrote. He gladly endured such foul spells for the glorious days of sunshine and the view of distant mountains standing large under brilliant skies. And he welcomed visitors. "The Macphersons made the climb up. Dinner by 8 P.M. in cheery lamplight; a happy day which made me forget all physical troubles & most of anxieties" (12 July).

"Dies mirabile. Received letter from Engert giving hope for Afghanistan support; from Cunningham with good news for the approaching Jalkot assembly; and from Wheeler [Colonel, later Sir Mortimer; then director-general of the Archaeological Survey] promising all needful help for Kohistan."

Kohistan or Afghanistan? Or, perhaps both. "I hope to carry out the intended [Kohistan] tour between the first week of October and the middle of November. I should then move to Kabul with as little delay as possible. If the 'tribal arrangements' cannot be secured in time, I propose to start for Kabul by the first half of October.

"For the present it may suffice to state that my immediate aim would be during the coming cold weather to move from Kabul to Kandahar and thence through the Helmand Valley to examine any ancient remains to be traced between the foot of the Hazara hills and the Baluchistan border in the south as far as Sistan, and thence up to Herat. I am anxious to devote to this long desired task as much time and work as the years ahead for me may permit."[14] That was Stein's program for the Promised Land. When, later, the Kabul trip was confirmed, including a grant, he noted "that the 1898 . . . 1943 plans were written on the same spot" (22 Aug.).

Kabul, good weather, welcome guests, and finally, toward the end of August, "Dash VII reincarnated, arrived. A dear jolly terrier with black head & all auspicious markings. A real joy. Walk up to ridge [on the Marg] with Dash VII who kept quietly by my side. He had during the day settled down at my feet or in his basket just as Dash VI did. A dear little dog, much pleased with chances for hunting rats, who has made me forget little sleep & dyspepsia. May he remain long with me & be blessed" (20 Aug.). "To rest by 10 P.M. to be awakened by attack of leopard beaten off by combination shouting & barking" (28 Aug.). "Was it the same killer who had caught Dash VI?" "Realized risks to Dash VII who keeps as lively as ever & is kept on a chain on evening walk." "Dash sharpens his teeth by cutting tent-rope" (6 Sept.), and "Dash indulged in biting off medicine bottle tops & absorbed chill-blain ointment" (19 Sept.).

The diary speaks across the years of that last summer on the Marg. Dyspepsia, Dash VII, guests, beating off an invasion by buffalo interlopers—all peripheral to writing: concentrated work on setting forth his field notes and sustaining the heavy flow of correspondence. He noted "the effect of successive cold weather tours on my old retainers" (24 April) but never considered that he, too, might be showing the same wear and tear. He might admit in his diary—"felt tired after poor night & rested in chair 2–4 P.M." (31 July)—and then add, "work on completing argument on Alex.'s route through Gedrosia" later the same day. His entries make his pattern of activity clear: squirrel-like, his cold-weather tours, necessarily short, were his time for gathering raw material —a hoard of notebooks and photographs, surveys and specimens—to be carried up to the Marg for writing fare. Indeed work kept him functioning as previously work had granted him tours of forgetfulness when deaths had devastated his heart.

With September and the approach of his Afghanistan tour, his mind turned to business matters. There was his will. (Drawn July 1934 in Oxford after Publius's death. Madam, Louis, and Kenneth Mason, Royal

Engineers Professor of Geography at Oxford, and old, valued friend from the Trigonometrical Survey, were named executors; Olga had been a witness. The first codicil, October 1936, executed at the British consulate, Kermanshah, took cognizance of Hetty's death. The second, July 1937, done at Oxford, revoked a bequest to Corpus Christi College; and the third, 9 October 1943, done at Srinagar with the Neves as witnesses, divided the income from his estate equally between Thesa and Ernst. On their deaths the money was to establish the Stein-Arnold Exploration Fund to be administered by the British Academy; Stein asked that qualified Hungarian scholars also be considered as candidates.) "Drafted codicil letting full income from capital go to Thesa and Ernst for lifetime. Disposal of same after expiry of life interests entirely to Brit. Academy. Poor night's rest; depressed" (2 Sept.).

Events of the war are noted: "Italy's capitulation & surrender of Italian fleet. Sic transit gloria—Romae. Escape of Mussolini: perhaps. Pleasant letter accepting Las Bela article [*Geographical Journal*]" (16 Sept.). The promise of peace filled him with optimism. "Wrote for passport renewal; suggest endorsement for Europe via Iraq & Turkey" (17 Sept.). When flowers, grasses, and falling leaves, the calendar of autumn, warned Stein it was time to leave, he used a few days to send "off Xmas cards to Harvard friends" (19 Sept.), and ordered the lorry to meet him at the foot of the Marg on 25 September.

"Descended from the Marg with loads on 59 men. On steep slope slipped & bruised shoulder-blade & back. Stony track amidst high grass caused slow progress. Glad for Abul's pony awaiting me" (25 Sept.). Stein did not go immediately to the Neves', but when the pain persisted he went to the hospital to have his shoulder X-rayed. "Macpherson finds no crack in shoulder. Posted MSS. & photos for G.J. [*Geographical Journal*] article on Alexander's route into Gedrosia" (30 Sept.). On 8 October he came to the Neves', and two days later, on Sunday, as usual "Went to Church at 11 A.M. At 6 P.M. farewell visit to Mrs. Carmichael [who had sent him Dash VI]. On clearing photo box opened Hypo bottle & its pungent smell caused fainting fit. Put to bed by Neve who finds pulse normal. Given whiskey; gives fresh strength. Dinner in bed restores feeling. Good Samaritans!" (10 Oct.).

The next morning, Monday, 11 October: "Rose fit at 6:30 A.M. Lorry arrives by 8:30. Baggage [26 pieces] loaded while we were at breakfast. Started saying goodbye to dear friends by 10:30." The morning of the 13th Stein was at Peshawar, guest of the Cunninghams. "Refreshed by bath; had breakfast alone; lunch with Lady Cunningham and walk after tea with H.E. [His Excellency] in garden." "Poor night. Taken to [visit with the] Wali of Swat at Dennis Hotel; as alert & eager as ever though his age of 60 shows. Quiet dinner with C. and his brother. Indigestion has troubled me all day" (14 Oct.). "Taken up with shopping. Posted airmail letter to Madam and fair copy (duplicate) of codicil to Solicitors. At 8 P.M. to dinner with Macphersons" (16 Oct.). And on Sunday, 17 October, the last entry in the diary: "Tent repairs checked. Baggage got ready."

Dr. Neve writes to Madam: "Yesterday's sea mail brought us your kind letter of Nov. 20th, and Oldham's able and full appreciation of the work of our dear friend, Aurel Stein. Our last view of Sir Aurel was when we walked with him down our garden path to the lorry and we bade him farewell. He had been with us a few days. The night before he had had an attack which gave some anxiety—a sudden attack of faintness, necessitating his lying down. After a small whiskey peg he recovered. We felt it a bad omen for a start on a journey, which, later, would be strenuous. I think that *he* felt doubtful, for, on the path, he mentioned your name and address to me, as executrix. The lorry journey, we hoped, would be no strain and if his health was doubtful, he would have advice and help available at Peshawar. All his plans were made and it was impossible at this stage to defer them.

"As you know our house was his 'pied à terre' in Srinagar. We were fond of him. His knowledge of History, Geography, archaeology and languages was amazing. Equally so was his simple, gentle, unassuming life amongst his friends.

"We used to have Morning Prayers, and usually read some passages from a work of devotion such for instance as Bishop Phillips Brooks' sermons and he liked this and enjoyed the privilege of going to All Saints' Church with us."[15]

Sir Francis Wylie's confidential, demi-official letter: 30 October 1943. The British minister to Afghanistan sent an extract to Madam: "I think that I should send you some account of Sir Aurel Stein's arrival here last week and of his sad death a few days later.

"He came up from Peshawar in the American Legation car and reached Kabul on Tuesday, 19th October. Engert brought him round to see me the next morning and he sat talking for about an hour. He was most happy to have arrived at last after, as he put it, three Viceroys had tried to arrange such a visit for him [for forty years!] and had failed. He was full of enthusiasm and eager to get the Afghan Government to let him spend the winter in the Helmand Valley. He saw the Minister of Foreign Affairs the same afternoon and was to meet other officials of the Afghan Government a little later. I saw him again on Saturday afternoon as he was walking in our garden but he had a slight cold and thought it wiser not to come to see the cinema show of 'Desert Victory' that evening. He was due to lunch with me the next day, Sunday [24th], but again had to ask to be excused as he was not too well. In the evening he was suddenly taken much worse, his heart began to give out and MacGregor [the doctor?] who went to see him thought it was unlikely that he would have strength enough to pull through. The same night he had a stroke and never really regained full consciousness. He was a little better on Tuesday but in the afternoon he suddenly failed and died at 5:30 P.M. It was very sad that he should have failed to achieve his ambition of exploration in Afghanistan but at the same time it was perhaps a good thing that he should have died here in Kabul in the house of a friend rather than in some lonely camp at the back of the beyond.

"He had realized on Sunday morning when he was first taken ill that he might not recover and had spoken quite happily to Engert about arrangements for his funeral, etc. He was content to have reached Kabul at last and to find himself in such good hands. He had particularly asked for a Church of England burial and we therefore wired at once to Peshawar asking if the padre could come up for the funeral. Some of his friends in Peshawar suggested we might send the body down for burial there, but Engert felt sure that he would have preferred to be buried in Kabul and we therefore carried on with the arrangements which we had been making.

"The Afghan Government who were, I think, genuinely pleased at his visit were most distressed and were very anxious to pay him after his death such tributes of respect as they had been unable to accord during his lifetime. They arranged for the road to the cemetery to be done up and watered, especially for the occasion, and the King, the Foreign Minister, and other departments, sent representatives to the funeral. The pall bearers consisted of members of the American and British Legation staffs (including Engert and myself) and Dr. Macpherson of the C. M. S. Mission in Peshawar, a very old friend of Sir Aurel Stein who came up with the padre. The Persian Ambassador, the Iraqi Minister, the Soviet Chargé d'Affaires and representative from other friendly legations also attended the funeral which took place on Friday morning the 29th October."[16]

Notes

CHAPTER 1

1. Heinrich Heine, *Works of Prose,* ed. Herman Kesten, trans. E. B. Ashton (New York: L. B. Fischer, 1943), p. 121.

2. "A letter from the Viceroy's hand announcing the K.C.I.E. . . . The telegrams address me in a queer medley as Sir Marcus, Sir Mark-Aurel etc. . . . I stick to the plain *Aurel* by which I was always called at home. It is true it does not readily recall *Aurelius* to English ears or eyes. But I cannot well help that, and after all fellow-scholars know me by that short form for a good number of years." Stein to Allen: 17. VI. 12. All letters to the Allens can be found in the Bodleian Library.

3. Stein to Ernst, 6 May 1877. All letters to Ernst and Hetty Stein can be found in the British Academy.

4. Stein to Ernst, 31 May 1889.

5. Anna Stein to Aurel, January 1894, Amagyar Tudományos Akadémia, Budapest.

6. Carl T. Keller to Stein, 22 April 1930, Allen file, Bodleian Library.

7. Stein to Ernst, 31 May 1889.

8. Stein to Ernst, 8 June 1889.

9. Stein to Ernst, 26 January 1891.

10. *Rájatarangini: A Chronicle of the Kings of Kasmir* (Delhi: Motilal Banarsidass, 1961), 1:xxiv.

11. C. E. A. W. Oldham, *Sir Aurel Stein, 1862–1943. Proceedings of the British Academy,* 29:19.

12. Fred H. Andrews, "Sir Aurel Stein: The Man," *Indian Arts and Letters,* 18, no. 2 (May 1944), 4–6.

13. *Ibid.*

14. E. D. MacLagan, "Marc Aurel Stein," *Hungarian Quarterly,* 4, no. 2 (Summer 1933), 274.

15. Andrews, "Sir Aurel Stein."

Chapter 2

1. Garland Cannon, *Oriental Jones* (London: Asia Publishing House, 1964), p. 141.

2. M. Aurel Stein, *In Memoriam Theodore Duka* (Oxford: privately printed, 1914), pp. 3, 25–29.

3. Arthur Waley, *The Secret History of the Mongols* (London: Allen & Unwin, 1963), pp. 28–29.

4. Stein, *Theodore Duka,* p. 25–26.

5. *Ibid.,* p. 27.

6. A. L. Basham, *The Wonder That Was India* (New York: Grove Press, 1959), p. 6.

7. *Rájatarangini,* 1:vii–viii, 45.

8. *Ibid.,* 1: 45.

9. Stein, *Theodore Duka,* p. 25.

10. "A dedication to Sir Henry Yule's memory has long been on my mind and will, I hope, appear appropriate to others. Anyhow it comes from feelings of true admiration. . . . Sir H. Y. wrote such splendidly simple dedications—and his daughter seems to love long ones." Stein to Allen, 10 January 1906.

11. Samuel Noah Kramer, *The Sumerians: Their History, Culture, and Character.* (Chicago: University of Chicago Press, 1963), p. 18.

12. V. Gordon Childe, *What Happened in History* (Harmondsworth: Penguin Books, 1952), p. 133.

13. William Marsden, *Marco Polo* (London, 1818), p. xxvii.

14. Herbert Butterfield, *Man on His Past: The Study of the History of Historical Scholarship* (Cambridge: At the University Press, 1955), p. 71, n. 2.

15. M. Aurel Stein, *Eine Ferienreise nach Srinagar, aus Briefen.* (Munich: Sonder-Abdruck aus der Allgemeinen Zeitung, 1889), pp. 1–37. Privately trans. by Brigitte Schaeffer; pagination mine. (Hereafter cited as *1888 Trip.*)

16. *Rájatarangini,* 1:45, 46.

17. "Did I ever tell you of my hunt after the learned Annotator [A$_2$ in *Rájatarangini,* 1:48–49] of the archetypus of the Kashmir Chronicle? I found his hand in many a Ms., but it took nearly 10 years before I pierced his incognito." Stein to Allen, 4 June 1905.

18. *1888 trip.*

Chapter 3

1. *Rájatarangini,* 2:350.

2. "Two boxes full of some 50 years' letters written by me to my brother & sister-in-law. My niece's devoted husband Gusti [Dr. Gustav Steiner] had spent his leisure for many evenings in sorting and arranging all these weekly epistles. What a flood of reminiscences of my life in Lahore, Kashmir & elsewhere they would revive if Time is ever granted to me to read through them." Stein to Allen, 5 January 1935.

3. Rudyard Kipling, *Kim* (1901), p. 10.

4. Benjamin Rowland, *Gandharan Sculpture from Pakistan Museums* (New York: Asia House, 1960), p. 11.

5. Andrews, *Stein: The Man,* pp. 2–3.

6. Richard N. Frye, *The Heritage of Persia* (New York and Toronto: New American Library, 1966), p. 48.

7. Basham, *The Wonder That Was India,* pp. 53–56.

8. Stein to Ernst, 7 June 1892.

9. *Rájatarangini,* 2:293, 287.

10. *Ibid.*

11. National Archives of India, Foreign Dept., Ext. B, February 1893, nos. 90–91. (Hereafter cited as NAI.)

12. MacLagan, *Hungarian Quarterly,* p. 274.

13. M. Aurel Stein, *Catalogue of the Sanskrit Manuscripts in the Raghunatha Temple Library of His Highness the Maharaja of Jammu and Kashmir* (Bombay, 1894).

14. *Rájatarangini,* vol. 1, First Book, para. 35. p. 8.

15. *Ibid.,* 2:273–75.

16. *Ibid.,* 2:277.

17. *Ibid.,* 2:279.

CHAPTER 4

1. M. Aurel Stein, "Notes on Inscriptions from Udyâna, presented by Major Harold Deane," *J.R.A.S.* (October 1899), pp. 895–903.

2. Basham, *The Wonder That Was India,* p. 66.

3. Stein, "Notes on Inscriptions from Udyâna."

4. NAI, Foreign Dept., Frontier B., January 1898, nos. 128–31.

5. M. Aurel Stein, "Alexander's Campaign on the North-West Frontier," *The Indian Antiquary,* 58 (1928), 15–17.

6. Andrews, *Stein: The Man,* p. 3.

7. Stein to Allen, 13 August 104.

8. H. R. Trevor-Roper, *Historical Essays* (New York: Harper & Row, 1966), p. 35.

9. Stein to Allen, 27 December 1926.

10. M. Aurel Stein, *In Memoriam Thomas Walker Arnold, 1864–1930* (London: Proceedings of the British Academy, 1932), p. 3.

11. *Ibid.,* p. 16.

12. Lionel Dunsterville, *Stalky's Reminiscences* (London, 1928), p. 110.

13. Basham, *The Wonder That Was India,* p. 396.

14. NAI, Dept. of Revenue & Agriculture, A & E, January 1899, nos. 1–2.

15. Stein to Ernst, 2 December 1898.

CHAPTER 5

1. Stein to Andrews, 11 January 1899. All letters to Andrews can be found in the Bodleian Library.

2. Sourindranath Roy, *The Story of Indian Archaeology, 1784–1947.* (New Delhi: Archaeological Survey of India, 1961), p. 78.

3. *Ibid.,* p. 80.

4. Stein to Ernst, 7 April 1898.

5. Roy, *Story of Indian Archaeology*, p. 84.

6. *Ibid.*, p. 14.

7. *Ibid.*

8. *Ibid.*, p. 26.

9. *Ibid.*, p. 25.

10. Henri Maspero, *La Chaise des langues et littérature chinoises et tartares-manchous*. Jubilee volume on the occasion of the fourth centenary of the Collège de France (Paris, 1932), pp. 355–65. I am most grateful to Professor Arthur Wright for this reference.

11. Wu Ch'êng en, *Monkey*, trans. Arthur Waley (New York: Grove Press, 1958), p. 7. Waley's *Monkey* is but a generous sampling of the *Hsi-yu chi*. Soon the whole fascinating adventure will be available to charm the West as it has China in a complete English translation by Anthony C. Yu under the title *The Journey to the West* (vol. 1, University of Chicago Press, 1977; vols. 2–4 in preparation).

12. Roy, *Story of Indian Archaeology*, p. 38.

13. *Ibid.*, p. 37.

14. *Ibid.*, pp. 48–49.

15. M. Aurel Stein, *Hatim's Tales: Kashmiri Stories and Songs*, ed. G. A. Grierson. Recorded with the assistance of Pandit Govind Kaul. (London: Murray, 1923), p. 14.

16. M. Aurel Stein, *Notes on an Archaeological Tour in South Bihar and Hazaribagh*. (Bombay: Educational Society, 1901).

17. NAI, Foreign Dept., Frontier B, May 1899, nos. 202–9.

CHAPTER 6

1. M. Aurel Stein, *Sand-buried Ruins of Khotan: Personal Narrative of a Journey of Archaeological and Geographical Exploration in Chinese Turkestan* (London: Unwin, 1903), 1:204.

2. *Ibid.*, 1:viii–ix.

3. M. Aurel Stein, *Ancient Khotan: Detailed Report of Archaeological Explorations in Chinese Turkestan.* Vol. 1, text; vol. 2, plates. (Oxford, 1907), 1:vii, xi.

4. M. Aurel Stein, *On Ancient Central-Asian Tracks: Brief Narrative of Three Expeditions in Innermost Asia and Northwestern Kansu.* (Chicago: University of Chicago Press, 1974), p. 3.

5. *Ibid.*, p. 4.

6. *Ibid.*, p. 5.

7. *Ibid.*, p. 7.

8. *Ibid.*, p. 5.

9. *Ibid.*, p. 6.

10. *Ibid.*, pp. 12–13.

11. *Ibid.*, p. 14.

12. *Sand-buried Ruins*, 1:276.

13. *Ibid.*, p. 2.

14. *Ibid.*, p. 7.

15. *Ibid.*, p. 2.

16. *Ibid.*, p. 7.

17. *Ibid.*, p. 8.
18. *Ibid.*, p. 12.
19. *Ibid.*
20. *Ibid.*, p. 13.
21. *Ibid.*, p. 14.
22. *Ibid.*, p. 15.
23. *Ibid.*, p. 16.
24. *Ibid.*, p. 17.
25. *Ibid.*, p. 18.
26. *Ibid.*, p. 20.
27. *Ibid.*, p. 21.
28. *Ibid.*, p. 23.
29. *Ibid.*, p. 22.
30. *Ibid.*, p. 24.
31. *Ibid.*, p. 27.
32. *Ibid.*, p. 28.
33. *Ibid.*, p. 30.
34. *Ibid.*, p. 44.
35. *Ibid.*, p. 45.
36. *Ibid.*, p. 31.
37. Frye, *The Heritage of Persia*, pp. 34–35.
38. *Sand-buried Ruins*, 1:36.
39. *Ibid.*, p. 39.
40. *Ibid.*
41. *Ibid.*, p. 43.
42. *Ibid.*, p. 45.
43. *Ibid.*, p. 46.
44. *Ibid.*, p. 47.
45. *Ibid.*, p. 45.
46. *Ibid.*, p. 47.
47. *Ibid.*, p. 48.
48. *Ibid.*, p. 50.
49. *Ibid.*, p. 51.
50. *Ibid.*, p. 52.
51. *Ibid.*, p. 53.
52. *Ibid.*, p. 54.
53. *Ibid.*, p. 56.
54. *Ibid.*, p. 58.
55. *Ibid.*, p. 59.
56. *Ibid.*, p. 71.
57. *Ibid.*, p. 60.
58. *Ibid.*, p. 62.
59. *Ibid.*, p. 64.
60. *Ibid.*, p. 68.
61. *Ibid.*, p. 69.
62. *Ibid.*, p. 70.
63. *Ibid.*, p. 72.

64. *Ibid.,* p. 73.
65. NAI, Foreign Dept., Frontier G, April 1901, no. 376.
66. *Sand-buried Ruins,* 1:77.
67. *Ibid.,* p. 81.
68. *Ibid.,* p. 84.
69. Gerald Morgan, *Ney Elias* (London: Allen & Unwin, 1971), p. 202.
70. *Sand-buried Ruins,* 1:86.
71. *Ibid.,* p. 87.
72. *Ibid.,* p. 88.
73. *Ibid.,* p. 90.
74. *Ibid.,* p. 94.
75. *Ibid.,* p. 99.
76. *Ibid.,* p. 100.
77. *Ibid.,* p. 101.
78. *Ibid.,* p. 102.
79. *Ibid.,* p. 107.
80. *Ibid.,* p. 109.
81. *Ibid.,* p. 112.
82. *Ibid.,* p. 114.
83. *Ibid.,* p. 115.
84. *Ibid.,* p. 116.
85. *Ibid.,* p. 117.
86. *Ibid.,* p. 118.
87. *Ibid.,* p. 119.
88. *Ibid.,* p. 120.

CHAPTER 7

1. C. P. Skrine and Pamela Nightingale, *Macartney at Kashgar: New Light on British, Chinese and Russian Activities in Sinkiang, 1890–1918* (London: Methuen, 1973), p. 3.
2. *Ibid.,* p. 4.
3. David Dilks, *Curzon in India* (London: Hart-Davis, 1969), 1:29.
4. Skrine and Nightingale, *Macartney at Kashgar,* p. 32.
5. NAI, Foreign Dept., Frontier B, April 1901, no. 383.
6. Skrine and Nightingale, *Macartney at Kashgar,* p. 39.
7. *Ibid.,* p. 102; quotation from Catherine Macartney, *An English Lady in Chinese Turkestan* (London, 1931).
8. *Sand-buried Ruins,* 1:122.
9. *Ibid.,* p. 121.
10. *Ibid.,* p. 123.
11. *Ibid.,* p. 124.
12. *Ibid.,* p. 125.
13. *Ibid.,* p. 126.
14. *Ibid.,* p. 130.
15. *Ibid.,* p. 127.
16. *Ibid.,* p. 128.
17. *Ibid.,* p. 127.

18. *Ibid.*, p. 132.
19. *Ibid.*, p. 139.
20. *Ibid.*, p. 148.
21. *Ibid.*, p. 157.
22. *Ibid.*, p. 165.
23. *Ibid.*, p. 151.
24. *Ibid.*, pp. 169–70.
25. *Ibid.*, p. 164.
26. *Ibid.*, p. 162.
27. *Ibid.*, p. 167.
28. *Ibid.*, p. 169.
29. *Ibid.*, p. 172.
30. *Ibid.*, p. 173.
31. *Ibid.*, p. 174.
32. *Ibid.*, p. 177.
33. *Ibid.*, p. 180.
34. *Ibid.*, p. 181.
35. *Ibid.*, p. 183–4.
36. *Ibid.*, p. 190.
37. *Ibid.*, p. 191.
38. *Ibid.*, p. 194–5.
39. *Ibid.*, p. 196.
40. *Ibid.*, p. 198.
41. *Ibid.*, p. 199.
42. *Ibid.*, p. 200.
43. *Ibid.*, p. 201.
44. *Ibid.*, p. 202.
45. *Ibid.*, p. 204.
46. *Ibid.*, p. 203.
47. *Ibid.*, p. 206.
48. *Ibid.*, p. 207.
49. *Ibid.*, p. 212.
50. *Ibid.*, p. 213–4.
51. NAI, Foreign Dept., Frontier B, April 1901, nos. 70–96.
52. *Sand-buried Ruins*, pp. 239–40.
53. *Ibid.*, p. 249.
54. *Ibid.*, p. 255.
55. *Ibid.*, p. 254.
56. *Ibid.*, p. 257.
57. *Ibid.*, p. 261.
58. *Ibid.*, p. 263.
59. *Ibid.*, p. 264.

CHAPTER 8

1. *Sand-buried Ruins*, p. 270.
2. *Ibid.*, p. 273.
3. *Ibid.*, p. 274.

4. *Ibid.*, p. 275.

5. *Ibid.*, p. 276.

6. *Ibid.*, p. 277.

7. This popular anodyne was a mixture of chloroform, morphia, Indian hemp, prussic acid, etc.

8. *Sand-buried Ruins*, p. 278.

9. *Ibid.*, p. 279.

10. *Ibid.*, p. 281.

11. *Ibid.*, p. 282.

12. *Ibid.*, p. 286.

13. *Ibid.*, p. 290.

14. *Ibid.*, p. 291.

15. *Ibid.*, p. 292.

16. *Ibid.*, p. 296.

17. *Ibid.*, p. 297.

18. *Ibid.*, p. 301.

19. *Ibid.*, p. 303.

20. *Ibid.*, p. 302.

21. *Ibid.*, p. 311.

22. *Ibid.*, p. 315.

23. *Ibid.*, p. 317.

24. NAI, Foreign Dept., Frontier B, April 1901, nos. 372–89.

25. *On Ancient Central-Asian Tracks*, p. 58.

26. Joseph Needham, with the research assistance of Wang Ling, *Science and Civilization in China* (Cambridge: At the University Press, 1954——), 1:185–86.

27. *On Ancient Central-Asian Tracks*, p. 58.

28. Sir Aurel Stein, *Innermost Asia: Detailed Report of Explorations in Central Asia, Ken-su and Eastern Iran*, 4 vols. (Oxford: Clarendon Press, 1928), 2:906.

29. Guy LeStrange, *Lands of the Eastern Caliphate* (Cambridge: At the University Press, 1930), p. 347.

30. Stein to Miss Maud Allen (sister of "Madam"), 4 January 1916, Bodleian Library.

31. *On Ancient Central-Asian Tracks*, p. 59.

32. *Ibid.*, p. 60.

33. *Ibid.*, p. 61.

34. *Sand-buried Ruins*, pp. 326–28.

35. *Ibid.*, p. 333.

36. *Ibid.*, pp. 329–30.

37. *Ibid.*, p. 333.

38. *Ibid.*, p. 337.

39. *Ibid.*, p. 342.

40. *Ibid.*, p. 343.

41. *Ibid.*, p. 360.

42. *Ibid.*, p. 361.

43. *On Ancient Central-Asian Tracks*, p. 72.

44. *Sand-buried Ruins*, p. 367.

45. *Ibid.*, p. 368.

46. Quoted in Glyn Daniel, *The Idea of Prehistory* (Cleveland and New York: World Publishing Co., 1963), p. 39.
47. *Sand-buried Ruins*, p. 374.
48. *On Ancient Centarl-Asian Tracks*, p. 76.
49. *Sand-buried Ruins*, p. 383.
50. *Ibid.*, p. 381.
51. *Ibid.*, p. 383.
52. *Ibid.*, p. 384.
53. *Ibid.*, p. 387.
54. *On Ancient Central-Asian Tracks*, p. 77.
55. *Sand-buried Ruins*, p. 388.
56. *On Ancient Central-Asian Tracks*, p. 78.
57. *Ibid.*, p. 79.
58. *Sand-buried Ruins*, p. 396.
59. *On Ancient Central-Asian Tracks*, p. 81.
60. *Sand-buried Ruins*, p. 409.
61. *On Ancient Central-Asian Tracks*, p. 84.

CHAPTER 9

1. *Sand-buried Ruins*, p. 410.
2. *Ibid.*, p. 413.
3. *On Ancient Central-Asian Tracks*, p. 94.
4. *Sand-buried Ruins*, p. 420.
5. *Ibid.*, p. 424.
6. *Ibid.*, p. 425.
7. *Ibid.*, p. 428.
8. *Ibid.*, p. 429.
9. *Ibid.*, p. 431.
10. *Ibid.*, p. 433.
11. *Ibid.*, p. 451.
12. C. P. Skrine, *Chinese Central Asia* (Boston: 1926), p. 116.
13. *Sand-buried Ruins*, p. 435.
14. *Ibid.*, p. 436.
15. *Ibid.*, p. 437.
16. *Ibid.*, p. 440.
17. *Ibid.*, p. 441.
18. *Ibid.*, p. 442.
19. *Ibid.*, p. 443.
20. *Ibid.*, p. 445.
21. *Ibid.*, p. 446.
22. *Ibid.*, p. 449.
23. *Ibid.*, p. 452.
24. *Ibid.*, p. 457.
25. *Ibid.*
26. *Ibid.*, p. 463.
27. *Ibid.*, p. 466.

28. *Ibid.,* p. 468.
29. *Ibid.,* p. 470.
30. *Ibid.,* p. 471.
31. *Ibid.,* p. 472.
32. *Ibid.,* p. 474.
33. *Ibid.,* p. 475.
34. *Ibid.,* p. 481.
35. *Ibid.,* p. 485.
36. *Ibid.,* p. 486.
37. *Ibid.,* p. 488.
38. *Ibid.,* p. 489.
39. *Ibid.,* p. 491.
40. NAI, Dept. of Revenue and Agriculture, A & E, May 1901, nos. 2–3.
41. *Ibid.;* telegram 18 May 1901 sent via Gilgit.
42. NAI, Foreign Dept., Frontier B, April 1901, nos. 372–89.
43. *Sand-buried Ruins,* p. 492.
44. *Ibid.,* p. 495.
45. *Ibid.,* p. 492.
46. *Ibid.,* p. 493.
47. *Ibid.,* p. 495.
48. Stein to Allen, 12 June 1901.
49. *Sand-buried Ruins,* p. 501.
50. *Ibid.,* p. 502.

CHAPTER 10

1. Stein to Ernst, 4 July 1901.
2. Stein to Ernst, 9 July 1901.
3. Stein to Ernst, 13 July 1901.
4. Stein to Ernst, 4 August 1901.
5. NAI, Home Dept. Education, 12 October 1901, nos. 30–31.
6. Stein to Ernst, 13 November 1901.
7. Stein to Ernst, 2 December 1901.
8. Stein to Ernst, 10 December 1901.
9. Stein to Ernst, 17 December 1901.
10. Stein to Ernst, 6 January 1902.
11. NAI, Dept. of Revenue and Agriculture, A & E, April 1902, nos. 1–5.
12. NAI, Roreign Dept., Frontier A, October 1902, nos. 23–26; Stein to Barnes, 30 June 1902.
13. NAI, Foreign Dept., Frontier A, May 1903, nos. 72–75.
14. *Ancient Khotan,* 1:xi.
15. Krishna Riboud, "A Study of Two Central-Asian Silk Specimens from the Han Dynasty," *Bulletin,* National Museum, New Delhi, no. 2 (1970), 1–13.
16. *Ancient Khotan,* 1:xii.
17. *Ibid.*
18. *Ibid.,* 1:xiii.

19. *Ibid.*, 1:xiv.
20. NAI, Foreign Dept., Exterior B, March 1904, no. 190.
21. Archives of the Archaeological Survey of India, New Delhi, Archaeology, October 1904, File 157, Series A, nos. 1–4. (Hereafter cited as ASI.)
22. NAI, Foreign Dept., External B, April 1904, no. 163.
23. ASI, Archaeology, October 1904, File 157, Series A, nos. 1–4.
24. Stein to Andrews, 23 September 1904.

CHAPTER 11

1. Stein to Allen, 12 October 1904.
2. London *Times*, 21 August 1958, p. 10.
3. ASI, 25 February 1904, p. 1.
4. ASI, Archaeology, October 1904, File 157, Series A, nos. 1–4.
5. NAI, Foreign Dept., Frontier B, June 1905, no. 199; see Stein's letter, 7 March 1905, pp. 1–3.
6. Stein to Andrews, 30 December 1905.
7. NAI, Foreign Dept., Frontier B, March 1905, nos. 73–78.
8. NAI, Home Dept. A & E, March 1906, nos. 14–22.
9. Stein to Allen, 2 October 1905.
10. Stein to Allen, 2 September 1905.
11. Stein to Allen, 2 October 1905.
12. Stein to Allen, 15 October 1905.
13. Stein to Allen, 30 December 1905.

CHAPTER 12

1. Aurel Stein, *Ruins of Desert Cathay: Personal Narrative of Explorations in Central Asia and Westernmost China*, 2 vols. (London: Macmillan, 1912), 1:117.
2. *Ibid.*, p. 116.

CHAPTER 13

1. *Ruins of Desert Cathay*, 1:337.
2. *Ibid.*, 1:6.
3. NAI, Foreign Dept., Frontier A, April 1906, nos. 38–43.
4. Stein to Allen, 17 February 1906.
5. Stein to Andrews, 27 January 1906.
6. NAI, Home Dept. A & E, February 1906, nos. 18–19.
7. In Allen file, Bodleian Library, no date.
8. *Ruins of Desert Cathay*, 1:21–23.
9. In Allen file, Bodleian Library, n.d.
10. NAI, Foreign Dept., Frontier B, June 1906, nos. 166–68.
11. *Ruins of Desert Cathay*, 1:52–69.
12. *Ibid.*, p. 57.
13. *Ibid.*, p. 58.
14. *Ibid.*, p. 60.

15. *Ibid.*, p. 64.
16. NAI, Foreign Dept., Frontier B, February 1907, nos. 21–26.
17. Stein to Allen, 21 May 1906.
18. Stein to Allen, 9 June 1906.
19. *Ruins of Desert Cathay*, 1:107–8.
20. *Ibid.*, p. 124.
21. Stein to Allen, 2 July 1906.
22. Stein to Allen, 10 July 1906.
23. Stein to Allen, 19 July 1906.
24. *Ruins of Desert Cathay*, 1:163.
25. *Ibid.*, p. 262.
26. Stein to Allen, 19 November 1906.
27. *Ruins of Desert Cathay*, 1:317.
28. Stein to Allen, 19 November 1906.
29. *Ruins of Desert Cathay*, 1:354–55.
30. Stein to Allen, 7 January 1907.
31. Stein to Andrews, 31 January 1907.
32. Stein to Allen, 17 February 1907.
33. Stein to Andrews, 31 January 1907.
34. Stein to Allen, 17 February 1907.
35. ASI, Archaeology, July 1907, File 174, Series 1.
36. Stein to Allen, 17 February 1907.
37. Stein to Allen, 16 March 1907.

Chapter 14

1. *Ruins of Desert Cathay*, 1:539.
2. *Ibid.*, 2:19.
3. Stein to Allen, 16 March 1907.
4. *Ruins of Desert Cathay*, 2:22–26.
5. *Ibid.*, pp. 28–31.
6. *Ibid.*, pp. 13–14.
7. *Ibid.*, p. 15.
8. *Ibid.*, p. 18.
9. *Ibid.*, pp. 33–34.
10. *Ibid.*, p. 41.
11. *Ibid.*, p. 155.
12. Stein to Allen, 26–28 April 1907.
13. *Ruins of Desert Cathay*, 2:159.
14. *Ibid.*, pp. 163–64.
15. *Ibid.*, pp. 164–82. I have chosen to handle the story so that Stein's actions and rationalizations move with speed. Hopefully, I have not given a biased account —only a shortened one.
16. *On Ancient Central-Asian Tracks*, p. 189.
17. *Ruins of Desert Cathay*, 2:164–82.
18. Stein to Allen, 9 June 1907.
19. *Ruins of Desert Cathay*, 2:232–33.

CHAPTER 15

1. Owen Lattimore, *Inner Asian Frontiers of China* (New York: American Geographical Society, 1940), p. 76n.
2. Stein to Allen, 2 December 1907.
3. *Ruins of Desert Cathay*, 2:355–56.
4. Stein to Allen, 11–23 January 1908.
5. Stein to Allen, 16 February 1908.
6. Stein to Allen, 12 May 1908.
7. Stein to Allen, 9–10 June 1908.
8. Stein to Allen, 11 April 1908.
9. Stein to Allen, 12 May 1908.
10. Stein to Allen, 9–10 June 1908.
11. Stein to Allen, 11 April 1908.
12. Stein to Allen, 27 April 1908.
13. Stein to Allen, 28 May 1908.
14. Stein to Allen, 9–10 June 1908.
15. R. B. Shaw, an Indian tea planter, visited Eastern Turkestan in a private capacity in 1868–69; G. W. Hayward, a private traveler, was one of the first Englishmen to visit Kashgar and Yarkand at the same time but independent of Shaw; Douglas Forsyth, a British senior officer, started a business venture in a diplomatic way in Central Asian trade—his first mission there was in 1870. A. Lamb, *The China-India Border* (London: Oxford University Press, 1964), pp. 82–84.
16. Stein to Allen, 27 April 1908.
17. *Ruins of Desert Cathay*, 2:441–42.

CHAPTER 16

1. *Ruins of Desert Cathay*, 2:435.
2. Lamb, *China-India Border*, p. 83.
3. *Ibid.*, p. 33.
4. *Ruins of Desert Cathay*, 2:441–42.
5. *Ibid.*, pp. 473–88.

CHAPTER 17

1. NAI, Home Dept., A & E, May 1909, nos. 13–15.
2. *Ibid.*, July 1909, nos. 2–3.
3. NAI, Foreign Dept., Secret F. June 1909, nos. 81–87.
4. *Ibid.*, August 1909, nos. 159–60.

CHAPTER 18

1. Quoted in N. Chaudhuri, *Scholar Extraordinary* (Oxford: Oxford University Press, 1974), p. 208.
2. NAI, Home Dept., A & E., February 1909, no. 33.
3. *Ibid.*, February 1910, nos. 11–17.

4. Roy, *Story of Indian Archaeology,* p. 102.

5. Stein to Andrews, 24 April 1909.

6. Stein to Andrews, 27 May 1909.

7. Stein to Allen, 19 August 1909.

8. Stein to Allen, 20 February 1910.

9. Stein to Allen, 21 February 1910.

10. Stein to Allen, 5 April 1910.

11. Stein to Andrews, 16 July 1915.

12. Stein to Allen, 3 March 1910.

13. Sir Aurel Stein, *Serindia, Detailed Report of Explorations in Central Asia and Westernmost China* (Oxford: At the Clarendon Press, 1921), 1:xlv.

CHAPTER 19

1. Stein to Allen, 3 January 1912.

2. Stein to Allen, 22 January 1912.

3. Stein to Andrews, 6 March 1912.

4. Stein to Allen, 3 April 1912.

5. NAI, Foreign Dept., Frontier B, October 1912, nos. 58–61.

6. Stein to Allen, 6 May 1912.

7. Stein to Allen, 20 May 1912.

8. Stein to Allen, 16 June 1912.

9. Stein to Barnett, 8 July 1912. Letters to Barnett can be found in the British Museum.

10. Stein to Allen, 14 October 1912.

11. Stein to Barnett, 1 December 1912.

12. "A drily pedantic work on Buddhist psychology and metaphysics, and of little interest except to specialists." Basham, *The Wonder That Was India,* p. 267.

13. Stein to Barnett, 2 March 1913.

14. Stein to Barnett, 31 March 1913.

15. Stein to Barnett, 2 March 1913.

16. Stein to Allen, 28 October 1912.

17. Stein to Campbell, assistant secretary to the chief commissioner of the Northwest Frontier Province, 23 November 1912; copy in Allen file, Bodleian Library.

18. NSI, Dept. of Education, A & E, July 1913, nos. 25–35.

19. Stein to Campbell, 23 November 1912.

20. *Innermost Asia,* 1:x.

21. Stein to Allen, 21 April 1913.

22. Stein to Allen, 2 August 1913.

23. Stein to Allen, 11 August 1913.

24. *Innermost Asia,* 1:29.

25. Stein to Allen, 20–22 September 1913.

26. Macartney to Stein, 12 February 1913; copy in Allen file, Bodleian Library.

27. Stein to Allen, 30 September 1913.

28. Stein to Allen, 9 October 1913.

29. Stein to Allen, 30 September 1913.

CHAPTER 20

1. Stein to Andrews, 2 November 1913.
2. Stein to Allen, 1 November 1913.
3. Stein to Allen, 30 December 1913.
4. Stein to Allen, 8 March 1914.
5. Stein to Allen, 21 February 1914.
6. Stein to Andrews, 21–24 February 1914.
7. Stein to Allen, 27 March 1914.
8. Ibid.
9. Stein to Allen, 13 April 1914.
10. Stein to Allen, 3 May 1914.
11. Stein to Allen, 14 May 1914.
12. Stein to Allen, 24 May 1914.
13. Stein to Allen, 6 June 1914.
14. Stein to Allen, 14 August 1914.
15. Stein to Allen, 27 September–5 October 1914.
16. Stein to Allen, 18 October 1914.
17. *Innermost Asia,* 2:557.
18. Stein to Allen, 29 October 1914.
19. Stein to Allen, 12 November 1914.
20. Stein to Andrews, 6 February 1915.
21. Stein to Andrews, 6 January 1915.
22. Stein to Andrews, 6 February 1915.
23. Stein to Allen, 2 February (?) 1915.
24. *Innermost Asia,* 2:644–45.
25. Stein to Allen, 15–25 March 1915.
26. Stein to Allen, 30 April 1915.
27. Stein to Allen, 10 May 1915.
28. Stein to Allen, 1 July 1915.
29. Stein to Allen, 16 July 1915.
30. Stein to Allen, 30 July 1915.

CHAPTER 21

1. *Innermost Asia,* 2:883.
2. Stein to Allen, 15 October 1915.
3. Stein to Allen, 10 November 1915.
4. Stein to Allen, 26 November–2 December 1915.
5. *Academy,* 16 May 1884, p. 349; reprinted in *The Indian Antiquary* (1886), p. 22.
6. Stein to Barnett, 10 January 1916.
7. Stein to Allen, 12 February 1916.
8. *Innermost Asia,* 2:980.
9. Stein to Allen, 3 March 1916.
10. Stein to Allen, 21 March 1916.
11. Stein to Allen, 29 March 1916.
12. Stein to Allen, 3 March 1916.

CHAPTER 22

1. ASI, Stein to Marshall, 10 May 1916.
2. Stein to Allen, 1 June 1916.
3. Stein to Allen, 5 September 1916.
4. Stein to Andrews, 8 November 1916.
5. Stein to Andrews, 14 November 1916.
6. Stein to Andrews, 8 November 1916.
7. Stein to Andrews, 13 December 1916.
8. Stein to Andrews, 8 November 1916.
9. Stein to Allen, 26 January 1917.
10. Stein to Allen, 3 February 1917.
11. Stein to Allen, 2 May 1917.
12. Stein to Barnett, 2 August 1917.
13. Stein to Barnett, 22 October 1917.
14. Stein to Allen, 27 November 1917.
15. Stein to Allen, 22 December 1917.
16. Stein to Allen, 1 February 1918.
17. Stein to Allen, 28 February 918.
18. Stein to Barnett, 11 December 1918.
19. Stein to Allen, 16 February 1918.
20. Stein to Allen, 25 February 1918.
21. Stein to Allen, 19 October 1918.
22. Stein to Allen, 8 September 1918.
23. Stein to Allen, 29 April 1918.
24. Stein to Barnett, 11 December 1918.
25. Stein to Allen, 1 May 1922.
26. Stein to Allen, 12 December 1918.
27. Stein to Allen, 8 January 1919.
28. Stein to Allen, 30 June 1919.
29. Stein to Allen, 16 May 1919.
30. Stein to Allen, 30 July 1919.
31. Stein to Allen, 16 May 1919.
32. Stein to Allen, 30 July 1919.
33. Stein to Allen, 11 October 1919.
34. Stein to Allen, 17 March 1919.
35. Stein to Allen, 16 July 1919.
36. Stein to Allen, 25 November 1919.
37. Stein to Allen, 20 January 1920.
38. Stein to Allen, 13 April 1920.
39. Stein to Barnett, 16 March 1920.
40. Stein to Allen, 29 December 1920.
41. Stein to Andrews, 2 June 1921.
42. Stein to Allen, 9 June 1921.
43. Stein to Allen, 25 April 1922.
44. Stein to Allen, 9 July 1922.
45. Stein to Allen, 1 October 1922.

46. Stein to Allen, 15 October 1922.
47. Stein to Allen, 26 June 1922.
48. Stein to Andrews, 28 June 1922.
49. Stein to Andrews, 22 September 1922.
50. Stein to Andrews, 20 September 1922.
51. Stein to Allen, 2–21 November 1922.
52. Stein to Allen, 17 December 1922.
53. Stein to Barnett, 1 April 1922.
54. Stein to Allen, 1 January 1923.
55. Stein to Allen, 10 June 1923.
56. Stein to Allen, 21 January 1924.
57. Stein to Allen, 2 March 1924.

CHAPTER 23

1. Stein to Allen, 2 March, 1924.
2. Stein to Allen, 4 April 1924.
3. Stein to Allen, 8 April 1924.
4. Stein to Allen, 24 April 1924.
5. ASI, Stein to Marshall, 23 September 1924.
6. Stein to Andrews, 29 October 1924.
7. Roy, *Story of Indian Archaeology,* pp. 107–8.
8. Walter A. Fairservis, Jr., *The Roots of Ancient India* (New York: The Macmillan Co., 1971), pp. 250–51.
9. Roy, *Story of Indian Archaeology,* p. 69.
10. ASI, Stein to Marshall, 11 August 1925.

CHAPTER 24

1. Stein to Allen, 10 December 1925.
2. Stein to Allen, 7 January 1926.
3. Stein to Allen, 21 January 1926.
4. Stein to Allen, 17 February 1926.
5. Stein to Allen, 10 March 1926.
6. The Chief commissioner justified the gifts supplied by sending Simla three brief reports Stein had written. His covering note mentioned that Stein "is having an extraordinarily interesting tour and from the point of view of Government in that part of the world, it is, I think, difficult to exaggerate the importance of it." NAI, F & P Dept., E.D.U.C. no. 1528, F&P d/– 11 April 1926.
7. Stein to Allen, 6 December 1926.

CHAPTER 25

1. Stein to Allen, 4 July 1926.
2. Stein to Allen, 21 September 1904.
3. Sir Aurel Stein: *The Indo-Iranian Borderlands: Their Prehistory in the Light of Geography and of Recent Explorations.* The Huxley Memorial Lecture for 1934. Reprinted from the *J.R.A.I.* 65 (July-December 1934):184.

4. Stein to Allen, 16 January 1927.
5. Stein to Allen, 30 January 1927.
6. ASI, 16 March 1927.
7. *Ibid.,* 27 September 1927.
8. Stein to Andrews, 13 April 1927.
9. Stein to Allen, 19 June 1927.
10. F. Nansen, the Nobel Prize humanitarian, wrote Stein an enthusiastic letter on his Swat book. A few months later, in the spring of 1928, Stein was nominated to be "an External Member of the Royal Norwegian Academy. I suppose I owe it to Nansen's kindness." Stein to Allen, 11 May 1928.
11. Stein to Allen, 11 December 1927.
12. Stein to Allen, 7 December 1927.
13. Stein to Allen, 11 December 1927.
14. Stein to Allen, 31 December 1927.
15. Stein to Allen, 17 January 1928.
16. Stein to Allen, 25 January 1928.
17. Stein to Allen, 24 February 1928.
18. Stein to Allen, 6 March 1928.
19. Fairservis, *Roots of Ancient India,* p. 215.
20. ASI, Stein to Marshall, 20 March 1928.
21. Stein to Allen, 25 March 1928.
22. Fairservis, *Roots of Ancient India,* p. 171.
23. Stein to Barnett, 20 May 1928.
24. Stein to Allen, 9 October 1928.
25. Stein to Andrews, 14 December 1928.
26. Stein to Andrews, 14 January 1929.
27. Stein to Andrews, 22 January 1929.
28. Stein to Andrews, 11 February 1929.
29. Stein to Allen, 5 March 1929.
30. Stein to Allen, 11 April 1929.
31. Stein to Allen, 25 July 1929.

Chapter 26

1. Stein to Allen, 5 August 1929.
2. Stein to Andrews, 12 August 1929.
3. Stein to Andrews, 17 October 1929.
4. Langdon Warner, *Buddhist Wall Paintings* (Cambridge, Mass.: Harvard University Press, 1938), pp. xiii–xv.
5. Stein to Allen, 13 December 1929.
6. Stein to Allen, 20 December 1929.
7. Stein, P.N. MS, China, pp. 7–10. 21 April 1930; pp. 1–21. 17 May 1930; pp. 1–35. Bodleian Library, Allen file.
8. Stein to Allen, 5 May 930.
9. Stein to Allen, 1 June 1930.
10. Stein to Barnett, 11 July 1930.
11. Stein to Allen, 29 August 930.

12. Stein to Allen, 9 October 1930.
13. Stein to Allen, 23 October 1930.
14. Stein to Barnett, 27 April 1931.
15. Stein to Allen, 3 February 1931.
16. Stein to Andrews, 16 May 1931.
17. Stein to Allen, 25 June 1931.
18. Stein to Allen, 3 July 1931.
19. Stein to Barnett, 25 July 1931.
20. Stein to Allen, 6 September 1931.

CHAPTER 27

1. Stein to Allen, 5 June 1931.
2. Stein to Allen, 26 October 1931.
3. Stein to Allen, 30 November 1931.
4. Stein to Allen, 8 December 1931.
5. Stein to Allen, 30 November 1931.
6. Stein to Allen, 4 March 1932.
7. Stein to Allen, 22 March 1932.
8. Stein to Allen, 4 April 1932.
9. Stein to Allen, 29 January 1932.
10. Stein to Allen, 4 March 1932.
11. *Indo-Iranian Borderland,* pp. 195–96.
12. Stein to Allen, 22 March 1932.
13. *Indo-Iranian Borderland,* pp. 195–96.
14. Stein to Allen, 4 April 1932.
15. Stein to Allen, 16 May 1932.
16. Stein to Allen, 25 May 1932.
17. Stein to Allen, 28 October 1932.
18. Stein to Allen, 25 November 1932.
19. Stein to Andrews, 27 November 1932.
20. Stein to Allen, 6 December 1932.
21. Stein to Allen, 13 December 1932.
22. Stein to Allen, 10 January–7 February 1933.
23. Stein to Allen, 1 April 1933.
24. Stein to Allen, 13 May 1933.
25. Stein to Mrs. Allen, 16 July 1933.
26. P.N. MS., 8 December 1933. Bodleian Library, Allen file.
27. Frye, *Heritage of Persia,* p. 32.
28. P.N. MS., 8 December 1933.
29. Stein, "Archaeological Tour in the Ancient Persis," *Geographical Journal,* 86, no. 6 (December 1935), 490.
30. P.N. MS., 8 December 1933.

CHAPTER 28

1. Stein to Mrs. Allen, 20 May, 1934.
2. Stein to Mrs. Allen, 9 April 1934.

3. Stein to Andrews, 14 September 1934.
4. Stein to Mrs. Allen, 11 October 1934.
5. Stein to Mrs. Allen, 16 October 1934.
6. Stein to Mrs. Allen, 5 January 1935.
7. Stein to Mrs. Allen, 28 January 1935.
8. Stein to Mrs. Allen, 6 March 1935.
9. Stein to Mrs. Allen, 16 March 1935.
10. Stein to Mrs. Allen, 28 July 1935.
11. Stein to Mrs. Allen, 31 August, 1935.
12. Stein to Mrs. Allen, 6 November 1935.
13. Stein to Mrs. Allen, 13 November 1935.
14. Bahman Karami, *Les Anciennes Routes de l'Iran: voyage en compagnie et pour la surveillance des opérations de Sir Aurel Stein . . . de Shiraz jusqu'à la dernière frontière des Kurdes de l'Azerbaidjan.* (Iran: Banque Melli, n.d.). Copy kindly lent by Mrs. David Lilienthal.
15. Stein to Mrs. Allen, 13 November 1935.
16. Stein to Mrs. Allen, 19 November 1935.
17. Stein, "An Archaeological Journey in Western Iran," *Geographical Journal,* 91, no. 4 (October 1938), 313–41.
18. *Ibid.,* p. 321.
19. *Ibid.*
20. Frye, *Heritage of Persia,* p. 210.
21. Stein, "An Archaeological Journey in Western Iran," p. 326.
22. Stein to Mrs. Allen, Easter Sunday, 1936.
23. Stein, "An Archaeological Journey in Western Iran," p. 332.
24. *Ibid.*
25. Stein to Mrs. Allen, 1 June 1936.
26. Stein, "An Archaeological Journey in Western Iran," p. 336.
27. Frye, *Heritage of Persia,* p. 183.
28. Stein to Andrews, 26 September 1936.
29. Stein, "An Archaeological Journey in Western Iran," p. 339.
30. *Ibid.,* p. 340.
31. *Ibid.,* p. 341.
32. Stein to Mrs. Allen, 29 October 1936.
33. Stein to Mrs. Allen, 2 December 1936.
34. Stein to Mrs. Allen, 20 December 1936.
35. Stein to Mrs. Allen, 14 November 1936.

CHAPTER 29

1. Oldham, "Sir Aurel Stein," *Proceedings of the British Academy,* 29:20.
2. *British Academy,* 16 May 1884, p. 349.
3. Jean Fairley, *The Lion River the Indus* (London: Allen Lane, 1975), p. 45.
4. Stein to Mrs. Allen, 17 April 1937.
5. Stein to Andrews, 13 June 1937.
6. Stein to Mrs. Allen, 15 December 1937.
7. Stein to Mrs. Allen, 28 December 1937.

8. Stein to Mrs. Allen, 5 December 1937.

9. Stein to Mrs. Allen, 15 December 1937.

10. Stein to Mrs. Allen, 10 January, 1938.

11. Stein to Mrs. Allen, 8 March 1938.

12. Stein to Mrs. Allen, 24 March 1938.

13. Stein to Mrs. Allen, 18 March 1938.

14. Stein to Mrs. Allen, 12 May 1938.

15. Stein to Mrs. Allen, 4 June 1938.

16. Stein to Mrs. Allen, 3 July 1938.

17. Stein to Mrs. Allen, 25 September 1938.

18. Stein to Mrs. Allen, 11 December 1938.

19. Stein to Mrs. Allen, 18 November 1938.

20. Stein to Mrs. Allen, 30 November 1938.

21. Stein to Mrs. Allen, 19 December 1938.

22. Stein to Mrs. Allen, 6 January 1939.

23. Stein, "Surveys on the Roman Frontier and Trans-Jordan," *Geographical Journal*, 95, no. 6 (June 1940), 432–33.

24. *Ibid.*

25. Stein to Mrs. Allen, 5 April 1939.

26. Stein to Mrs. Allen, 23 April 1939.

27. Stein to Mrs. Allen, 10 April 1939.

28. Stein to Mrs. Allen, 5 May 1939.

CHAPTER 30

1. Stein to Mrs. Allen, 6 September 1939.

2. Stein to Mrs. Allen, 20 September 1939.

3. Stein to Mrs. Allen, 6 September 1939.

4. Stein to Mrs. Allen, 15 November 1939.

5. Stein to Mrs. Allen, 25 November 1939.

6. Stein to Andrews, 10 December 1939.

7. Stein to Mrs. Allen, 23 December 1939.

8. Stein to Mrs. Allen, 24 January 1940.

9. Stein to Mrs. Allen, 30 December 1939.

10. Stein to Andrews, 13 February 1940.

11. Fairley, *Lion River,* p. 76.

12. *Ibid.,* p. 70.

13. Stein, "Surveys on the Roman Frontier and Trans-Jordan."

14. Stein to Mrs. Allen, 14 June 1940.

15. Stein, "Notes on Alexander's Crossing of the Tigris and the Battle of Arbela, vol. C[100], no. 4 *Geographical Journal* (October 1942).

16. Stein to Mrs. Allen, 9 October 1940.

17. Stein to Mrs. Allen, 5 November 1940.

18. Stein to Mrs. Allen, 22 November 1940; he is referring to "The Ancient Trade Route past Hatra and the Roman Posts," *J.R.A.S.,* October 1941.

19. Stein to Mrs. Allen, 14 November 1940.

20. Stein to Mrs. Allen, 22 November 1940.

21. Stein to Mrs. Allen, 17 January 1941.

22. Stein to Mrs. Allen, 27 January 1941.

23. Stein to Mrs. Allen, 16 February 1941.

24. Stein to Mrs. Allen, 24 February 1941.

25. Stein, "A Survey of Ancient Sites along the 'Lost' Sarasvati River," *Geographical Journal,* 99, no. 4 (April 1942), 180.

26. Stein to Mrs. Allen, 18 March 1941.

27. Stein to Mrs. Allen, 29 March 1941.

28. Stein to Mrs. Allen, 15 April 1941.

29. Stein to Andrews, 29 June 1941.

30. Stein to Mrs. Allen, 24 September 1941.

CHAPTER 31

1. Stein to R. Ghirshman, 17 April 1942. I am grateful to Professor Ghirshman for this letter.

2. Stein to Mrs. Allen, 27 June 1942.

3. Stein to Mrs. Allen, 26 July 1942.

4. Stein to Mrs. Allen, 19 August 1942.

5. Stein to Mrs. Allen, 18 September 1942.

6. Linthithgow to Stein, 28 November 1942. Dunsterville to Stein, same date. Both in Allen file, Bodleian Library.

7. Stein to Mrs. Allen, 10 November 1942.

8. Stein to Mrs. Allen, 19 December 1942.

9. ASI, Krishna Deva to Dikshit, 24 January 1943.

10. Stein to Dikshit, 4 March 1943.

11. Deva to Dikshit, 24 March 1943.

12. Fairservis, *Roots of Ancient India,* p. 194.

13. Stein to Mrs. Allen, 8 April 1943.

14. Stein to Wheeler, 16 July 1943. In Allen file, Bodleian Library.

15. Neve to Mrs. Allen, 14 January 1944. In Allen file, Bodleian Library.

16. Sir Francis Wylie, British Minister to Afghanistan, to Mrs. Allen. Extract from demi-official letter no. 536/43 dated 30 October 1943. In Allen file, Bodleian Library.

Index

Ram Singh, Naik, 303-4